Vascular Endothelium

Responses to Injury

NATO ASI Series

Advanced Science Institutes Series

A series presenting the results of activities sponsored by the NATO Science Committee, which aims at the dissemination of advanced scientific and technological knowledge, with a view to strengthening links between scientific communities.

The series is published by an international board of publishers in conjunction with the NATO Scientific Affairs Division

A	**Life Sciences**	Plenum Publishing Corporation
B	**Physics**	New York and London
C	**Mathematical and Physical Sciences**	Kluwer Academic Publishers
D	**Behavioral and Social Sciences**	Dordrecht, Boston, and London
E	**Applied Sciences**	
F	**Computer and Systems Sciences**	Springer-Verlag
G	**Ecological Sciences**	Berlin, Heidelberg, New York, London,
H	**Cell Biology**	Paris, Tokyo, Hong Kong, and Barcelona
I	**Global Environmental Change**	

PARTNERSHIP SUB-SERIES

1. Disarmament Technologies	Kluwer Academic Publishers
2. Environment	Springer-Verlag
3. High Technology	Kluwer Academic Publishers
4. Science and Technology Policy	Kluwer Academic Publishers
5. Computer Networking	Kluwer Academic Publishers

The Partnership Sub-Series incorporates activities undertaken in collaboration with NATO's Cooperation Partners, the countries of the CIS and Central and Eastern Europe, in Priority Areas of concern to those countries.

Recent Volumes in this Series:

Volume 278 — Obesity Treatment: Establishing Goals, Improving Outcomes, and Reviewing the Research Agenda
edited by David B. Allison and F. Xavier Pi-Sunyer

Volume 279 — Organization of the Early Vertebrate Embryo
edited by Nikolas Zagris, Anne Marie Duprat, and Antony Durston

Volume 280 — Etiology of Hodgkin's Disease
edited by Ruth F. Jarrett

Volume 281 — Vascular Endothelium: Responses to Injury
edited by John D. Catravas, Allan D. Callow, and Una S. Ryan

Series A: Life Sciences

Vascular Endothelium

Responses to Injury

Edited by

John D. Catravas

Medical College of Georgia
Augusta, Georgia

Allan D. Callow

Boston University School of Medicine
Boston, Massachusetts

and

Una S. Ryan

T Cell Sciences, Inc.
Needham, Massachusetts

Plenum Press
New York and London
Published in cooperation with NATO Scientific Affairs Division

Proceedings of a NATO Advanced Study Institute on
Vascular Endothelium: Responses to Injury,
held June 18–27, 1994,
in Limin Hersonissou, Greece

NATO-PCO-DATA BASE

The electronic index to the NATO ASI Series provides full bibliographical references (with keywords and/or abstracts) to about 50,000 contributions from international scientists published in all sections of the NATO ASI Series. Access to the NATO-PCO-DATA BASE is possible in two ways:

—via online FILE 128 (NATO-PCO-DATA BASE) hosted by ESRIN, Via Galileo Galilei, I-00044 Frascati, Italy

—via CD-ROM "NATO Science and Technology Disk" with user-friendly retrieval software in English, French, and German (©WTV GmbH and DATAWARE Technologies, Inc. 1989). The CD-ROM also contains the AGARD Aerospace Database.

The CD-ROM can be ordered through any member of the Board of Publishers or through NATO-PCO, Overijse, Belgium.

Library of Congress Cataloging in Publication Data

Vascular endothelium: responses to injury / edited by John D. Catravas, Allan D. Callow, and Una S. Ryan.
p. cm.—(NATO ASI series. Series A, Life sciences: vol. 281)
"Proceedings of a NATO Advanced Study Institute on Vascular Endothelium: Responses to Injury, held June 18–27, 1994, in Limin Hersonissou, Greece"—T.p. verso.
"Published in cooperation with NATO Scientific Affairs Division."
Additional material to this book can be downloaded from http://extra.springer.com.

ISBN-13:978-1-4613-8013-9 e-ISBN-13:978-1-4613-0355-8
DOI: 10.1007/978-1-4613-0355-8

1. Vascular endothelium—Pathophysiology—Congresses. 2. Vascular endothelium—Physiology—Congresses. I. Catravas, John D. II. Callow, Allan D. III. Ryan, Una S. IV. North Atlantic Treaty Organization. Scientific Affairs Division. V. NATO Advanced Study Institute on Vascular Endothelium: Responses to Injury (1994: Limin Hersonissou, Greece) VI. Series: NATO ASI series. Series A, Life sciences; v. 281.
RC691.4.V384 1996 95-50886
616.1'3—dc20 CIP

ISBN-13:978-1-4613-8013-9

Softcover reprint of the hardcover 1st edition 1996

A Division of Plenum Publishing Corporation
233 Spring Street, New York, N. Y. 10013

10 9 8 7 6 5 4 3 2 1

PREFACE

This monograph contains the contributions to the NATO Advanced Studies Institute on *"Vascular Endothelium: Responses to Injury"*, which took place in Crete, Greece from June 18 to 27, 1994. This was the fourth in the series of NATO-supported ASIs in selected specific areas of endothelial cell biology, which began in 1988. Each time, it has combined clinical with basic scientists and renowned experts with novices in the field, in an attractive setting conducive to dissemination of knowledge, high quality discussions and exchange of ideas.

This, as well as previous ASIs, reflects the hard work of numerous people. As Co-Directors, we have been particularly fortunate to benefit from the expert advice of the Organizing Committee, which included Elizabetta Dejana, Alberto Mantovani and Magdi Yacoub. Their insightful suggestions helped formulate the scientific program of the ASI. We are also grateful for the work and dedication of the local organizing committee of Thomie Douraki, Stylianos Orfanos, Panayotis Behrakis, Michael Maragoudakis and Lydia Argyropoulos who tirelessly looked after the main details necessary to ensure that all of us enjoyed the meeting. Similarly, we wish to thank Jim Parkerson, Connie Snead, Andreas Papapetropoulos and Nandor Marczin for their assistance in various aspects of the many preparations for the conference. A very special thanks goes to Annie Cruz, the ASI Co-ordinator, for her efficient, solicitous and pleasant demeanor during the months prior to the ASI and especially during the ten days of the conference. Finally, this volume would not have been possible without the meticulous work of Aretha Bogan whose high editorial standards helped maintain the quality of these monographs.

We wish to acknowledge the generous contributions of T-Cell Sciences, Inc., Pfizer Ltd., and Texas Biotechnology Corporation. Their support was germaine to the success of the ASI.

John Catravas (Augusta)
Allan Callow (Boston)
Una Ryan (Boston)

CONTENTS

Abbreviations ix

I. TRANSPLANTATION AND TISSUE DAMAGE

Soluble Complement Receptor and Hyperacute Rejection 3
Scott K. Pruitt, William M. Baldwin, R. Randall Bollinger, Henry C. March, Jr., and Fred Sanfillippo

Early Detection of Lung Endothelial Dysfunction in Man 11
John D. Catravas, Attila Cziraki, Adrian Chester, Anna F. Maizza, Samad Tajkaremi, Asgar Kaghani, and Magdi Yacoub

Interactions of Superoxide Dismutases with Nitric Oxide 23
Brian Day, Tim Oury, and James D. Crapo

II. RESTENOSIS

Clinical Profile of Restenosis 33
Allan D. Callow

Role of Thrombin and Plasminogen Activators, Essential Proteases of the Coagulation and Fibrinolysis Cascades, in Fibrin Homeostasis and Vascular Cell Activation 45
Victor W. M. van Hinsbergh

Complement and Mechanisms of Restenosis 61
George B. Zavoico and Una S. Ryan

Genetic Factors Contributing to Cardiovascular Disease That May Affect Endothelial Structure and Function: The Role of Proteins Involved in Lipoprotein Transport 69
Vassilis I. Zannis, Maria Laccotripe, Savvas C. Makrides, Dimitris Kardassis, and Eleni Zanni

III. INFLAMMATION

Acute Inflammatory Lung Injury: Mechanisms and Interventions 131
Peter A. Ward, Andreas Seekamp, and Michael S. Mulligan

Cytokine Activation of Endothelium: Cloning and Characterization of a New IL-1 Inducible Gene 139
M. Introna, V. Vidal Alles, G. Picardi, A. Basile, C. Matteucci, B. Bottazzi, G. Peri, E. D'Aniello, F. Breviario, and A. Mantovani

Pro-Inflammatory Adhesins 145
Dorian O. Haskard

IV. ANGIOGENESIS/VASCULOGENESIS

Angiogenesis: An Overview of Regulation and Potential Clinical Application 157
Michael E. Maragoudakis

Molecular Organization of Endothelial Cell to Cell Junctions 167
Elisabetta Dejana and Jean-Luc Plantier

V. PULMONARY HYPERTENSION

The Cellular Basis of the Pathophysiology and Treatment of Pulmonary Hypertension 175
Stuart Rich

Inhaled Nitric Oxide in Acute Respiratory Failure 181
Jean-Jacques Rouby

Tissue Angiotensin Converting Enzyme 205
Suzanne Oparil, Qingcheng Meng, Shuang-dan Sun, Yiu-Fai Chen, and Louis J. Dell'Italia

VI. ABSTRACTS OF ORAL AND POSTER PRESENTATIONS 241

VII. PARTICIPANTS 347

VIII. INDEX 355

ABBREVIATIONS

ACE	angiotensin converting enzyme
ANG	angiotensin
ARDS	adult respiratory distress syndrome
bFGF	basic fibroblast growth factor
CAM	chorioallantoic membrane
CR1	complement receptor 1
CVF	cobra venom factor
DAF	decay accelerating factor
DDC	diethyldethiocarbonate
EC	endothelial cells
EC-SOD	extra cellular superoxide dismutase
ECM	extracellular matrix
EGF	epidermal growth factor
GAG	glycosaminoglycan
HAR	hyperacute rejection
Hb	hemoglobin
HHC	human heart chymase
HR	heart rate
HTS	hypothalamus-thalamus-septal
HUVEC	human umbilical vein endothelial cells
ICAM-1	intercellular adhesion molecule
IFNγ	interferon gamma
IFPI	issue factor pathway inhibitor
IGF-I	insulin-like growth factor-I
IL-1	interleukin-1
LPS	lipopolysaccharide
MAC	membrane attack complex
MAP	mean arterial pressure
MMP	matrix metalloproteinases
MPAP	mean pulmonary arterial pressure
MPO	myeloperoxidase

NO	nitric oxide
NOS	nitric oxide synthase
NTS	nuclear tractus solitarius
PA	plasminogen activators
PAF	platelet-activating factor
PAP	pulmonary arterial pressure
PAPD	pulmonary arterial pressure diastolic
PAPS	pulmonary arterial pressure systolic
PCR	polymerase chain reaction
PDGF	platelet-derived growth factor
PHA	phyto hemagglutinin
PKC	protein kinase C
PMA	phorbol myristate acetate
PMN	polymorphonuclear
PPH	primary pulmonary hypertension
PTCA	percutaneous transluminal coronary angioplasty
PVRI	pulmonary vascular resistance index
Q.s./Qt	pulmonary shunt
RBC	red blood cell
sCR1	soluble complement receptor 1
SHR	spontaneously hypertensin rats
SHRSP	spontaneously hypertensin rats stroke prone
SMC[15]	smooth muscle cell
SOD	superoxide dismutase
t-PA	tissue type plasminogen activator
TGF-b	transforming growth factor-b
TNF	tumor necrosis factor
TXA_2	thromboxane A_2
VEGF	vascular endothelial growth factor
VPF	vascular permeability factor
VSMC	vascular smooth muscle cell
Xg	Xenograft

I. TRANSPLANTATION AND TISSUE DAMAGE

SOLUBLE COMPLEMENT RECEPTOR AND HYPERACUTE REJECTION

Scott K. Pruitt, M.D., Ph.D, William M. Baldwin, III, M.D., Ph.D., R. Randall Bollinger, M.D., Ph.D., Henry C. March, Jr., Ph.D., and Fred Sanfillippo, M.D., Ph.D.

Duke University Medical Center
Durham, North Carolina USA

John Hopkins School of Medicine
Baltimore, Maryland USA

T Cells, Inc.
Needham, Massachusetts USA

ABSTRACT

Hyperacute rejection (HAR) of vascularized organ transplants occurs in both presensitized recipients of allografts and in discordant xenograft (Xg) recipients. HAR of allografts is currently avoided by excluding from transplantation those patients with high levels of alloreactive antibodies, since no clinically effective treatment exists to prevent this process. These patients are then added to ever expanding waiting lists, and many die while awaiting organ transplantation. Additionally, these lists continue to expand because of the increasing shortage of organs available for clinical transplantation. The use of Xgs from distantly related species might relieve this problem of donor availability but is also prevented by HAR.

HAR is mediated by the binding of recipient antibodies to the endothelium of the graft, with subsequent activation of the classical pathway of complement (C). In some models, alternative pathway C activation may also play a role in this process. In an effort to prevent the C-mediated injury of HAR, our laboratory has investigated the use of soluble complement receptor type 1 (sCR1) in several animal models.

In an ACI-to-Lewis rat presensitized cardiac allograft model, the administration of sCR1 significantly prolonged allograft survival in comparison to saline treated controls. In both the guinea pig-to-rat and the pig-to-cynomolgus monkey species combinations *in vivo*, and in an *ex vivo* pig-to-human model, the administration of sCR1 significantly prolonged cardiac Xg survival and function. These studies suggest that sCR1 may be a useful agent for both allotransplantation in the presensitized recipient and xenotransplantation.

ALLOTRANSPLANTATION IN THE PRESENSITIZED RECIPIENT

HAR of allografts occurs when elevated levels of antibodies directed against the donor organ are present in the recipient at the time of transplantation. Upon allograft reperfusion in these patients, the alloantibodies bind to the endothelium of the graft, with subsequent C activation via the classical pathway, platelet activation, cellular infiltration, and allograft loss within minutes to hours.

Production of these alloantibodies is stimulated by prior blood transfusions, pregnancies, or previous allografts, and no clinically effective therapy exists which will inhibit their production once they have been elicited. Most centers prevent HAR by avoiding transplantation in sensitized patients. These individuals are then added to the ever growing waiting lists, where many will die while awaiting a closely matched organ *(Evans et al., 1992)*.

Vascular Endothelium: Responses to Injury
Edited by John D. Catravas *et al.*, Plenum Press, New York, 1996

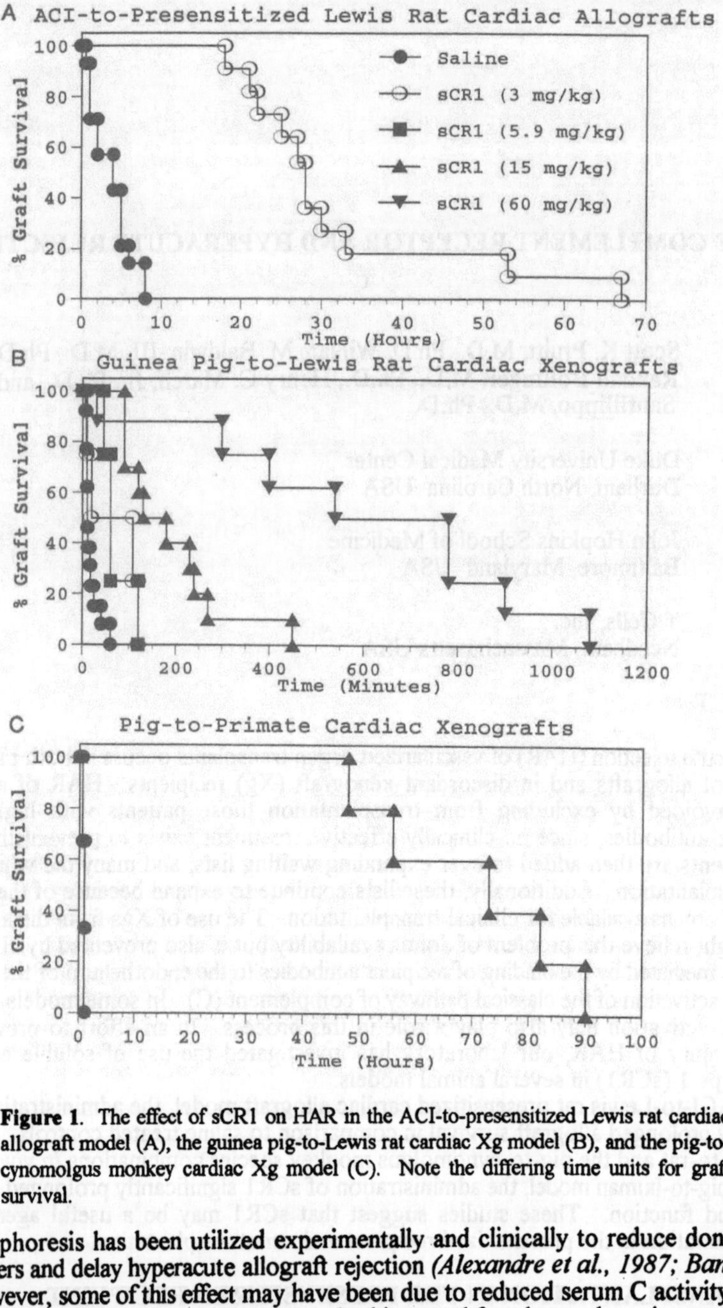

Figure 1. The effect of sCR1 on HAR in the ACI-to-presensitized Lewis rat cardiac allograft model (A), the guinea pig-to-Lewis rat cardiac Xg model (B), and the pig-to-cynomolgus monkey cardiac Xg model (C). Note the differing time units for graft survival.

Plasmaphoresis has been utilized experimentally and clinically to reduce donor-specific antibody titers and delay hyperacute allograft rejection *(Alexandre et al., 1987; Bannett et al., 1989)*. However, some of this effect may have been due to reduced serum C activity, since the C activating properties of the membranes and tubing used for plasmaphoresis are well known *(Henderson and Chenoweth, 1987; Kimikawa et al., 1992)*.

The critical role for C activation in hyperacute rejection has been demonstrated by experiments in which C depletion using cobra venom factor (CVF) delays this process *(Knechtle et al., 1985)*. However the clinical administration of CVF seems unlikely for several reasons. First, administration of CVF has been reported to result in neutropenia, thrombocytopenia, increased pulmonary vascular permeability, and death *(Orr and Warner, 1990; Regal, 1990)*. Second, as a foreign protein, CVF might be expected to elicit a humoral immune response resulting in the production of neutralizing antibodies as has been reported with the administration of the foreign mouse IgG OKT3 in human transplant recipients *(Hammond et al., 1993)*. As a recombinant form of a ubiquitously expressed human protein, sCR1 might not be as immunogenic as foreign proteins in transplant patients.

With the development of soluble C receptor type 1 (sCR1), a more clinically applicable method of C inhibition may now be available. sCR1 is a recombinant human protein which acts to inhibit both the alternative and classical C pathways at multiple points during the activation process *(Weisman et al., 1990)*. Specifically, sCR1 inhibits both the classical and alternative C pathways by binding to proteins (C3b and C4b) of the multi-subunit C3 and C5 convertases and promoting the dissociation of the catalytic subunits (C2a and Bb) in each case. By serving as a cofactor for the serum protease Factor I, sCR1 further acts to promote the degradation of activated C3 (C3b) to inactive forms iC3b and C3dg and of activated C4 (C4b) to inactive iC4b *(Weisman et al., 1990; Yeh et al., 1991)*.

We hypothesized that C inhibition with sCR1 would inhibit HAR of allografts in the presensitized recipient. Based on studies performed in the rat showing that the administration of sCR1 reduced C-mediated myocardial reperfusion injury *(Weisman et al., 1990; Weisman et al., 1990)*, we utilized the well-established ACI-to-presensitized Lewis rat cardiac allograft model in an effort to prove this hypothesis.

THE EFFECT OF sCR1 ON HYPERACUTE ALLOGRAFT REJECTION

In this study *(Pruitt and Bollinger, 1991)*, male Lewis rats were sensitized with three successive full thickness ACI rat skin grafts at approximately two week intervals. Two to three weeks later, after high levels of anti-ACI antibodies had been confirmed, these presensitized rats received heterotopic cardiac allografts from an ACI donor. Immediately prior to allograft reperfusion, the recipients received either a bolus intravenous injection of sCR1 (3mg/kg, n=11) or an equivalent volume of PBS as a control (n=10).

In the PBS treated controls, HAR occurred, with allograft survival averaging 3.25±0.81 hours. The single 3 mg/kg dose of sCR1 significantly prolonged allograft survival to 32.0±4.47 hours ($p<0.0001$, Figure 1A), and no adverse effects were noted. This dose of sCR1 was found to reduce serum total C hemolytic activity to between 30 and 40% of preoperative levels over the first 3.25 hours following allograft reperfusion (Figure 2A). At the time of rejection in the sCR1 treated recipients, serum C activity had risen to levels comparable to those seen in the PBS treated recipients at rejection.

When the allografts were examined histologically, significant histologic differences between treatment groups were observed. sCR1 prevented the formation of intravascular platelet thrombi within the coronary arteries and arterioles of the grafts which were observed occluding the vessel lumena in the allografts from the saline treated controls.

sCR1 therapy also prevented the neutrophil infiltration seen in the controls, but interestingly, the allografts from the sCR1 treated recipients at the time of rejection demonstrated marked infiltration with 0X42+ (rat CR3+) mononuclear cells. Since sCR1 functions as a cofactor for factor I mediated degradation of C3b deposited on the endothelial cell surface, treatment with sCR1 might be expected to result in hemolytically inactive iC3b and C3dg remaining deposited on these cells. Mononuclear cells might then interact with the graft through their receptors for iC3b (CR3).

XENOTRANSPLANTATION

In addition to the growing number of presensitized patients that are excluded from transplantation, the number of patients awaiting organs for transplantation continues to increase each year because of a shortage of donor organs. Xenotransplantation offers a potentially unlimited supply of transplantable organs to meet this clinical need *(Reemstma, 1992)*. "Concordant" xenotransplantation, where naturally occurring xenoreactive antibodies that are associated with hyperacute rejection are not detectable in the recipient, has been performed clinically but with limited success. In the 1960's, Reemstma and colleagues performed twelve chimpanzee to human concordant renal Xgs, with one patient surviving 9 months with a functioning chimpanzee kidney before dying of infection *(Reemstma et al., 1964)*. More recently, a neonate survived 4 weeks with a concordant baboon cardiac Xg *(Bailey et al., 1985)*.

While this limited clinical experience with concordant xenotransplantation suggests that monkeys and apes might be a logical choice for transplantable Xgs, the use of primates as donors for human transplantation creates several practical problems. First of all, primates are in short supply and breed very slowly in captivity. Secondly, primates can carry a variety of viruses that are pathogenic to humans *(Michaels and Simmons, 1994)*. Finally, the use of

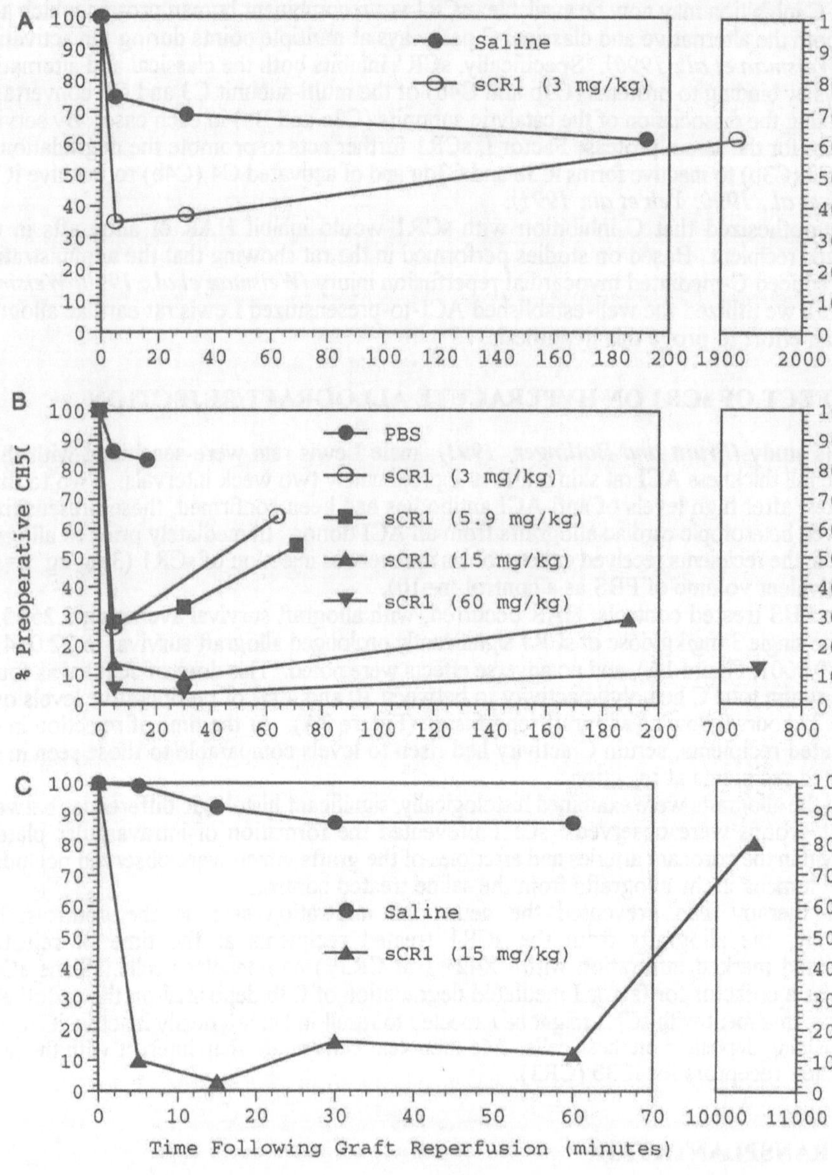

Figure 2. The effect of sCR1 on total serum hemolytic activity in the ACI-to-presensitized Lewis rad cardiac allograft model (A), the guinea pig-to-Lewis rat cardiac Xg model (B), and the pig-to-cynomolgus monkey cardiac Xg model (C). Values are expressed as the percentage of preoperative CH50 for each recipient and were determined using the standard antibody sensitized SRBC hemolysis assay.

otherwise normal primates as organ donors raises complicated ethical issues and public policy concerns. For these reasons, primate organs are unlikely to meet the need for clinically transplantable organs.

The use of organs from more distantly related species, many of which are already used for food and other resources, overcomes these problems. However, hyperacute rejection, which occurs within minutes to hours of Xg reperfusion and results in immediate graft loss, is currently a barrier to successful organ transplantation between these distantly related discordant species. Unfortunately, while numerous immunosuppressants are available to inhibit the cellular immune response to Xgs *(Auchincloss, 1990)*, no clinically effective immunosuppressive regimens yet exist that will safely prevent this humorally mediated Xg rejection process. As donor organ shortages continue to worsen, the clinical need for such a regimen becomes increasingly apparent.

The process of hyperacute Xg rejection is thought to be initiated by the binding of xenoreactive naturally occurring antibodies (NAbs) to the endothelium of the xenotransplanted organ *(Platt et al., 1990)*. Studies, in which depletion of NAbs using plasmaphoresis and pretransplantation xenogeneic organ perfusion have led to prolonged survival of experimental discordant Xgs, support the concept that NAb binding to the xenogeneic organ is an important component of the hyperacute rejection response *(Platt et al., 1990; Flores et al., 1992)*. However, C consumption during these procedures may also have contributed to the observed prolongation Xg survival in these studies.

Following the binding of NAbs to the endothelium of the Xg, the classical C pathway is activated *(Platt et al., 1990)*. Alternative C pathway activation may then help amplify classically initiated C activation. In some experimental models, however, activation of the alternative pathway by the xenogeneic organ, in the absence of measurable NAbs, may initiate hyperacute Xg rejection *(Johnston et al., 1993)*. The importance of C activation in the process of hyperacute Xg rejection has been demonstrated by several studies in which depletion of recipient C using CVF administration *in vivo* has led to marked prolongation of discordant Xg survival in several animal models *(Adachi et al., 1987; Miyagawa et al., 1988)*. Because of potentially serious side effects as described previously, the clinical administration of CVF remains unacceptable. Other inhibitors of C have also been shown experimentally to delay hyperacute rejection of discordant Xgs *(Miyagawa et al., 1993)*.

After C has been activated, various cellular elements such as platelets or red blood cells may contribute to irreversible damage of the Xg *in vivo (Han et al., 1990)*. Makowka and colleagues have reported that the administration of the platelet activating factor antagonist SRI 63-441 to recipients significantly prolongs Xg survival in the discordant guinea pig-to-rat model *(Makowka et al., 1990)* suggesting that platelet activation may contribute to discordant Xg damage, at least in the rodent. However, unpublished data from this laboratory using an *ex vivo* isolated heart preparation has demonstrated in a variety of species combinations including guinea pig-to-rat, rat-to-human, and rat-to-chicken, that xenogeneic C plus NAb, in the absence of cellular elements, is sufficient to cause hyperacute Xg rejection of a discordant Xg. Thus, modulation of NAb activity, C activity, or both would be expected to significantly modulate the HAR response to discordant Xgs.

Recipients with even very low titers of xenoreactive natural antibodies still hyperacutely reject discordant Xgs *(Kaplan et al., 1990; Leventhal, et al., 1992)*. Because the currently available methods for xenoreactive antibody depletion do not fully eliminate NAbs from recipient serum, it is unlikely that these methods alone will prevent hyperacute Xg rejection. For this reason, our laboratory has chosen to investigate methods to reduce C activity *in vivo* in an effort to inhibit or delay hyperacute Xg rejection. Because sCR1 was effective in inhibiting HAR of allografts in our rodent model, we performed a variety of experiments to evaluate the effectiveness of sCR1 in delaying HAR of Xgs.

THE EFFECT OF sCR1 ON HAR IN A SMALL ANIMAL MODEL

To test the effectiveness of sCR1 in preventing HAR of discordant Xgs, we first utilized the well established guinea pig-to-Lewis rat heterotopic cardiac xenotransplantation model *(Pruitt et al., 1991; Pruitt et al., 1992)*. The dose (3 mg/kg) which was effective in significantly prolonging allograft survival in the presensitized rat did not prolong guinea pig cardiac Xg survival (Figure 1B).

The failure of this dosage, which effectively prevented HAR of an allograft, to prolong discordant guinea pig cardiac Xg survival might be attributed to several factors. First, work in our laboratory has demonstrated that in the guinea pig-to-Lewis model, the guinea pig cardiac Xg can activate rat C via the alternative pathway *(Pruitt et al., 1993)*. Second, the C regulatory proteins present on the graft endothelium differ. While the rat homologous restriction factor, decay accelerating factor, membrane cofactor protein, and CD59 may help protect the allograft from the C mediated injury, the xenogeneic guinea pig forms of these regulatory proteins may not all interact with rat serum C components, making the Xg more susceptible to C mediated injury *(Parker, 1992)*.

However, as we increased the dose of sCR1 administered to the recipients, we were able to demonstrate a direct correlation between sCR1 dosage and Xg survival (Figure 1B). The highest dose of sCR1 (60 mg/kg) prolonged Xg survival to over 12 hours, in contrast to an average of 17 minutes in PBS treated controls. sCR1 administration significantly inhibited serum total C activity in a parallel dose dependent fashion (Figure 2B). It is of note that at the time of Xg rejection in the 60 mg/kg sCR1 group, minimal total C activity was present. Rejection in this setting might be attributable to the cumulative effects of continuous low levels of C activation, suggesting that continuous inhibition of C with an infusion of sCR1 might be beneficial. Alternatively, the release of C3a and C5a even in the face of sCR1 levels that significantly inhibit total hemolytic serum C activity *(Yeh et al., 1991)* might contribute to Xg damage *(Jose et al., 1981; Bjork et al., 1985; del Balzo et al., 1985)*.

THE EFFECT OF sCR1 ON HAR IN AN *EX VIVO* PIG-TO-HUMAN MODEL

Because of the species specificity of the HAR response to Xgs, we next evaluated sCR1 using an isolated working pig heart preparation in which hearts from porcine donors were perfused with AB+ normal human whole blood *(Kirk et al., 1993)*. This preparation allowed measurement of cardiac functional parameters throughout the four hour period of observation. Blood from each of three human volunteers was tested with PBS, sCR1 (70 µg/ml), and sCR1 (300 µg/ml) added.

In a dose dependent response, the addition of sCR1 to the whole blood perfusate improved porcine cardiac Xg cardiac performance *(Pruitt et al., 1994)*. In contrast to in vivo studies in which either the presence or absence of cardiac contraction has been used to assess Xg survival or rejection, respectively, this study was the first to document actual preservation of Xg function with an immunomodulatory agent. This study thus indicated that sCR1 was effective in inhibiting human C and in delaying the detrimental effects of HAR on porcine cardiac Xgs. Because the limitations of the ex vivo model prevented the continuation of the experiments beyond 4 hours duration, we next evaluated the effect of sCR1 in a preclinical pig-to-primate in vivo model of Xg HAR.

THE EFFECT OF sCR1 ON HAR OF PIG-TO-PRIMATE CARDIAC Xgs

In these studies *(Pruitt et al., 1994)*, male cynomolgus monkeys received heterotopic cardiac Xgs from porcine donors, and immediately prior to Xg reperfusion received either sCR1 (15 mg/kg, n=5) or an equivalent volume of saline (n=3) intravenously. This sCR1 dosage was chosen to yield serum levels of approximately 300 µg/ml, comparable to those obtained in the ex vivo perfusion study. As shown in Figure 1C, this single dose of sCR1 prolonged Xg survival from 1 hour or less for the saline controls to 2-3 days. Importantly, no detrimental effects were observed with sCR1 administration.

This prolongation achieved with a single bolus of sCR1 is similar to that achieved by Leventhal and colleagues using repeated doses of CVF to deplete serum C activity in a pig-to-baboon model *(Leventhal et al., 1993)*. In contrast to our studies where a minimal cellular infiltrate was observed in the rejected cardiac Xgs after sCR1 therapy, this group noted a significant mononuclear cell infiltrate within the Xgs. These differences might be due to the large amounts of C3a and C5a released when serum C is consumed by CVF. These anaphylatoxins are able to activate neutrophils and increase their adhesion to endothelial cells *Tonneson et al., 1989)*, and both neutrophils and macrophages have receptors for C5a which can stimulate chemotaxis and cytokine upregulation.

The administration of sCR1 markedly inhibited serum alternative and classical pathway C activity (Figure 2C) over the first hour following Xg reperfusion. In the saline treated controls, total C hemolytic activity was unchanged, but alternative pathway decreased, suggesting that alternative pathway activation might play a role in HAR in the pig-to-primate model. C activity was returning toward preoperative levels at the time of Xg rejection. Because HAR occurred when C activity was no longer maximally inhibited, studies are currently underway in the pig-to-cynomolgus monkey model in which recipients receive a continuous intravenous infusion of sCR1 to maintain C inhibition.

Anti-porcine IgM and IgG antibody levels fell over the initial 2-3 days of Xg perfusion, but had begun to return toward or even exceed preoperative levels at the time of Xg rejection. Whether these NAbs are detrimental to the Xg is currently unknown, but this rise in anti-porcine NAb levels may be difficult to inhibit given the currently available immunosuppressive agents.

Immunofluorescence microscopic evaluation of Xg biopsies at 1 hour following reperfusion (and at rejection in saline treated controls) demonstrated significant IgM and IgG deposition in all Xgs, indicating that sCR1 treatment had no effect on NAb deposition *in vivo*. C3b/iC3b deposition was noted in Xgs from saline treated recipients but was absent in sCR1 treated recipients both at 1 hr and at rejection. C3d deposition, as detected with a mAb recognizing C3d as both an isolated fragment and as part of C3b/iC3b, was present in all Xgs. These immunohistologic findings are consistent with the cofactor activity of sCR1.

SUMMARY

The administration of sCR1 significantly prolongs cardiac allograft survival in presensitized rats. Likewise, sCR1 significantly prolongs the survival and function of cardiac xenografts between discordant species (the guinea pig-to-rat and the pig-to-cynomolgus monkey species combinations *in vivo*, and in an *ex vivo* pig-to-human model). These studies suggest that sCR1 may be a useful agent for both allotransplantation in the presensitized recipient and xenotransplantation.

REFERENCES

Adachi, H., Rosengard, B.R. et al., 1987, Effects of cyclosporine, aspirin, and cobra venom factor on discordant cardiac xenograft survival in rats. *Transplant Proc.* 19:1145.

Alexandre, G.P., Sqifflet, J.P. et al., 1987, Present experiences in a series of 26 ABO-incompatible living donor renal allografts. *Transplant Proc.* 19(6):4538-42.

Auchincloss, H., 1990, Xenografting: A review. *Transplant Rev.* 4:14.

Bailey, L.L.,, Nehlsen-Cannarella, S.L. et al., 1985, Baboon-to-human cardiac xenotransplantation in a neonate. *JAMA* 254:3321.

Banett, A.D., McAlack, R.F. et al., 1989, ABO incompatible renal transplantation: a qualitative analysis of native endothelial tissue ABO antigens after transplantation. *Transplant Proc.*

Bjork, J., Hugli, T.E. et al., 1985, Microvascular effects of anaphylatoxins C3a and C5a. *J. Immunol.* 234(2):1115-9.

del Balzo, U., Levi, R. et al., 1985, Cardiac dysfunction caused by purified human C3a anaphylatoxin. *Proc. Natl. Acad. Sci. USA* 82(3):886-90.

Evans, R.W., Orians, C.E. et al., 1992, The potential supply of organ donors. An assessment of the efficacy of organ procurement efforts in the United States. *JAMA* 267(2):239-246.

Fores, H.C., Leventhal, J.R. et al., 1992, Natural antibody production can be inhibited by 15-deoxyspergualin in a discordant xenograft model. *Transplantation Proceedings* 24(2):

Forty, J., White, D.J. et al., 1993, Activation of the alternative pathway of complement in hyperacute xenograft rejection of rabbit hearts by human blood. *J. Heart and Lung Transplant.* 12:(2):283-6.

Hammond, E.A., Yowell, R.L. et al., 1993, Prevention of adverse clinical outcome by monitoring of cardiac transplant patients for murine monoclonal CD3 antibody (OKT3) sensitization. *Transplantation* 55(5):1061-3.

Han, L., Henry, M.L., 1990, Modification of hyperactive xenograft rejection by ex vivo xenoantibody adsorption. *Current Surgery* 47(1):15-7.

Henderson, L.W. and Chenoweth, D., 1987, Biocompatibility of artificial organs: an overview. [Review]. *Blood Purif.* 5(2-3):100-11.

Johnston, P.S., Wang, M.W. et al., 1992, Discordant xenograft rejection in an antibody-free model. *Transplantation* 54(4):573-6.

Jose, P., Forrest, M.J. et al., 1981, Human C5a des Arg increases vascular permeability. *U. Immunol.* 127(6):2376-80.

Kaplan, E., Dresdale, A.R. et al., 1990, Total lymphoid irradiation and discordant cardiac xenografts. *J. Heart Transplant* 9(1):11-3.

Kimikawa, M., Agishi, T. et al., 1992, Prolongation of cardiac xenograft survival by double filtration plasma pheresis and ex vivo xenoantibody adsorption. *Transplantation Proceedings* 24(2):447.

Kirk, A.D., Heinle, J.S. et al., 1993, Ex vivo characterization of human anti-porcine hyperacute cardiac rejection. *Transplantation* 56(4):785-93.

Knechtle, S.J., Halperin, E.C. et al., 1985, The effect of cyclosporine, total lymphoid irradiation, and cobra venom factor on hyperacute rejection. *J. Heart Transplant* 4(5):541-5.

Leventhal, J.R., Dalmasso, A.P. et al., 1993, Prolongation of cardiac xenograft survival by depletion of complement. *Transplantation* 55(4):857-65.

Leventhal, J.R., Flores, H.C. et al., 1992, Evidence that 15-deoxyspergualin inhibits natural antibody production but fails to prevent hyperacute rejection in a discordant xenograft model. *Transplantation* 54(1):26-31.

Makowka, L., Chapman, F.A. et al., 1990, Platelet-activating factor and hyperacute rejection. The effect of a platelet-activating factor antagonist, SRI 63-441, on rejection of xenografts and allografts in sensitized hosts. *Transplantation* 50(3):359-65.

Michaels, M.G. and Simmons, R.L., 1994, Xenotransplant-associated zoonoses. Strategies for prevention. [Review]. *Transplantation* 57(1)1-7.

Miyagawa, S., Hirose, H. et al., 1988, The mechanism of discordant xenograft rejection. *Transplantation* 46(6):825-30.

Miyagawa, S., Shirakura, R. et al., 1993, Prolonging discordant xenograft survival with anticomplement reagents K76COOH and FUT175. *Transplantation* 55(4):709-13.

Orr, F.W. and Warner, D.J., 1990, Effects of systemic complement activation and neutrophil-mediated pulmonary injury on the retention and metastasis of circulating cancer cells in mouse lungs. *Lab Invest.* 62(3):331-8.

Parker, C.J., 1992, Regulation of complement by membrane proteins: an overview. [Review]. *Curr. Top. Microbiol. Immunol.* 178(1):1-6.

Platt, J.L., Vercellotti, G.M. et al., 1990, Transplantation of discordant xenografts: a review of progress. [Review]. *Immunology Today* 11(12):450-6.

Pruitt, S.K., Baldwin, W. et al., 1991, The effect of soluble complement receptor type 1 on hyperacute xenograft rejection. *Transplantation* 52(5):868-73.

Pruitt, S.K., Baldwin, W. et al., 1992, Effect of soluble complement receptor type 1 on natural antibody levels during xenograft rejection. *Transplant Proc.* 24(2):477-8.

Pruitt, S.K., Baldwin, W. et al., 1993, The effect of xenoreactive antibody and B cell depletion on hyperacute rejection of guinea pig-to-rat cardiac xenografts. *Transplantation* 56:1318-1324.

Pruitt, S.K. and Bollinger, R.R., 1991, The effect of soluble complement receptor type 1 on hyperacute allograft rejection. *J. Surg. Res.* 50(4):350-5.

Pruitt, S.K., Kirk, A.D. et al., 1994, The effect of soluble complement receptor type 1 on hyperacute rejection of porcine xenografts. *Transplantation* 57(3):363-70.

Reemstma, K., McCracken, B.H. et al., 1964, Heterotransplantation of the kidney: Two clinical experiences. *Science* 143:700.

Reemstma, K., 1992, Xenografts. *Transplantation Proceedings* 24(5)

Regal, J.F., 1990, Enhancement of antigen-induced bronchoconstriction in the guinea pig after intravascular complement activation with cobra venom factor. *Int. Arch. Allergy Appl. Immunol.* 91(1):86-94.

Tonnesen, M.G. Anderson, D.C. et al., 1989, Adherence of neutrophils to cultured human microvascular endothelial cells. Stimulation by chemotactic peptides and lipid mediators and dependence upon the Mac-1, LFA-1, p150, 95 glycoprotein family. *J. Clin. Invest.* 83(2):637-46.

Weisman, H.F., Bartow, T. et al., 1990, Recombinant soluble CR1 suppressed complement activation, inflammation, and necrosis associated with reperfusion of ischemic myocardium. *Trans. Assoc. Am. Physicians* 103(64):64-72.

Wiesman, H.F., Bartow, T. et al., 1990, Soluble human complement receptor type 1: in vivo inhibitor of complement suppressing post-ischemic myocardial inflammation and necrosis. *Science* 249(4965):146-51.

Yeh, C., Marsh, H.J. et al., 1991. Recombinant soluble human complement receptor type 1 inhibits inflammation in the reversed passive arthus reaction in rats. *J. Immunol.* 146(1):250-6.

EARLY DETECTION OF LUNG ENDOTHELIAL DYSFUNCTION IN MAN

John D. Catravas, Attila Cziraki, Adrian Chester, Anna F. Maizza, Samad Tajkaremi, Asgar Kaghani, and Magdi Yacoub

Vascular Biology Center
Medical College of Georgia
Augusta, Georgia, U.S.A.

Heart Science Center
Harefield Hospital
Harefield, Middlesex, U.K.

INTRODUCTION

Pulmonary vascular endothelial cells express a variety of transporter and enzyme proteins which are known or suspected modulators of cardiovascular homeostasis and are susceptible to injury (*Ryan, U.S., 1987*). Over the past decade, studies by us and other investigators have accumulated considerable experimental evidence suggesting that the activity of pulmonary capillary endothelium-bound angiotensin converting enzyme (ACE), a prominent endothelial ectoenzyme which is uniformly distributed throughout the vasculature, can be assayed repeatedly, accurately and reproducibly, *in vivo* (*Ryan, J.W., 1977, 1991; Catravas et al, 1981, 1983*). Reduced endothelium-bound ACE activity is the earliest yet known manifestation of impeding lung injury in various animal models of lung damage, *in vivo* (e.g., hyperoxia, irradiation, bleomycin, activated neutrophil, etc.) (*Bakhle, 1979; Catravas et al, 1984, 1989; Gillis et al, 1986; McCormick et al, 1987; Toivonen et al, 1981*). Based on these findings, we have hypothesized that monitoring of changes in pulmonary capillary endothelium-bound ACE activity may be used as a diagnostic test to aid in the early detection of lung injury in man.

Recently, we have accumulated evidence from a population of lung transplant patients indicating that patients with pulmonary dysfunction have dramatically lowered pulmonary capillary endothelium-bound ACE activity and that ACE activity in lung donors may be a useful parameter in the screening of lung grafts.

Angiotensin Converting Enzyme

Angiotensin converting enzyme (kininase II) is a carboxypeptidase found in most mammalian tissues. In the lung, ACE is primarily located along the luminal surface of the endothelial cells and is responsible for the processing of circulating angiotensin I and bradykinin. *In vivo* assays of ACE activity utilizing either angiotensin I or bradykinin as

Vascular Endothelium: Responses to Injury
Edited by John D. Catravas *et al.*, Plenum Press, New York, 1996

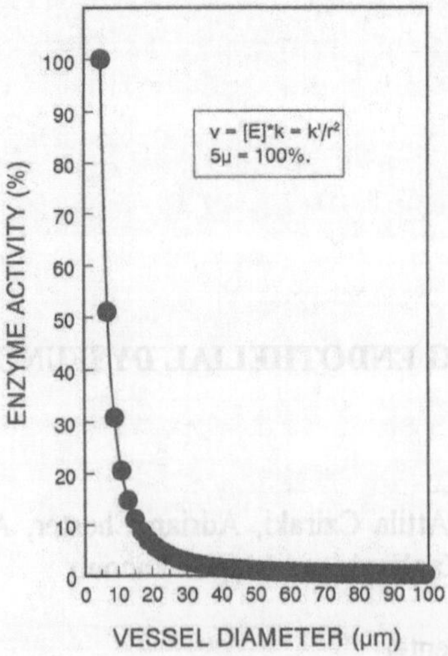

Figure 1. Correlation between vessel diameter and vessel wall enzyme activity assuming cylindrical vessels with 100% activity at 5µ diameter. Enzyme activity is estimated as: $v=V_{max}=[E]\bullet k_{cat}=[E]\bullet k$ or $v=V_{max}/K_m=[E]\bullet k_{cat}/K_m=[E]\bullet k$. In either case $[E]\bullet k=E\bullet 1/\pi r^2\bullet k=k^1/r^2$.

substrates are complicated by the fact that compounds are also metabolized by endogenous aminopeptidases and endopeptidases. In contrast, certain synthetic substrates of ACE show low affinity for enzymes other than ACE and yield products that are easily distinguished from the parent compound. One such substrate, benzoyl-Phe-Ala-Pro (BPAP) is specific for blood, lung and urine ACE. In cultured endothelial cells, conversion of BPAP is completely inhibited by captopril at concentrations (10^{-6}M) known to preferentially inhibit ACE without affecting other peptidases. In the presence of ACE, BPAP is converted to benzoyl-Phenylalanine and Alanyl-Proline; *in vitro,* the affinity (K_m) of BPAP for ACE is slightly higher than that for bradykinin, but lower than that for angiotensin I. *In vivo*, BPAP has been shown to extensively metabolized during a single transpulmonary passage in rabbits, dogs, cats, minipigs, rats, guinea pigs and sheep (*Ryan, J.W. et al, 1978; Catravas, 1984, 1986*).

Measurement of BPAP Hydrolysis by Pulmonary Capillary Endothelium-Bound ACE

For transpulmonary measurements of capillary endothelium-bound ACE activity in man, the specific ACE substrate ^{3}H-BPAP (20 µCi; 26 Ci/mmol) is rapidly injected as a 0.7 ml bolus into the right atrium through the proximal port of a Swan-Ganz catheter. At the same time, arterial blood is collected from the radial artery over a 60 second period by means of a peristaltic pump (operating at 30 ml/min) into a fraction collector equipped with tubes containing 3.5 mM EDTA and 3.4 mM 8-hydroxyquinoline in saline (to prevent further substrate hydrolysis by blood ACE) and advancing at the rate of one tube per 2.4 sec.

All sample tubes are centrifuged at 3,000 rpm for 10 minutes, to separate cells from plasma. Aliquots (0.5 ml) from the supernatant of each tube are transferred into two sets of 7 ml polyethylene scintillation vials. Total ^{3}H activity is measured in one set of vials in the presence

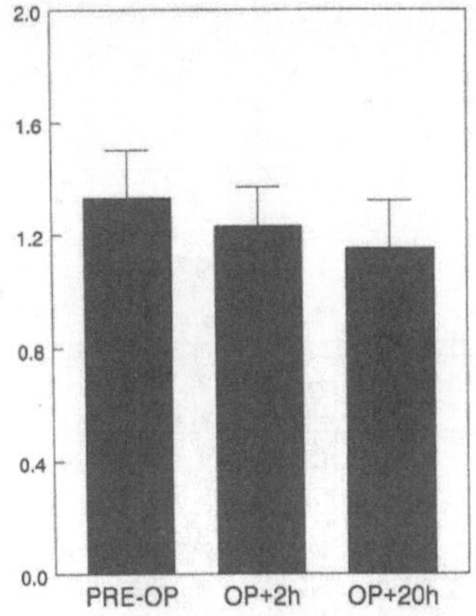

Figure 2. *CABG Patients (N=12) Lung Ace Activity.* ^{3}H-BPAP hydrolysis (v) by pulmonary capillary endothelium-bound ACE in patients undergoing surgery for coronary arterial bypass graft. Three measurements were performed in each patient, one after induction of anesthesia, but before surgery (PRE-OP), a second approximately two hours after completing cardiopulmonary bypass (OP+2), and a third 20 hours later in the intensive care unit (OP+20). Means ±SE of 12 patients.

of 5 ml Ecoscint A scintillation cocktail by a liquid scintillation spectrometer. The other 0.5 ml aliquot in the second set of vials was mixed with 2.5 ml of 0.12 N hydrochloric acid and radioactivity is estimated in the presence of 3 ml of 4 g/L Omnifluor in toluene, after 48 hours of undisturbed equilibration in the dark. In this way, less than 8% of the substrate and 62% of the metabolite ([^{3}H]BPhe) is extracted into the toluene (counting) phase (the precise fraction is determined by concurrently assayed pure [^{3}H]BPAP and [^{3}H]-BPhe, see below). The amount of metabolite radioactivity in each sample is calculated according to:

$$[^3H]BPhe = (toluene\ ^3H - f_s \bullet plasma\ ^3H)/(f_p - f_s) \quad (1)$$

where f_p is the fractional extraction of product [^{3}H]BPhe into the counting phase (toluene) of the second set of vials and f_s is the fractional extraction of the substrate [^{3}H]BPAP into the counting phase. Ten microliters of injectate are added to five additional blood sample tubes to determine $f_s \bullet f_p$ is similarly determined by adding 10 µl of previously synthesized [^{3}H]BPhe into five different blood sample tubes and processing them as all the other samples. Metabolite disintegrations per minute (dpm) per milliliter of plasma were calculated for each sample collected.

The single-pass transpulmonary substrate hydrolysis of ^{3}H-BPAP (v) is calculated by applying the integrated Henri-Michaelis-Menten equation, under first order reaction conditions, to each sample and then integrating over the entire 60 sec interval, as proposed by Ryan and Catravas (*Ryan et al, 1978; Catravas et al, 1981*):

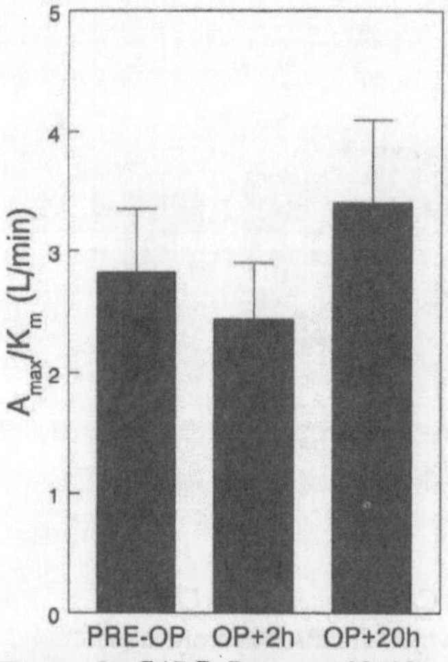

Figure 3. *CABG Patients (N=12) Capillary Surface Area.* Surface area index of dynamically perfused lung capillaries inferred from A_{max}/K_m values of ^{3}H-BPAP for pulmonary capillary endothelium-bound ACE in patients undergoing surgery for coronary arterial bypass graft. Three measurements were performed in each patient, one after induction of anesthesia, but before surgery (PRE-OP), a second approximately two hours after completing cardiopulmonary bypass (OP+2), and a third 20 hours later in the intensive care unit (OP+20). Means ±SE of 12 patients.

$$v = \ln([S_0]/[S]) = [E] \bullet t_c \bullet k_{cat}/K_m \quad (2)$$

with [E], t_c, and k_{cat} being the capillary enzyme concentration, reaction time (capillary mean transit time), and catalytic rate constant , respectively, while K_m is the Henri-Michaelis-Menten constant. [S_0] is the initial substrate concentration ([BPAP] + [BPhe]) and [S] is the surviving substrate [BPAP] concentration in the effluent arterial plasma, estimated in dpm/ml.

The Modified First Order Parameter A_{max}/K_m : Capillary Surface Area Index

Angiotensin converting enzyme is distributed homogeneously over the pulmonary endothelial luminal surface (*Ryan, U.S. et al, 1976*). Therefore, changes in the amount of ACE available to interact with a circulating substrate reflect changes in the dynamically perfused capillary surface area. Under first order reaction conditions, the modified first order parameter A_{max}/K_m is proportional to enzyme mass and thus proportional to the dynamically perfused capillary surface area; it is calculated from the integrated Henri-Michaelis-Menten equation:

$$A_{max}/K_m = E \bullet k_{cat}/K_m = \dot{Q} \bullet \ln([S_0]/[S])$$

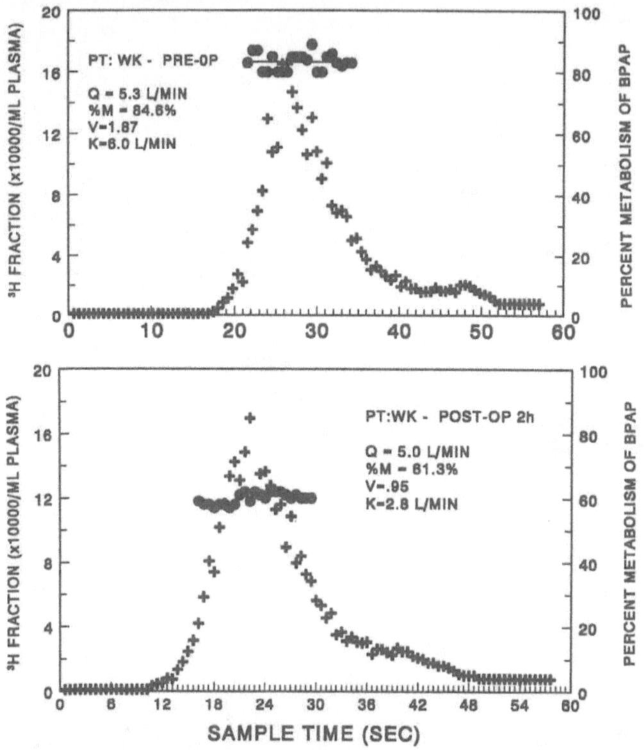

Figure 4. Top Panel: Arterial plasma fractional concentrations of tritium (crosses) and percent ^{3}H-BPAP metabolism (%M) by pulmonary capillary endothelium-bound ACE (solid circles) in a patient undergoing surgery for coronary arterial bypass graft after induction of anesthesia, but before surgery. Notice the stable sample-to-sample %M values, despite widely ranging substrate concentrations, reflecting the existence of first order reaction condition, as well as homogeneous capillary transit times. Bottom Panel: Arterial plasma fractional concentrations of tritium (crosses) and percent ^{3}H-BPAP metabolism (%M) by pulmonary capillary endothelium-bound ACE (solid circles) in the same patient two hours after completion of cardiopulmonary bypass. Notice that $k=(A_{max}/K_m)$, an index of dynamically perfused capillary surface area) is approximately half of the value measured pre-operatively. Subsequent chest x-ray identified a collapsed right lung.

where $\dot{Q}$ is plasma flow (calculated from the indicator dilution, arterial effluent ^{3}H concentration curve), $[S_0]$ and $[S]$ are the initial and surviving substrate concentrations, respectively, A_{max} is the product of enzyme mass and k_{cat}; (the catalytic rate constant) and K_m is the Michaelis constant.

During a single transpulmonary passage, metabolism of ^{3}H-BPAP takes place almost exclusively (>90%) at the capillaries (vessel diameter < 20μ) where the ratio of vessel wall surface area to plasma volume, and hence luminal enzyme concentration, is the highest. This is demonstrated in Figure 1 which depicts the relationship between vessel diameter and enzyme activity (v). For this relationship, it is assumed that capillaries are perfect tubes and that a 5μ diameter capillary expresses 100% enzyme activity.

Pulmonary Capillary ACE Activity in CABG Subjects

Pulmonary capillary endothelium-bound ACE activity was determined form the single pass transpulmonary hydrolysis of ^{3}H-BPAP in twelve patients undergoing coronary arterial bypass graft (CABG) surgery without manifest lung disease. Three measurements were performed

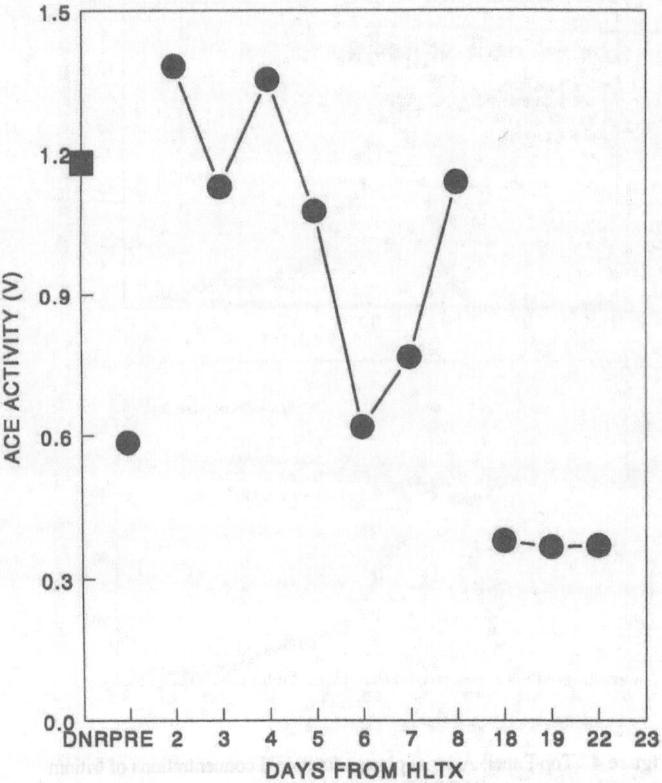

Figure 5. *HLTX Subject: C.S.* Pulmonary capillary endothelium-bound ACE activity (v; solid circles) before and after heart and lung transplant(HLTX) in a 28-year old female. Solid square shows graft ACE activity in the donor prior to harvesting. Unconnected solid circle shows ACE activity in the recipient before the transplant.

in each patient; one after induction of anesthesia, but before surgery, a second approximately two hours after completion of the CABG procedure and the third 20 hours later in the intensive care unit. Figure 2 shows that BPAP hydrolysis (v) by pulmonary capillary endothelium-bound ACE in these subjects was approximately 1.3, a value similar to what has been previously reported in various mammalian species (*Ryan and Catravas, 1991; Orfanos et al, 1994*). Furthermore, on average, v remained statistically unchanged during the first 20 hours postoperatively. Similar results were observed with the modified first order parameter A_{max}/K_m (index of the dynamically perfused pulmonary capillary surface area) which also remained stable around 3 l/min (Figure 3). Pulmonary blood flow, calculated by the indicator dilution method using total ^{3}H as the intravascular indicator, remained unchanged during the same period and, on average, fluctuated around 3.5 l/min.

Figure 4, top panel, presents a typical arterial outflow blood indicator dilution curve of total tritium and percent metabolism of ^{3}H-BPAP measured preoperatively in one subject. In this patient, approximately 85% of ^{3}H-BPAP was hydrolyzed by pulmonary capillary endothelium-bound ACE (v=1.87) and the parameter A_{max}/K_m (referred to here as k) was 6.0 l/min. Figure 4, bottom panel, shows the same parameters in the same patient two hours after completion of CABG surgery. Notice the dramatic decrease in percent ^{3}H-BPAP metabolism (60%), as well as in the A_{max}/K_m value (2.8 l/min), the latter suggesting approximately 50% decrease in the dynamically perfused capillary surface area of the lungs. Subsequent chest x-ray confirmed this biochemical observation and revealed that the deterioration was due to a temporarily collapsed right lung .

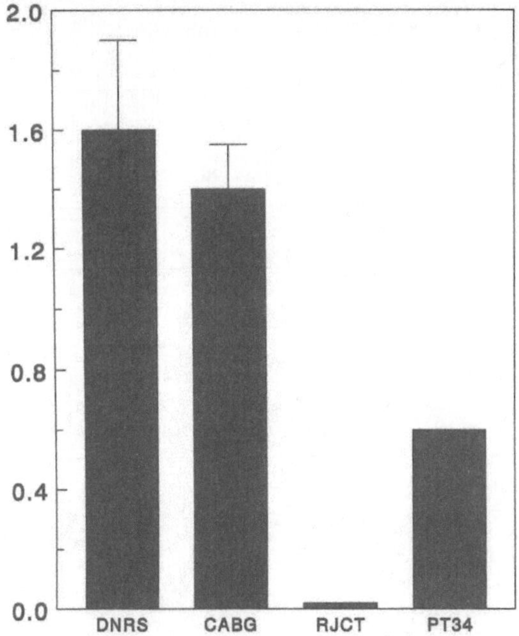

Figure 6. *ACE Activity in Donor Lungs*. Pulmonary capillary endothelium-bound ACE activity (expressed as V) in donors of lung grafts (DNRS, N=8), in patients undergoing surgery for coronary arterial bypass graft (CABG; N=12), in one donor whose lungs were rejected because of poor compliance and gas exchange (RJCT) and in a donor whose lungs were transplanted in a recipient who died with pulmonary dysfunction within 24 hours of transplant. (PT 34). Means ±SEM.

Pulmonary Capillary ACE Activity in Lung Transplantation

Previous experimental data had suggested that alterations in pulmonary capillary endothelial ACE activity may be predictive of existing or impending lung injury (*Dobuler et al, 1982; Catravas et al, 1983; 1988*). We investigated this hypothesis in patients scheduled for lung transplantation: recipients as well as donors.

Lung graft recipients frequently exhibit signs of pulmonary dysfunction and/or rejection which, on occasion, prove fatal. If altered capillary ACE activity were predictive of pulmonary dysfunction and/or rejection, then routine monitoring could aid in early diagnosis, avoid the use of transbronchial biopsies and help improve the outcome of the recipient. Figure 5 demonstrates the time-course of pulmonary capillary endothelium-bound ACE activity in a 28-year old subject undergoing heart and lung transplantation. The solid square symbol represents ACE activity of the graft prior to harvesting ; the activity is normal and comparable to that in normal controls (see Figure 2). Solid circles represent recipient ACE activities, the first one immediately prior to transplant, reflecting activity approximately 50% below normal. Notice the return to normal ACE activity immediately after transplant. Subsequent fluctuations with nadirs at day 6 and 18 -22 post-transplant preceded incidents of rejection. Thus, data from this ongoing investigation strongly suggest that changes in pulmonary endothelium-bound ACE activity may prognosis of certain pathological conditions.

In lung donors, we examined whether monitoring of capillary endothelial ACE activity may aid in the screening for suitable grafts. Pulmonary capillary endothelium-bound ACE activity was determined in eight donors, prior to removal of the graft. The results, expressed as ^{3}H-

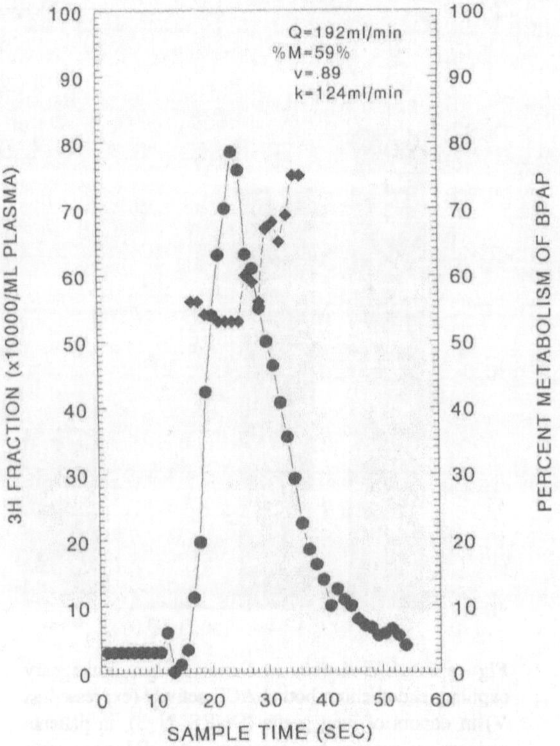

Figure 7. *Coronary ACE Activity.* Coronary sinus plasma fractional concentrations of tritium (solid circles) and percent ^{3}H-BPAP metabolism (%M) by coronary endothelium-bound ACE (solid diamonds) during pump-perfusion of the coronary bed in a patient undergoing surgery for aortic valve replacement.

BPAP hydrolysis (v) are shown in Figure 6. In six subjects (DNRS) v levels were comparable to those of normal controls (CABG); in a seventh potential donor (RJCT), the graft was not harvested because of poor compliance and impaired gas exchange; pulmonary capillary ACE activity in this subject was virtually absent. In another donor (PT34) ACE levels were well below control values; the recipient exhibited post-transplant pulmonary dysfunction and died within 24 hours. These data confirm that pulmonary endothelium-bound ACE activity could be useful as a quantifiable indicator of existing, subtle lung injury in man.

Measurement of Coronary Endothelium-Bound ACE Activity and Dynamically Perfused Capillary Surface Area

We have applied similar techniques to measure the activity of coronary capillary-bound ACE and the area of the perfused coronary capillary vascular bed in patients undergoing corrective or replacement valve surgery with or without coronary artery bypass grafting. ^{3}H-BPAP (2μCi or 0.1 nmol) was injected as a bolus into the inflow coronary perfusion catheter. Blood withdrawal began immediately from a retrograde coronary sinus cannula connected to a fraction collector equipped with tubes advancing at the rate of one every 2.4 sec. for 60 sec.

Figure 7 shows data from a typical transcoronary measurement in a patient undergoing surgery for aortic valve replacement. Fractional concentration of tritium in the effluent

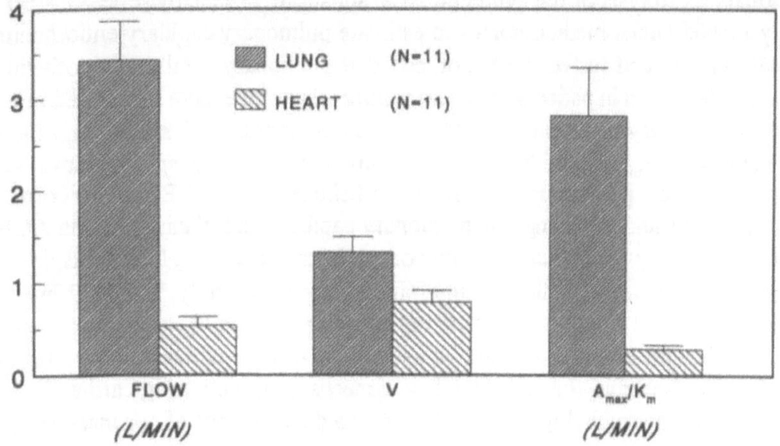

Figure 8. *Human Pulmonary v. Coronary ACE.* Comparison of flow, v and A_{max}/K_m, (k) values between the human coronary and pulmonary vascular beds. Means±SEM. Notice that despite almost 10-fold greater flow and capillary surface area index values in the lung compared to the heart, ACE activity (v) per capillary is relatively similar between the two beds.

coronary sinus plasma, percent ^{3}H-BPAP metabolism (%M), substrate hydrolysis (v), coronary perfused plasma flow and the modified first order parameter A_{max}/K_m (k) values are presented. Unlike the lung, there is significantly sample-to-sample variation in substrate metabolism, probably suggestive of perfusion heterogeneity (variable capillary transit times) in this subject. Average values from eleven such measurements are shown in Figure 8 and are compared to those obtained in the pulmonary capillary bed of the same subjects. Notice a remarkably similar degree of substrate hydrolysis, suggesting a comparable endothelial ACE concentration in pulmonary v. coronary capillaries. A_{max}/K_m (k) values, however, demonstrate an approximately ten-fold greater area in the perfused pulmonary capillary bed, compared to the coronary bed.

SUMMARY AND IMPLICATIONS

Endothelial cell injury is the initiating step in numerous pathologies. It has long been recognized that the development of markers of endothelial integrity would be of considerable benefit for the early detection of injury to the lung and possibly other organs. Dysfunction of pulmonary vascular endothelium has been suggested as one of the major and early events in lung injury (*Doubler et al, 1982; McCormick et al, 1987; Witschi et al, 1984*). The clinical relevance of these concepts is continuously evolving.

Approximately twenty years ago, it was proposed that monitoring of certain key pulmonary endothelial biochemical functions might serve as a basis for a test of impeding lung injury (*Hammond et al, 1977*). Subsequent investigation into changes in the transpulmonary uptake of proprenolol, prostaglandin E_2 or serotonin by subjects suffering from adult respiratory distress syndrome (ARDS) confirmed the hypothesis: all three processes were depressed in ARDS patients, however a significant overlap in the values between normal and ARDS subjects was observed, rendering a test based on such measurements of limited diagnostic value (*Morel et al, 1985; Gillis et al, 1986*).

More and more data appear to suggest that there is a need to develop a feasible, clinically useful method for identifying impending lung injury. Over the past decade, studies by us and other investigators have accumulated considerable experimental evidence suggesting that the activity of pulmonary microvascular endothelium-bound ACE, an endothelial ectoenzyme which is uniformly distributed throughout the vasculature, can be assayed repeatedly, accurately, and reproducibly, *in vivo*. Methods now exist for the *in vivo* assay of pulmonary capillary endothelium-bound ACE activity in laboratory animals via the single pass transpulmonary hydrolysis of the synthetic ACE substrate ^{3}H-benzoyl-Phe-Ala-Pro (BPAP). We recently applied similar methodologies to estimate pulmonary capillary endothelium-bound ACE activity in man and tested the hypothesis that pulmonary capillary endothelium-bound ACE activity is depressed in patients with severe lung injury. We have succeeded in measuring percent transpulmonary metabolism (%M), transpulmonary hydrolysis (v), and the modified first order parameter (A_{max}/K_m) of ^{3}H-BPAP for pulmonary capillary endothelium-bound ACE. We demonstrated that pulmonary capillary endothelium-bound ACE activity correlates with diagnosed lung injury and the changes in pulmonary capillary endothelium-bound ACE activity may assist in predicting the clinical outcome of patients undergoing lung transplantation.

We have also developed indicator dilution techniques to study ^{3}H-BPAP hydrolysis by coronary capillary-bound ACE in patients with heart disease. It is likely that determinations of coronary endothelium-bound ACE activities will provide significant contribution to the diagnosis of cardiac ischemia, the understanding of mechanisms underlying cardiac hypertrophy, failure and perhaps other pathologies, as well as in the development of coronary-specific ACE inhibitors.

REFERENCES

Bakhle, Y.S., Hartiala, J., Toivonen, H.J., and Uotila, P., 1979, Effects of cigarette smoke on the metabolism of vasoactive hormones in rat isolated lungs. *Br. J. Pharmacol.* 65:495-499.

Brigham, K.L., 1991, The clinical profile of sepsis and the adult respiratory distress syndrome, in: Vascular Endothelium: Physiological Basis of Clinical Problems, ed Catravas, John D., Ryan, U.S., Gillis, C.N., and Callow, A.D., Plenum Press, New York, pp. 3-11.

Catravas, J.D. and Gillis, C.N., 1981, Metabolism of ^{3}H-benzoyl-phenylanyl-alanyl-proline by pulmonary angiotensin converting enzyme, *in vivo*: effects of bradykin, SQ14225 or acute hypoxia. *J. Pharmacol. Exp. Ther.* 217:263-270.

Catravas, J.D., 1986, Michealis-Menten kinetics of pulmonary endothelial angiotensin converting enzyme in the conscious rabbit. *Adv. Exp. Med.* 445-451.

Catravas, J.D. and White, R.E., 1984, Kinetics of pulmonary angiotensin converting enzyme and 5′ nucleotidase, *in vivo*. *J. Appl. Physiol.* 57:1173-1181.

Catravas, J.D., Burch, S.E., Spurlock, B.O., and Mills, L.R., 1988, Early effects of ionizing radiations on pulmonary endothelial angiotensin converting enzyme and 5′-nucleotidase, *in vivo*. *Tox. And Appl. Pharm.* 94:342-355.

Catravas, J.D., Lazo, J.S., Dobuler, K.D., Mills, L.R., and Gillis, C.N., 1983, Pulmonary endothelial destruction in the presence or absence if interstitial injury by intratracheal injection of bleomycin to rabbits. *Amer. Rev. Resp. Dis.* 128:740-746.

Catravas, J.D., Lazo, J.S., and Gillis, C.N., 1981, Biochemical markers of bleomycin toxicity: clearance of ^{14}C-5-hydroxytryptamine and ^{3}H-norepinephrine by rabbit lung, *in vivo*. *J. Pharmacol. Exper. Ther.* 217:524-529.

Dobuler, K.J., Catravas, J.D., and Gillis, C.N., 1982, Early detection of oxygen-induced lung injury in concious rabbits. *Amer. Rev. Resp. Dis.* 126:534-539.

Gillis, N.C., Pitt, B.R., Wiedemann, H.P., and Hammond, G.L., 1986, Depressed prostaglandin E_1 and 5-hydroxytryptamine removal in patients with adult respiratory distress syndrome. *Am. Rev. Respir. Dis.* 134:2371-2375.

Hammond, G.L., Cronau, L.H., Whittaker, D., and Gillis, N.C., 1977, Fate of prostaglandins E_1 and A_1 in the human pulmonary circulation. *Surgery* 81:716-722.

Howell, R.E., Moalli, and Gillis, C.N., 1985, Analysis of rabbit pulmonary angiotensin converting enzyme kinetics, *in vivo*. *Biochem. Pharmacol.* 34:2371-2375.

McCormick, J.R., Chrzanowski, R., Andreani, J., and Catravas, J.D., 1987, Early pulmonary endothelilal enzyme dysfunction after phorbol ester in concious rabbits. *Am. Pysiosl. Soc.* 1972-1978.

Morel, D.R., Dargent F., Bachman, M., Suter, P.M., Junod, A.F., 1985, Pulmonary extraction of serotinin and propranol in patient with adult respiratory distress syndrome. *Am. Rev. Respir. Dis.* 134:479-484.

Orfanos, S.E., Chen, X-L, Ryan, J.W., Chung, A.Y.K., Burch, S.E., and Catravas, J.D., 1994, Assay of Pulmonary Microvascular Endothelial Angiotensin-Converting Enzyme *in vivo*: Comparison of three probes. *Toxicol. Appl. Pharmacol.* 124, 99-111.

Rinaldo, J.E. and Rogers, R.M., 1982, Adult respiratory distress syndrome. *The New England Journal of Medicine*.900-907,

Ryan, J.W., Roblero, J., and Stewart, J.M., 1968, Inactivation of bradykinin in the pulmonary circulation. *Biochem. J.* 110:795-797.

Ryan, J.W., Chung, H.S., Ammons, C., and Carlton, M.L., 1977, A simple radioassay for angiotensin converting enzyme. *Biohem. J.* 167:501-504.

Ryan, J.W. and Catravas, J.D., 1991, Angiotensin converting enzyme as an index of pulmonary mirovascular function, in: *In Focus on Pulmonary Pharmacology and Toxicology,* ed M. Hollinger, CRC Press, Boca Raton, pp. 183-210.

Ryan, J.W., Chung, H.S., Martin, L.C., and Ryan, U.S., 1978, New substrates for the radioassay of angiotensin converting enzyme of endothelial cells in culture. *Tissue & Cell* 10:555-562.

Ryan, U.S., Ryan, J.W., Whittaker, C., and Chiu, A., 1976, Localization of angiotensin converting enzyme (kininase II). Immunocytochemstry and immunofluorescence. *Tissue & Cell* 8:125-146.

Ryan, U.S., 1987, *Pulmonary Endothellium in Health and Disease.* Marcel Dekker, Inc., New York.

Toivonen, H.J., Hitiala, J., and Bakhle, Y.S., 1981, Effects of high oxygen tension on the metabolisam of vasoactive hormones in isolated perfused rat lungs. *Act. Physiol. Scand.* 111:185-192.

Witschi, H.P. and Lindenschmidt, R.C., 1984, Pathogenesis of acute and chronic lung injury induced by foreign compounds. *Clin. Physiol. Biochem.* 3:135-146.

INTERACTIONS OF SUPEROXIDE DISMUTASES WITH NITRIC OXIDE

Brian Day, Ph.D., Tim Oury, M.D., Ph.D., and James D. Crapo, M.D.

Pulmonary and Critical Care Medicine
Duke University Medical Center
Durham, North Carolina 27710, USA

INTRODUCTION

Superoxide (O_2^-) is an important component of a variety of vascular inflammatory disorders including reperfusion injuries, hemorrhagic shock, oxygen toxicities, infections, neoplasia and apoptosis. Extracellular superoxide dismutase (EC-SOD) has been found in the extracellular spaces and fluids where it is thought to defend against O_2^- produced by inflammatory cells. Nitric oxide (NO) is also produced during inflammatory reactions and has some anti-inflammatory properties. In the extracellular spaces the primary species that will react with NO is O_2^-. The product of this reaction is peroxynitrite which is a strong proinflammatory mediator. Therefore, to control inflammation, the reaction of O_2^- with NO must be prevented. High EC--SOD levels in the extracellular spaces may act to preserve NO and thus create strong anti-inflammatory microenvironments.

REACTIONS INITIATED BY REACTIVE O_2 SPECIES

Oxygen free radicals are produced normally as an inevitable byproduct of oxidative metabolism *(Cohen and Cederbaum, 1979)*. In usual quantities these radicals can be eliminated by native antioxidant defenses. With hyperoxia, free radical generation is increased in proportion to the elevated oxygen concentrations in the tissues *(Turrens et al., 1982; Turrens et al., 1982; Freeman and Crapo, 1982)*. As the injury progresses, inflammatory cells including neutrophils and mononuclear phagocytes, are sequestered within the alveolar septa *(Barry and Crapo, 1985)*. These cells are capable of producing oxygen free radicals and release other products that serve to amplify the resultant injury *(Martin, 1984)*.

Biochemical consequences of unchecked free radical production include protein denaturation, cross linkage and breakage with subsequent loss of enzymatic activity or structural integrity. Nucleic acid chains undergo strand breakage and nucleotide alteration with resultant degradation of genetic information. Other consequences of radical exposure include alteration of cell surface receptor glycoproteins with disruption of trans-plasmalemmal communication, cross linking and polymerization of plasma membrane lipid constituents resulting in impairment of normal membrane fluidity and permeability *(Freeman and Crapo, 1982)*.

Cellular defenses against oxygen radicals exist and include antioxidant enzymes which catalyze reactions yielding less reactive species. One important enzymatic defense is superoxide dismutase (SOD). SOD exists in three forms in primate tissue; as a cytoplasmic dimeric enzyme with a CuZn active site *(McCord and Fridovich, 1969)*, as a tetrameric, largely mitochondrial enzyme with an active site containing Mn *(McCord et al., 1977)*, and in a more recently described form occurring primarily in the intravascular and extracellular spaces (*Marklund*, 1982). All forms of SOD catalyze the reaction: $2O_2^- + 2H^+ \rightarrow H_2O_2 + O_2$. The

Vascular Endothelium: Responses to Injury
Edited by John D. Catravas *et al.*, Plenum Press, New York, 1996

Superoxide Dismutases

$$2O_2^- + 2H^+ \rightarrow H_2O_2 + O_2$$

Isozyme	Characteristics	Location
CuZn SOD	32,000 M.W. Dimer	Cytoplasm Nucleus
Mn SOD	86,000 M.W. Tetramer	Mitochondria
EC-SOD	135,000 M.W. Tetramer	Extracellular Matrix Extracellular Fluids Cell Surfaces

Figure 1. The known mammalian superoxide dismutases.

hydrogen peroxide produced is scavenged by other important antioxidant enzymes, including catalase and peroxidases.

THE SUPEROXIDE DISMUTASES

Superoxide dismutases (SOD) are a class of enzymes which catalyze the dismutation of two superoxide radicals to form hydrogen peroxide and oxygen. Three distinct types of superoxide dismutase have been found in mammalian tissues (Figure 1). The rate constants of the superoxide dismutases are high ($>10^9\ M^{-1}\bullet s^{-1}$) (10), allowing scavenging of O_2^- at rates which are almost diffusion limited. Saturation of these enzymes with substrates is difficult *in vivo*. The superoxide dismutases obey first order kinetics, resulting in a steady state concentration of O_2^- in lung tissues that varies directly with the rate of O_2^- generation and inversely with the tissue concentration of scavenging enzymes.

Intracellular SODS

The first SOD to be discovered, a copper and zinc containing enzyme (CuZn SOD), was originally isolated and characterized from bovine erythrocytes by McCord and Fridovich in 1969 *(McCord and Fridovich)*. It is a stable molecule of approximately 32,000 molecular weight, consisting of two identical subunits — each containing one atom of copper and one atom of zinc. The second, a manganese containing superoxide dismutase (Mn SOD), was first isolated from the bacterium *E. coli* and found in both pro- and eukaryotic organisms *(McCord et al., 1977)*. The eukaryotic Mn SOD is usually associated with mitochondria, has a molecular weight of 93,000 and consists of four identical subunits *(McCord et al)*.

The extracellular superoxide dismutase

The third (EC-SOD) is predominantly found in the extracellular spaces, has a molecular weight of 135,000, contains copper and zinc in four equal subunits and has a strong affinity for heparin *(Marklund, 1984; Marklund, 1984; Sandström et al., 1993)*. This protein was first described by Marklund *et al.*, and was named EC-SOD after the location where it was first detected (i.e., extracellular fluids). Its pattern of tissue distribution is distinctly different from the CuZn SOD and Mn SOD which are highest in tissues with high metabolic activity *(Marklund, 1984a; Marklund, 1984b)*. The EC-SOD activity is particularly high in lung tissue (Figure 2). This suggests a unique role for EC-SOD in protecting against O_2^- produced in

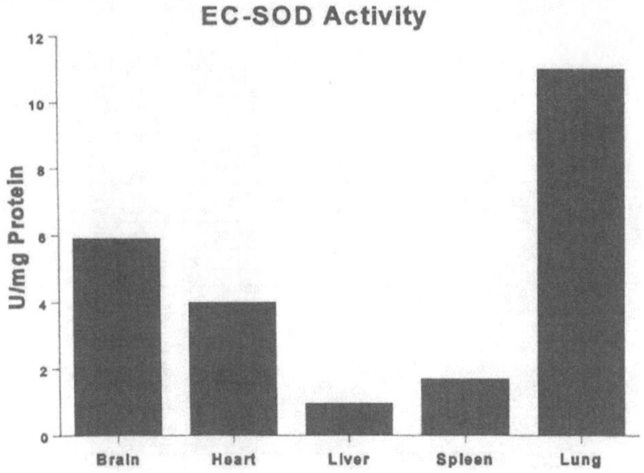

Figure 2. EC-SOD Activity is high in human lung. (Modified from 11).

extracellular spaces of the lung. The enzyme has hydrophobic properties *(Marklund 1984a)* and contains a cluster of positively charged amino acids near the carboxy terminal of the protein which give this protein its heparin or g lycosaminoglycan (GAG) binding properties *(Sandström et al., 1992; Marklund and Karlsson, 1990; Sandström et al., 1993)*. EC-SOD has been speculated to partition between cell surfaces and extracellular fluids where it is thought to be a primary defense against O_2^- produced by inflammatory cells *(Marklund and Karlsson, 1990; Sandström et al., 1993)*.

We have done immunolocalization studies using polyclonal antibodies against recombinant human EC-SOD and found that it is particularly located in the interstitial spaces surrounding airways and vasculature smooth muscle cells *(Oury et al., 1994)* (Table 1). Localization of EC-SOD in association with type I collagen suggests an important function for EC-SOD in protecting this extracellular matrix protein from reactive oxygen species. Collagen has previously been demonstrated to be susceptible to attack by reactive oxygen species such as O_2^- *(Creenwald and Moy, 1979; Borel and Monboisse, 1985; Monboisse et al., 1988)*. In addition, O_2^- may be capable of activating latent collagenases from polymorphonuclear leukocytes (PMN) *(Burkhardt et al., 1986; Sorsa et al., 1989)*, which can lead to further collagen degradation. Consequently, the association of EC-SOD with collagen may be important in preventing O_2^- mediated degradation of collagen and therefore represent a means of controlling inflammatory responses.

Recent findings by our group, using transgenic mice expressing high levels of EC-SOD in the brain, suggest that the function of EC-SOD is to regulate the availability of the powerful smooth muscle relaxant NO· *(Oury et al., 1992)*. When transgenic mice were exposed to 6 atm of oxygen for 25 min, they demonstrated a lower survival (17%) than nontransgenic litter-mates (67%) (Figure 3). Pretreatment with diethyldithiocarbamate (DDC), which inhibits both EC-SOD and CuZn SOD activity, increased resistance to CNS O_2 toxicity, in terms of both survival (100% in transgenics) and resistance to seizures (4-fold increase in seizure latency in both transgenic and nontransgenic mice, $P < 0.05$). Thus, O_2^- apparently protects against CNS O_2 toxicity. We hypothesized that O_2^- decreased toxicity by inactivating NO·. To test this, we inhibited NO synthase with N^{ω}-nitro-L-arginine (LNNA) to determine whether NO contributes to enhanced CNS O_2 toxicity in transgenic mice. LNNA protected both transgenic and nontransgenic mice against CNS O_2 toxicity (100% survival and a 4-fold delay in time to first seizures; $P < 0.05$). These results implicate NO as an important mediator in CNS O_2 toxicity and suggest that EC-SOD increases CNS O_2 toxicity by inhibiting O_2^--mediated inactivation of NO *(Oury et al., 1992)*. Our mice containing the EC-SOD transgene were also resistant to vasogenic edema by a mechanism thought to be due to the inhibition of peroxynitrite formation *(Oury et al., 1993)*.

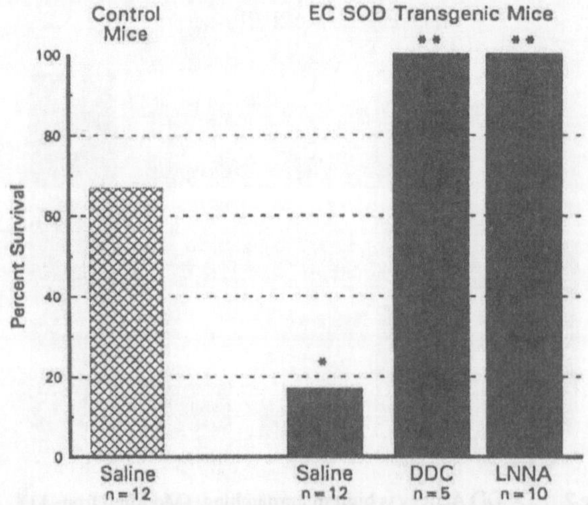

Figure 3. Percent survival in hyperbaric oxygen. EC-SOD transgenic animals had lower survival (*p<0.05) and this effect was completely reversed by either inhibiting SOD activity with Ddc or inhibiting NO activity with LNNA (**p>0.05). Modified from 23.

NITRIC OXIDE SYNTHASES (NOS)

NO is synthesized by oxidative deamidation of L-arginine to citrulline by a calcium calmodulin dependent (constitutive) and a calcium independent (inducible) form of nitric oxide synthase (NOS). Both the constitutive and inducible forms of NOS are homologous to cytochrome P450 reductase and require NADPH, heme, tetrahydrobiopterin, flavins, L-arginine, thiols and molecular oxygen as cofactors *(White and Marletta, 1992)*. Both forms of NOS have similar kinetic profiles; however, due to differences in their regulation, the constitutive forms have a lower output of NO than the inducible forms *(Fostermann et al., 1991)*. The constitutive forms of NOS produce NO in a tightly regulated manner that is dependent on intracellular calcium levels *(Bredt and Snyder, 1990)*. This form is thought to be important in the maintenance of basal smooth muscle tone and neurotransmission *(Toda and O kamura, 1992)*. The inducible NOS forms contain tightly coupled calmodulin and, as a result, continuously produce high amounts of NO *(Pufahl et al., 1992)*. The inducible forms of NOS are only present in cells after appropriate cytokine stimulation (i.e. IL-1, TNFα, interferon g amma) and/or inflammatory responses such as those induced by endotoxin *(Radomski et al., 1990)*. The constitutive and inducible forms of NOS from a number of species have been cloned and sequenced including the endothelial (types Ib, II, III) smooth muscle cell (type II) and macrophage enzymes (types II, IV) *(Fostermann et al., 1991)*. The membrane associated forms of NOS (III & IV) are myristilated which is thought to give them this property; however, the role these forms play in cellular biology is not known. The nomenclature for identifying the various NOS isoforms is evolving with one approach given in Table 2. The three major gene products are thought to be Types Ia, II and III.

REACTIONS OF NO and O_2^-

As would be expected for two small free radicals, O_2^- and NO interact at a diffusion-limited rate ($k_2 = 6x10^9 M^{-1}s^{-1}$ at 25°C) *(Huie and Padmaja, 1993)*, and the product of this reaction is the powerful oxidant peroxynitrite. O_2^- can thus limit the half-life, and, therefore, the diffusion range, of NO. The high rate constant for reactions of NO with O_2^- means that superoxide will be one of the primary scavengers of NO in the extracellular spaces. Therefore, production of O_2^- extracellular spaces will dramatically reduce the extracellular concentrations of NO except in domains rich in EC-SOD *(Huie and Padmaja, 1993; Gryglewski et al., 1986; Radi et al., 1991)*.

Table I. Distribution of EC-SOD in Human Lung (modified from 7).

Cell Surfaces	**Labeling Density**
Endothelial cell	---
Type I cell	---
Type II cell	---
Smooth muscle cell	++
Bronchial epithelial cell	+
Extracellular Matrix	
Type I collagen	++++
Elastin	---
Cartilage	---
Unidentified matrix elements	++++

Table II. Classification of currently recognized forms of NOS (Modified from 26).

Type	Expression	Subcellular Site	Localization
Ia	constitutive	cytosolic	neurons
Ib	constitutive	cytosolic	endothelial cells (soluble form of Type III)
Ic	constitutive	cytosolic	PMN
II	inducible	cytosolic	macrophages, type II cells, hepatocytes, fibroblasts, smooth muscle & endothelial cells
III	constutive	membrane	endothelial cells
IV	inducible	membrane	macrophage

THE DARK SIDE OF NO

Nitric oxide reacts with O_2^- to produce the peroxynitrite anion ($ONOO^-$) *(Radi et al., 1991; Radi et al., 1991; Blough and ZaFlriou, 1985; Beckman, 1991)*. $ONOO^-$ is a toxic species which when protonated disassociates into OH^- and NO_2. $ONOO^-$ has significantly less smooth muscle relaxing properties than does NO and is thought to be a strong pro-inflammatory mediator *(Beckman, 1991)*. Thus it is critical that reactions of NO with O_2^- in the alveolar septal spaces be prevented. This is likely one of the teleologic reasons for the development of high levels of EC-SOD in the lung.

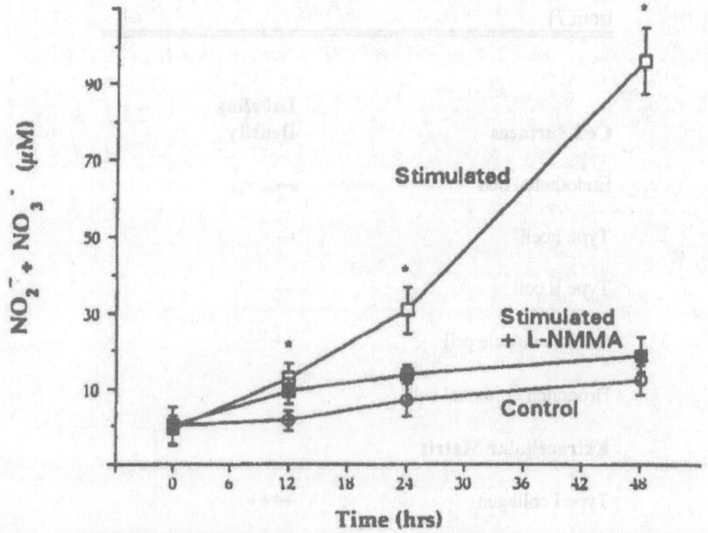

Figure 4. Time course of NO production (measured as nitrite + nitrate) by cultured type II cells stimulated with zymosan-activated serum, LPS, and cytokines. The NOS inhibitor, L-NNMA, inhibited the release verifying that NO was the original species. (Modified from 38).

SOURCES OF O_2^- IN THE ALVEOLAR SEPTAL EXTRACELLULAR SPACES

The normal levels of O_2^- in lung extracellular spaces are expected to be extremely low, as they are in other tissues *(Halliwell and Gutteridge, 1990)*. The relatively high EC-SOD levels found in the lung *(Marklund, 1984a; Marklund, 1984b)* suggest both that there is a source of extracellular O_2^- and that it is controlled by its enzymatic scavenger. The NADPH oxidase on the cell membranes of neutrophils and macrophages is a well established source of extracellular O_2^- *(Freeman and Crapo, 1982)*. It is also possible that O_2^- may reach lung extracellular spaces by passing through anion channels in alveolar septal cell membranes or being produced by an oxidase on these cell membranes. These postulated sources of extracellular O_2^- have not yet been established.

THE ROLE OF NO AND O_2^- IN ALVEOLAR INFLAMMATION

Figure 4 illustrates that type II cells can produce large amounts of NO, particularly when stimulated by cytokines *(Gutierrez et al., 1993)*. NO· has been shown to inhibit neutrophil adhesion *(Kubes and Granger, 1992; Kubes et al., 1991)* and to inhibit the NADPH oxidase on neutrophil membranes. Thus, NO has the potential to act as a powerful anti-inflammatory agent in the alveolar interstitium or air spaces. NO may exert similar effects on alveolar macrophages, thus helping provide an environment in alveolar surface lining layers that will suppress amplification signals for inflammation. EC-SOD would play a critical role in this environment by preventing the reaction of NO with superoxide to form the peroxynitrite molecule. Thus EC-SOD may control the critical balance of whether NO acts as a regulator of inflammation in the alveolar surface or is transformed to a strong pro-inflammatory agent (Figure 5).

SUMMARY

Superoxide is chemically reactive, and it can contribute to the production of even more reactive entities, such as hydroxyl radical (OH·). There is increasing evidence showing that superoxide is an important component of a variety of vascular inflammatory disorders including reperfusion injuries, hemorrhagic shock, oxygen toxicities, infections, neoplasia and apoptosis. EC-SOD has been speculated to partition between cell surfaces and extracellular fluids where

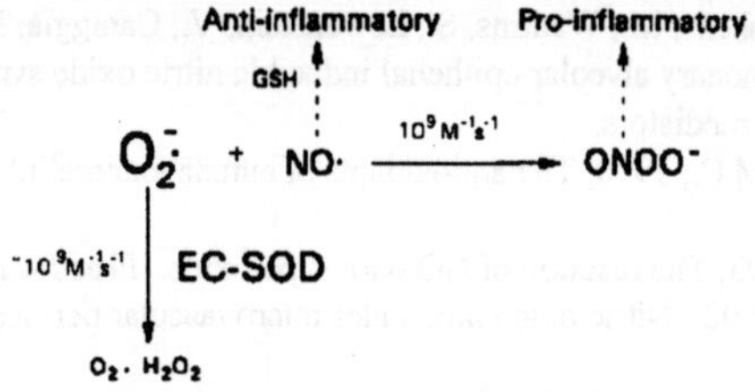

Figure 5. EC-SOD and O_2^- mediate the anti-inflammatory/pro-inflammatory balance in the alveolar septum. NO released by septal cells acts to suppress inflammation unless it reacts with O_2^- to form $ONOO^-$. By scavenging O_2^- EC-SOD tips the balance in the alveolar septum against inflammation. Significant amounts of $ONOO^-$ will form only when EC-SOD is deficient or when there is greatly increased O_2^- release.

it is thought to be a primary defense against superoxide produced by inflammatory cells. Nitric oxide is also produced during inflammatory reactions and has anti-inflammatory effects under certain conditions. In the extracellular spaces the primary species that will react with nitric oxide is superoxide. This reaction leads to the formation of peroxynitrite which has less smooth muscle relaxing properties and is a strong proinflammatory mediator. Therefore, it is critical that where inflammation is to be controlled, the reaction of superoxide with nitric oxide must be prevented. High EC-SOD levels in the extracellular spaces may act to preserve nitric oxide and, thus, create strong anti-inflammatory microenvironments.

REFERENCES

Barry, B.E. and Crapo, J.D., 1991, Patterns of accumulation of platelets and neutrophils in rat lungs during exposure to 100% and 85% oxygen. *Am. Rev. Respir. Dis.* 132:548-555.

Beckman, J.S., 1991, The double-edged role of nitric oxide in brain function and superoxide-mediated injury. *J. Dev. Phys.* 15:53-59.

Borel, J. P. and Monboisse, J. C, 1985, Collagenolysis. In: *CRC Handbook of Methods for Oxygen Research*, pp 407-412.

Blough, N.V. and ZaFlriou, O.C., 1985, Reaction of superoxide with nitric oxide to form peroxynitrite in alkaline aqueous solution. *Inorg. Chem.* 24:3502-3504.

Bredt, D.S. and Snyder, S.H., 1990, Isolation of nitric oxide synthase, a calmodulin requiring enzyme. *Proc.Natl. Acad. Sci. USA* 87:682-685.

Burkhardt, H., Hartmann, F. and Schwingel, M.L., 1986, Activation of latent collagenase from polymorphonuclear leukocytes by oxygen radicals. *Enzyme* 36:221-231.

Clancy, R.M., Leszczynska-Piziak, J. and Abramson, S.B., 1992, Nitric oxide, an endothelial cell relaxation factor, inhibits neutrophil superoxide anion production via a direct action on the NADPH oxidase. *J. Clin. Invest.* 90:1116-1121.

Cohen, G. and Cederbaum, A., 1979, The production of hydroxyl radicals by rat liver microsomes. *Science* 204:66-68.

Forman, H.J. and Fridovich, I., 1973, Superoxide dismutase: A comparison of rate constraints. *Arch. Biochem. Biophys.* 396-400.

Fostermann, U., Schmidt, H.H., Pollock, H.H.H.W., Sheng, J.S., Mitchell, H., Warner, J.A., Nakane, T.D., and Murad, F., 1991, Isoforms of nitric oxide synthase: Characterization and purification from different cell types. *Biochem. Pharmacol.* 42:1849-1857.

Freeman, B.A. and Crapo, J.D., 1982, Biology of disease: free radicals and tissue injury. *Lab. Invest.* 47:412-426.

Greenwald, R.A. and Moy, W.W., 1979, Inhibition of collagen gelation by action of the superoxide radical. *Arthritis Rheum.* 22:251-359.

Gryglewski, R.J., Palmer, R.M.J. and Moncada, S., 1986, Superoxide anion is involved in the breakdown of endothelium-derived relaxing factors. *Nature* 320:454-456.

Gutierrez, H.H., Pitt, B.R., Brookens, M., Watkins, S., Lowenstein, V., Caniggia, I., Chumley, P. and Freeman, B.A., submitted 1993, Pulmonary alveolar epithelial inducible nitric oxide synthase gene expression: regulation by inflammatory mediators.

Halliwell, B. and Gutteridge, J.M.C., 1990, The antioxidants of human extracellular fluids. *Arch. Biochem. & Biophys.* 280:1-8.

Huie, R.E. and Padmaja, S., 1993, The reaction of NO with superoxide. *Free Rad. Res. Commun.* 18:195-199.

Kubes, P. and Granger, D.N., 1992, Nitric oxide modulates microvascular permeability. *Am. J. Physiol.* 262:H611-H615.

Kubes, P., Suzuki, M. and Granger, D.N., 1991, Nitric oxide: an endogenous modulator of leukocyte adhesion. *Proc. Natl. Sci. USA* 88:4651-4655.

Marklund, S.L., 1984a, Properties of extracellular superoxide dismutase from human lung. *Biochem.* J220:269-272.

Marklund, S.L., 1984b, Extracellular superoxide dismutase and other superoxide dismutase isoenzymes in tissues from nine mammalian species. *Biochem. J.* 222:649-655.

Marklund, S. and Karlsson, K., 1990, Extracellular superoxide dismutase, distribution in the body and therapeutic applications. In: *Antioxidants in Therapy and Preventive Medicine*, Emerit *et al.* (eds), Plenum Press, New York pp 1-4.

Marklund, S.L., 1984, Extracellular superoxide dismutase in human tissues and human cell lines. *J. Clin. Invest.* 74:1398-1403.

Marklund, S.L., 1982, Human copper-containing superoxide dismutase of high molecular weight. *Proc. Natl. Acad. Sci. USA* 79:7634-7638.

Martin, W.J., 1984, Neutrophils kill pulmonary endothelial cells by a hydrogen peroxide-dependent pathway: An *in vivo* model of neutrophil-mediated lung injury. *Am. Rev. Respir. Dis.* 130:209-213.

McCord, J.M. and Fridovich, I., 1969, SOD: an enzymatic function for erythrocuprein chemocuprein. *J. Biol. Chem.* 244:6049-6055.

McCord, J.M., Boyle, J.A., Day Jr, E.D., Rizzolo, L.J. and Salin, M.L., 1977, A managanese-containing superoxide dismutase from human liver. In: *Superoxide and Superoxide Dismutase*, (A.M. Mitchelson, J.M. McCord and I. Fridovich, eds), 129-138.

Monboisse, J-C, Gardes-Albert, M., Randoux, A., Borel, J-P and Ferradini, C., 1988, Collagen degradation by superoxide anion in pulse and gamma radiolysis. *Biochim. Biophys. Acta* 965:29-35.

Oury, T.D., Chang, L, Marklund, S.L., Day, B.J. and Crapo, J.D., 1994, Immunocytochemical localization of extracellular superoxide dismutase in human lung. *Lab Invest* (in press).

Oury, T.D., Ho, Y.S. Piantadosi, C.A., and Crapo, J.D., 1992, Extracellular superoxide dismutase, nitric oxide and central nervous system O_2 toxicity. *Proc. Natl. Acad. Sci. USA* 89:9715-9719.

Oury, T.D., Piantadosi, C.A. and Crapo, J.D., 1993, Cold-induced brain edema in mice: Involvement of extracellular superoxide dismutase and nitric oxide. *J. Biol. Chem.*, (in press).

Pufahl, R.A., Nanjappan, P.G., Woodard, R.W. and MA Marletta, M.A., 1992, Mechanistic probes of N-hydroxylation of L-arginine by the inducible nitric oxide synthase from murine macrophages. *Biochem.* 31:6822-6828.

Radi, R., Beckman, J.S. and Freeman, B.A., 1991, Peroxynitrite oxidation of sulfhydryls: the cytotoxic potential of endothelial-derived superoxide and nitric oxide. *J. Biol. Chem.* 266:4244-4250.

Radi, R., Beckman, J.S., Buch, K.M. and Freeman, B.A., 1991, Peroxynitrite-induced membrane lipid peroxidation: the cytotoxic potential of superoxide and nitric oxide. *Arch. Biochem. Biphys.* 288:481-487.

Radomski, M.N., RMJ Palmer, R.M.J. and Moncada, S., 1990, Glucocorticoids inhibit the expression of an inducible, but not the constitutive, nitric oxide synthase in vascular endothelial cells. *Proc. Natl. Acad. Sci. USA* 37:10043-10047.

Sandström, J., Karlsson, K., Edlund, A. and Marklund, S.L., 1992, The heparin-binding domain of extracellular superoxide dismutase C and formation of variants with reduced affinity. *J. Biol. Chem.* 267:18205-18209.

Sandström, J, Karlsson, K., Edlund, A., and Marklund, S.L., 1993, Proteolytic modification of heparin-binding affinity of extracellular superoxide dismutase. *Biochem. J.* 290:623-626.

Sorsa, T., Saari, H., Konttinen, Y.T., Suomalainen, K., Lindy, S. and Uitto, V-J., 1989, Non-proteolytic activation of latent human neutrophil collagenase and its role in matrix destruction in periodontal diseases. *Inter. J. Tissue React.* 11:153-159.

Toda, N. and Okamura, T., 1992, Regulation of nitroxidergic nerves of arterial tone. News *Physiol. Sci.* I:148-152.

Turrens, J.F., Freeman, B.A., Levitt, J.G. and Crapo, J.D., 1982, The effect of hyperoxia on superoxide production by lung submitochondrial particles. *Arch. Biochem. Biophys.* 217:401-410.

Turrens, J.R., Freeman, B.A. and Crapo, J.D., 1982, Hyperoxia increases H_2O_2 release by lung mitochondria and microsomes. *Arch. Biochem. Biophys.* 271:411-421.

White, K.A. and Marletta, M.A., 1992, Nitric oxide synthase is a P-450 type hemoprotein. *Biochemistry* 31:6627-6631.

II. RESTENOSIS

CLINICAL PROFILE OF RESTENOSIS

Allan D. Callow, M.D., Ph.D.

Boston University
School of Medicine
Boston, MA

The actual number of arterial interventions of one type or another performed each year in the United States is estimated to be between 550,000 and 600,000. This includes bypass grafts, endarterectomies, and a variety of intraluminal procedures which blast, burn or break the plaque. They are about equally divided between the coronary and the rest of the circulation but unequally distributed as to procedure. In the carotid artery surgical endarterectomy is the overwhelming favorite with balloon angioplasty under cautious trial in a handful of centers. In the coronary circulation angioplasty may enjoy a slight edge for the present, but more repeat procedures are being referred for bypass surgery. The renal artery is preferentially treated by angioplasty and, over the years, has yielded the best results for this procedure in any arterial territory. Bypass grafts dominate methods to improve lower extremity circulation and are often combined with balloon angioplasty to relieve a proximal stenosis. The early results with all these techniques are generally excellent. Morbidity and mortality are low, and functional performance is satisfactory in most cases. Over the long term, as short as a few months and as long as ten years, results are disappointing due to recurrence of symptoms and the need for reintervention. In round numbers this amounts to approximately 50% of patients. Secondary procedures as compared to primary are approaching 50% of the case load in many clinics. At least 50% of all recurrences, and much more than this number of early recurrences, are due to an exuberant overgrowth of the myointimal cell and an excessive accumulation of extracellular matrix at the site of the angioplasty or the graft anastomosis. There are two noteworthy exceptions to these generalizations: the internal mammary artery as an aorta-coronary conduit and the synthetic graft in the aorta-iliac position.

Three decades ago, peripheral vascular surgeons realized that arterial reconstructions were not immortal irrespective of the care with which patients were selected and the operation performed. Two decades ago cardiac coronary bypass surgeons experienced the same disappointment. Balloon and other interventionalists had to learn the same lesson: restenosis compromises the best technology. At present restenosis is unpredictable as to patient susceptibility, time of onset, severity and symptom likelihood. There is no reliable method of prevention. There is no consistently reliable treatment short of repeat intervention.

Restenosis of an arterial segment, following a mechanical intervention, is known by a variety of names. Recurrent stenosis, anastomotic hyperplasia, neointimal and myointimal hyperplasia are the same. Custom selects anastomotic hyperplasia to identify the lesion by location in the arterial wall: that is, formation of a new intima, neo, and the predominant cell type, the myofibroblast. With slight variations of no clinical importance, the lesions are the same. All are the response to injury of the cellular and molecular elements of the arterial wall.

Hyperplasia as a cause of restenosis varies by location and procedure. Approximate figures are:

Angioplasty - coronary	30-50%
Angioplasty - femoral	30-50%

Vascular Endothelium: Responses to Injury
Edited by John D. Catravas *et al.*, Plenum Press, New York, 1996

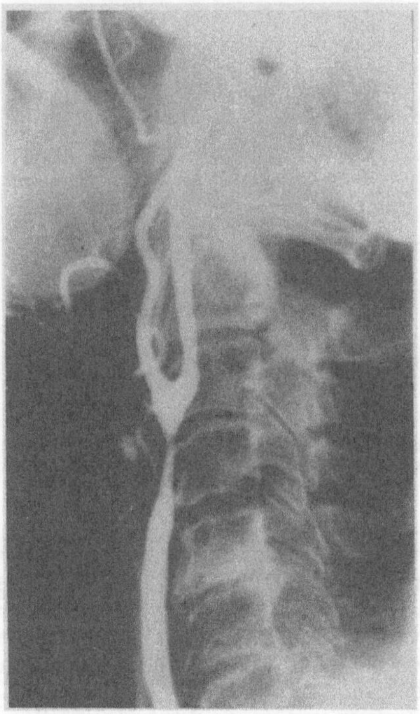

Figure 1. Stenosis of saphenous vein graft in aorta graft in aorta coronary position 2 years after implantation. Note animal residual lumen(arrow), organized thrombus clefts of cholesterol crystals and cellula proliferation(Human material).

Endarterectomy - carotid		20% (early)
	Vein bypass - coronary	10-30%
	Vein bypass - femoral distal tibial	10-20%

The Carotid Artery

Restenosis following carotid endarterectomy is reported to be from 3 to 23% of operations. The spread is based on whether one is reporting the incidence of abnormalities in the lumen as detected by arteriography or noninvasive studies, which tend to be high, or on the presence of symptoms associated with the area of stenosis, which tends to be low. The spread is also influenced by the intensity of the search, the sensitivity of the method, and the degree of narrowing the reviewer considers significant. For example, review of our own carotid endarterectomy follow-up data resulted in the identification of two sets of patients: those with symptomatic recurrence and those without. This generated the next quesion: for how long should patients with an asymptomatic carotid restenosis lesion be followed? For how long is the treated artery vulnerable? What is the expense? What is the yield?

Based on the reappearance of neurologic symptoms we found an incidence of recurrent carotid stenosis of 3.6% in a time interval from operation of 2 to 24 months *(Cossman et al., 1982)*. Two other series of patients followed for the same postoperative period noted symptomatic recurrent stenosis in 1.3% and 1.5% *(Hertzer et al., 1979; Stoney and String, 1976)*. By contrast, the incidence of restenosis as determined by noninvasive tests (which are of increasing sensitivity in recent years) ranges between 10 to 30% *(Zierler et al., 1982; Cantelmo et al., 1981; DeGroote et al., 1987)*. An important byproduct of these studies was the discovery that the postoperative carotid endarterectomy site undergoes remodeling with the passage of time, and single or time limited examinations may not give a true evaluation of

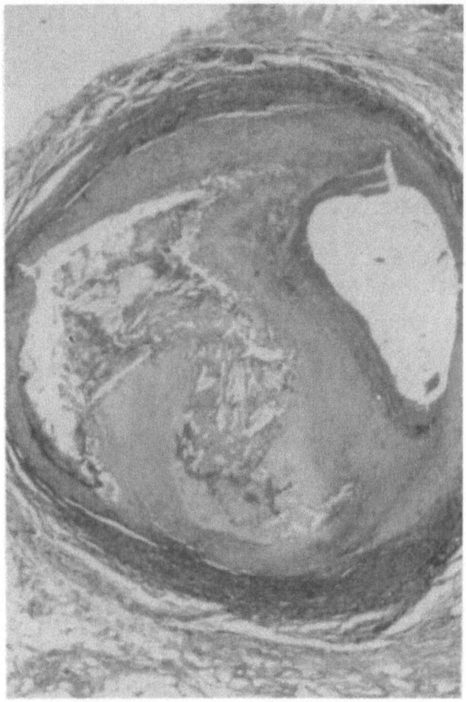

Figure 2. Atreriogram showing hour-glass constriction of the lumen of the common carotid artery characteristic of an early cellular proliferative restenosis lesion.

the process. Serial tests by duplex scanning showed that some early lesions with severe stenosis underwent regression on repeated examination *(Zierler et al., 1982; Healy et al., 1989)*.

Available data are inadequate to identify with reasonable certainty the risk factors for developing restenosis. The risk factors generally accepted for atherosclerosis seem to hold true for the lesions developing more than two years after operation. Late lesions, on gross and microscopic examination, display arterial wall changes consistent with atherosclerosis. Early lesions, arbitrarily set as occurring within two years of operation, are of the myointimal proliferative variety with little if any of the characteristics of atherosclerosis. These early lesions are the ones that may undergo remodeling, but identifying which one is impossible. The stenosis of an early lesion is usually smooth on the luminal surface: thus, it is less likely to attract platelets and less likely to develop thrombi which may become a source of emboli. Whatever the circumstances, the early proliferative restenotic lesion is far less productive of symptoms than the atherosclerotic plaque of the first operation. Many vascular specialists believe it is a waste of resources to keep these carotid endarterectomy patients under surveillance for more than one year *(Nitzberg, 1991)*, not so with the late or atherosclerotic lesion. These must be followed as any plaque in the internal carotid artery. They are more likely to become symptomatic and, apparently, for the same reasons: development of an ulcerated surface with secondary thrombus formation and embolization.

Although requiring a little more time and more meticulous dissection than at the initial operation, reoperation on most carotid restenoses is not particularly difficult. It is not a guarantee, however, that restenosis will not again recur. Indeed, in our experience, we have had a few patients who required four procedures to control the problem. Recurrences did not cease until the cartid bifurcation and the old operative sites had been excised and replaced with a vein interposition graft. The accepted combined morbidity and mortality for satisfactory risk patients undergoing first time carotid endarterectomy for symptomatic disease is 2.5-3% *(Callow et al., 1988)*. For operations on the recurrent lesion this figure jumps to 10%. The patient is exposed to a substantially greater risk than at the first operation.

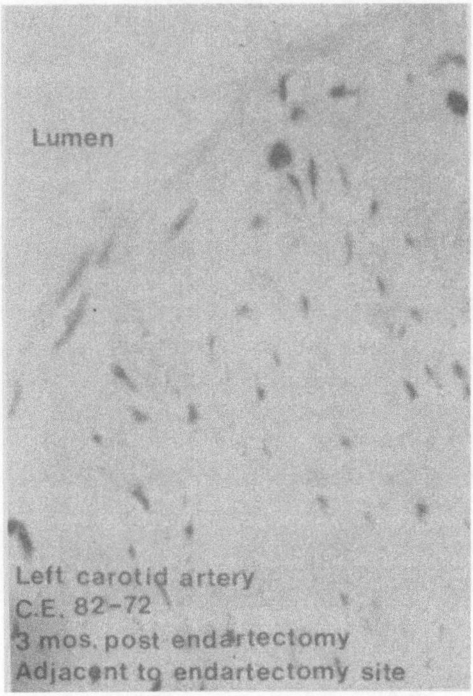

Figure 3. Microscopic demonstration (H and E stains) of a human cellular proliferative type of restenosis lesion, human, 6 months after initial coarotid endarterectomy.

THE CORONARY ARTERY

The recurrent stenosis rate for the saphenous vein bypass is approximately 50% at ten years. The internal mammary artery recurrence rate is 10% at ten years in the same position. Internal mammary artery bypasses require more time and care to perform and are not available in the length: and thus, the number of bypasses available with the saphenous vein. Stripping the sternum of its blood supply, especially when both internal mammary arteries are used, leads to ischemia and wound breakdown. These problems have largely been overcome by sometimes using vein as well as artery.

Reoperation in the coronary position is more difficult than the carotid, and, because coexisting coronary disease is the leading cause of morbidity and mortality in vascular surgery, the risk of reoperation is substantial.

Balloon angioplasty has a worse record of failure-approximately 30 to 50% in five years. The risk of reangioplasty is much less than that of redo coronary bypass grafts, but second and third redos are not uncommon. Most of these patients eventually come to bypass surgery.

THE AORTA, ILIAC, AND LOWER EXTREMITY VESSELS

The failure rate of balloon angioplasty, surgical endarterectomy and bypass grafts are all much lower in the proximal portion of this system than anywhere else. The aorta, iliac and common femoral arteries are large vessels of low resistance. The continued patency rate of aorta-femoral bypass grafts using synthetic material is in the 90% range. It has been established over the past three decades as the most durable reconstruction of all available modalities and of all anatomic locations. The most common cause of failure at this high location is progression of atherosclerosis in the inflow and outflow vessels. Below the knee, these figures fall off dismally. In distal tibial vessels, the posterior and anterior tibials and the peroneal arteries, continuing patency three years after the operation is in the 15 to 30% range, and very

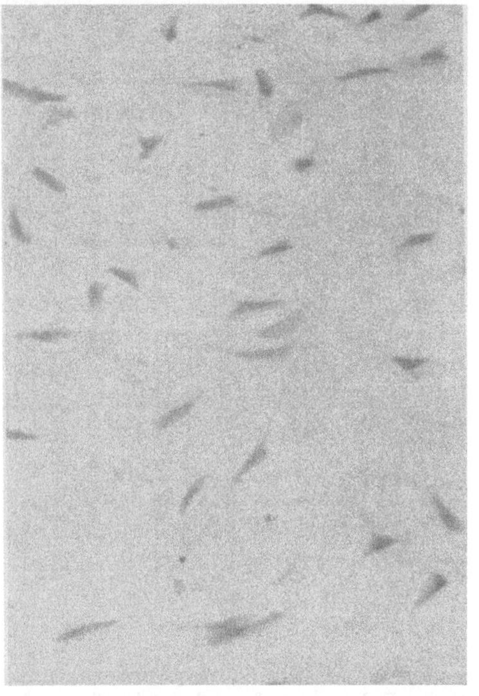

Figure 4. Microscopic display of early cellular proliferative type of restenotic lesion. The absence of reendothelialization is the important point.

few reach five years. These are small vessels with high resistance and limited outflow. Reoperation is common. Both the cellular hyperplastic lesion of early restenosis and progression of atherosclerosis are seen at the line of anastomosis and beyond.

Microscopic examination of the hyperplastic lesion biopsied at reoperation does not permit identification of the vessel of origin. The lesions are quite uniform from wherever they are obtained. On gross inspection the surface is smooth, glistening, and pearly white. Ulcerations, calcification and the yellowish stain of cholesterol are absent. If not pearly white, it is pale. The consistency of the material is firm and even rubbery. Dissection of this material from the adventitia of the underlying arterial wall is quite difficult and requires cutting of the lesion from the wall, quite unlike removing a typical atherosclerotic plaque where a cleavage plane between the lesion and the adventitia is easily developed. The difficulty of separation of a hyperplastic lesion is simply because it involves the adventitia. It is a lesion of the media and the intima, unlike the intimal and subintimal lesion of atherosclerosis. This involvement of both media and intima was the reason for originally describing the predominant cell as a myointimal or myofibroblast cell. The smooth muscle cell is considered the predominant proliferating cell. Underscoring this statement is the lack of a new endothelial lining beyond the first few millimeters of the anastomosis or the ends of the endarterectomy site in recovered human specimens. We have looked without success for many years for evidence of endothelial regeneration from our material and from material of numerous colleagues.

Microscopically the early hyperplastic lesion consists of stellate or spindle-shaped cells surrounded by a liberal amount of clear, nonstaining, homogenous matrix. It has been described as fibromyomatous connective tissue. No hemosiderin, red cell remnants or thrombi are seen. Only a single cell type is found which resembles the smooth muscle cell in many respects. These cells demonstrate the typical morphology of SMCs; they grow in clumps and masses and are without contact inhibition in tissue culture; they stain positively for smooth muscle cell specific actin chains and for sulfate glycosaminoglycans. These cells have myofilaments, dense bodies and external lamina. They satisfy the requirements of smooth muscle cells. In all likelihood, they originate in the media.

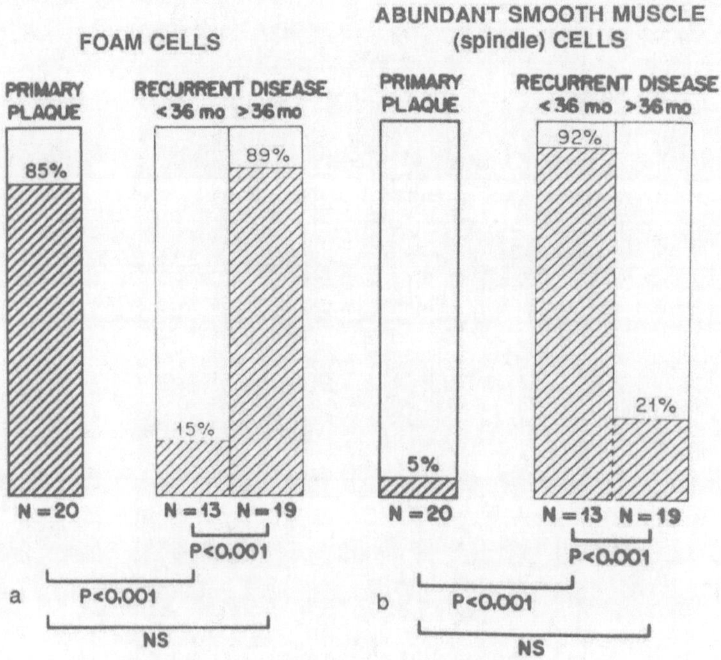

Figure 5. A: Frequency of foam cell in initial (ASO) plaque removed at initial operation compared to frequency in recurrent lesion. B: Frequency of smooth muscle cells in early versus late restenotic lesions.

The temporal profile of the cellular proliferative lesion sheds some light on the response to injury of the arterial wall. At least, this is so in the experimental animal model which uses endothelial cell denudation as the injury. The sequence of events which enlist the participation of the cellular and molecular players is similar to a domino scenario in which the next event depends on the one or ones immediately preceding it. Following denudation there is migration of smooth muscle cells from the media through the internal elastic lamina into the intima. Proliferation of smooth muscle cells occurs both in the media of the cell's origin and among the migrated cells. Whether migration or proliferation is the first event is a matter of discussion and depends to some extent on the animal species, rat or rabbit. Once in the intima and subintimal region , the cells proliferate rapidly, undergo a change of phenotype from contractile to secretory and proceed to manufacture extracellular matrix. Early in the process the bulk of the lesion is made up of migrated and proliferating cells. Later, the extracellular matrix contributes to the mass. Endothelial cells are not present, and in the human, unlike other animals, never completely cover the area of injury or the surface of the inserted synthetic graft. It is for this reason that grafts in the human are said to never reach a stage of complete healing. The normal controls over smooth muscle proliferation and its maintenance of the normal physiologic contractile phenotype are missing. In the normal state the smooth muscle cell resides in the media surround by an extracellular matrix of collagen and elastin. Its rate of turnover is very low. Absence of the endothelium appears to initiate a complex series of events, the sum of which results in great activity of the smooth muscle cell.

Platelets are rapidly attracted to the subendothelial layer which is exposed after endothelial injury. Platelet adhesion is mediated via platelet membrane glycoprotein receptors, von Willebrand Factor and fibronectin. Platelet activation occurs through the arachidonic acid pathway with additional platelet recruitment and thromboxane release. Additional players are the membrane receptors GPIIb and GPIIIa. Platelet activation is followed by release of Platelet Derived Growth Factor (PDGF) which is not only a smooth muscle mitogen but also a chemoattractant of major importance in smooth muscle cell migration.

It was the observation that the endothelial cell was unable to replicate, or to do so with sufficient speed after injury, that led us to a series of experiments with Vascular Endothelial Growth Factor (VEGF), also known as Vascular Permeability Factor (VPF). Using a rabbit model, the endothelium was stripped from the carotid artery with a balloon catheter after which

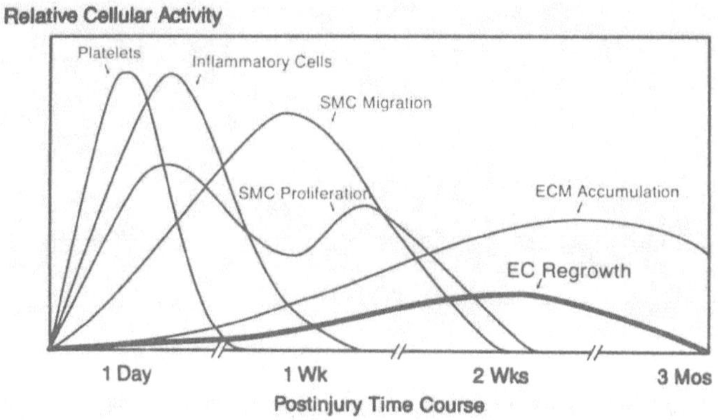

Figure 6. *Comparative Cellular Activity Following Injury*. Detection of PDGF-like material in biopsy specimens of human restenosis lesions in human lower extremity bypass grafts (3) removed at second operation.

VPF was administered. VPF is an extremely potent polypeptide mitogen for endothelial cells but not for vascular smooth muscle cells *(Connolly et al., 1989)*. We noted that rapid reendothelialization appeared to reduce the proliferative SMC response in a rabbit model *(Stevens et al., 1992)*. We administered VPF intravenously for two weeks following balloon denudation of the endothelium and observed marked acceleration of reendothelialization. Cellular proliferation was markedly reduced, in comparison to that in animals treated with control solution, the vehicle for the VPF only, as measured by the intimal area to medial area ratio *(Callow et al., 1995; Stevens et al., 1992; Ryan et al., 1979; Clowes et al., 1989; Jawien et al., 1992)*.

The lack of human material for study is illustrated by the absence of reports in the literature except for our 1989 report on cell culture studies of specimens obtained at reoperation of failed by-pass lower extremity grafts in humans. These biopsies placed in culture expressed a PDGF-like mitogen as well as a gene for a PDGF receptor.

Human material is difficult to obtain even by the operating surgeon interested in the problem. Recurrent stenoses are treated by balloon dilatation which, in most cases, does not provide an opportunity for inspection and biopsy, or by bypass graft which precludes opening the area of constriction. Thus, the overwhelming predominance of *in vitro* data from cell culture and *in vivo* data from animal models are not comparable to the human lesion beyond its cellular proliferative portion *(Painter, 1991)*.

Graft failure must be examined from several aspects: was the graft of autologous tissue, e.g., saphenous vein, or was it synthetic? Did failure occur in the immediate postoperative period or later? Was failure due to an excessive burden of atherosclerosis in the inflow and the outflow vessels? Was there a technical error in the construction of the graft, such as twisting of the vein or a constricting "last stitch" place in an attempt to control bleeding? Was patient selection improper because of poor outflow? All of these relate to early failure. Late failure is commonly due to atherosclerosis in adjacent vessels and to anastomotic hyperplasia. In synthetic grafts, whether teflon or dacron, stenosis occurs at the line of anastomosis, which is where the rigid non-yielding graft with its line of sutures anchoring it to the less rigid recipient artery may create a continuing stimulus to the elements of the arterial wall. The resulting rebellion to this intrusion on the natural environment of the arterial wall in which the smooth muscle cell and the endothelial cell exist in quiet harmony with each other cannot be surprising. Porous synthetic grafts develop an endothelial lining throughout the length, probably because of ingrowth of these cells with the granulation tissue which enter the wall of the graft through the interstices in the material. Such porous grafts cannot be used in the human because of unacceptable bleeding. In the non-porous graft used in the human, the lining of the graft is compacted fibrin, and cells are presumed to be of mesenchymal origin. In the injured artery of the experimental model, intimal thickening associated with the healing process stops when and where endothelial repaving is complete. In both instances the graft, vein or synthetic, develops fibromuscular stenosing constrictions at the suture lines of the anastomoses. In both types of grafts the stenosing lesion consists of smooth muscle cells and extracellular matrix. In summary, experimental animal models demonstrate smooth muscle cell migration and proliferation, extracellular matrix formation and accumulation and some degree, usually

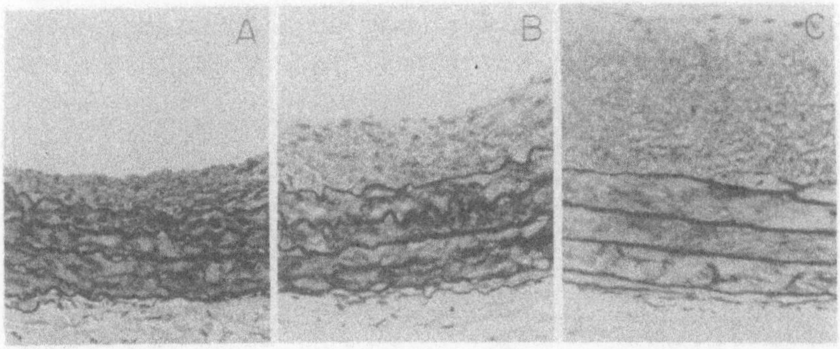

Figure 7. Endothelin administration to a rat balloon injury model showing stimulation of proliferative restenosis.

incomplete, of endothelial regeneration. The human lesion, if longer than two or three centimeters, and with few or no side branches to the injured artery which could provide a source of endothelial cells along the length of the artery, has a very limited ability to replace the missing or damaged endothelial layer. If the arterial injury includes insertion of a nonporous synthetic graft, endothelial regeneration is limited to a centimeter or so beyond the anastomotic line at which distance the host or recipient artery cannot supply endothelial cells. Even at this location where limited endothelial repaving has occurred, smooth muscle cells continue to proliferate under the endothelium and matrix continues to build up. This phenomenon suggests autocrine and paracrine pathways; that is, the endothelium is a source of smooth muscle growth factors. The evidence involving the platelet and its growth factors is now extensive and indicate that the platelet serves as source of growth factor in the injured artery and probably largely regulates the intimal hyperplastic response *(Ross and Glomset, 1976)*. Platelet factors also regulate migration of smooth muscle cells, an observation that has been confirmed in balloon injury animal models *(Fingerle et al., 1989; Choi et al., 1994)*. Platelet derived growth factor induces chemotaxis by focal redistribution of $\alpha_v\beta_3$ integrin along the leading edge of migrating smooth muscle cells. The lack of human material is more severe with angioplasties. Biopsies are rarely obtained either at the initial or the repeat procedure. A 1991 report describes the restenosis lesions in seven patients who died within 20 days of a percutaneous transluminal coronary balloon angioplasty. The arteries in which the angioplasties were performed were sectioned serially at 1 mm intervals from the proximal origin of the artery to the most distal epicardial site to ensure that the total length of the artery was examined. The authors note that distinct differences in the nature of the response to the angioplasty exist in relation to the type of tissue injured, whether it is an atheromatous plaque or a fibromusculo-elastic lesion. Injury to an atheroma plaque rupture, leads to exposure of atheromatous debris. The consequence is almost instant mural thrombosis which may be so severe as to result in occlusion of the artery. The possibility of a fatal outcome is high. On the other hand angioplasty of a fibromusculo-elastic lesion results in laceration of the media or the musculo-elastic layers with deposition of a fibrin-platelet thrombus, soon followed by the appearance of the spindle shaped cells typical of the cellular proliferative type recurrent lesion. This type of response is not instant and requires weeks to months, or more, to reach a symptomatic size. The report is based on a very small series, and it describes changes limited to the first twenty days following angioplasty. Small as it is, it is the beginning of data accumulation of human material *(Ueda et al., 1991)*.

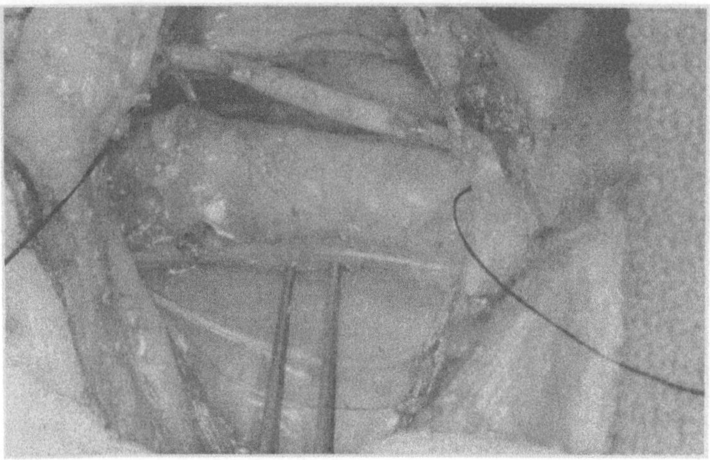

Figure 8. Restenosis is markedly increased by the synergistic effect of hypercholesterolemia in a rabbit model of balloon injury to a carotid artery. The animal had been on a high cholesterol diet for 12 weeks and endothelial denudation was performed at the onset of the diet. The uppermost carotid demonstrates the effect of the high cholesterol diet plus balloon denudation. The lower artery shows the effect of high cholesterol alone.

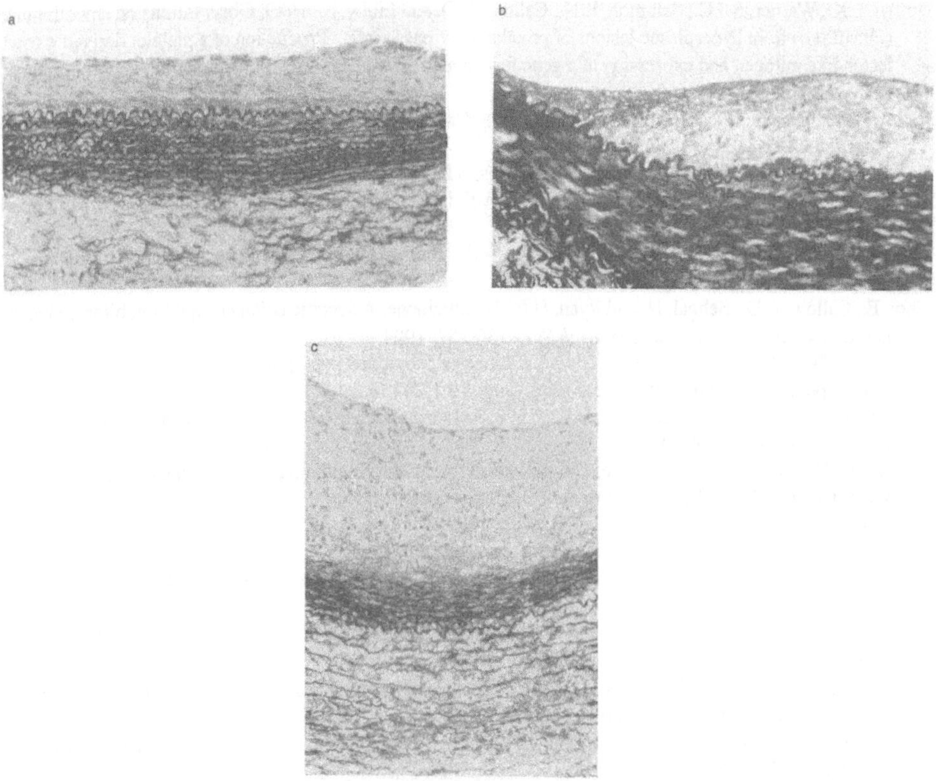

Figure 9. The synergistic effect of hypercholesterolemia and endothelial injury in a rabbit model using the common carotid artery. All specimens harvested at the same time point of 12 weeks after onset of diet and/or endothelial injury. A: Effect of endothelial denudation alone. B: Effect of administering high cholesterol diet alone. C: Effect of high cholesterol diet plus endothelial injury.

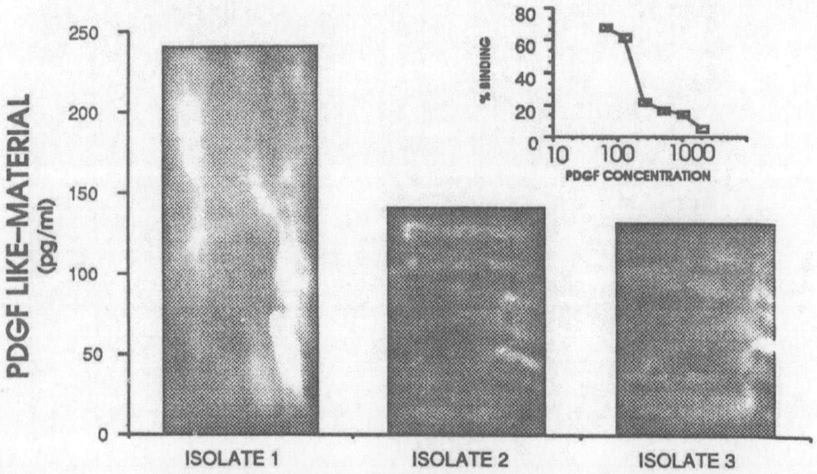

Figure 10. A temporal profile of the many events occurring in an animal model (rat) of response to balloon denudation injury. This is probably not a reasonable profile for the human restenosis lesion because of the blood lipid evaluations usually not found in this animal model injury.

REFERENCES

Birinyi, L.K., Warner, S.J.C., Salomon, R.N., Callow, A.D. and Libby, P., 1989, Observations on smooth muscle cell cultures from hyperplastic lesions of prosthetic by-pass grafts: Production of a platelet derived g rowth factor-like mitogen and expression of a gene for a platelet-derived growth factor receptor - A preliminary study. *J. Vasc. Surg.* 10:160.

Callow, A.D., Caplan, L.R., Correll, J.W., et al., 1988, Carotid endarterectomy: What is its current status? *Am. J. Med.* 85:835-838.

Callow, A.D., Choi, E.T., Trachtenberg, J.D., Stevens, S.L., Connolly, D. and Ryan U.S., in press, Vascular permeability factor accelerates endothelial regrowth in vivo following balloon angioplasty. *Growth Factors* 10:223-228,1994.

Cantelmo, N.L., Cutler, B.S., Wheeler, H.B. et al., 1981, Noninvasive detection of carotid stenosis following endarterectomy. *Arch. Surg.* 116:1005.

Choi, E., Callow, A.D., Sehgal, N. and Ryan, U.S., Halofuginone, A specific collagen type I inhibitor, reduces anastomotic intimal hyperplasmia. *Arch Surg* 130:257, 1994.

Choi, E., Callow, A.D., Engel, L. et al., 1994, Inhibition of neointimal hyperplasia by blocking $\alpha_v\beta_3$ integrin with a small peptide GpenGRGDSPCA. *J. Vasc. Surg.* 19:125-134.

Clagett, G.P., Graor, R.A. and Salzman, E.W., 1992, Antithrombotic therapy in peripheral arterial occlusive disease. *Chest* 102:516S-28S.

Clowes, A.W., Clowes, M.M., Fingerle, J. and Reidy, M.A., 1989, Regulation of smooth muscle cell growth in injured artery. *J. Cardiovasc. Pharmacol.* 14(Suppl 6): S12-S15.

Connolly, D.T., Heuvelman, D.M., Nelson, R. et al., 1989, Tumor vascular permeability factor stimulates endothelial growth and angiogenesis. *J. Clin. Invest.* 84:1470-1478.

Cossman, D., Callow, A.D., Stein, A., Matsumoto, G., 1982, Early restenosis after carotid endarterectomy. *Arch. Surg.* 113:1408.

DeGroote, R.D., Lynch, T.G., Jamil, Z. and Hobson, R.W. III., 1987, Carotid restenosis: Long term noninvasive follow up after carotid endarterectomy. *Stroke* 18:1031.

Fingerle, J., Johnson, A.W., Clowes, A.W. et al., 1989, Role of platelets in smooth muscle cell proliferation and migration after vascular injury in rat carotid artery. *Proc. Natl. Acad. Sci. USA* 86:8412.

Healy, D.A., Zierler, R.E., Nicholls, S.C. et al., 1989, Long term follow up and clinical outcome of carotid restenosis. *J. Vasc. Surg.* 10:662, 1989.

Hertzer, N.R., Martinez, B.D. and Bevan, E.G., 1979, Recurrent stenosis after carotid endarterectomy. *Surg. Gynecol Obstet.* 149:360.

Jawien, A., Bowen-Pope, D.F., Lindner, V. et al., 1992, Platelet-derived growth factor promotes smooth muscle cell migration and intimal thickening in a rat model of balloon angioplasty. *J. Clin. Invest.* 89:407-511.

Nitzberg, R.S., Mackey, W.C., Prendiville, E., Foster, E., Kastrinakis, W., Belkin, M., O'Donnell, T.F., Jr. and Callow, A.D., 1991. Long-term follow-up of patients operated on for recurrent carotid stenosis. *J. Vasc. Surg.* 13:121-126.

Painter, T.A., 1991, Myointimal hyperplasia: Pathogenesis and implications.*In vitro* Characteristics Artifical Organs 15:42-55.

Ryan, U.S., Whitaker, C., Hart, M.A. et al., 1979, Structural interaction between endothelial and smooth muscle cells. *J. Cell Biol.* 79:207a.

Ross, R. and Glomset, J.A., 1976, The pathogenesis of atherosclerosis. *N. Eng. J. Med.* 295:369-377.

Stevens, S.L., Choi, E.T., Callow, A.D. et al., Vascular permeability factor speeds endothelial repaving following arterial injury. *Surg. Forum XLIII* 358-360.

Stevens, S.L., Hilgarth, K., Ryan, U.S. and Callow, A.D., 1992, The synergistic effects of hypercholesterolemia and mechanical injury on intimal hyperplasia. *Ann. Vasc. Surg.* 6:55-62.

Stoney, R.J. and String, S.T., 1976, Recurrent carotid stenosis. *Surgery* 80:705.

Ueda, M., Becker, A.E., Fujimoto, T. and Tsukada, T. The early phenomena of restenosis following percutaneous transluminal coronary angioplasty. *European Heart J.* 12:937-945.

Zierler, R.E., Bandyk, D.F., Thiele, B.L. and Strandness, D.E. Jr., 1982, Carotid artery stenosis following endarterectomy. *Arch. Surg.* 117:1408.

ROLE OF THROMBIN AND PLASMINOGEN ACTIVATORS, ESSENTIAL PROTEASES OF THE COAGULATION AND FIBRINOLYSIS CASCADES, IN FIBRIN HOMEOSTASIS AND VASCULAR CELL ACTIVATION

Victor W.M. van Hinsbergh

Gaubius Laboratory TNO-PG
Leiden
The Netherlands

INTRODUCTION

Coagulation and fibrinolysis regulate the formation and removal of fibrin, the temporary repair matrix in wounded areas of the body. The fibrin matrix acts as a sealing meshwork enforcing the hemostatic plug and preventing further blood loss after wounding, and provides a scaffold into which new microvessels or other invading cells can grow during wound healing. The formation of the fibrin matrix is initiated by proteolytic conversion of fibrinogen to fibrin by thrombin, a serine protease formed by the coagulation pathway . The matrix is strengthened by crosslinking of the fibrin subunits in fibrin fibers by factor XIII. Thrombin, the eventual product of the activated coagulation system, not only initiates fibrin formation and platelet activation, but also acts as a paracrine hormone at the vessel wall. On the endothelium, it interacts with two receptors and thereby enhances anticoagulant properties of the endothelium. The first receptor, thrombomodulin, catalyzes thrombin-dependent activation of protein C; upon thrombin binding, the other receptor activates the endothelial cell, by which acute release of t-PA and production of prostacyclin and NO is induced, and the cell is activated to synthesize new proteins. Thrombin also acts on smooth muscle cells, in particular in the thickened intima, via the thrombin receptor, and may stimulate the growth of smooth muscle cells. Hence, thrombin initiates and regulates fibrin formation; and activates vascular cells.

A similar combination of functions can be ascribed to the proteases that regulate the breakdown of fibrin, i.e. plasmin and plasminogen activators (PAs). Like thrombin, the plasminogen activation system also plays a dual role in the vascular tree. Firstly, it causes the proteolysis of fibrin (fibrinolysis), both by the rapid degradation of intravascular fibrin, necessary to prevent anoxia and damage to the distal tissue; and by removal of a wound matrix, which proceeds slowly and precisely timed to prevent rebleeding. Secondly, PAs, in particular u-PA, are involved in cell migration and in angiogenesis (pericellular proteolysis affecting cell-matrix contacts and/or matrix proteins). These processes also require proper timing, otherwise scar tissue may form. In this chapter we shall successively review the role of thrombin and PAs in the formation and degradation of a fibrin matrix; and in vascular cell activation. Both the fibrin matrix and vascular cell activation play an important role in the control of tissue repair.

Vascular Endothelium: Responses to Injury
Edited by John D. Catravas *et al.*, Plenum Press, New York, 1996

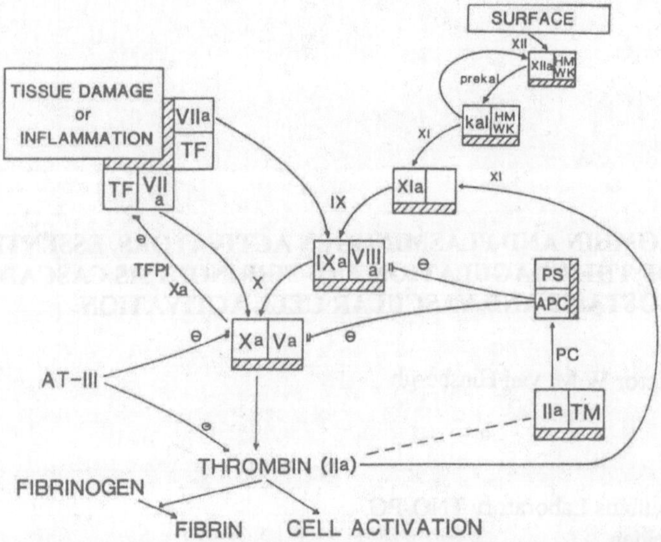

Figure 1. Schematic representation of the coagulation cascade. Coagulation factors are indicated by Roman figures. Activated coagulation factors by the addition "a".II: prothrombin; IIa: thrombin; TF: tissue factor; TFPI: tissue factor pathway inhibitor; ATIII: antithrombin III; PC: protein C; APC: activated PC; TM: thrombomodulin; PS: protein S; kal: kalikrein; HMWK: high molecular weight kininogen.

However, if these processes proceed uncontrolled, thrombosis, (re)bleeding, scar tissue formation or restenosis after angioplasty may occur.

THROMBIN GENERATION: THE COAGULATION CASCADE

A fibrin matrix is formed after proteolytic conversion of fibrinogen into fibrin monomer, followed by spontaneous fibrin polymerisation and crosslinking of the fibrin subunits in the fibrin fibers by the transglutaminase factor XIII. Proteolytic conversion of fibrinogen into fibrin is initiated by the serine protease thrombin, which is formed from its zymogen prothrombin by the activation of a cascade of proteolytic events, the coagulation cascade (*Davie, Fujikawa, and Kisiel, 1991*). A schematic diagram of the coagulation cascade is depicted in Figure 1. Usually the coagulation cascade is initiated by the exposure of tissue factor to the blood, leading to activation of the extrinsic pathway of coagulation. In special conditions, in particular when blood is exposed to artificial negatively-charged surfaces, the intrinsic pathway can be activated via activated factor XII. However, the current opinion is that activation of the intrinsic pathway usually occurs via activation of factor XI by thrombin, and that the physlological role of factor XII in coagulation is obscure. Hence, the intrinsic pathway acts as an amplification mechanism of thrombin generation. The same holds true for the activation of factor IX. While tissue factor-bound factor VIIa rapidly activates both factor X and IX, the activated factors Xa and VIIa form a complex with TFPI (tissue factor pathway inhibitor) (*Rapaport, 1991*), which blocks tissue factor and further initiation of the intrinsic coagulation pathway. Continuation of thrombin formation by factor Xa then depends on the availability of factor IXa, which was initially formed by tissue factor-bound factor VIIa and can be generated via thrombin-activated factor XIa.

Like nearly all proteolytic steps in the coagulation cascade, the activation of prothrombin proceeds enzymatic cleavage in a complex that consists of negatively charged phospholipids, calcium ions, a nonenzymatic cofactor protein, an active serine protease and the substrate, prothrombin. In this prothrombinase complex, factor Va represents the cofactor protein and factor Xa converts prothrombin in α-thrombin via two proteolytic cleavages. If thrombin is generated on the surface of the endothelium, the first proteolytic cleavagege is preferentially at Arg^{323}-Ile^{324}. The then formed intermediate, meizothrombin, activates preferentially anticoagulant properties on the endothelium (*Tijberg, van Heerde, Leenhouts, Hessling, Bouma, and de Groot, 1991; Doyle and Mann, 1990)*. If the first cleavage is at the Arg^{274}-Thr^{275} position, the intermediate molecule is prothrombin-2, which has strong procoagulant properties. Both intermediates are rapidly converted in the procoagulant α-thrombin (further indicated as thrombin).

Although the formation of a hemostatic plug and a fibrin meshwork are necessary to prevent bleeding after wounding and to ensure proper healing of a wound, intravascular fibrin formation may seriously impair blood flow and cause ischemia and damage to the distal tissue. To prevent vascular occlusion, the inner lining of the blood vessel, the endothelium, has several properties which counteract coagulation and platelet deposition. The major anticoagulant properties of the endothelium are: (a) The synthesis and exposure of specific proteoglycans, which bind antithrombin III, the major thrombin inhibitor in the lumenal compartment (*Marcum and Rosenberg, 1987*), (b) The synthesis and exposure of thrombomodulin, a receptor on which thrombin activates protein C (*Esmon, 1993*). Activated protein C inhibits the coagulation pathway by proteolytic inactivation of the factors Va and VIIIa, (*Esmon, 1989*),the cofactors needed for thrombin generation and factor IXa-dependent factor Xa generation. Protein S is the cofactor involved in this inhibitory activity of protein C (*Walker, 1980).* Homozygous deficiencies of protein C or protein S cause serious thrombotic complications. (*Griffin, Evatt, zimmerman, Kleiss, and Wideman, 1981; Bertina, Broekmans, van der Linden, and Mertens, 1982; Comp, Nixon, Cooper, and Esmon, 1984)*. The risk for thrombosis in heterozygous patients varies within families. In this respect the recent discovery of a generally occurring mutation in factor V, which causes a decreased sensitivity for proteolytic cleavage by activated protein C, (*Bertina, Koeleman, Koster, Rosendaal, Dirven, de Ronde, van der Velden, and Reitsma, 1994; Dahlbäck and Hildebrand, 1994)* is of interest. This type of mutation was highly associated with the risk of obtaining thrombosis, (*Bertina, Koeleman, Koster, Rosendaal, Dirven, de Ronde, van der Velden, and Reitsman, 1994*) and may act in concert with heterozygous protein C deficiency. (c) Furthermore, endothelial cells contribute to the synthesis of TFPI, (*Rapaport, 1991)* and in culture they can produce annexin V, (*Reutlingsperger, Hornstra, and Hemker, 1985; Funakoshi, Heimark, Hendrickson, McMullen, and Fujikawa, 1987*) a phospholipid binding protein, which interferes with the formation of the prothrombinase complex.

Endothelial cells also regulate hemostasis by preventing platelet activation by the synthesis of prostacyclin(*Marcus, Weksler, Jaffe, and Broekman, 1980*), prostaglandin E_2 and nitric oxide (NO) (*Radomski and Moncada,* 1993*)*, and by the production of ectonucleotidases (*Marcus, Safier, Hajjar, Ullman, Islam, Broekman, and Eiroa, 1991*), which rapidly degrade the strong platelet-activating agent ADP. Furthermore, endothelial proteoglycans contribute to the prevention of platelet adhesion on the surface of endothelial cells. If not withstanding all these regulatory mechanisms fibrin formation still occurs, endothelial cells have a third mechanism to prevent vascular occlusion. They provide the regulatory protein tissue-type plasminogen activator (t-PA) (*Rijken, Wijngaards, and Welbergen, 1980; van Hinsbergh, 1992),* which activates fibrinolysis, the process by which fibrin is dissolved. This aspect will be dealt with after discussing the effect of thrombin on cell activation.

THROMBIN RECEPTORS AND CELL ACTIVATION

In addition to its pivotal role in coagulation, thrombin also plays a crucial role in the activation of platelets and other cells. Platelet activation is necessary to form a hemostatic plug after wounding. Thrombin stimulates endothelial cells to control the extent of coagulation by release and synthesis of a number of antithrombotic products. It induces mitogenesis in fibroblasts and smooth muscle cells, which become involved in a subsequent healing process.

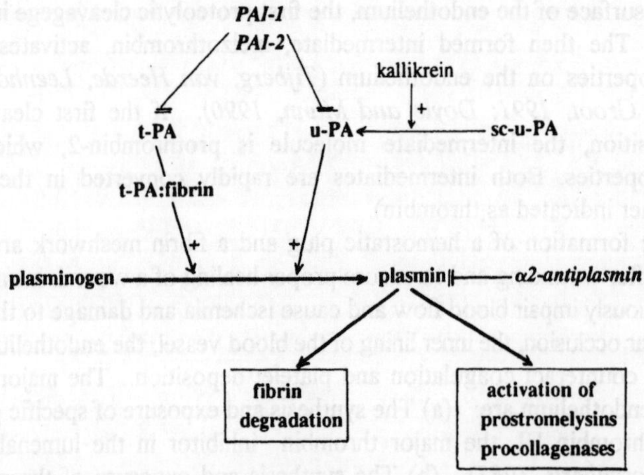

Figure 2. Schematic representation of the plasminogen activation system. +: activation; -: inhibition. Inhibitors are indicated in italics. PA: plasminogen activator; t-PA: tissue-type PA; u-PA: urokinase-type PA; sc-u-PA: single chain u-PA; PAI: PA inhibitor.

These reactions are induced by interaction of thrombin with a cellular signal-transducing receptor.

Several proteins have been reported to act as binding sites for thrombin on platelets. Among them, are glycoprotein Ib (GpIb) and a recently cloned receptor, further indicated as the thrombin receptor (*Vu, Hung, Weaton, and Coughlin, 1991; Rasmussen, Vouret-Craviari, and Jallet et.al.,1991*), which binds thrombin with high affinity. Gplb can also bind thrombin with high affinity, but it is currently believed that Gplb only plays an indirect role in thrombin-induced platelet activation via modulation of the interaction of thrombin with the thrombin receptor (*Jandrot-Perrus, Clemetson, Huisse, and Guillin, 1992*). On endothelial cells, both the thrombin receptor and thrombomodulin act as a thrombin receptors. However, while thrombomodulin acts as the cofactor needed for protein C activation, it does not generate second messengers in the cell. On the other hand, the thrombin receptor is a member of the family of receptors which have seven transmembrane domains (*Vu, Hung, Weaton, and Coughlin, 1991; Coughlin, 1993*), and is able to activate platelets and other cells via G-protein-mediated generation of second messengers (*Van Obberghen-Schilling and Pouysségur*, 1993; *Brass, Hozie, and Manning, 1993*). Hence, this receptor is the only second messenger-generating thrombin receptor presently known.

The thrombin receptor has a rather long extracellular N-terminal end, seven transmembrane domains connected with alternating intra- and extracellular loops and a C-terminal cytoplasmatic domain (*Coughlin, 1993*). Similar to other members of the seven transmembrane domain receptor family, the activated thrombin receptor interacts with G-proteins, which transduce the activation signal to second messenger generating systems, such as phospholipase C (via an activating G-protein) and adenylate cyclase (in platelets via an inhibiting G-protein) (*Van Obberghen-Schilling and Pouysségur, 1993; Brass, Hozie, and Manning, 1993*). The N-terminal extracellular domain harbours a binding domain for thrombin, which has close similarity to the thrombin binding sequence of hirudin, and a so-called tethered ligand domain. The tethered ligand domain becomes the new N-terminus after proteolytic cleavage of the N-terminal part by receptor-bound thrombin, by which a stretch of 41 amino acids is removed. This new N-terminus acts as the activating ligand for the thrombin receptor itself. When a synthetic peptide identical to the tethered ligand domain is added to platelets or cells, this synthetic peptide, called thrombin-receptor activating protein or TRAP, is also able to activate the thrombin receptor.

Thrombin is a strong stimulator of endothelial cells. By G-protein-mediated activation of phospholipase C, it increases the generation of inositoltrisphosphate, which subsequently induces an increase of the cytoplasmic calcium ion concentration (*Berridge, 1993*), and of diacylglycerol, a potent stimulator of protein kinase C, which phosphorylates a myriad of proteins on serine and threonine residues (*Nishizuka, 1992*). In addition thrombin induces phosphorylation of a number of proteins (*Mackie, Nairn, Hampel, Lam, and Jaffe, 1989*), and can activate phospholipase A_2 and D (*Hong and Deykin, 1982; Garcia, Fenton, and Natarajan, 1992*), As a consequence the endothelium responds immediately by releasing prostacyclin (*Weksler, Ley, and Jaffe, 1978*), nitric oxide (*Boulanger and Lüscher, 1990*), t-PA (*Emeis, 1992*) and von Willebrand factor (*Levine, Harlan, Harker, Joseph, and Counts, 1982*) and by exposing P-selectin on its surface (*Hattori, Hamilton, Fugate, McEver, and Sims, 1989*), and - in postcapillary venules - by contraction causing vascular leakage(*Lum, Andersen, Siflinger-Birnboim, Tiruppathi, Goligorsky, Fenton, and Malik, 1993*). Several hours after exposure cultured endothelial cells to thrombin, a number of genes are induced, which results in the enhanced synthesis of t-PA (*Levin, Marzec, Anderson, and Harker, 1984; van Hinsbergh, Sprengers, and Kooistra, 1987*), u-PA (*van Hinsbergh, Sprengers, and Kooistra, 1987*) PAI-1 (*van Hinsbergh, Sprengers, and Kooistra, 1987; Gelehrter and Sznycer-Laszuk, 1986*), endothelin (*Boulanger and Lüscher, 1990*) and many other proteins.

Thrombin is also a potent mitogen for fibroblasts (*Pohjanpelto, 1978; Glenn, Carney, Fenton, and Cunningham, 1980*) and smooth muscle cells (*McNamara, Sarembock, Gimple, Fenton, Coughlin, and Owens, 1993*). This effect is mediated via the thrombin receptor. Probably related to this effect, activation of the thrombin receptor by thrombin or TRAP enhances the rate of wound healing *in vivo* (*Carney, Mann, Redin, Pernia, Berry, Heggers, Hayward, Robson, Christie, Annable, Fenton, and Glenn, 1992*). In this perspective it is of interest to note that smooth muscle cells in the thickened intima express considerable amounts of thrombin receptors, whereas smooth muscle cells in the media express little or no thrombin receptors (*Nelken, Soifer, O'Keefe, Vu, Charo, and Coughlin, 1992*). In addition, the thrombin receptor expression increases markedly after wounding the thickened intima (*Nelken, Soifer, O'Keefe, Vu, Charo, and Coughlin, 1992*). Combined with the observation that inhibition of thrombin by a chemical thrombin inhibitor over a 1 hour period caused a reduced adhesion of platelets in a baboon arterectomie model (*Harker, Kelly, and Hanson, 1991*), these observations suggest a possible role for thrombin as a smooth muscle mitogen or activator. The foregoing recent observations indicate that thrombin can act as a paracrine agent modulating repair processes. On the one hand it induces platelet aggregation and generates the repair matrix fibrin, which prevent blood loss after wounding; on the other hand it activates vascular cells to become involved in the subsequent healing process. Sometimes, such as is the case in atherosclerosis and restenosis, this repair process overreacts and becomes lifethreatening in itself.

FIBRINOLYSIS AND PLASMINOGEN ACTIVATION

While the coagulation cascade results in the formation of the temporary fibrin matrix, another cascade of proteases regulates the breakdown of the latter. This breakdown is performed by the fibrinolytic system, in which the broadly acting protease plasmin is generated from its zymogen plasminogen by plasminogen activators (PAs) (Figure 2). Proof for the importance of plasminogen ctivators in fibrin clearance was recently obtained with transgenic animals missing t-PA and/or u-PA (*Carmeliet, Schoonjans, Kieckens, Ream, Degen, Bronson, de Vos, van den Oord, Collen, and Mulligan, 1994.*) The dissolution of a fibrin matrix requires precise timing. Fibrin formed at the luminal site of the endothelium has to be dissolved immediately to maintain blood vessel patency. Fibrin in the wound should be dissolved in adjusted pace with the healing process, otherwise rebleeding or scar tissue formation will occur. Furthermore, fibrin acts as a repair matrix, along which cells enter the wound during the healing process. It is therefore, of biological interest that PAs, which regulate fibrinolysis by

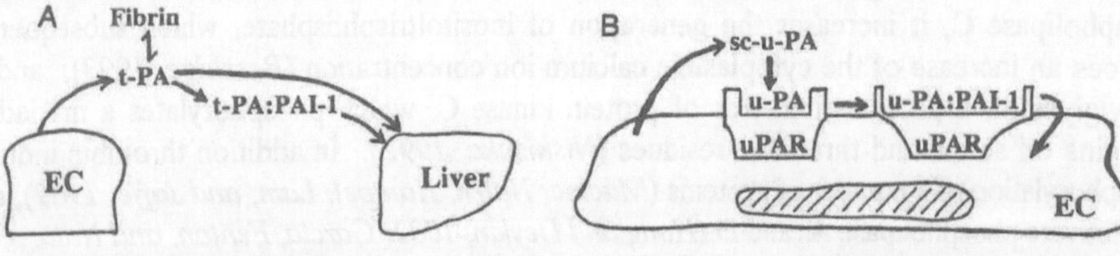

Figure 3. Schematic presentation of the synthesis and cellular uptake of plasminogen activators (PAs). Left: fibrinolysis; right: pericellular proteolysis. After secretion the PA binds to a matrix or cellular receptor where it becomes active and performs its proteolytical activity until it is inhibited by its inhibitor PAI-1. The PA:PAI-1 complex is subsequently internalized by the same or another cell. PAs that do not interact with a suitable matrix or receptor are cleared by the liver. EC: endothelial cell; uPAR: u-PA receptor.

converting plasminogen to plasmin, also play an important role in the pericellular proteolysis involved in cell migration and angiogenesis. Both in fibrinolysis and in pericellular proteolysis, PAs are secreted, e.g. by endothelial cells, and are taken up by other or the same cells after their action. This action occurs at a surface, i.e. on fibrin in the case of fibrinolysis or on a cell-bound receptor in the case of pericellular proteolysis. Figure 3 depicts this schematically. Protection against unwanted fibrin deposition occurs in the bloodstream by the continuous and/or stimulated secretion of t-PA by the endothelium, and by clearance of it by the liver. In pericellular proteolysis, a PA is secreted by the cell itself or an adjacent cell and uptake occurs in association with a cellular receptor on the same cell at the surface of which the PA performed its function. After a survey of the components of the fibrinolytic system, we shall subsequently discuss the cellular regulation of plasminogen activator synthesis and release, and how the various aspects of plasminogen activation are orchestrated in time and space by specific inhibitors and cellular receptors.

COMPONENTS OF THE FIBRINOLYTIC SYSTEM

Fibrin degradation is accomplished by the serine protease plasmin, which is formed from its zymogen plasminogen by PAs. Two types of PAs are presently known: tissue-type plasminogen activator (t-PA) and urokinase-type plasminogen activator (uPA) (*Bachman, 1987; Wallén, 1987*) (Figure 2). The genes of plasminogen, t-PA and u-PA have been characterized. All three proteins are synthesized as a single polypeptide chain, which is converted by specific proteolytic cleavage to a molecule with two polypeptide chains connected by a disulphide bond. The carboxy-terminal parts of the molecules (the so called B-chain) contain the proteolytically active site, whereas the amino-terminal parts of the molecule (the A-chain) are built up of domains that determine the interaction of the proteases with matrix proteins and cellular receptors. The cleavage of plasminogen and single-chain u-PA is necessary to disclose the proteolytic active site and to activate the molecule. In contrast, generation of t-PA activity does not depend on its conversion in a two-chain form, but on its interaction with a specific substrate, in particular fibrin. Once bound to this substrate, both the single-chain form and the two-chain form of t-PA are active.

The actual activity of the PAs is regulated not only by their concentration, but also by their interaction with PA inhibitors (PAIs) (*Sprengers and Kluft, 1987; Kruithof, 1988; Loskutoff, 1991*), cellular receptors (*Miles and Plow, 1988; Barnathan, 1992; Hajjar, 1993*) and, as indicated above, matrix proteins. Table 1 summarizes some properties of the proteins involved in plasminogen activation.

The three proteases of the fibrinolytic system are counterregulated by potent inhibitors, which are members of the serine protease inhibitor (serpin) superfamily. Plasmin is instantaneously inhibited by α2-antiplasmin (*Holmes, Nelles, Lijnen, and Collen, 1987*), but this reaction is attenuated when plasmin is bound to fibrin. The predominant regulators of t-PA and u-PA are PAI-1 and PAI-2 (*Kruithof, 1988*). PAI-1 is a 50 kD glycoprotein present in blood platelets and synthesized by endothelial cells, smooth muscle cells and many other cell types in culture (*Loskutoff, 1991*). PAI activity in human plasma is normally exclusively PAI-1. PAI-2 is produced by monocytes/macrophages and can be found as a glycosylated secreted molecule and as an intracellular molecule (*Wohlwend, Belin, and Vassalli, 1987*).

Regulation of fibrinolytic activity also occurs by cellular receptors. These recptors direct the action of PAs and plasmin to focal areas on the cell surface. High affinity binding sites for plasminogen (*Miles, Levin, Plescia, Collen, and Plow, 1988; Nachman, 1992*), t-PA (*Hajjar, 1991*) and u-PA (*Barnathan, 1992*) are found on various types of cells including endothelial cells. A specific u-PA receptor has been identified and cloned, which binds both single-chain u-PA and two-chain u-PA via their growth factor domains (*Roldan, Cubellis, Masucci, Behrendt, Lund, Danø, Appella, and Blasi, 1990*). The u-PA receptor is heavily glycosylated and proteolytically processed at its carboxy-terminal end; the receptor with the new carboxy-terminus is anchored in the membrane by a glycosylphosphatidylinositol moiety (*Plough, Behrendt, Løber, and Danø, 1992*). Its function will be discussed in a later paragraph. The nature of the t-PA receptor and the plasmin(ogen) receptor(s) on endothelial cells is less clear. The lipoprotein Lp(a), which has strong structural homology with a large part of the plasminogen molecule, can compete for plasminogen binding to endothelial cells (*Nachman, 1992; Miles, Fless, Levin, Scanu, and Plow, 1989*). In addition, clearance receptors exist on liver hepatocytes and liver endothelial cells, which can clear t-PA (*Kuiper, Otter, Rijken, and Van Berkel, 1988*) and plasmin-α2-antiplasmin and PA:PAI-1 complexes (*Orth, Madison, Gething, Sambrook, and Herz, 1992; Bu, Williams, Strickland, and Schwartz, 1992*) from the circulation.

REGULATION OF t-PA PRODUCTION: PREVENTION OF INTRAVASCULAR FIBRIN DEPOSITION

The fibrinolytic activity in blood is largely determined by the concentration of t-PA, which is synthesized in the endothelium (*Rijken, Wijngaards, and Welbergen, 1980; Wun and Capuano, 1985*). The concentration of t-PA in the circulation can change rapidly. This is due to the short half life time of t-PA in the circulation, which is 5 to 10 minutes in man, and to the ability of endothelial cells to release rapidly a relatively large amount of t-PA. Clearance of t-PA occurs in the liver. Consequently, changes in the liver blood flow affect t-PA clearance and the plasma t-PA concentration. The acute release of t-PA from a storage pool in the vessel wall can be induced by vasoactive substances, such as bradykinin, platelet activating factor and thrombin (*Emeis, 1992*). This mechanism makes it possible to enhance the t-PA concentration exclusively at those places where fibrin generation occurs. Hence, it contributes to the local protection against an emerging thrombus. If a generalized stimulation of the endothelium occurs, for example by catecholamines, the acute release mechanism causes a rapid temporary increase in the blood t-PA concentration.

The release of t-PA by endothelial cells also depends on the t-PA synthesis rate in the cells. The t-PA synthesis rate is different in various types of blood vessels, e.g. veins produce more t-PA than arteries. Furthermore, t-PA synthesis can be enhanced pharmacologically or physiologically by various mediators. During recent years insight has been gained regarding the regulation of the synthesis of t-PA by using endothelial cells *in vitro*. Activation of protein kinase C has been implicated in the regulation of transcription of the t-PA gene and t-PA synthesis in human endothelial cells. The stimulation of t-PA synthesis by histamine and thrombin is caused by this process. The induction of t-PA by protein kinase C activation is potentiated by a simultaneous increase of the cellular cAMP concentration (*Levin, Marotti, and Santell, 1989; Kooistra, Bosma, Toet, Cohen, Griffioen, Van den Berg, Le Clercq, and Van Hinsbergh, 1991*). It has been suggested that the protooncogenes c-fos and c-jun, which can form a heterodimer called AP1, are involved in the regulation of the t-PA gene in endothelial cells by interacting with one or more AP1-binding site(s) of the t-PA promoter (*Kooistra, Bosma, Toet, Cohen, Griffoen, Van den Berg, Le Clercq, and Van Hinsbergh, 1991*). In favor

Table 1. Properties of protein involved in plasminogen activation.

Protein	MW (kD)	Number of amino acids	Plasma concentration (mg/l)	Produced by
Plasminogen	92	791[a]	200	hepatocytes
t-PA	68	530[a]	0.005	endothelium, mesothelium
scu-PA	54	411	0.008[b]	monocytes/mø, renal tubuli, activated endothelial cells
α2-Antiplasmin	70	452	70	hepatocytes
PAI-1	50	379	0.05	smooth muscle cells, activated endothelium, ?liver
PAI-2	46/56[c]	393	<0.005	macrophages, placenta, mesothelial cells
u-PA Receptor	55	313/282/284[d]	-	present on monocytes-mø, endothelial cells, and other cells
t-PA Receptor	40	?	-	endothelial cells

[a] The numbering of amino acid residues is usually based on a total of 790 residues for plasminogen (Lys-plasminogen) and of 527 residues for t-PA.
[b] Sum of scu-PA and u-PA: inhibitor complexes.
[c] 46kD: intracellular non-glycosylated form, 60kD excreted glycosyated form.
[d] 313: postulated transmembrane form; 282-284: PI-linked form, which has been processed on its COOH-terminal side.
Abbreviations: mø: macrophages; MW: molecular weight.

of this suggestion is an experiment of nature, in which a single mutation in a AP-1 binding site of the t-PA promoter (-TGACATCA-) alter this PMA-responsive element in a cAMP-responsive element (-TGACGTCA-) (*Feng, Ohlsson, and Ny, 1990*). In contrast to human cells, rat endothelial cells enhance their t-PA synthesis markedly upon stimulation of the cAMP generation alone, whereas rat t-PA synthesis does not respond to the sole addition of PMA (*Emeis and Kooistra, 1993*).

Other pharmacologically interesting agents that can enhance t-PA synthesis are retinoids (*Kooistra, Opdenberg, Toet, Hendriks, Van den Hoogen, and Emeis, 1991; Thompson, Nelles, and Collen, 1991*) and certain benzodiazepines (*Kooistra, Toet, Kluft, Von Voigtlander, Ennis, Aiken, Boadt, and Erickson, 1993*). Retinoic acid and vitamin A have been demonstrated also to enhance t-PA synthesis in the rat, an elevation which is still present after a feeding period of six weeks (*Van Bennehum, Emeis, Kooistra, and Hendricks, 1993*). As these components have little effect on the production of PAI-1, vitamin A derivatives are interesting candidates for pharmacological enhancement of t-PA synthesis. They probably act via specific nuclear receptors in the endothelial cell. Furthermore, shear forces have been reported to increase the production of t-PA by endothelial cells (*Diamond, Eskin, and McIntire, 1989*). A short promoter sequence has recently been identified, which is present in a mumber of genes that respond to shear forces in endothelial cells (*Resnick, Collins, Atkinson, Bonthron, Dewey, and Gimbrone, 1993*).

INFLAMMATORY ACTIVATION OF THE ENDOTHELIUM: EFFECT ON PAI-1 AND u-PA SYNTHESIS

In vivo, changes of the plasma levels of t-PA and its main inhibitor PAI- 1 often occur in the same direction. Nevertheless, the synthesis of PAI-1 is independently regulated from that of t-PA albeit that certain mediators, such as thrombin, can induce both t-PA and PAI-1 synthesis. In a number of diseases predominantly PAI-1 is elevated in the blood (sepsis, maturity onset diabetes, postoperative thrombosis) and/or in tissues (arteriosclerosis, sepsis). Several factors involved in inflammatory and vascular diseases, such as the cytokines tumor necrosis factor-α (TNFα) and interleukin-1 (IL-1), endotoxin (LPS), transforming growth factor-β (TGF-β), oxidized lipoproteins and thrombin, can stimulate PAI-1 production by endothelial cells *in vitro*. Also *in vivo*, administration of TNFα, IL-1, LPS or thrombin causes an increase in circulating PAI-1. After infusion of LPS in animals, PAI-1 mRNA increased in vascularized tissues and PAI-1 mRNA was elevated in the endothelium of various organs (*Quax, Van den Hoogen, Verheijen, Padró, Zeheb, Gelehrter, Van Berkel, Kuiper, and Emeis, 1990; Keeton, Eguchi, Sawdey, Ahn, and Loskutoff, 1993*). Administration of LPS or TNFα to patients or healthy volunteers caused after about 2 hours, a large increase in circulating PAI-1, which was preceded by a rapid and sustained increase in circulating t-PA (*Suffredini, Harpel, and Parrillo, 1989; Van Hinsbergh, Bauer, Kooistra, Kluft, Dooijewaard, Sherman, and Nieuwenhuizen, 1990; Van Deventer, Büller, Ten Cate, Aarden, Hack, and Sturk, 1990*). The mechanism underlying the stimulation of t-PA synthesis *in vivo* by LPS or TNFα is still unresolved. It may be indirect via the generation of another mediator generated by the LPS- or TNFα- infusion; and/or it may be a direct effect of TNFα on microvascular endothelial cells (*Van Hinsbergh, Kooistra, Van den Berg, Princen, Fiers, and Emeis, 1988*). The large increase in PAI-1 induced two hours after TNFα- or LPS-administration far exceeds the production of t-PA (*Suffredini, Harpel, and Parrillo, 1989; Van Hinsbergh, Bauer, Kooistra, Kluft, Dooijewaard, Sherman, and Nieuwenhuizen, 1990; Van Deventer, Büller, Ten Cate, Aarden, Hack, and Sturk, 1990*). This may result after an initial raise in fibrinolytic activity in a prolonged attenuation of the fibrinolysis process. It is generally believed that induction of PAI-1 by inflammatory mediators may contribute to the thrombotic complications in endotoxinemia and sepsis.

However, the effect of inflammatory mediators TNFα, IL-1 and LPS on plasminogen activation is probably more complex. Simultaneous with the increase in PAI-1, these inflammatory mediators induce the synthesis of u-PA in human endothelial cells (*Van Hinsbergh, Van den Berg, Fiers, and Dooijewaard, 1990*). Induction of u-PA by TNFα is associated by an increased degradation of matrix proteins (*Niedbala and Stein Picarella, 1992*). The enhanced secretion of u-PA occurs entirely towards the basolateral side of the cell, whereas the secretion of t-PA and PAI-1 proceed equally to the luminal and basolateral sides of the cell (*Van Hinsbergh, Van den Berg, Fiers, and Dooijewaard, 1990*). The polar secretion of u-PA underlines the suggestion that u-PA may be involved in local processes causing the remodelling of the basal matrix of the cell. Therefore, the increase of PAI-1 induced by inflammatory mediators may represent, in addition to a role in the modulation of fibrinolysis, a protective mechanism of the cell against uncontrolled u-PA action.

THE u-PA RECEPTOR

The site, where local u-PA activity is localized, is propably directed by the u-PA receptor, a GPI-anchored glycoprotein of about 45kD (*Plough, Behrendt, Løber, and Danø, 1992; Blasi, 1993*). Human endothelial cells *in vitro* contain about 40,000 u-PA receptors per cell. Upon secretion, single-chain u-PA binds via its growth factor domain to the receptor, and can subsequently be converted to two-chain u-PA, by which it becomes proteolytically active. As the endothelial cell contains also plasmin(ogen) receptors, an interplay between receptor-bound u-PA and receptor-bound plasmin(ogen) is likely to happen. The then generated plasmin can degrade a number of matrix proteins. On the other hand, a direct plasmin-independent action of u-PA on matrix proteins may also occur, as has been reported by Quigley *et al* (*Quigley, Gold, Schwimmer, and Sullivan, 1987*) for u-PA-dependent degradation of avian fibronectin. In this respect it is of interest to note that in various cell types including endothelial cells the u-PA receptor has been localized in the focal attachment sites, which host integrin-matrix interactions, and in cell-cell contact areas (*Pöllänen, Hedman, Nielsen, Danø, and Vaheri,*

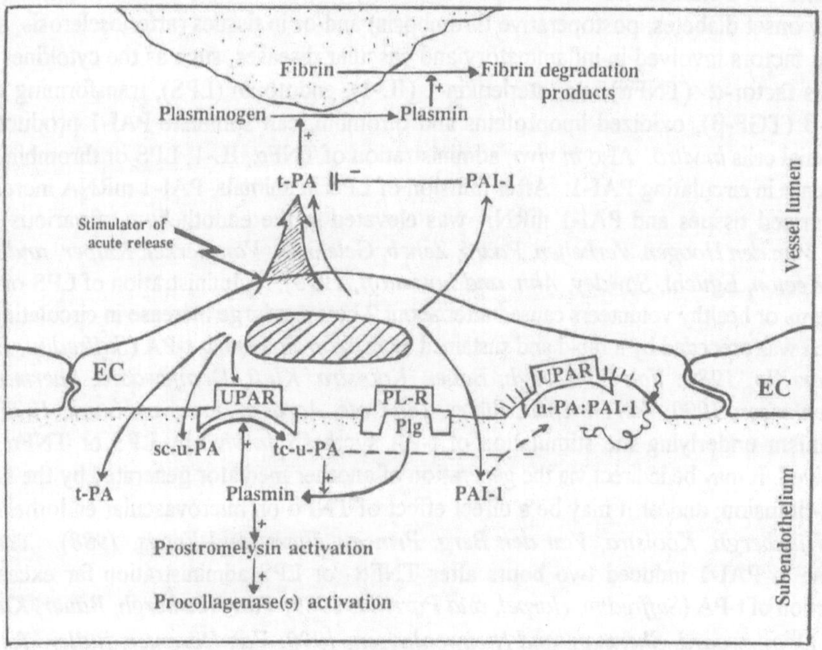

Figure 4. Schematic representation of postulated aspects of the involvement of endothelial cell plasminogen activators in fibrinolysis and local proteolysis. Abbreviations: PA: plasminogen activator; t-PA: tissue-type PA; u-PA: urokinase-type PA; sc-u-PA: single chain u-PA; tc-u-PA: two chain u-PA; Plg: plasminogen; UPAR: u-PA receptor; PL-R: plasminogen receptor; PAI-1: PA inhibitor-1; EC: endothelial cell; +: stimulation; -: inhibition.

1988; Conforti, Dominguez-Jimenez, Rønne, Høyer-Hansen, and Dejana, 1994). The activity time of u-PA is probably rather short. Receptor-bound two-chain u-PA is also subject to inhibition by PAI-1. The formed u-PA:PAI-1 complex is internalized (*Olson, Pöllänen, Høyer-Hansen, Rønne, Sakaguchi, Wun, Appella, Danø, and Blasi, 1992*), in contrast to receptor-bound free u-PA. This occurs probably after interaction with (an)other receptor(s), such as the LDL-receptor-related protein (LRP) (*Nykjær, Petersen, Møller, Jensen, Moestrup, Holtet, Etzerodt, Thøgersen, Munch, Andreasen, and Gliemann, 1992*). After internalization, the u-PA: PAI-1 complex is dissociated from the receptor and degraded in the lysosomes, whereas the u-PA receptor returns to the plasma membrane.

The number of u-PA receptors on human and bovine endothelial cells is enhanced by activation of protein kinase C and elevation of the cellular cAMP concentration (*Van Hinsbergh, 1992; Langer, Kuo, Kariko, Ahuja, Klugherz, Ivanics, Hoxie, Williams, Liang, Cines, and Barnathan, 1993*), as well as by several angiogenic growth factors, including basic fibroblast growth factor (b-FGF) (*Mignatti, Mazzieri, and Rifkin, 1991*) and vascular endothelial cell growth factor (VEGF) (*Pepper, Ferrara, Orci, and Montesano, 1992*). Although b-FGF enhances u-PA production in bovine endothelial cells, it is unable to induce u-PA production in various types of human endothelial cells. Therefore, the simultaneous exposure of human endothelial cells to TNFα, which induces u-PA synthesis, and to b-FGF and VEGF, which enhance the expression of u-PA receptors, are needed for a maximal increase in local u-PA activity on human endothelial cells *(Koolwijk and Van Hinsbergh)*.

LOCAL PERICELLULAR PROTEOLYSIS INVOLVED IN CELL MIGRATION

Migrating and invading cells, such as monocytes and tumor cells, express u-PA activity bound to u-PA receptors on their cellular protrusions and on focal attachment sites, which suggest that PA activity is involved in cell migration and invasion. Such a mechanism is also likely to be involved in smooth muscle and endothelial cell migration and in the formation of new blood vessels (angiogenesis). Inhibition of plasminogen activation interferes with smooth muscle cell migration *in vitro* (*Schleef and Birdwell, 1982*) and affects smooth muscle migration and proliferation *in vivo* (*Clowes, Clowes, Au, Reidy, and Belin, 1990*). Proteolysis of the basement membrane of endothelial cells is a prerequisite for endothelial cell migration and angiogenesis. A direct correlation between the expression of PA activity and the migration and formation of capillary sprouts by bovine microvascular endothelial cells was demonstrated *in vitro* by Pepper and Montesano (*Pepper, Vassalli, Montesano, and Orci, 1987; Pepper, Belin, Montesano, Orci, and Vassalli, 1990; Montesano, 1992; Pepper, Sappino, Stöcklin, Montesano, Orci, and Vassalli, 1993*). The outgrowth of tubular structures was increased by b-FGF which increases u-PA activity in bovine endothelial cells, and is counteracted by transforming growth factor-β (TGF-β), which enhances predominantly PAI-1 and inhibits PA activity. VEGF can also stimulate bovine adrenal microvascular endothelial cells to form tubular structures, and acts cooperatively with b-FGF in this induction (*Pepper, Ferrara, Orci, and Montesano, 1992*). Preliminary studies in our laboratory (*Koolwijk and Van Hinsbergh*) indicate that no or a limited number of capillary sprouts grow from a monolayer of human endothelial cells into a fibrin matrix after exposure to b-FGF and VEGF. However, many tubular structures are obtained after the simultaneous exposure of the cell monolayers to TNFα, b-FGF and VEGF. Whether this marked proliferation of tubular structures is due to an enhanced local u-PA activity by the induction of u-PA (by TNFα) and u-PA receptors (by b-FGF and VEGF) remains to be elucidated. Alternative or additional mechanisms may be involved. However, in favor of the hypothesis that u-PA is involved in or associated with angiogenesis are *in vivo* observations, which demonstrate the expression of u-PA during neovascularization of ovarian follicles, the corpus luteum and the maternal decidua (*Bacharach, Itin, and Keshet, 1992*).

The regulation of proteolytic activation and activity may be more complex than depicted above. It is not known whether interaction of u-PA with its receptor may affect cell metabolism. This may occur by a direct signal transduction into the cell (*Nusrat and Chapman, 1991; Del Rosso, Anichini, Pedersen, Blasi, Fibbi, Pucci, and Ruggiero, 1993; Dumler, Petri, and Schleuning, 1993*) or by an indirect signal generated by a proteolytic event in the focal attachment site, which is connected via a complex of proteins with the actin cytoskeleton of the cell. Furthermore, the local proteolytic activity is not limited to plasminogen activation. Endothelial cells *in vitro* can produce a number of matrix metalloproteinases (MMPs) (*Matrisian, 1992; Murphy, Atkinson, Ward, Gavrilovic, and Reynolds, 1992*). They are all secreted in a zymogen form and depend on zinc and calcium ions for activity. Activation of these proteins can proceed by the action of proteases, of which plasmin and stromelysin (MMP-3) have received particular attention (*Murphy, Atkinson, Ward, Gavrilovic, and Reynolds, 1992*). This results in the loss of a pro-peptide and the reduction of the molecular weight by about 8 to 10 kD. The MMPs are inhibited by tissue inhibitors of metalloproteinases, the so-called TIMPs (*Liotta, Steeg, and Stetler-Stevenson, 1991; Docherty and Murphy, 1990*). Two mammalian TIMPs have been characterized: TIMP-1 and TIMP-2, the mRNAs of which are both found in human endothelial cells *in vitro*. Hence, MMP activities are regulated by (a) the activation of zymogen, (b) the presence and interaction with TIMPs and (c) by the presence of binding and competing substrates. Recent data (*Hanemaaijer, Koolwijk, Le Clercq, De Vree, and Van Hinsbergh, 1993*) have indicated that the synthesis of several MMPs, including stromelysin, is enhanced or induced by the inflammatory mediator TNFα, the same mediator that induces the synthesis of u-PA. This sets the scene for a coordinate expression of matrix remodeling proteases in endothelial cells in inflammation.

From the foregoing it is clear that endothelial cells use PAs both for fibrinolysis and in pericellular proteolysis during cell migration. Figure 4 depicts schematically the involvement of endothelial PAs in fibrinolysis, which can be stimulated by the acute release of t-PA after exposure to vasoactive substances, and in cell-associated proteolysis via the u-PA receptor. It remains to be elucidated whether cell surface bound t-PA also contributes to pericellular proteolysis.

SUMMARY

The coagulation system helps the body to limit blood loss after wounding and provides a scaffold for the subsequent repair process. Formation and timely disappearance of a fibrin matrix plays a central role in these processes. As surveyed in this chapter, several key proteases involved in the generation and dissolution of fibrin have a role in cell activation, simultaneously. Thrombin interacts with a signal-transducing receptor on many cell types, including platelets, othelial cells and smooth muscle cells. Thrombin interaction with the endothelium activates several anticoagulant properties of the endothelial cells; activation of smooth muscle cells and fibroblasts by thrombin can induce cell proliferation. The proteases that are involved in the regulation of fibrinolysis, the plasminogen activators t-PA and u-PA, also interact with cells and, in addition to their regulatory role in fibrinolysis, stimulate certain cell functions, such as migration. The combined actions of thrombin and PAs on fibrin metabolism and cell activation enables the body to orchestrate a number of reactions, which prevent tissue damage and eventually result in healing. If this coordination does not proceed properly, i.e. proceeds too slowly or overreacts, blood loss, vascular occlusion and tissue damage, scar tissue formation, or restenosis after angioplasty or thrombolytic intervention may occur.

REFERENCES

Bacharach, E., Itin, A., and Keshet, E., 1992, *In Vivo* patterns of expression of urokinase and its inhibitor PAI-1 suggest a concerted role in regulating physiological angiogenesis. *Proc. Natl. Acad. Sci. U.S.A.* 89:10686.

Bachman, F., 1987, Fibrinolysis, in: *"Thrombosis and Haemostasis 1987"*, M. Verstraete, J. Vermylen, R. Lijnen, and J. Arnout, eds., Leuven University Press, Leuven.

Baranthan, E.S., 1992, Characterization and regulation of the urokinase receptor on human endothelial cells. *Fibrinolysis.* 6, Suppl.1:1.

Berridge, M.J., 1993, Inositol trisphosphate and calcium signalling. *Nature.* 361:315.

Bertina, R.M., Broekmans, A.W., van der Linden, I. K., and Mertens, K., 1982, Protein C deficiency in a Dutch family with thrombotic disease. *Thromb. Haemostas.* 48:1.

Bertina, R.M., Koeleman, B.P.C., Koster, T., Rosendaal, F.R., Dirven, R.J., de Ronde, H., van der Velden, P.A., and Reitsma, P.H., 1994, Mutation in blood coagulation factor V associated with resistance to activated protein C. *Nature.* 369:64.

Blasi, F., 1993, Urokinase and urokinase receptor: a paracrine/autocrine system regulating cell migration and invasiveness. *BioEssays,* 15:105.

Boulanger, C. and Lüscher, T.F., 1990, Release of endothelin from the porcine aorta. Inhibition by endothelium-derived nitric oxide. *J. Clin. Invest.* 85:587.

Brass, L.F., Hozie, J.A., and Manning, D.R., 1993, Signalling through G-proteins and G-protein-coupled receptors during platelet activation. *Thromb. Haemostas.* 70.217.

Bu, G., Williams, S., Strickland, D.K., and Schwartz, A.L., 1992, Low density lipoprotein receptor-related protein/α_2-macroglobulin receptor is an hepatic receptor for tissue-type plasminogen activator. *Proc. Natl. Acad. Sci. U.S.A.* 89:7427.

Carmeliet, P., Schoonjans, L., Kieckens, L., Ream, B., Degen, J., Bronson, R., de Vos, R., van den Oord, J.J., Collen, D., and Mulligan, R.C., 1994, Physiological consequence of loss of plasminogen activator gene function in mice. *Nature.* 368:419.

Carney, D.H., Mann, R., Redin, W.R., Pernia, S.D., Berry, D., Heggers, J.P., Hayward, P.G., Robson, M.C., Christie, J., Annable, C., Fenton II, J.W., and Glenn, K.C., 1992, Enhancement of incisional wound healing and neovascularization in normal rats by thrombin and synthetic thrombin receptor-activating peptides. *J. Clin. Invest.* 89:1469.

Clowes, A.W., Clowes, M.M., Au, Y.P.T., Reidy, M.A., and Belin, D., 1990, Smooth muscle cells express urokinase during mitogenesis and tissue-type plasminogen activator during migration in injured rat carotid artery. *Circ. Res.* 67:61.

Comp, P.C., Nixon, R.R., Cooper, M.R., and Esmon, C.T., 1984, Familial protein S deficiency is associated with recurrent thrombosis. *J. Clin. Invest.* 74:2082.

Conforti, G., Dominguez-Jimenez, C., Rønne, E., Høyer-Hansen, G., and Dejana, E., 1994, Cell-surface plasminogen activation causes a retraction of *in vitro* cultured human umbilical vein endothelial cell monolayer. *Blood.* 83:994.

Coughlin, S.R., 1993, Thrombin receptor structure and function. *Thromb Haemostas.* 66:184.

Dahlbäck, B. and Hildebrand, B., 1994, Inherited resistance to activated protein C is corrected by anticoagulant cofactor activity found to be a property of factor V. *Proc. Natl. Acad. Sci. U.S.A.* 91;1396.

Davie, E.W., Fujikawa, K., and Kisiel, W., 1991, The coagulation cascade: initiation, maintenance, and regulation. *Biochemistry.* 30:10363.

Del Rosso, M., Anichini, E., Pedersen, N., Blasi, F., Fibbi, G., Pucci, M., and Ruggiero, M., 1993, Urokinase-urokinase receptor interaction: non-mitogenic signal transduction in human epidermal cells. *Biochem. Biophys. Res. Comm.* 190.347.

Diamond, S.L., Eskin, S.G., and McIntire, L.V., 1989, Fluid flow stimulates tissue plasminogen activator secretion by cultured human endothelial cells. *Science.* 243:1483.

Docherty, A.J.P. and Murphy, G., 1990, The tissue metalloproteinase family and the inhibitor TIMP: a study using cDNAs and recombinant proteins. *Ann. Rheumatic Diseases.* 49:469.

Doyle, M.F., and Mann, K.G., 1990, Multiple active forms of thrombin IV Relative activities of meizothrombins. *J. Biol. Chem.* 265:10693.

Dumler, I., Petri, T., and Schleuning, W.D., 1993, Interaction of urokinase-type plasminogen activator with its cellular receptor (u-PAR) induces phosphorylation on tyrosine of a 38kD protein. *FEBS lett.* 322:37.

Emeis, J.J., 1992, Regulation of the acute release of tissue-type plasminogen activator from the endothelium by coagulation activation products. *Ann. N. Y. Acad. Sci.* 667:249.

Emeis, J.J. and Kooistra, T., 1993, Animal models and experimental procedures to study the synthesis and acute release of tissue-type plasminogen activator. *Fibrinolysis* 7, Suppl.1:31.

Esmon, C.T., 1989, The roles of protein C and thrombomodulin in the regulation of blood coagulation. *J. Biol. Chem.* 246;4743.

Esmon, C.T., 1993, Molecular events that control the protein C anticoagulant pathway. *Thromb. Haemostas.* 70:29.

Feng, P., Ohlsson, M., and Ny, T., 1990, The structure of the TATA-less rat tissue-type plasminogen activator gene. Species-specific sequence divergences in the promoter predict differences in regulation of gene expression. *J. Biol. Chem.* 265:2022.

Funakoshi, T., Heimark, R.L., Hendrickson, L.E., McMullen, B., and Fujikawa, K., 1987, Human placental anticoagulant protein: isolation and characterization. *Biochemistry.* 26:6238.

Garcia, J.G., Fenton II, J.W., and Natarajan, V., 1992, Thrombin stimulation of human endothelial cell phospholipase D activity. Regulation by phospholipase C, protein kinase C and cyclic adenosine 3',5'-monophosphate. *Blood.* 79:2056.

Gelehrter, T.D. and Sznycer-Laszuk, R., 1986, Thrombin induction of plasminogen activator inhibitor in cultured human endothelial cells. *J. Clin. Invest.* 77:165.

Glenn, K.C., Carney, D.H., Fenton II, J.W., and Cunningham, D.D., 1980, Thrombin active site regions required for fibroblast receptor binding and initiation of cell division. *J. Biol. Chem.* 255:6609.

Griffin, J.H., Evatt, B., Zimmerman, T.S., Kleiss, A.J., and Wideman, C., 1981, Deficiency of protein C in congenital thrombotic disease. *J. Clin. Invest.* 68:1370.

Hajjar, K.A., 1991, The endothelial cell tissue plasminogen activator receptor. Specific interaction with plasminogen. *J. Biol. Chem.* 266:21962.

Hajjar, K.A., 1993, Homocysteine-induced modulation of tissue plasminogen activator binding to its endothelial cell membrane receptor. *J. Clin. Invest.* 91:2873.

Hanemaaijer, R., Koolwijk, P., Le Clercq, L., De Vree, W.J.A., and Van Hinsbergh, V.W.M., 1993, Regulation of matrix-degrading metalloproteinases (MMPs) expression in human vein and microvascular endothelial cells. Effects of TNFα, IL-1 and phorbol ester. *Biochem. J.* 296:803

Harker, L.A., Kelly, A.B., and Hanson, S.R., 1991, Experimental arterial thrombosis in nonhuman primates. *Circulation.* 83:IV41.

Hattori, R., Hamilton, K.K., Fugate, R.D., McEver, R.P., and Sims, P.J., 1989, Stimulation of endothelial von Willebrand factor is accompanied by rapid redistribution to the cell surface of the intracellular granule membrane protein GMP-140. *J. Biol. Chem.* 264:7768.

Holmes, W.E., Nelles, L., Lijnen, H.R., and Collen, D., 1987, Primary structure of human α_2-antiplasmin, a serine protease inhibitor (Serpin), *J. Biol. Chem.* 262:1659.

Hong, S.L. and Deykin, D., 1982, Activation of phospholipases A_2 and C in pig aortic endothelial cells synthesizing prostacyclin. *J. Biol. Chem.* 257:7151.

Jandrot-Perrus, M., Clemetson, K.G., Huisse, M.G., and Guillin, M.C., 1992, Thrombin interaction with platelet glycoprotein Ib. Effect of glycocalicin on thrombin specificity. *Blood.* 80:2781.

Keeton, M., Eguchi, Y., Sawdey, M., Ahn, C., and Loskutoff, D., 1993, Cellular localization of type 1 plasminogen activator inhibitor messenger RNA and protein in murine renal tissue. *Am. J. Pathol.* 142:59.

Kooistra, T., Bosma, P.J., Toet, K., Cohen, L.H., Griffioen, M., Van den Berg, E., Le Cercq, L., and Van Hinsbergh, V.W.M., 1991, Role of protein kinase C and cyclic adenosine monophosphate in the regulation of tissue-type plasminogen activator, plasminogen activator inhibitor-1, and platelet-derived growth factor mRNA levels in human endothelial cells. Possible involvement of proto-oncogenes c-jun and c-fos. *Arterioscler. Thrombos.* 11:1042.

Kooistra, T., Opdenberg, J.P., Toet, K., Hendriks, H.F.J., Van den Hoogen, R.M., and Emeis, J.J., 1991, Stimulation of tissue-type plasminogen activators synthesis by retinoids in cultured human endothelial cells and rat tissues *in vivo*. *Thromb. Haemostas.* 65:565.

Kooistra, T., Toet, K., Kluft, C., Von Voigtlander, P.F., Ennis, M.D., Aiken, J.W., Boadt, J.A., and Erickson, L.A., 1993, Triazolobenzodiazepines: a new class of stimulators of tissue-type plasminogen activator synthesis in human endothelial cells. Biochem. Pharmacol. 46:61.

Koolwijk, P. and Van Hinsbergh, V.W.M., unpublished results.

Kruithof, E.K.O., 1988, Plasminogen activator inhibitor type 1: biochemical, biological and clinical aspects. *Fibrinolysis* 2, Suppl. 2:59.

Kuiper, J., Otter, M., Rijken, D.C., and Van Berkel, T.J.C., 1988, Characterization of the interaction *in vivo* of tissue type plasminogen activator with liver cells. *J. Biol. Chem.* 263:18220.

Langer, D.J., Kuo, A., Kariko, K., Ahuja, M., Klugherz, B.D., Ivanics, K.M., Hoxie, J.A., Williams, W.V., Liang, B.T., Cines, D.B., and Barnathan, E.S., 1993, Regulation of the endothelial cell urokinase-type plasminogen activator receptor. Evidence for cyclic AMP-dependent and protein kinase C-dependent pathways. *Circ. Res.* 72:330.

Levin, E.G., Marotti, K.R., and Santell, L., 1989, Protein kinase C and the stimulation of tissue plasminogen activator release from human endothelial cells. Dependence on the elevation of messenger RNA. *J. Biol. Chem.* 264:16030.

Levin, E.G., Marzec, U., Anderson, J., and Harker, L.A., 1984, Thrombin stimulates tissue plasminogen activator release from cultured human endothelial cells. *J. Clin. Invest.* 74:1988.

Levine, J.D., Harlan, J.M., Harker, L.A., Joseph, M.L., and Counts, R.B., 1982, Thrombin-mediated release of factor VIII antigen from human umbilical vein endothelial cells in culture. *Blood.* 60:531.

Liotta, L.A., Steeg, P.S., and Stetler-Stevenson, W.G., 1991, Cancer metastasis and angiogenesis: an imbalance of positive and negative regulation. *Cell.* 64:327.

Loskutoff, D.J., 1991, Regulation of PAI-1 gene expression. *Fibrinolysis.* 5:197.

Lum, H., Andersen, T.T., Siflinger-Birnboim, A., Tiruppathi, C., Goligorsky, M.S., Fenton II, J.W., and Malik, A.B., 1993, Thrombin receptor peptide inhibits thrombin-induced increase in endothelial permeability by receptor desensitization. *J. Cell. Biol.* 120:1491.

Mackie, K.P., Nairn, A.C., Hampel, G., Lam, G., and Jaffe, E.A., 1989, Thrombin and histamine stimulate the phosphorylation of elongation factor 2 in human umbilical vein endothelial cells. *J. Biol. Chem.* 264:1748.

McNamara, C.A., Sarembock, I.J., Gimple, L.W., Fenton II, J.W., Coughlin, S.R., and Owens, G.K., 1993, Thrombin stimulates proliferation of cultured rat aortic smooth muscle cells by a proteolytically activated receptor. *J. Clin. Invest.* 91:94.

Marcum, J.A. and Rosenberg, R.D., 1987, Anticoagulantly active heparan sulfate proteoglycan and the vascular endothelium. *Semin. Thromb. Haemost.* 13:464.

Marcus, A.J., Safier, L.B., Hajjar, K.A., Ullman, H.L., Islam, N., Broekman, M.J., and Eiroa, A.M., 1991, Inhibition of platelet function by an aspirin-insensitive endothelial cell ADPase. Thromboregulation by endothelial cells. *J. Clin. Invest.* 88:1690.

Marcus, A.J., Weksler, B.B., Jaffe, E.A., and Broeman, M.J., 1980, Synthesis of prostacyclin from platelet-derived endoperoxides by cultured human endothelial cells. *J. Clin, Invest.* 66:979.

Matrisian, L.M., 1992, The matrix-degrading metalloproteinases. *BioEssays* 14:455.

Mignatti, P., Mazzieri, R., and Rifkin, D.B., 1991, Expression of the urokinase receptor in vascular endothelial cells is stimulated by basic fibroblast growth factor. *J. Cell. Biol.* 113:1193.

Miles, L.A., Fless, G.M., Levin, E.G., Scanu, A.M., and Plow, E.F., 1989, A potential basis for the thrombotic risks associated with lipoprotein (a). *Nature.* 339:301.

Miles, L.A., Levin, E.G., Plescia, J., Collen, D., and Plow, E.F., 1988, Plasminogen receptors, urokinase receptors, and their modulation on human endothelial cells. *Blood.* 72:628.

Miles, L.A. and Plow, E.F., 1988, Plasminogen receptors: ubiquitous sites for cellular regulation of fibrinolysis. *Fibrinolysis.* 2:61.

Montesano, R., 1992, Regulation of angiogenesis *in vitro*. *Eur. J. Clin. Invest.* 22:504.

Murphy, G., Atkinson, S., Ward, R., Gavrilovic, J., and Reynolds, J.J., 1992, The role of plasminogen activators in the regulation of connective tissue metalloproteinases. *Ann. N. Y. Acad. Sci.* 667:1.

Nachman, R.L., 1992, Thrombosis and atherogenesis: molecular connections. *Blood.* 79:1897.

Nelken, N.A., Soifer, S.J., O'Keefe, J., Vu, T.K.H., Charo, S.R., Coughlin, S.R., 1992, Thrombin receptor expression in normal and atherosclerotic human arteries. *J. Clin. Invest.* 90:1614.

Niedbala, M.J. and Stein Picarella, M., 1992, Tumor necrosis factor induction of endothelial cell urokinase-type plasminogen activator mediated proteolysis of extracellular matrix and its antagonism by γ-interferon. *Blood.* 79:678.

Nishizuka, Y., 1992, Intracellular signaling by hydrolysis of phospholipids and activation of protein kinase C. *Science.* 258:607.

Nusrat, A.R. and Chapman Jr., H.A., 1991, An autocrine role for urokinase in phorbol ester-mediated differentiation of myeloid cell lines. *J. Clin. Invest.* 87:1091
Nykjær, A., Petersen, C.M., Møller, B., Jensen, P.H., Moestrup, S.K., Holtet, T.L., Etzerodt, M., Thøgersen, H.C., Munch, M., Andreasen, P.A., and Gliemann, J., 1992, Purified α_2-macroglobulin receptor/LDL receptor-related protein binds urokinase-plasminogen activator inhibitor type-1 complex. Evidence that the α_2-macroglobulin receptor mediates cellular degradation of urokinase receptor bound complexes. *J. Biol. Chem.* 267:14543.
Olson, D., Pöllänen, J., Høyer-Hansen, G., Rønne, E., Sakaguchi, K., Wun, T.C., Appella, E., Danø, K., and Blasi, F., 1992, Internalization of the urokinase-plasminogen activator inhibitor type-1 complex is mediated by the urokinase receptor. *J. Biol. Chem.* 267:9129.
Orth, K., Madison, E.L., Gething, M.J., Sambrook, J.F., and Herz, J., 1992, Complexes of tissue-type plasminogen activator and its serpin inhibitor plasminogen-activator inhibitor type 1 are internalized by means of the low density lipoprotein receptor-related protein/α_2-macroglobulin receptor. *Proc. Natl. Acad. Sci. U.S.A.* 89:7422.
Pepper, M.S., Belin, D., Montesano, R., Orci, L., and Vassalli, J.D., 1990, Transforming growth factor-beta 1 modulates basic fibroblast growth factor-induced proteolytic and angiogenic properties of endothelial cells *in vitro*. *J. Cell. Biol.* 111:743.
Pepper, M.S., Ferrara, N., Orci, L., and Montesano, R., 1992, Potent synergism between vascular endothelial growth factor and basic fibroblast growth factor in the induction of angiogenesis *in vitro*. *Biochem. Biophys. Res. Commun.* 189:824.
Pepper, M.S., Sappino, A.P., Stöcklin, R., Montesano, R., Orci, L., and Vassalli, J.D., 1993, Upregulation of urokinase receptor expression on migrating endothelial cells. *J. Cell. Biol.* 122:673.
Pepper, M.S., Vassalli, J.D., Montesano, R., and Orci, L., 1987, Urokinase-type plasminogen activator is induced in migrating capillary endothelial cells. *J. Cell. Biol.* 105:2535.
Plough, M., Behrendt, N., Løber, D., and Danø, K., 1992, Protein structure and membrane anchorage of the cellular receptor for urokinase-type plasminogen activator. *Semin. Thrombos. Haemostas.* 17:183.
Pohjanpelto, P., 1978, Stimulation of DNA synthesis in human fibroblasts by thrombin. *J. Cell. Physiol.* 95:189.
Pöllänen, J., Hedman, K., Nielson, L.S., Danø, K., and Vaheri, A., 1988, Ultrastructural localization of plasma membrane-associated urokinase-type plasminogen activator at focal contacts. *J. Cell. Biol.* 106:87.
Quax, P.H.A., Van den Hoogen, C.R., Verheijen, J.H., Padró, T., Zeheb, R., Gelehrter, T.D., Van Berkel, T.J.C., Kuiper, J., and Emeis, J.J., 1990, Endotoxin induction of plasminogen activator and plasminogen activator inhibitor type 1 mRNA in rat tissues *in vivo*. *J. Biol. Chem.* 265:15560.
Quigley, J.P., Gold, L.I., Schwimmer, R., and Sullivan, L.M., 1987, Limited cleavage of cellular fibronectin by plasminogen activator purified from transformed cells. *Proc. Natl. Acad. Sci. U.S.A.* 84:2776.
Radomski, M.W., and Moncada, S., 1993, Regulation of vascular homeostasis by nitric oxide. *Thromb. Haemostas.* 70:36.
Rapaport, S.I., 1991, The extrinsic pathway inhibitor: a regulator of tissue factor dependent blood coagulation. *Thromb. Haemostas.* 66:6.
Rasmussen, U.B., Vouret-Craviari, V., and Jallat, S., et al., 1991, cDNA cloning and expression of a hamster α-thrombin receptor coupled to Ca^{2+} mobilization. *FEBS Lett.* 288:123.
Resnick, N., Collins, T., Atkinson, W., Bonthron, D.T., Dewey Jr., C.F., and Gimbrone Jr., M.A., 1993, Platelet-derived growth factor B chain promoter contains a cis-acting fluid shear-stress-responsive element. *Proc. Natl. Acad. Sci. U.S.A.* 90:4591.
Reutlingsperger, C.P.M., Hornstra, G., and Hemker, H.C., 1985, Isolation and partial purification of a novel anticoagulant from arteries of human umbilical cord. *Eur. J. Biochem.* 151:625.
Rijken, D.C., Wijngaards, G., and Welbergen, J., 1980, Relationship between tissue plasminogen activator and the activators in blood and vascular wall. *Thromb. Res.* 18:815.
Roldan, A.L., Cubellis, M.V., Masucci, M.T., Behrendt, N., Lund, L.R., Danø, K., Appella, E., and Blasi, F., 1990, Cloning and expression of the receptor for human urokinase plasminogen activator, a central molecule in cell surface, plasmin dependent proteolysis. *EMBO J.* 9:467.
Schleef, R.R. and Birdwell, C.R., 1982, The effect of proteases on endothelial cell migration *in vitro*. *Exp. Cell. Res.* 141:503
Sprengers, E.D. and Kluft, C., 1987, Plasminogen activator inhibitors. *Blood.* 69:381.
Suffredini, A.F., Harpel, P.C., and Parrillo, J.E., 1989, Promotion and subsequent inhibition of plasminogen activation after administration of intravenous endotoxin to normal subjects. *N. Engl. J. Med.* 320:1165.
Thompson, E.A., Nelles, L., and Collen, D., 1991, Effect of retinoic acid on the synthesis of tissue-type plasminogen activator and plasminogen activator inhibitor-1 in human endothelial cells. *Eur. J. Biochem.* 201:627.
Tijberg, P.M., van Heerde, W.L., Leenhouts, H.M., Hessling, M., Bouma, B.N., and de Groot, P.G., 1991, Formation of meizothrombin as intermediate in factor Xa-catalyzed prothrombin activation on endothelial cells. The influence of thrombin on the reaction mechanism. *J. Biol. Chem.* 266:4017.

Van Bennekum, A.M., Emeis, J.J., Kooistra, T., and Hendriks, H.F.K., 1993, Modulation of tissue-type plasminogen activator by retinoids in rat plasma and tissues. *Am. J. Physiol.* 264:R931.

Van Deventer, S.J.H., Büller, H.R., Ten Cate, J.W., Aarden, L.A., Hack, E., and Sturk, A., 1990, Experimental endotoxemia in humans: analysis of cytokine release and coagulation, fibrinolytic, and complement pathways. *Blood.* 76:2520.

Van Hinsbergh, V.W.M., 1992, Impact of endothelial activation on fibrinolysis and local proteolysis in tissue repair. *Ann. N.Y. Acad. Sci.* 667:151.

Van Hinsbergh, V.W.M., Bauer, K.A., Kooistra, T., Kluft, C., Dooijewaard, G., Sherman, M.L., and Nieuwenhuizen, W., 1990, Progress of fibrinolysis during tumor necrosis factor infusion in humans. Concomitant increase of tissue-type plasminogen activator, plasminogen activator inhibitor type-1, and fibrin(ogen) degredation products. *Blood.* 76:2284.

Van Hinsbergh, V.W.M., Kooistra, T., Van den Berg, E.A., Princen, H.M.G., Fiers, W., and Emeis, J.J., 1988, Tumor necrosis factor increases the production of plasminogen activator inhibitor in human endothelial cells *in vitro* and in rats *in vivo. Blood.* 72:1467.

Van Hinsbergh, V.W.M., Sprengers, E.D., and Kooistra, T., 1987, Effect of thrombin on the production of plasminogen activators and PA inhibitor-1 by human foreskin microvascular endothelial cells. *Thromb. Haemostas.* 57:148.

Van Hinsbergh, V.W.M., Van den Berg, E.A., Fiers, W., and Dooijewaard, G., 1990, Tumor necrosis factor induces the production urokinase-type plasminogen activator by human endothelial cells. *Blood.*75:1991.

Van Obberghen-Schilling, E. and Pouysségur, J., 1993, Signalling pathways of the thrombin receptor. *Thromb. Haemostos* 70:163

Vu, T.K.H., Hung, D.T., Weaton, V.I., and Coughlin, S.R., 1991, Molecular cloning of a functional thrombin receptor reveals a novel functional proteolytic mechanism of receptor activation. *Cell.* 64:1057.

Walker, F.J., 1980, Regulation of activated protein C by a new protein. *J. Biol. Chem.* 255:5521.

Wallén, P., 1987, Structure and Function of Tissue Plasminogen Activator and Urokinase, in: "*Fundamental and Clinical Fibrinolysis,*" Castellino, P.J., Gaffney, P.J., Samama, M.M., and Takada, A., eds., pp. 1-18, Elsevier, Amsterdam.

Weksler, B.B., Ley, C.W., and Jaffe, E.A., 1978, Stimulation of endothelial cell prostacyclin production by thrombin, trypsin, and the ionophore A 23187. *J. Clin. Invest.* 62:923.

Wohlwend, A., Belin, D., and Vassili, J.D., 1987, Plasminogen activator-specific inhibitors produced by human monocyte/macrophages. *J. Exp. Med.* 165:320.

Wun, T.C. and Capuano, A., 1985, Spontaneous fibrinolysis in whole human plasma. Identification of tissue activator-related protein as the major plasminogen activator causing spontaneous activity *in vitro. J. Biol. Chem.* 260:5061.

COMPLEMENT AND MECHANISMS OF RESTENOSIS

George B. Zavoico and Una S. Ryan

T Cell Sciences, Inc.
115 4th Ave.
Needham, MA 02194

The Whitaker Cardiovascular Institute
Boston University School of Medicine
Boston, MA 02118

Percutaneous transluminal coronary angioplasty (PTCA) is a highly successful procedure for increasing the luminal diameter of atherosclerotic coronary arteries in order to improve distal perfusion of the myocardium. However, PTCA induces a series of events that involves activation of the coagulation and complement cascades, platelet activation and aggregation, thrombus formation, leukocyte adherence, activation of the inflammatory response, vascular smooth muscle (VSM) migration and proliferation, and deposition of extracellular matrix (ECM) (*Popma et al., 1991; Ferns et al., 1992; Wilcox 1993; Herrman et al., 1993a; Schwartz et al., 1993; O'Brien & Schwartz 1994; Casscells et al., 1994*). In up to 50% of patients these events lead to restenosis, the narrowing or blockage of the lumen of the coronary artery, and the necessity for a repeat PTCA or a coronary artery bypass graft.

PTCA failure within a day or two of the procedure is due to acute closure of the arterial lumen by the formation of a platelet and thrombotic plug, blockage caused by separation of the atherosclerotic plaque from the vessel wall, elastic recoil of the stretched coronary artery, vasospasm, or a combination of the above. Acute closure occurs in up to 10% of patients (*Landau et al., 1994*). Research efforts aimed at preventing acute closure center on mechanical devices, such as stents, and the prevention of platelet aggregation and thrombosis. PTCA failure within 6 months of the procedure is due to intimal hyperplasia that causes restenosis of the newly expanded arterial lumen. The high frequency of delayed failure, 30-50% of patients, has provided the impetus for the research effort to find a therapeutic regimen that will effectively prevent its occurrence. In numerous clinical trials using drugs that inhibit one or several events that contribute to restenosis in humans, no drug has been conclusively shown to prevent restenosis (*Popma et al., 1991; Ferns et al., 1992; Herrman et al., 1993a*). In this review we will argue that inhibition of the complement cascade warrants investigation as a therapeutic strategy due to the effect of activated complement on the initiation and progression of many of the humoral and cellular events that lead to restenosis.

Vascular Endothelium: Responses to Injury
Edited by John D. Catravas *et al.*, Plenum Press, New York, 1996

CONVENTIONAL WISDOM

Restenosis begins with injury to the arterial wall caused by expansion of the balloon catheter, and progresses through at least three phases (*Currier et al., 1992; Schwartz et al., 1993; O'Brien & Schwartz 1994*). The first phase, lasting up to 3 days, is largely a normal tissue response to injury, involving the coagulation cascade, platelet activation, and deposition of neutrophils and monocytes. The release of numerous growth factors and cytokines in this phase is believed to initiate the subsequent phases. In the second phase, smooth muscle cells are stimulated to proliferate in the media and begin to migrate to the intima. In the third phase, the smooth muscle cells populate the intima and begin to synthesize ECM, which constitutes the bulk of the restenotic lesion.

In the first phase, inflation of the balloon catheter injures the coronary artery by causing deendothelialization, disruption and fracture of the atherosclerotic plaque and internal elastic lamina, and stretching and rupturing of vascular smooth muscle cells and remaining endothelial cells. One of the first events is activation of the coagulation cascade via the extrinsic pathway (*Shirotani et al., 1993*). Plaque rupture and fissuring exposes tissue factor, expressed by macrophages, to serum coagulation factors causing their activation. Thrombin is formed, leading to fibrin deposition and contributing to platelet activation. Plaque rupture also exposes collagen, which can directly activate platelets (*Ferns et al., 1992*). Platelets adhere to the exposed subendothelial matrix and may also aggregate. Peak platelet deposition occurs within 2 hours of the PTCA procedure (*Currier et al., 1992*). Thromboxane A2 (TXA2), synthesized and secreted by activated platelets, and ADP and serotonin, released by degranulation of activated platelets, are platelet activating agents that serve to amplify the platelet response. Together with thrombin, these agents offer considerable potential for the growth of an obstructive thrombus, which may contribute to acute closure in a small but significant percentage of patients. In most patients, natural antithrombotic regulatory elements, such as antithrombin III and activated protein C, as well as flow dilution of the activating agents, will limit the growth of the thrombus and platelet aggregate.

Neutrophils, lymphocytes, and monocytes are recruited to the site of injury by chemotactic factors and adhesion molecules that appear on the surface of platelets and endothelial cells following degranulation, and that are upregulated in activated and injured endothelial cells (*Shirotani et al., 1993; Herrman et al., 1993a; Schwartz et al., 1993*). Likely chemotactic factors are platelet-derived growth factor (PDGF) and CSA, and the adhesion molecules include P-selectin, E-selectin and ICAM-1. Neutrophils and monocytes are activated by platelet-activating factor (PAF), thrombin, PDGF, C5a, and leukotrienes to release proteases and oxygen free radicals that may exacerbate the injury, and mitogens to promote smooth muscle cell proliferation (*Ricevuti et al., 1991; Shirotani et al., 1994*). A significant, though small, fraction (1-11%) of the cells in the enlarging lesion are leukocytes, and the particular role of each type of leukocyte remains unclear (*Wilcox 1993; Ferns et al., 1992*).

The proliferation of vascular smooth muscle cells and endothelial cells begins in the first phase, in response to thrombin, basic fibroblast growth factor (bFGF), angiotensin II (AII), insulin-like growth factor-I (IGF-I), epidermal growth factor (EGF), and PDGF (*Shirotani et al., 1993; O'Brien & Schwartz 1994*). Thrombin directly stimulates smooth muscle cell proliferation in vitro and is present in regenerating arteries for days (*Majesky, 1994*). There is an intracellular pool of bFGF that is released by cell injury, lysis and death (*Lindner & Reidy, 1991*). AII is formed by the local renin-angiotensin system. Degranulating platelets release growth factors, including PDGF, IGF-I, and EGF. However, the induction of medial smooth muscle proliferation can occur without platelets, indicating that the release and expression of factors from other sources are sufficient (*Fingerle et al., 1989*). Activated monocytes also contribute several growth factors and cytokines, including PDGF, bFGF, and IL-1. IL-1 maintains the activated states of endothelial cells, smooth muscle cells, and monocytes which result in the continued synthesis and release of mitogens. The primary effect of these agents is to induce a phenotypic change in smooth muscle cells from a quiescent, non-secreting and contractile phenotype to a proliferative and secretory phenotype (*Currier et al., 1992; Wilcox, 1993*). This change perpetuates the proliferative response since smooth muscle cells begin to secrete PDGF and bFGF, thus stimulating their own proliferation in an autocrine fashion. However, by the end of the first phase most, but not all, smooth muscle cells return to a quiescent state (*Majesky, 1994*). The return to a quiescent state is attributed to reendothelialization of the luminal surface, which may occur within several days (*Schwartz et al., 1993; Shirotani et al., 1993*). Reendothelialization restores the normal balance of intercellular signals that prevents smooth muscle cell activation.

The second phase of restenosis, beginning on the first to third day and lasting up to 2 weeks, is characterized by the migration from the media to the intima of the subpopulation of smooth muscle cells that escaped growth inhibition in the first phase (*Majesky, 1994*). PDGF, in addition to bFGF and AII, are believed to mediate this phase (*O'Brien & Schwartz, 1994*). Smooth muscle cells and endothelial cells begin to synthesize and secrete ECM proteins, including fibronectin, heparan sulfate proteoglycans, tenascin, thrombospondin, and collagen. In addition, transforming growth factor-β (TGF-β), which stimulates the synthesis of ECM proteins, is produced (*Wilcox, 1993; Shirotani et al., 1993; Majesky, 1994; Gibbons & Dzau, 1994*).

In the third phase, beginning after about 7 days and continuing for several months, neointimal proliferation of smooth muscle is substantially reduced, but the deposition of matrix continues. The bulk of the neointima is composed of newly synthesized ECM proteins (*Casscells et al., 1994; Gibbons & Dzau, 1994; Majesky, 1994; O'Brien & Schwartz et al., 1994; Shirotani et al., 1994*). The matrix undergoes remodeling and becomes organized. By 6 months all the smooth muscle cells have returned to their contractile phenotype (*Nobuyoshi et al., 1991*). The issue of the contribution and time course of changes in smooth muscle proliferation is controversial. Recent work has suggested that there is no peak in smooth muscle proliferation; instead proliferation seems to occur at a steady, but low (≤1% of smooth muscle cells), rate for more than 9 months following PTCA (*O'Brien & Schwartz, 1994*). The significance of this finding on therapeutic drug design remains to be determined.

In patients that do not develop restenosis, the third phase completes the healing process. This desirable outcome of the events following PTCA, re-endothelialization, recovery of vascular tone, clearance of cellular debris, replacement of dead cells by new cells, vascular remodeling, formation of scar tissue and maintenance of vessel patency, occurs by the processes described above. In patients developing restenosis, neointimal proliferation and matrix deposition continues, probably mediated by the presence of mitogens and other mediators. Thus, the progression of events that are a normal response to injury also lead to restenosis. This abnormal response has been called the fourth phase or wave (*O'Brien & Schwartz, 1994*) and leads to the view that restenosis is an overly vigorous response to injury.

If this view is correct, then certain mediating factors must have sufficient influence to convert a controlled response to injury leading to vessel patency into continued proliferation and matrix deposition leading to restenosis. These factors may be pre-existing, they may occur as a result of a problematic procedure that causes severe injury, or they may emerge later. Attempts to define the risk factors that are predictive of restenosis have, for the most part, been inconclusive (*Landau et al., 1994*). Several studies have been unable to correlate serum lipoprotein, lipid and/or cholesterol levels to restenosis (*Califf et al., 1990; Austin, 1992*), while others have shown a correlation (*Reis et al., 1991; Shah & Amin, 1992*). Some have proposed that the greater the severity of arterial injury, the greater the neointimal thickening and risk of restenosis (*Schwartz et al., 1993; Casscells et al., 1994*). The goal of therapy to prevent restenosis is therefore not to prevent the response to injury, but to control the response such that the benefits of a healed coronary artery are realized without causing restenosis.

FAILED STRATEGIES

The lack of defined risk factors that predispose patients to restenosis has deprived researchers and clinicians of an obvious target. However, the complex nature of restenosis provides ample targets for therapeutic intervention. Consequently, many different strategies have been attempted and many of these have proven efficacious in various animal models of restenosis. However, human clinical trials have largely been disappointing (*Popma et al., 1991; Ferns et al., 1992; Herrman et al. 1993a*). In the belief that inhibition of the early events will prevent progression to restenosis, many therapeutic strategies have targeted coagulation, platelets, inflammation, and the generation of free radicals. Another strategy has been to target classical risk factors of atherosclerosis, in the belief that the mechanisms promoting neointimal hyperplasia in restenosis are similar to those in atherosclerosis. This rationale has supported the testing of lipid lowering drugs, vasodilators, and antihypertensive drugs.

Recent advances in the development of novel drug delivery technologies and in the understanding of the initiating mechanisms of smooth muscle proliferation have resulted in other approaches (*Herrman et al., 1993b*). Success in animal models by targeting smooth muscle cell proliferation directly with antisense oligonucleotides targeting cdc2 kinase and proliferating-cell nuclear antigen (*Morishita et al., 1993*), the cellular oncogenes *c-myc* or *c-myb* (*Simons et al., 1992; Epstein et al., 1993*) have been reported. In a gene therapy approach, balloon injured porcine iliofemoral arteries were exposed to adenovirus vectors encoding thymidine kinase, and then the animals were treated with ganciclovir (*Ohno et al., 1994*). Transduction of porcine arterial smooth muscle cells was confirmed, and three weeks following injury neointimal hyperplasia was significantly decreased in animals treated with ganciclovir. In several rat and rabbit studies a direct inhibitor of proliferation, angiopeptin, has also proven effective in inhibiting restenosis following balloon injury (*Conte et al., 1989; Lundergan et al., 1989; Hong et al., 1993*). These approaches have not yet been tested in clinical trials.

ANTICOMPLEMENT AGENTS AS INHIBITORS OF RESTENOSIS

PTCA would be expected to lead to immediate complement activation. Many types of cellular, tissue, and organ injuries, including burns (*Gelfand et al., 1983*), pulmonary inflammation (*Mulligan et al., 1992a, 1992b*), hyperacute transplant rejection (*Pruitt et al., 1991, 1994*), and myocardial, skeletal muscle, and hepatic ischemia (*Rubin et al., 1990; Weisman et al., 1990; Rossen et al., 1985, 1988, 1994; Kagiyama et al., 1989; Smith et al., 1993; Jaeschke et al., 1993*) are associated with complement activation. When cells and tissues are damaged, complement is activated by the release of subcellular components from injured and ischemic cells, including mitochondrial proteins (*Rossen et al., 1985, 1988*) that may be associated with cardiolipin (*Rossen et al., 1994*), a phospholipid abundant in mitochondrial membranes, or cardiolipin alone (*Kovacsovics et al., 1985*). These components bind and activate C1q, the first component of the classical pathway of complement activation. Evidence for complement activation following myocardial ischemia in dogs includes the detection of C1q in cardiac lymph (*Rossen et al., 1988, 1994*) and the deposition of C1q in reperfused myocardium (*Rossen et al., 1985*). Complement-activating lipid particles, composed of cholesterol, cholesterol esters, and phospholipids, have also been extracted from human atherosclerotic lesions (*Seifert et al., 1990*). In addition, activation of the alternative pathway in ischemia-reperfusion injury of skeletal muscle has been reported (*Rubin et al., 1990*). Complement may also be activated by plasmin, through the activation of C1r and C3 (*Ratnoff and Naff, 1967; Ward, 1967*). Plasmin is formed from plasminogen by the release of t-PA from thrombin-activated endothelial cells and its generation in animal models of balloon injury is increased (*Gibbons and Dzau, 1994*). Moreover, plasmin-mediated complement activation has been reported in patients undergoing thrombolytic therapy (*Frangi et al., 1994*). Reperfusion of ischemic myocardium and localized cell and tissue injury by inflation of the balloon catheter and coronary artery recanalization should, therefore, provide ample stimuli for complement activation.

Complement activation results in the formation of the anaphylatoxins, C3a, C4a, and C5a, and the formation and deposition of the membrane attack complex (MAC) on cell membranes. In the setting of PTCA-mediated injury, each of these products would be expected to contribute to the normal response to injury. In addition, they could exacerbate the normal response and contribute to the development of restenosis. The anaphylatoxins are chemotactic for neutrophils, inducing them to cross the endothelial cell barrier if it still exists. Moreover, the anaphylatoxins can also directly activate neutrophils, causing degranulation and the release of oxygen free radicals and proteolytic enzymes, highly toxic products that may contribute to additional cell injury. This, in turn, causes the release of more subcellular components and continued activation of complement. The formation and deposition of the MAC, a pore forming complex of multiple complement components (C5b,6,7,8,9), on cell membranes causes the collapse of the electrochemical gradient leading to ion and water fluxes that may ultimately lead to cell death by lysis. Sub-lytic deposition of the MAC stimulates mitogenesis on fibroblasts and smooth muscle cells (*Halperin et al., 1993; J.A. Halperin, personal communication*), both alone and synergistically with PDGF. The MAC also stimulates release of bFGF and PDGF from endothelial cells resulting in smooth muscle cell proliferation (*Benzaquen et al., 1994*). Endothelial cells are also stimulated to degranulate, releasing von Willebrand factor and exposing P-selectin, and to assemble the prothrombinase complex, promoting coagulation, by the MAC (*Hattori et al., 1989; Hamilton et al., 1990*). Experimental evidence in support of a role for complement in vascular disease has been obtained by the study of atherosclerotic plaques. C5b-9, CR1, CR3, decay accelerating factor (DAF), CD59, factor H, C3c, C3d, and S-protein have been found in plaques, suggesting a contributory role of complement in the development of atherosclerosis (*Niculescu et al., 1987; Seifert and Hansson, 1989a, 1989b; Seifert et al., 1992*). In addition to plaques, complement deposition and depletion of complement regulatory proteins have been reported in infarcted human myocardium (*Schäfer et al., 1986; Hugo et al., 1990; Zimmermann et al., 1990; Väkevä et al., 1992*).

The potentially important role for complement activation in restenosis described above requires testing. Activated complement may exacerbate the response to injury and may lead to restenosis for the following reasons: (1) Anaphylatoxins are chemotactic for neutrophils and monocytes, leading to the recruitment of these cell types to the injury site. (2) Anaphylatoxins activate neutrophils and monocytes, causing the release of oxygen free radicals and proteases that contribute to tissue injury, and growth factors and cytokines that may prolong the proliferative response. (3) The formation and deposition of sub-lytic amounts of the MAC activate cells at the injury site, stimulating proliferation and coagulation via prothrombinase assembly, and depleting complement regulatory proteins that might otherwise prevent complement activation and lysis of cells. These events occur early in the response to injury, and if they could be attenuated by a complement inhibitor, then it is possible that restenosis might be prevented.

Soluble complement receptor 1 (sCR1) is a recombinant protein that lacks the cytoplasmic and transmembrane domains of complement receptor 1 (CR1) (*Weisman et al., 1990*). CR1 is a naturally occurring membrane-bound inhibitor of C3 and C5 convertases and a co-factor for the degradation of C3b and C4b by factor I that is found primarily on erythrocytes and leukocytes. Substantial evidence indicates that sCR1 retains all of the functions of CR1 and suggests that it may counteract many of the initiating and continuing processes that lead to restenosis. In models of myocardial ischemia-reperfusion injury, sCR1 significantly decreased myeloperoxidase activity, a marker for neutrophil accumulation, in infarcted myocardium and reduced infarct size (*Weisman et al., 1990; Smith et al., 1993*). In a model of skeletal ischemia-reperfusion injury, sCR1 prevented leukocyte adherence and increased the number of reperfusing capillaries (*Pemberton et al., 1993*). In various models of dermal and lung inflammation, sCR1 has demonstrated potent antiinflammatory effects (*Mulligan et al., 1992a, 1992b; Rabinovici et al., 1992*). In a hyperacute xenograft rejection model, in which complement is rapidly activated by xenoreactive antibodies present on the surface of the endothelium of transplanted hearts, sCR1 prolonged xenograft survival, in part by reducing the number of occlusive platelet thrombi and cellular infiltrate (*Pruitt et al., 1991, 1994*). Although these examples do not include models of restenosis, the processes are largely comparable in regard to the role of complement.

The complex nature of restenosis has frustrated numerous attempts to inhibit its progression in human trials. None have targeted complement activation. It would be informative to use a potent and effective complement inhibitor, such as sCR1, in animal models of restenosis to determine its effectiveness. The failure of other inhibitors of the early events of restenosis in clinical trials, including antiplatelet drugs, anticoagulants, and antioxidants, suggests either that the most critical early event, perhaps complement activation, was not targeted, or that combination therapy may ultimately be necessary to prevent restenosis.

REFERENCES

Austin, G.E., 1992, Lipids and vascular restenosis. *Circ.* 85:1613-1615.

Benzaquen, L.R., Nicholson-Weller, A., and Halperin, J.A., 1994, Terminal complement proteins C5b-9 release basic fibroblast growth factor and platelet-derived growth factor from endothelial cells. *J. Exp. Med.* 179:985-992.

Califf, R.M., Ohman, E.M., Frid, D.J., Fortin, D.F., Mark, D.B., Hlatky, M.A., Herndon, J.E., and Bengton, J.R., 1990, Restenosis: the clinical issues. In: Textbook of Interventional Cardiology (Topol, E.J., ed.), W.B. Saunders, Philadelphia, pp. 363-394.

Casscells, W., Engler, D., and Willerson, J.T., 1994, Mechanisms of restenosis. *Texas Heart Instit. J.* 21:68-77.

Conte, J.V., Foegh, M.L., Calcagno, D., Wallace, R.B., and Ramwell, P.W., 1989, Peptide inhibition of proliferation following angioplasty in rabbits. *Transplant. Proc.* 21:3686-3688.

Currier, J.W., Haudenschild, C., and Faxon, D.P., 1992, Pathophysiology of restenosis: clinical implications. In: Strategies in Primary and Secondary Prevention of Coronary Artery Disease (Ischinger, T. and Gohlke, H., eds), W. Zuckschwerdt Verlag GmbH, New York, pp. 181-192.

Epstein, S.E., Speir, E., and Finkel, T., 1993, Do antisense approaches to the problem of restenosis make sense? *Circ.* 88:1351-1353.

Ferns, G.A.A., Stewart-Lee, A.L., and Änggård, E.E., 1992, Arterial response to mechanical injury: balloon catheter de-endothelialization. *Atherosclerosis* 92:89-104.

Fingerle, J., Johnson, R., Clowes, A.W., Majesky, M.W., and Reidy, M.A., 1989, Role of platelets in smooth muscle cell proliferation and migration after vascular injury in rat carotid artery. *Proc. Natl. Acad. Sci. USA* 86:8412-8416.

Frangi, D., Gardinali, M., Conciato, L., Cafaro, C., Pozzoni, L., and Agostoni, A., 1994, Abrupt complement activation and transient neutropenia in patients with acute myocardial infarction treated with streptokinase. *Circ.* 89:76-80.

Gelfand, J.A., Donelan, M., and Burke, J.F., 1982, Preferential activation and depletion of the alternative complement pathway by burn injury. *Ann. Surg.* 198:58-62.

Gibbons, G.H. and Dzau, V.J., 1994, The emerging concept of vascular remodeling. *New Eng. J. Med.* 330:1431-1438.

Halperin, J.A., Taratuska, A., and Nicholson-Weller, A., 1993, Terminal complement complex C5b-9 stimulates mitogenesis in 3T3 cells. *J. Clin. Invest.* 91:1974-1978.

Hamilton, K.K., Hattori, R., Esmon, C.T., and Sims, P.J., 1990, Complement proteins C5b-9 induce vesiculation of the endothelial plasma membrane and expose catalytic surface for assembly of the prothrombinase enzyme complex. *J. Biol. Chem.* 265:3809-3814.

Hattori, R., Hamilton, K.K., McEver, R.P., and Sims, P.J., 1989, Complement proteins C5b-9 induce secretion of high molecular weight multimers of endothelial von Willebrand factor and translocation of granule membrane protein GMP-140 to the cell surface. *J. Biol. Chem.* 264:9053-9060.

Herrman, J-P.R., Hermans, W.R.M., Vos, J., and Serruys, P.W., 1993a, Pharmacological approaches to the prevention of restenosis following angioplasty. The search for the Holy Grail? (Part I). *Drugs* 46:18-52.

Herrman, J-P.R., Hermans, W.R.M., Vos, J., and Serruys, P.W., 1993b, Pharmacological approaches to the prevention of restenosis following angioplasty. The search for the Holy Grail? (Part II). *Drugs* 46:249-262.

Hong, M.K., Bhatti, T., Matthews, B.J., Stark, K.S., Cathapermal, S., Foegh, M.L., Ramwell, R.W., and Kent, K.M., 1993, The effect of porous infusion balloon-delivered angiopeptin on myointimal hyperplasia after balloon injury in the rabbit. *Circ.* 88:638-648.

Hugo, F., Hamdoch, T., Mathey, D., Schäfer, H, and Bhakdi, S., 1990, Quantitative measurement of SC5b-9 and C5b-9(m) in infarcted areas of human myocardium. *Clin. Exp. Immunol.* 81:132-136.

Jaeschke, H., Farhood, A., Bautista, A.P., Spolarics, Z., and Spitzer, J.J., 1993, Complement activates Kupffer cells and neutrophils during reperfusion after hepatic ischemia. *Am. J. Physiol.* 264:G801-G809.

Kagiyama, A., Savage, H.E., Michael, L.H., Hanson, G., Entman, M.L., and Rossen, R.D., 1989, Molecular basis of complement activation in ischemic myocardium: Identification of specific molecules of mitochondrial origin that bind human C1q and fix complement. *Circ. Res.* 64:607-615.

Kovacsovics, T., Tschopp, J., Kress, A., and Isliker, H., 1985, Antibody-independent activation of C1, the first component of complement, by cardiolipin. *J. Immunol.* 135:2695-2700.

Landau, C., Lange, R.A., and Hillis, L.D., 1994, Percutaneous transluminal coronary angioplasty. *New Eng. J. Med.*330:981-993.

Lindner, V. and Reidy, M.A., 1991, Proliferation of smooth muscle cells after vascular injury is inhibited by an antibody against basic fibroblast growth factor. *Proc. Natl. Acad. Sci. USA* 88:3739-3743.

Lundergan, C., Foegh, M.L., Vargas, R., Eufemio, M., Bormes, G.W., Kot, P.A. and Ramwell, P.W., 1989, Inhibition of myointimal proliferation of the rat carotid artery by the peptides angiopeptin and BIM 23034. *Atherosclerosis* 80:49-55.

Majesky, M.W., 1994, Neointima formation after acute vascular injury. Role of counteradhesive extracellular matrix proteins. *Texas Heart Instit. J.* 21:78-85.

Morishita, R., Gibbons, G.H., Ellison, K.E., Nakajima, M., Zhang, L., Kaneda, Y., Ogihara, T., and Dzau, V.J., 1993, Single intraluminal delivery of antisense cdc2 kinase and proliferating-cell nuclear antigen oligonucleotides results in chronic inhibition of neointimal hyperplasia. *Proc. Natl. Acad. Sci. USA* 90:8474-8478.

Mulligan, M.S., Warren, J.S., Smith, C.W., Anderson, D.C., Yeh, C.G., Rudolph, A.R., and Ward, P.A., 1992a, Lung injury after deposition of IgA immune complexes. Requirements for CD18 and L-arginine. *J. Immunl.* 148:3086-3092.

Mulligan, M.S., Yeh, C.G., Rudolph, A.R., and Ward, P.A., 1992a, Protective effects of soluble CR1 in complement-and neutrophil-mediated tissue injury. *J. Immunol.* 148:1479-1485.

Niculescu, F., Rus, H.G., and Vlaicu, R., 1987, Immunohistochemical localization of C5b-9, S-protein, C3d and apolipoprotein B in human arterial tissues with atherosclerosis. *Atherosclerosis* 65:1-11.

Nobuyoshi, M., Kimura, T., Ohishi, H., Horiuchi, H., Nosaka, H., Hamasaki, N., Yokoi, H., and Kim, K., 1991, Restenosis after percutaneous transluminal coronary angioplasty: Pathologic observations in 20 patients. *J. Amer. Coll. Cardiol.* 17:433-441.

O'Brien, E.R. and Schwartz, S.M., 1994, Update on the biology and clinical study of restenosis. *Trends Cardiovsc. Med.* 4:169-178.

Ohno, T., Gordon, D., San, H., Pompili, V.J., Imperiale, M.J., Nabel, G.J., and Nabel, E.G., 1994, Gene therapy for vascular smooth muscle cell proliferation after arterial injury. *Science* 265:781-784.

Pemberton, M., Anderson, G., Vetvicka, V., Justus, D.E., and Ross, G.D., 1993, Microvascular effects of complement blockade with soluble recombinant CR1 on ischemia/reperfusion injury in skeletal muscle. *J. Immunol.* 150:5104-5113.

Popma, J.J., Califf, R.M., and Topol, E.J., 1991, Clinical trials of restenosis after coronary angioplasty. *Circ.* 84:1426-1436.

Pruitt, S.K., Baldwin III, W.M., Marsh, Jr., H.C., Lin, S.S., Yeh, C.G., and Bollinger, R.R., 1991, The effect of soluble complement receptor type 1 on hyperacute xenograft rejection. *Transplantation* 52:868-873.

Pruitt, S.K., Kirk, A.D., Bollinger, R.R., Marsh, Jr., H.C., Collins, B.H., Levin, J.L., Mault, J.R., Heinle, J.S., Ibrahim, S., Rudolph, A.R., Baldwin III, W.M., and Sanfilippo, F., 1994, The effect of soluble complement receptor type 1 on hyperacute rejection of porcine xenografts. *Transplantation* 57:363-370.

Rabinovici, R., Yeh, C.G., Hillegass, L.M., Griswold, D.E., DiMartino, M.J., Vernick, J., Fong, K-L.L., and Feuerstein, G., 1992, Role of complement in endotoxin/platelet-activating factor-induced lung injury. *J. Immunol.* 149:1744-1750.

Ratnoff, O.D. and Naff, G.B., 1967, The conversion of C'1s to C'1 esterase by plasma and trypsin. *J. Exp. Med.* 125:337-358.

Reis, G.J., Kuntz, R.E., Silverman, D.I., and Pasternak, R.C., 1991, Effects of serum lipid levels on restenosis after coronary angioplasty. *Amer. J. Cardiol.* 68:1431-1435.

Ricevuti, G., Mazzone, A., Pasotti, D., de Servi, S., and Specchia, G., 1991, Role of granulocytes in endothelial injury in coronary heart disease in humans. *Atherosclerosis* 91:1-14.

Rossen, R.D., Michael, L.H., Hawkins, H.K., Youker, K., Dreyer, W.J., Baughn, R.E., and Entman, M.L., 1994, Cardiolipin-protein complexes and initiation of complement activation after coronary artery occlusion. *Circ. Res.* 75:546-555.

Rossen, R.D., Michael, L.H., Kagiyama, A., Savage, H.E., Hanson, G., Reisberg, M.A., Moake, J.N., Kim, S.H., Self, D., Weakley, S., Giannini, E., and Entman, M.L.; 1988, Mechanism of complement activation after coronary artery occlusion: Evidence that myocardial ischemia in dogs causes release of constituents of myocardial subcellular origin that complex with human C1q in vivo. *Circ. Res.* 62:572-584.

Rossen, R.D., Swain, J.L., Michael, L.H., Weakley, S., Giannini, E., and Entman, M.L., 1985, Selective accumulation of the first component of complement and leukocytes in ischemic canine heart muscle. A possible initiator of an extra myocardial mechanism of ischemic injury. *Circ. Res.* 57:119-130.

Rubin, B.B., Smith, A., Liauw, S., Isenman, D., Romaschin, A.D., and Walker, P.M., 1990, Complement activation and white cell sequestration in postischemic skeletal muscle. *Am. J. Physiol.* 259:H525-H531.

Schwartz, R.S., Edwards, W.D., Huber, K.C., Antoniades, L.C., Bailey, K.R., Camrud, A.R., Jorgenson, M.A., and Holmes, Jr., D.R., 1993, Coronary restenosis: Prospects for solution and new perspectives from a porcine model. *Mayo Clin. Proc.* 68:54-62.

Seifert, P.S. and Hansson, G.K., 1989a, Decay-accelerating factor is expressed on vascular smooth muscle cells in human atherosclerotic lesions. *J. Clin. Invest.* 84:597-604.

Seifert, P.S. and Hansson, G.K., 1989b, Complement receptors and regulatory proteins in human atherosclerotic lesions. *Arteriosclerosis* 9:802-811.

Seifert, P.S., Hugo, F., Tranum-Jensen, J., Zähringer, U., Muhly, M., and Bhakdi, S., 1990, Isolation and characterization of a complement-activating lipid extracted from human atherosclerotic lesions. *J. Exp. Med.* 172:547-557.

Seifert, P.S., Roth, I., Schmiedt, W., Oelert, H., Okada, N., Okada, H., and Bhakdi, S., 1992, CD59 (homologous restriction factor 20), a plasma membrane protein that protects against complement C5b-9 attack, in human atherosclerotic lesions. *Atherosclerosis* 96:135-145.

Shäfer, H., Mathey, D., Hugo, F., and Bhakdi, S., 1986, Deposition of the terminal C5b-9 complement complex in infarcted areas of human myocardium. *J. Immunol.* 137:1945-1949.

Shah, P.K. and Amin, J., 1992, Low high density lipoprotein level is associated with increased restenosis rate after coronary angioplasty. *Circ.* 85:1279-1285.

Shirotani, M., Yui, Y., and Kawai, C., 1993, Restenosis after coronary angioplasty: Pathogenesis of neointimal thickening initiated by endothelial loss. *Endothelium* 1:5-22.

Simons, M., Edelman, E.R., DeKeyser, J-L., Langer, R., and Rosenberg, R.D., 1992, Antisense c-*myb* oligonucleotides inhibit intimal arterial smooth muscle cell accumulation *in vivo*. *Nature* 359:67-70.

Smith III, E.F., Griswold, D.E., Egan, J.W., Hillegass, L.M., Smith, R.A.G., Hibbs, M.J., and Gagnon, R.C., 1993, Reduction of myocardial reperfusion injury with soluble complement receptor 1 (BRL 55730). *Eur. J.Pharmacol.* 236:477-481.

Väkevä, A., Laurila, P., and Meri, S., 1992, Loss of expression of protectin (CD59) is associated with complement membrane attack complex deposition in myocardial infarction. *Lab. Invest.* 67:608-616.

Ward, P., 1967, A plasmin-split fragment of C3 as a new chemotactic factor. *J. exp. Med.* 126:189-206.

Weisman, H.F., Bartow, T., Leppo, M.K., Marsh, Jr., H.C., Carson, G.R., Concino, M.F., Boyle, M.P., Roux, K.H., Weisfeldt, M.L., and Fearon, D.T., 1990, Soluble human complement receptor type 1: In vivo inhibitor of complement suppressing post-ischemic myocardial inflammation and necrosis. *Science* 249:146-151.

Wilcox, J.N., 1993, Molecular biology: Insight into the causes and prevention of restenosis after arterial intervention. *Amer. J. Cardiol.* 72:88E-95E.

Zimmermann, A., Gerber, H., Nussenzweig, V., and Isliker, H., 1990, Decay-accelerating factor in the cardiomyocytes of normal individuals and patients with myocardial infarction. *Virchows Archiv A Pathol. Anat.* 417:299-304.

GENETIC FACTORS CONTRIBUTING TO CARDIOVASCULAR DISEASE THAT MAY AFFECT ENDOTHELIAL STRUCTURE AND FUNCTION: THE ROLE OF PROTEINS INVOLVED IN LIPOPROTEIN TRANSPORT

Vassilis I. Zannis+*, Maria Laccotripe*, Savvas C. Makrides*, Dimitris Kardassis+, and Eleni Zanni*

* Section of Molecular Genetics, Boston University Medical Center
+ The University of Crete Medical School, Institute of Molecular Biology and Biotechnology of Crete

GENERAL INTRODUCTION

PROTEIN STRUCTURE

Lipoproteins are macromolecular complexes of lipids and proteins which originate mainly from the liver and intestine, and are involved in the transport and redistribution of lipids in the body. The plasma lipoproteins are spherical particles with cores of nonpolar neutral lipid consisting of cholesteryl ester and triglycerides and coats of relatively polar materials consisting of phospholipid, free cholesterol, and proteins[1,2] (Figure 1A). The plasma lipoproteins have traditionally been grouped in four major lipoprotein classes: chylomicrons, very low density lipoproteins (VLDL), low density lipoproteins (LDL), and high density lipoproteins (HDL). A lipoprotein class of intermediate density (IDL) found between VLDL, LDL and several subpopulations of VLDL, LDL and HDL have been described. Finally, lipoprotein particles (Lp) with defined lipid and apolipoprotein composition have also been isolated from plasma and the media of cell cultures [3,4]. The protein components of lipoproteins are called apolipoproteins and have been designated apoA-I, apoA-II, apoA-IV, apoB, apoCI, apo CII, apoCIII, apoD, and apoE [5]. The lipid, apolipoprotein composition and the properties of the human plasma lipoproteins are shown in Table I.

LIPOPROTEIN BIOSYNTHESIS AND CATABOLISM

Lipoprotein metabolism can be viewed as a complex biological pathway, which is shown schematically in Figure 1B. A variety of proteins participate in this pathway, including the apolipoproteins, plasma enzymes, lipid transfer proteins and the lipoprotein receptors. Table II

Vascular Endothelium: Responses to Injury
Edited by John D. Catravas *et al.*, Plenum Press, New York, 1996

TABLE I: PROPERTIES AND COMPOSITION OF HUMAN PLASMA LIPOPROTEINS[a]

Properties and Composition	Chylomicrons	VLDL	LDL	HDL
Size (Å)	750-12,000	300-700	180-300	50-120
Density (g/ml)	0.94	0.94-1.006	1.019-1.063	1.063-1.21
Triglycerides (% wt)	80-95	45-65	4-8	2-7
Phospholipids (% wt)	3-6	15-20	18-24	26-32
Free cholesterol (% wt)	1-3	4-8	6-8	3-58
Esterified cholesterol (% wt)	2-4	16-22	45-50	15-20
$S_f°$	400[b]	20-400[b]	0-12[b]	0-9[c]
Electrophoretic mobility	Origin (cathode)	Pre-β	β	α
Proteins (% wt)	1-2	6-10	18-22	45-55
Major apolipoproteins	A-I,A-IV,B, CI,CIII,E	B,E,CI,CII,CIII, A-I,A-II,A-IV	B	A-I,A-II,E
Minor apolipoproteins	A-II,CII	A-I,A-II,A-IV	CI,CII,CIII,E	C I,CII,CIII, D,E

a Modified from Ref. 2.
b Corrected flotation rate at a density of 1.063 g/ml, expressed in svedbergs [10^{-13}sec].
c $F_{1.20}°$ designates corrected flotation at a density of 1.20 g/ml.

summarizes our knowledge on the structure/function, sites of synthesis, genetics, and physiological functions of the apolipoprotein moieties [1]. Following synthesis, apolipoproteins assemble with lipids to form lipoproteins either inside the cell as is the case of VLDL, or outside the cell, as is the case of HDL. The nascent lipoproteins formed are altered in plasma by the action of plasma enzymes and lipid transfer proteins to become mature lipoproteins which are then recognized and catabolized by lipoprotein receptors. The sites of synthesis, structure/function relationships, and disease correlations of these proteins, which are involved in lipoprotein remodeling and catabolism, are shown in Table III. The association of the proteins involved in cholesterol transport with human diseases is shown in Tables II and III.

Figure 2A shows how a cell can maintain cholesterol homeostasis by the action of lipoproteins and lipoprotein receptors. The cells can take up cholesterol by the interaction of lipoproteins containing apoB such as LDL, IDL or apoE (βVLDL, IDL) with the LDL receptor or by the interaction of lipoproteins containing apoE with the apoE/α_2 macroglobulin receptor found in all cell types [1,6,7]. When LDL accumulates in plasma, it is modified and is taken up by the modified LDL receptor of monocyte-macrophages. Such an uptake may lead to the formation of foam cells which are found in the atherosclerotic lesions [7] (Figure 2B). To maintain cholesterol homeostasis, the cell must have the ability to lose excess cholesterol. This is achieved by the interaction of HDL via apoA-I and possibly other small apolipoproteins with cell membrane binding sites through a mechanism which is not well understood at the present time (Figure 2A). This process has been termed reverse cholesterol transport. The importance of the lipoprotein and lipoprotein receptor system for the maintenance of cholesterol

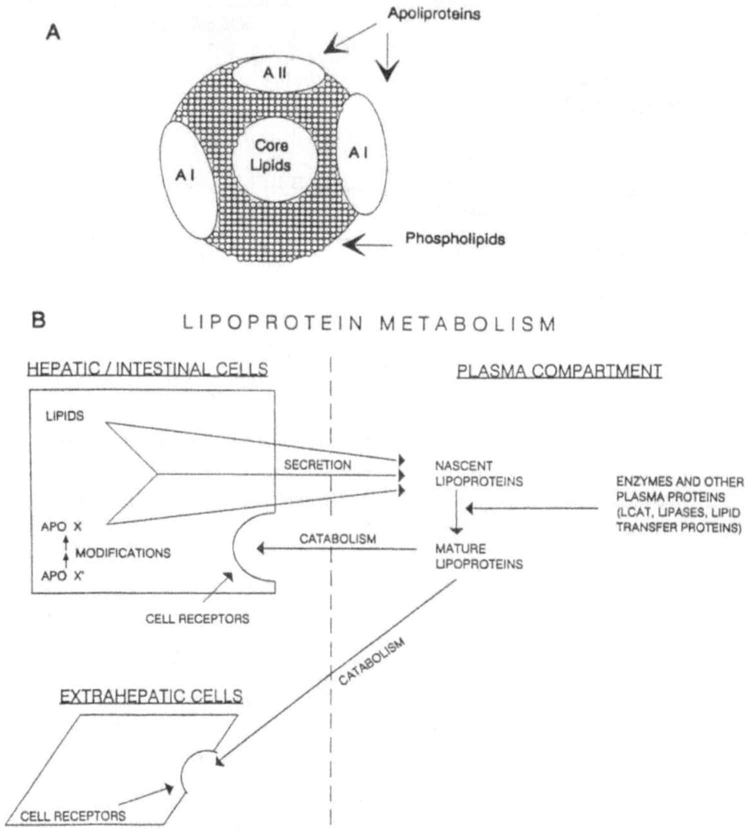

Figure 1. Panel A: Schematic representation of a spherical HDL particle containing apoA-I and apoA-II as the major apolipoproteins. Panel B: Simplified representation of the pathway of lipoprotein metabolism showing biosynthesis, assembly, secretion, modifications, and catabolism of apolipoproteins and lipoproteins.

homeostasis has been further supported by numerous genetic mutations in humans which alter plasma lipid and lipoprotein concentrations [1] as well as by transgenic and gene inactivation experiments [8-12]. This review will provide a summary of the structure, function, genetic variation, and the physiological importance of the proteins of the cholesterol transport system with emphasis on two key apolipoproteins, apoA-I and apoE. As shown in Figure 2A, these proteins are involved in the efflux and influx of cholesterol out of and into the cells, respectively, and thus play an important role in the maintenance of cellular cholesterol homeostasis.

Importance of the Lipoprotein System

Lipoproteins and apoprotein levels can predict the risk for coronary heart disease (CHD); in fact, the momentum for the study of lipoproteins has been provided by epidemiological and genetic data which demonstrate clearly that increases in LDL or apoB levels and decreases in HDL or apoA-I levels are associated with an increased risk for CHD [13,14]. The risk factors which contribute to CHD are age, total cholesterol, HDL cholesterol, cigarette smoking, diabetes and left ventricular hypertrophy as determined by electrocardiography [15]. The contribution of the total plasma cholesterol and HDL cholesterol to the risk of CHD is shown in Table IV A, B [15]. The probability of developing heart disease increases as the sum of the risk points from different causes increases (Figure 3).

A CELLULAR CHOLESTEROL HOMEOSTASIS

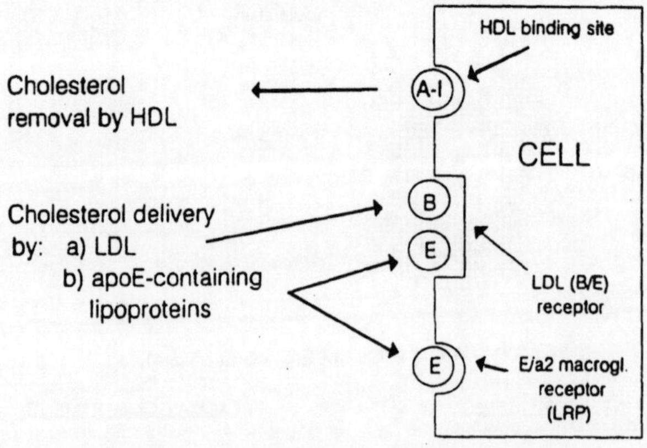

1) LDL(B), IDL(B) deliver cholesterol to cells through the LDL receptor

2) Chylomicron remnants (E), βVLDL(E) deliver cholesterol throu the LDL receptor and the E/α2 macroglobulin receptor

3) HDL reacts with the cell membrane and removes cholesterol from cells.

B CHOLESTEROL ACCUMULATION IN MONOCYTES-MACROPHAGES

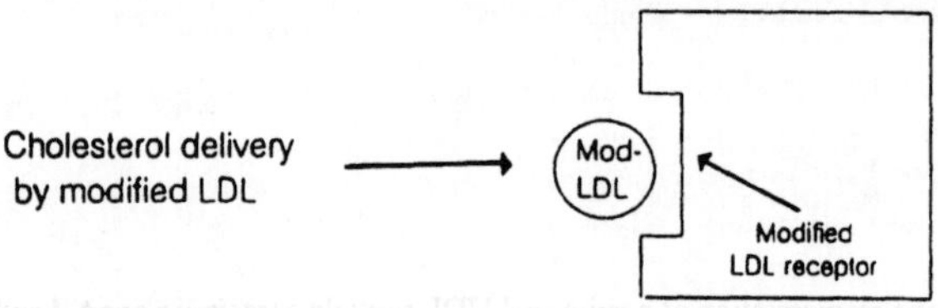

1) LDL can be modified in plasma by products of lipid peroxidation

2) Uptake of modified LDL by the modified LDL receptor of monocytes-macrophages leads to the formation of foam cells that are found in the atherosclerotic lesions.

Figure 2. Panel A shows a schematic representation illustrating potential mechanisms of cholesterol homeostasis. Panel B shows a schematic representation illustrating cholesterol accumulation in macrophages caused by the uptake of the modified LDL by the modified LDL receptor.

The Role of Apolipoprotein A-I and HDL

The plasma HDL consists of spherical particles 50 to 100Å in diameter which float in the density range 1.063 to 1.21 g/ml. The apoprotein and the lipid moieties each make up approximately 50% of the weight of HDL. The major protein component of HDL is apoA-I followed by apoA-II (Figure 1A and Table I), whereas apoE and C peptides are minor constituents. ApoA-I is synthesized predominantly by the liver and/or the intestine of all mammalian species [16]. In plasma and the media of lipoprotein secreting cells apoA-I is found predominantly in two populations of lipoprotein particles containing either apoA-I (LpA-I) or both apoA-I and apoA-II (LpA-I:A-II) [3,17].

TABLE II: APOLIPOPROTEINS: STRUCTURE, FUNCTION, GENETICS AND ASSOCIATION WITH HUMAN DISEASES.[a]

Name of Apo-protein	Sites of Synthesis	Gene size in Kb (exons)	Chromosomal localization	Amino acids in the mature protein	Function	Association with disease
A-I	Liver, intestine, minor sites[b]	2 (4)	11(11q13→qter)	243	Activates LCAT, binds to LCAT[f]	Deletion of apoA-I, CIII, A-IV loci and inversion of the apoA-I, CIII, A-IV loci are associated with atherosclerosis; some are associated with lipoprotein disorders
A-II	Liver	1.3 (4)	1(1p21→qter)	77		
A-IV	Intestine, liver[c]	2.8 (3)	11(11q13→qter)	376	Activates LCAT	Gene deletion (see comments for apoA-I)
B-100	Liver, intestine	43 (29)	2(2p23→p24)	4536	Receptor-mediated catabolism of LDL	Abetalipoproteinemia, hypobetalipoproteinemia, LDL-receptor binding defects
B-48	Liver, intestine			2152		
CI	Liver, adrenal gland[d]	4.7 (4)	19	57	Activates LCAT (moderately), inhibits the catabolism of apoE-containing lipoproteins by B/E and E/α^2M receptor	
CII	Liver, intestine	3.3 (4)	19	79		
CIII	Liver	3.2 (4)	11(11q13→qter)	79	Activates lipoprotein lipase. Inhibits catabolism of apoE-containing lipoproteins by the LDL receptor	Familial type I hyperlipoproteinemia. Deletion of apoA-I, CIII, A-IV loci; inversion of apoA-I, CIII loci (see comments for apoA-I)
D	Adrenal gland, kidney, pancreas, brain, intestine, testis, minor sites[e]			169		
E	Liver, peripheral tissues	3.6 (4)	19	299	Receptor-mediated catabolism of apoE-containing lipoproteins	Familial type III hyperlipoproteinemia. ApoE deficient humans & mice develop premature atherosclerosis. ApoE4 predominates in patients with late onset of Alzheimer's disease
J	Liver			427		

a For supporting references see ref. 1.
b Minor sites of apoA-I mRNA synthesis in fetal human tissues are the adrenal gland, kidney, gonads, ovaries, heart, and stomach, whereas apoE and apoJ mRNA are present in every tissue analyzed.
c Liver is a minor site of apoA-IV synthesis in primates and rats, and the major site of synthesis in rabbits. Intestine is a minor site of synthesis of apoCII and apoCIII.
d V.I. Zannis, unpublished.
e Minor sites of apoD mRNA synthesis are liver, spleen, lung, and heart.
f LCAT, lecithin cholesterol acyltransferase

TABLE III: LIPOPROTEIN RECEPTORS AND PLASMA PROTEINS: STRUCTURE, FUNCTION, GENETICS AND ASSOCIATION WITH HUMAN DISEASES

Name of protein	Sites of synthesis	Gene size in Kb (exons)	Chromosomal localization	Amino acids	Function	Association with disease
LDL Receptor	Ubiquitous synthesis	45 Kb (18)	Short arm of chromosome 19	860 (including a 21 residue signal peptide)	Binds and internalizes lipoproteins which contain apoB and apoE on their surface (LDL, βVLDL, IDL, HDL with apoE) is subject to feedback regulation by cholesterol or other sterols.	Familial hypercholesterolemia
ApoE/α_2 Macroglobulin Receptor	Ubiquitous synthesis		Chromosome 12 (q13-14)	4544 (including a 19 residue signal peptide)	Binds and internalizes apoE containing lipoproteins such as βVLDL as well as activated α_2/macroglobulin	
Modified LDL Receptor	Monocytes Macrophages			Type I: 453 Type II: 349	Binds and internalizes modified LDL. Is not subject to feedback regulation by the intracellular cholesterol. Is involved in the formation of foam cells.	
VLDL Receptor	Heart, muscle, adipose tissue (liver) (Tissues with active fatty acid metabolism)	40 Kb (19)	Chromosome 9	846 (including a 27 residue signal peptide)	Binds VLDL, βVLDL and IDL (does not bind LDL)	
Lipoprotein Lipase	Adipocytes, skeletal & cardiac muscle cells, macrophages	30 Kb (10)	Chromosome 8 (8p22)	475 (including a 27 residue signal peptide)	Hydrolyzes TAGs of chylomicrons and VLDL	Type I Hyperlipoproteinemia
Hepatic Lipase	Liver	35 Kb (9)	Chromosome 15 (15q21)	499 (including a 23 residue signal peptide)	Hydrolyzes TAGs and PL of IDL and HDL_2 phospholipids	Hepatic lipase deficiency
Lecithin Cholesterol Acyltransferase	Liver (brain, testis)	4.2 Kb (6)	Chromosome 16 (16q22)	440 (including a 24 residue signal peptide)	Esterifies HDL and LDL cholesterol in plasma (αLCAT and βLCAT)	Familial LCAT deficiency (fish eye disease)
Cholesteryl Ester Transfer Protein (Lipid Transfer Protein I)	Liver, small intestine, spleen, adipose tissue, muscle, adrenal (kidney)	25 Kb (16)	Chromosome 16	493 (including a 17 residue signal peptide)	Catalyzes the transfer of cholesteryl ester and TAGs among lipoproteins (mainly HDL→VLDL & HDL→LDL)	CETP deficiency syndrome
Lipid Transfer Protein II	Adipose tissue, muscle				Facilitates the exchange and transfer of phospholipid from VLDL to HDL	
Lipid Transfer Inhibitory Protein					Inhibits the cholesteryl ester, TAG, phospholipid transfer activities of CETP and LTPII	
88 KDa subunit of the microsomal triglyceride transfer protein	Liver Intestine			894	Catalyzes the transfer of triglycerides, phospholipids and cholesteryl esters among membranes. Participates in the assembly of LDL.	Abetalipoproteinemia

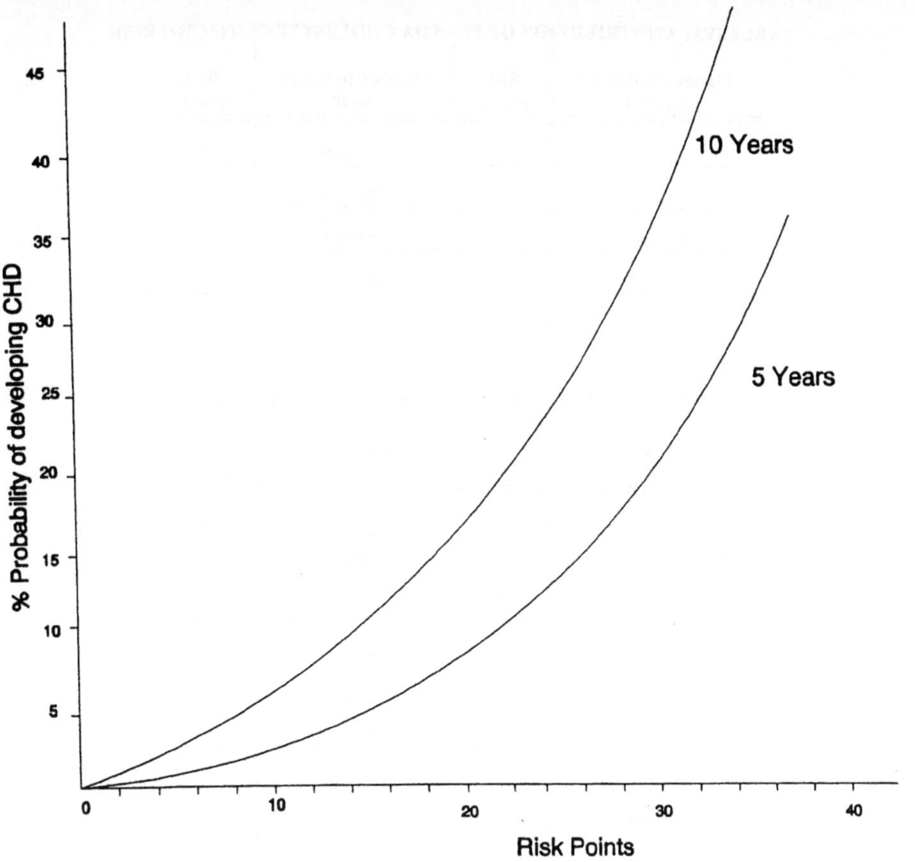

Figure 3. *Contribution of risk factors to the development of CHD.* Diagram showing the contribution of all the known risk factors to the development of CHD. From ref 15, with permission.

It has been proposed that the LpA-I particles are beneficial while LpA-I:A-II particles inhibit the functions of LpA-I particles and may be detrimental [18,19]. In general, epidemiological and genetic data have shown convincingly that low levels of HDL or apoA-I are associated with increased risk of developing CHD [13-15]. In contrast, high HDL levels are protective against CHD and confer longevity to humans [13,15,20-22].

The Role of Apolipoprotein E and ApoE-Containing Lipoproteins

ApoE is a 34.2 KDa protein which is synthesized in the liver and the peripheral tissues. Following synthesis, apoE is incorporated into lipoproteins such as VLDL, IDL and a subfraction of HDL and directs their catabolism by the LDL receptor, the VLDL receptor, and the apoE/α_2 macroglobulin receptor [1]. There is a common genetic polymorphism of apoE in humans which generates six common apoE phenotypes that affect the lipid and lipoprotein levels in the population. Mutations in apoE which affect its recognition by the LDL receptor are associated with type III hyperlipoproteinemia. Mutations leading to type III hyperlipoproteinemia as well as apoE deficiency in humans and animal models result in premature atherosclerosis [1]. Finally, a common apoE mutation which results in the apoE4 phenotype occurs with increased frequency in patients with late onset of Alzheimer's disease[23].

TABLE IVA: CONTRIBUTION OF PLASMA CHOLESTEROL TO CHD RISK

Plasma Cholesterol (mg/dl)	Risk Points	Plasma Cholesterol (mg/dl)	Risk Points
139-151	-3	220-239	2
152-166	-2	240-262	3
167-182	-1	263-288	4
183-199	0	289-315	5
200-219	1	316-330	6

TABLE IVB: CONTRIBUTION OF HDL CHOLESTEROL TO CHD RISK

HDL Cholesterol (mg/dl)	Risk Points	HDL Cholesterol (mg/dl)	Risk Points
25-26	7	51-55	-1
27-29	6	56-60	-2
30-32	5	61-66	-3
33-35	4	67-73	-4
36-38	3	74-80	-5
39-42	2	81-87	-6
43-46	1	88-96	-7
47-50	0		

From ref. 15, with permission.

APOA-I AND ITS ROLE IN CHOLESTEROL HOMEOSTASIS

ApoA-I Structure/Function

ApoA-I is the major protein component of high density lipoprotein HDL. The mature plasma form of apoA-I consists of 243 amino acids [24] and is generated by removal of the six N-terminal residues of the nascent proapoA-1 form which is secreted by the liver and the intestine[25,26].

Both the amino acid [27-29] and the nucleotide [30] sequence analyses of apoA-I showed that the protein and the gene sequence contain repeated units. The protein repeat consists of 22 amino acids and contains two symmetrical 11 residue long segments with *grossly* similar arrangements of charged and hydrophobic residues (Table V). Similar repeated units have been found in the other exchangeable apolipoproteins [32]. It has been proposed that these repeated units form amphipathic α-helices which have a unique distribution of apolar and polar amino acids. On one side of the helix are the hydrophobic amino acids which can associate with the hydrophobic fatty acyl chains of the phospholipids. On the other side are the hydrophilic amino acids which can associate with the hydrophilic phospholipid head groups and the aqueous phase[32]. A variety of evidence outlined below suggests that these amphipathic α-helices are important for lipid binding and the biological properties of apoA-I. The concept of amphipathic α-helices has been verified recently with the derivation of the three-dimensional structure of the amino terminal portion of the human apolipoprotein E (apo E) [33-35] and the insect

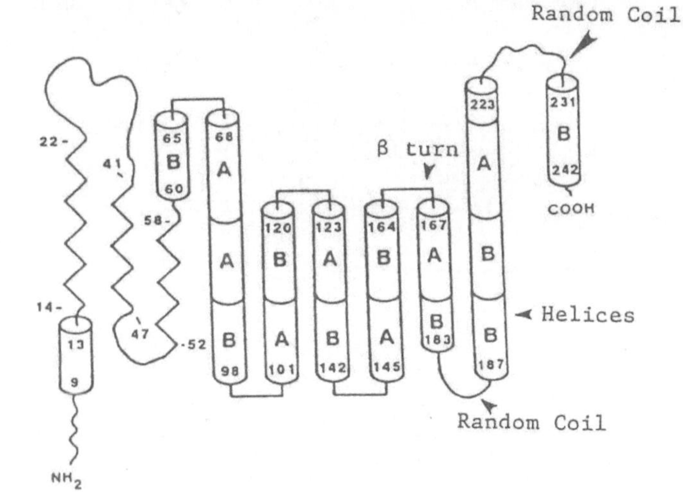

Helix	1	2	3	4	5	6	7	8	9
Repeat		1	2&3	4	5	6	7	8&9	10

Figure 4. *Apolipoprotein A-I.* Secondary structure model of apoA-I. Helical sections denoted as cylinders, β structures are as ∧ ∧ ∧, and turns as ⊓. Sections of random coil are shown as irregular curves. From ref. 37, with permission.

apolipoprotein III (apoLpIII) [36]. Both molecules fall into the category of α-helical bundles, and apoLpIII exists as a five helix bundle. It has been proposed that apoLpIII on the surface of lipophorin acquires an unfolded conformation by movement of two of the helices out of the bundle conformation which exists in solution [36]. A probable model of secondary structure and the relative arrangement of the repeated units of apoA-I was proposed recently by Nolte & Atkinson. This model took into account the homologies of the repeated units, and is based on several structure prediction calculations and the circular dichroism spectra of apoA-I-lipid complexes [37]. This model (Figure 4) predicts the existence of 9 antiparallel helices 1 to 9 and is consistent with the formation of an up and down α-helical bundle structure within the carboxy terminal region of apoA-I. The proposed relationship of the putative helices with the repeated units and the bundle formation is shown in Figure 4 and Table V.

The analysis of the apoA-I structure by monoclonal antibodies showed that the epitopes of the lipid-free protein in the region of the repeated units 4 and 5 (helix 4 and 5) recognize one but not both of these repeats. This suggests that these helices may be continuous or at an angle[38]. In contrast, antibodies recognizing repeated units 6 and 7 have overlapping epitopes suggesting that these helices may have antiparallel configuration [38]. Other monoclonal antibodies have discontinuous epitopes which recognize residues 45 to 51, 83 to 92, 119 to 126, and 136 to 143 suggesting that these apoA-I regions are in close proximity and may have conformational flexibility. There were no monoclonal antibodies recognizing the carboxy terminal repeats 9 and 10 [38]. Three major classes of amphipathic helices designated A, G and Y are found in apoA-I and other apolipoproteins which are different from those found in globular proteins [39]. It has been proposed that the lipid affinity of an apolipoprotein depends on the number of class A α-helices it contains [39]. Utilization of synthetic peptide fragments has identified the lipid-binding domains in the 66 to 120 region (hinged domain) [40] 147 to 168 region (repeats 6 and 7) [41] and the 200 to 243 region repeats 8 to 10 [42].

Binding of ApoA-I to Lipids, Formation of HDL

ApoA-I can be associated with phospholipids. This association leads to the formation of discoidal particles (Figure 5) and increases the helical content of apoA-I from 55 to 75% [43,44].

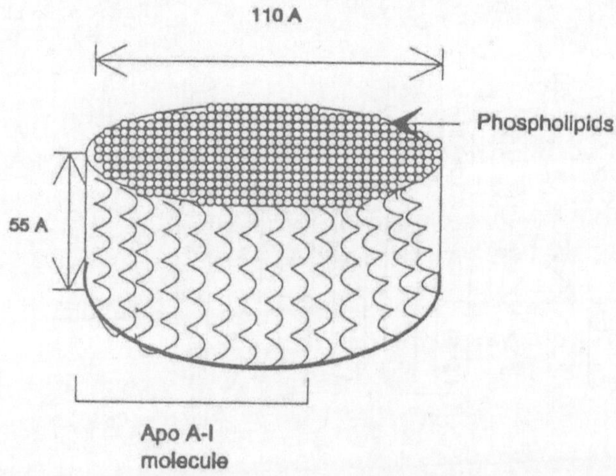

Figure 5. Schematic representation of a discoidal particle formed by mixing apo A-I with phospholipids. The diagram shows the polar phospholipid heads on top and the repeated units of apo A-I covering the hydrocarbon chains of the fatty acids on the perimeter of the cylinder. From ref 47, 48.

The diameter of the particles decreases from 150 to 75Å as the ratio of phospholipid to protein decreases [47]. X-ray and neutron scattering analysis showed that at 2.5:1 DMPC phospholipid to apoA-I ratio the discs have 110Å diameter with a bilayer thickness of 55Å (Figure 5). The apoA-I covers the hydrocarbon chains of the phospholipids on the perimeter of the disc [47,48]. The interaction of apoA-I with liposomes increases in the presence of small amounts of cholesterol [49]. Jonas and colleagues have suggested that six to eight helical segments of apoA-I participate in the formation of discoidal recombinant HDL particles [50]. In the discoidal particles the helix axis is parallel to the phospholipid side chains [51,52]. Fluorescence polarization studies have shown that the generation of recombinant particles which contain the same number of apoA-I molecules but have different size is caused by the ability of the amino terminal segment of apoA-I to associate reversibly with lipids [53]. Methylation of Met residues 86 and 112 found in the polar face of helices 2 and 3 which introduces a positive charge inhibits the ability of apoA-I to cause stepwise increases in apoA-I phospholipid recombinants [39]. The difference in apoA-I conformation is also suggested by the physicochemical and electrokinetic analysis of discoidal and spherical HDL particles [54,55] as well as by the immunochemical heterogeneity of HDL particles [38,56-58]. Based on the conformational flexibility of the aminoterminal region of apoA-I it has been postulated that the region 66-120 of apoA-I undergoes a hinged movement which allows it either to bind to lipids or be exposed to the aqueous phase [39,59]. It has been suggested that this movement may be relevant to the efflux of cholesterol from cells, to the activation of LCAT, and to the size of the lipoprotein particle [39,59].

Activation of the Enzyme Lecithin Cholesterol Acyl Transferase (LCAT)

HDL is the site of cholesterol esterification by LCAT as shown below [60].

$$\text{HDL Chol} \xrightarrow{\alpha\ \text{LCAT}} \text{HDL Chol Ester}$$

$$\text{LDL Chol} \xrightarrow{\beta\ \text{LCAT}} \text{LDL Chol Ester}$$

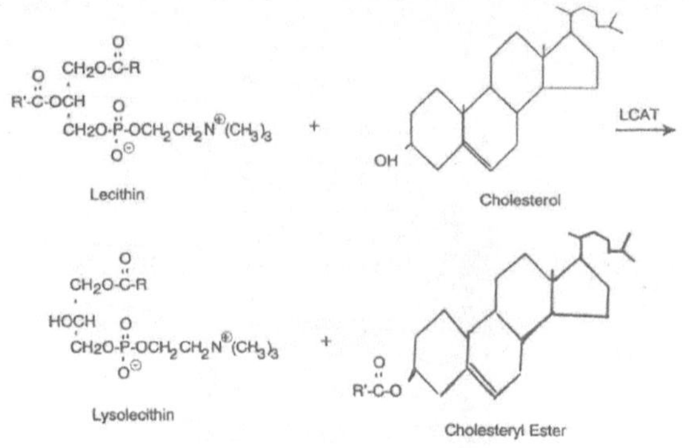

TABLE V. APOA-I REPEATED UNITS

No. of Repeats	Helix	Helix Class	Residue	Predicted topologically distinct positions a to g in the α-helical bundle residue[&]	
				a+ d e f g a b c d e f g a	
1	2	A	44	LKLLDNWDSVTSTFSKLREQLG	65
2	3	A	66	PVTQEFWDNLEKETEGLRQEMS	87
3	3	Y	88	KD*LEEVKAKVQ	98
4	4	Y	99	PYLDD*FQKK*WQE*EMELYRQKVE	120
5	5	A	121	PLRAE LQEGARQKLHE*LQE*KLS	142
6	6	A	143	P*LGEE*MRDRARAHVDA*LRTHLA	164
7	7	A	165	P*YSDE*LRQR*LAAR*LEALKENGG	186
8	8	A	187	ARLAEYHAKATE*HLSTLSEKAK	208
9	8	A	209	PALED*LRQGLL	219
10	9	Y	220	PVLESFKVSFLSALEEYTKKLN	241

& The hydrophobic positions a and d in the predicted bundle region are indicated. indicates that there are 4 rather than three residues between the position a and the next position d. This can occur if we assume 90° helix turn per residue in this segment of the helix. No assignment has been made to the carboxy and the amino terminal residues of the repeats which may participate in the formation of β turns. * Reported natural amino acid substitutions. Underlined sequences indicate lipid binding regions. A and Y indicate the class of the amphipathic α-helices.

ApoA-I is an effective cofactor for LCAT activity [61,62] particularly with physiological lecithins, which typically are long-chain with 1-saturated and 2-unsaturated acyl groups. ApoA-I also stimulates the efflux of cholesterol from peripheral cells, providing substrate for the LCAT reaction [63]. As a result of this activity, apoA-I may play an important role in regulating the cholesterol content of peripheral tissues through the putative reverse cholesterol transport pathway [60,63]. The 96Å recombinant discs formed by binding of apoA-I to phospholipid vesicles containing cholesterol are efficient substrates for LCAT [53]. Discoidal HDL particles containing apoA-I are detected in peripheral lymph. These particles are most likely lipid bilayer

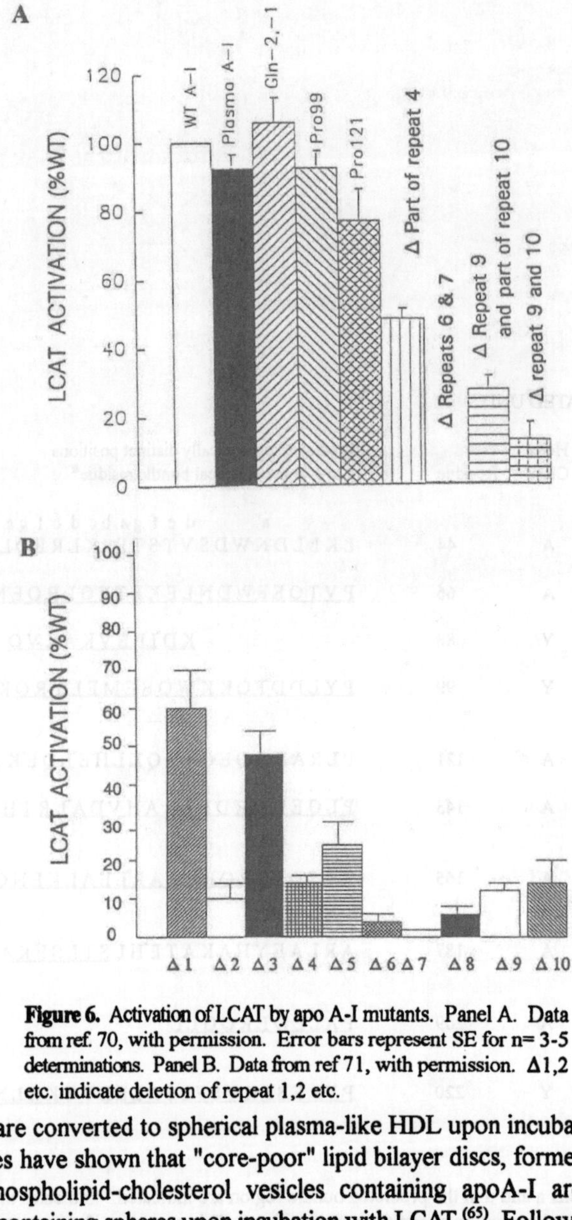

Figure 6. Activation of LCAT by apo A-I mutants. Panel A. Data from ref. 70, with permission. Error bars represent SE for n= 3-5 determinations. Panel B. Data from ref 71, with permission. Δ1,2 etc. indicate deletion of repeat 1,2 etc.

in structure, and are converted to spherical plasma-like HDL upon incubation with LCAT [(64)]. Numerous studies have shown that "core-poor" lipid bilayer discs, formed *in vitro* by mixing apoA-I with phospholipid-cholesterol vesicles containing apoA-I are transformed into cholesteryl-ester-containing spheres upon incubation with LCAT [(65)]. Following the introduction of the amphipathic α-helix hypothesis in 1974 [(32)], several studies have examined the ability of peptide fragments of apoA-I and a variety of synthetic peptides to associate with phospholipid vesicles. This association requires a minimum length of approximately 20 amino acids (5.5 helical turns) and sequence organization into distinct polar and nonpolar faces [(66)]. Several laboratories have shown that synthetic peptides with these characteristics bind phospholipid and activate LCAT, although enzyme activation appears to have been in all cases significantly less effective than that obtained with native apoA-I in reactions with physiological lecithins [(65)]. Specifically, LCAT activation, approximately 25-30% of native apoA-I, has been attained with the amino (1 to 85) and carboxyl (149 to 243) terminal cyanogen bromide fragments derived from native apoA-I [(66)] and synthetic peptides corresponding to the 145 to 185 and 121 to 164 apoA-I sequences [(42,67)]. Studies by Sparrow and Gotto using synthetic peptides showed that the 148 to 185 region can both bind phospholipids and activate LCAT whereas the carboxy

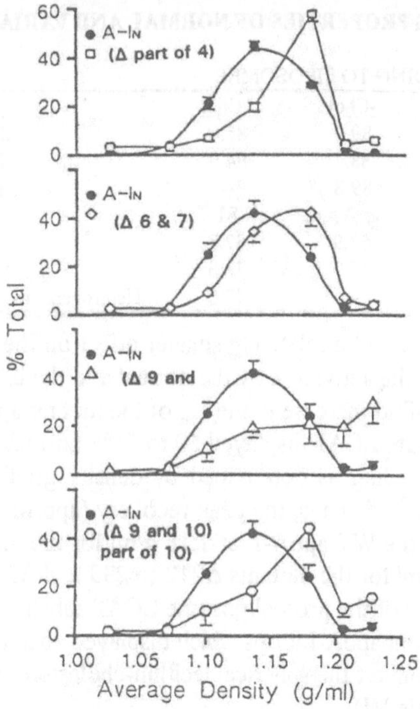

Figure 7. Binding of apo A-I mutants to HDL. 2ml [35S]-labeled culture media from permanent cell lines expressing different forms of apo-A-I was mixed with 100 μg of HDL, adjusted to a density of 1.21 g/ml with KBr and subjected to discontinuous density gradient ultracentrifugation. Gradient fractions were analyzed by 12% SDS-PAGE and autoradiography. Bands containing apo A-I were excised from the gel, solubilized, and counted. Analysis shows the distribution of nascent apo A-I forms to VLDL, LDL, and HDL In all panels, distribution of the mutant apo A-I form is compared to the wild type apo A-I secreted by C127 cells. Error bars: S.E.M. (n=3). Data from ref. 70 with permission.

terminal region 200-243 can bind to phospholipids but cannot activate LCAT [40]. Activation of LCAT was also achieved with synthetic peptides corresponding to the 121 to 164 region [67]. Finally, substitution of Pro-143 or Pro-165 for arginine results in LCAT activation which is 40 to 70% of normal [68]. To assess the importance of the repeated units of apoA-I for lipid binding and LCAT activation, we have mutagenized the human apoA-I gene and have generated cell lines which express normal and mutant apoA-I forms. Point mutations were introduced which changed Gln-1, Gln-2 to Arg, Arg, Pro-99 to His, and Pro-121 to His. In addition, the following amino acid deletions (Δ) were generated: Δ113 to 124, Δ148 to 186, Δ212 to 233, and Δ212 to 243 [69,70]. The apoA-I form isolated from the culture medium of C127 cells was analyzed for its ability to activate LCAT as well as to bind to phospholipid vesicles and to HDL. Compared to the wild type (WT) apoA-I, the relative activation of LCAT achieved by the point mutations Gln-1, Gln-2→Arg,Arg, Pro-99→His, and Pro-121→His was 106±7, 92±6, and 77±9%, respectively. Furthermore, the activation achieved by the internal deletion mutants Δ113 to 124, Δ148 to 186, Δ212 to 233 and Δ212 to 243 was 47±3, 0.5±0.4, 28±4, and 13±5% [71] respectively (Figure 6A). In another study mutant forms of apoA-I were generated by deletion of each of the repeated units 1 to 10 of apoA-I. Analysis of the ability of the variant apoA-I forms to activate LCAT showed that deletions of repeats 6, 7 and 8 had the most severe effects on the activation of LCAT followed by deletions of r epeats 2, 4, 9 and 10. Finally,

TABLE VI: LIPID-BINDING PROPERTIES OF NORMAL AND VARIANT APOA-I FORMS

	% BINDING TO LIPOSOMES		EM ANALYSIS	
A-I form	-Chol	+Chol	-Chol	+Chol
Plasma A-I	89.0	85.0	Discs	Discs
Gln-1-2→Arg,Arg	88.7	94.9	Discs	Discs
Pro-121→His	89.8	--	Discs	Discs
Δ113-124		81	A	Discs
Δ213-233	67.9	87.8	Few discs + vesicles	
Δ213-243	57.7	72.5	Few discs + vesicles	
Δ148-186	--	43.0	Heterogeneous size particles <10 nm	

deletions of repeats 1, 3 and 5 had a relatively smaller effect on the activation of LCAT [71] (Figure 6B). Kinetic analysis of the activation by the normal and the Δ212 to 243 mutant form showed similar V_{max} but a 15-fold increase in the K_m of the mutant apoA-I form [70]. Mutants deficient in their ability to activate LCAT displayed 50 to 75% normal binding to liposomes as well as alterations in HDL binding as determined by density gradient ultracentrifugation analysis of the culture medium [70,72]. Thus, the peak recovery (approximately 50%) of apoA-I was at density 1.14 g/ml for the WT apoA-I, at 1.18 g/ml for the mutants Δ113 to 124 and Δ148 to 186, and at d>1.21 g/ml for the mutants Δ212 to 233 and Δ212 to 243 (Figure 7 and Table VI). Electron microscopy of the proteoliposome LCAT substrate generated by different apoA-I forms showed that mutant apoA-I forms which displayed aberrant binding to HDL also displayed reduced ability to convert the spherical lecithin-cholesterol vesicles into discs compared to WT (Figure 8 & Table VI).

The findings suggest that a) the ability of apoA-I to activate LCAT is mediated by its ability to bind to lipid and/or to form discoidal substrate for LCAT; b) the interaction of several domains of apoA-I is required for the activation of LCAT. The present data also do not rule out direct involvement of apoA-I in the catalytic mechanism of LCAT.

Other studies suggest the importance of the hinged domain region of the apoA-I molecule for LCAT activation. Thus, deletion of Lys-107 results in LCAT activation which is approximately 40 to 60% of normal [73]. In addition, monoclonal antibodies which bind to the 95 to 121 apoA-I sequence inhibit the activation of LCAT by apoA-I [74]. Fluorescence spectroscopy and CD studies suggest that the conformation of the amino terminal domain apoA-I in the small particles which are poor activators of LCAT may be globular and may not interact directly with lipids [50,53].

The preceding analysis suggests that structural alterations in apoA-I which interfere with the formation of the nascent discoidal apoA-I containing particles or spherical HDL particles could in principle contribute to low apoA-I and HDL levels (see also Table VII).

Binding of HDL and ApoA-I Proteoliposomes to Cell Surfaces, Reverse Cholesterol Transport

It was shown earlier that apoA-I, as a component of HDL, promotes cholesterol efflux from cells, and through this mechanism might be important in maintaining cellular cholesterol homeostasis [75]. Three different mechanisms have been proposed to explain this phenomenon which include a) HDL receptor mediated efflux, b) retroendocytosis of HDL, and c) passive diffusion of cholesterol through the aqueous phase between cholesterol-poor membrane region and HDL or other acceptor molecules. a) Several early studies demonstrated binding of HDL to primary cultures of hepatocytes [76-78], nonparenchymal cells [77], cultured human fibroblasts [79], and adrenal gland [80] as well as to membrane preparations of rat and dog liver [81,82] and rat testis [83]. The number of HDL binding sites per hepatocyte and nonparenchymal cell has been estimated to be 10^5-10^6 [76-78]. The dissociation constant of the HDL-receptor complex is in the

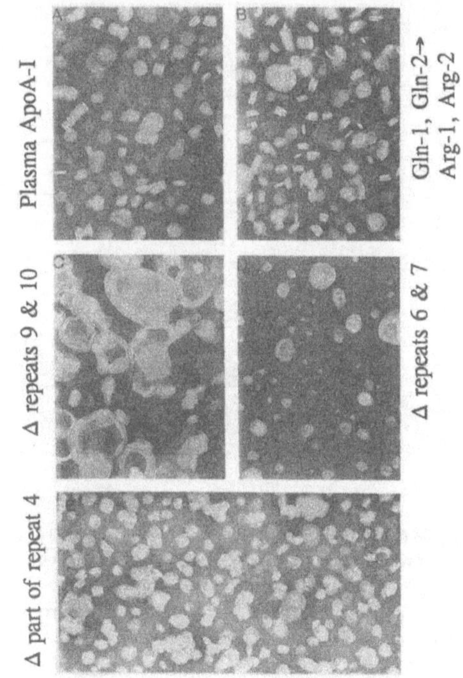

Figure 8. Electron miroscopy of proteoliposome generated by mixing normal and mutant apo A-I forms with phospholipids (DMPC). Panel A: Plasma apo A-I discoidal particles of relatively uniform size. Similar results were observed with WT apoA-I. Panel B: Point mutation A-I (Gln-2, Gln-1→ Arg, Arg) forms discs of similar size together with a small amount of larger liposomes. Panel C: Deletion mutant A-I (Δ213-243:lacking repeats 9 & 10) forms multilamellar liposomes of larger size (100-200nm diameter) than original liposomes (30-40nm), together with a few discoidial particles. Similar results were observed with A-I (Δ212-233: lacking repeat 9 and part of repeat 10). Panel D: Deletion mutant A-I (Δ148-186:lacking repeats 6 & 7) formed no visible discs, but created lipid particles heterogeneous in size, many of which were 10 nm or smaller. Panel E: Deletion mutant A-I (Δ113-124 lacking part of repeat 4) generated discoidal particles of relatively uniform size similar to those observed for plasma A-I (Panel A). Data from ref. 70.

range of 10^{-6}-10^{-7}M [78,81]. The binding is Ca^{2+} -independent [78,81] and is not affected by acetylation or reductive methylation of HDL [78,82] or pronase treatment of the membrane preparations or of cell cultures [78,79,81,82]. Various lipoproteins including HDL_3, HDL_2, HDL with apoE [82], LDL, and VLDL [81,82] as well as liposomes containing apoA-I [84-86], apoA-II and apoC's [85,86], apoA-IV [87], and phospholipid vesicles alone [88] compete to varying degrees for the HDL binding site. Human fibroblasts and arterial smooth muscle cells contain specific, high-affinity sites for HDL free of apoE (maximum saturation 20 μg HDL/ml) [89]. HDL bound to these sites was not internalized or degraded by these cells; instead it caused an increase in LDL receptor activity, sterol synthesis, and cholesterol efflux. These processes were saturated at

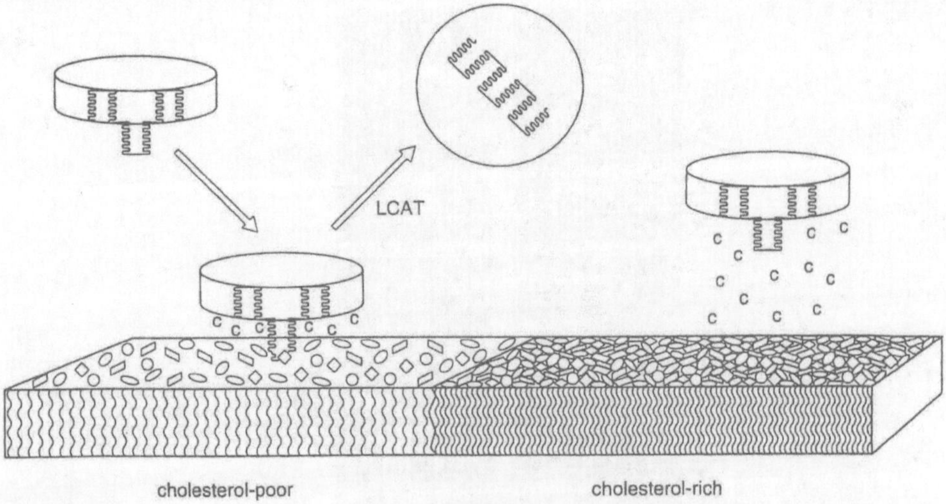

Figure 9. Model for the efflux of cholesterol from cholesterol-poor and cholesterol-rich plasma membrane domains. The model assumes that the hinged region of apoA-I on the HDL disc interacts with cholesterol-poor domain of the plasma membrane. Cholesterol molecules (C) move from membrane to the acceptor. Through the LCAT action the disc is converted to a spherical HDL. The model assumes that flux of cholesterol from the cholesterol-rich domain occurs through a large unstirred water layer. From ref 59, with permission.

HDL concentrations of 20 μg/ml, as was the specific, high-affinity HDL binding (89). Subsequent studies by the same group of investigators suggested that binding of HDL to cell surface receptors promotes translocation of cholesterol from intracellular membranes to cell surfaces (90,91). The translocated cholesterol can then be removed from the plasma membrane by cholesterol acceptors. *b)* Other studies suggested that HDL binds to cell surface receptors of human monocytes (92). Following binding to normal monocytes, apoA-I is internalized, transported through cytoplasmic organelles, and finally exocytosed undegraded. The hypothesis advanced by these studies was that regardless of the subsequent events, binding of HDL to a cell surface receptor via apoA-I promotes efflux of cholesterol from cells which is subsequently delivered to the liver for excretion or reutilization. This process has been termed reverse cholesterol transport (93). Recent studies have also shown that in addition to the low affinity, high capacity binding site of HDL described above, hepatic cells contain another binding site with 100-fold higher affinity and 15-fold lower capacity. This site can bind both HDL as well as lipid-free apoA-I (94). Other studies utilizing monoclonal antibodies showed that the binding of HDL protein and cholesterol esters to HepG2 cells is mediated by non-exclusive cooperative binding of several apoA-I regions containing repeated units (95). Numerous studies also showed that, in addition to HDL a variety of cholesterol acceptors which do not bind to cells, such as nitrosylated HDL, albumin-phospholipid complexes, nitrosylated HDL and phospholipid vesicles alone, and intact erythrocytes, efficiently promote cholesterol efflux from cells, thus suggesting bidirectional equilibration of membrane and HDL cholesterol (96-100). To date, numerous efforts to clone and sequence the putative HDL receptor remain inconclusive. Ligand blotting indicated that molecular mass of an HDL binding protein varies from 70-110 kD (91,101), whereas radiation inactivation studies showed that the subunit molecular mass of the HDL binding site is 10 to 16 kD (102). As discussed previously, the putative hinged domain of apoA-I which consists of two helical regions around residue 100 has the flexibility depending on the packing of apoA-I on a lipoprotein particle to be either associated with the lipid on a spherical HDL particle or extend out of the lipoprotein particle into the aqueous phase when apoA-I is associated with a nascent HDL particle (39). It has recently been proposed by Rothblat and colleagues that transient association of this domain with cholesterol-poor membrane regions might result in the efflux of cholesterol from cells (59). Subsequent esterification of these

TABLE VII: APOA-I VARIANTS WITH ABNORMAL LIPID PROFILES OR APOA-I PROPERTIES[a]

Variant form	ApoA-I and HDL levels	ApoA-I properties	Atherosclerosis	Other
Pro-3±Arg (Munster-3), Pro-3→His	—	Reduced proapoA-I to apoA-I conversion	No	—
Gly-26→Arg (Iowa)	Low	—	—	Familial amyloidotic neuropathy
Gln-84→Stop	Deficiency	—	Mild	—
Lys-107 deletion (Marburg, Munster-2)	—	40-60% LCAT activation	—	—
Gln-110→Lys (Fukuoka)	Low	—	No	—
Pro-143→Arg (Giessen)	Low	60-70% LCAT activation	No	—
Pro-165→Arg	Low (55-60%)	50-60% LCAT activation	—	—
Arg-173→Cys (Milano)	Low (33-60%) abundance HDL^{3b}-type particles	Low HDL and apoA-I levels	No	Reduced LCAT mass and activity
Thr-202→frameshift addition 27 amino acids	Deficiency	—	—	Partial LCAT deficiency, corneal opacities
Deletion of 45 nucleotides encoding aa 146 to 160		15% HDL and apoA-I levels		
ApoAI-CIII deficiency	Deficiency of HDL, apoA-I, CIII	—	Severe	Partial LCAT deficiency
ApoAI, CIII, A-IV deficiency	Deficiency of HDL, apoA-I, CIII, A-IV	—	Severe	Partial LCAT deficiency

[a] For supporting references, see ref. 1.

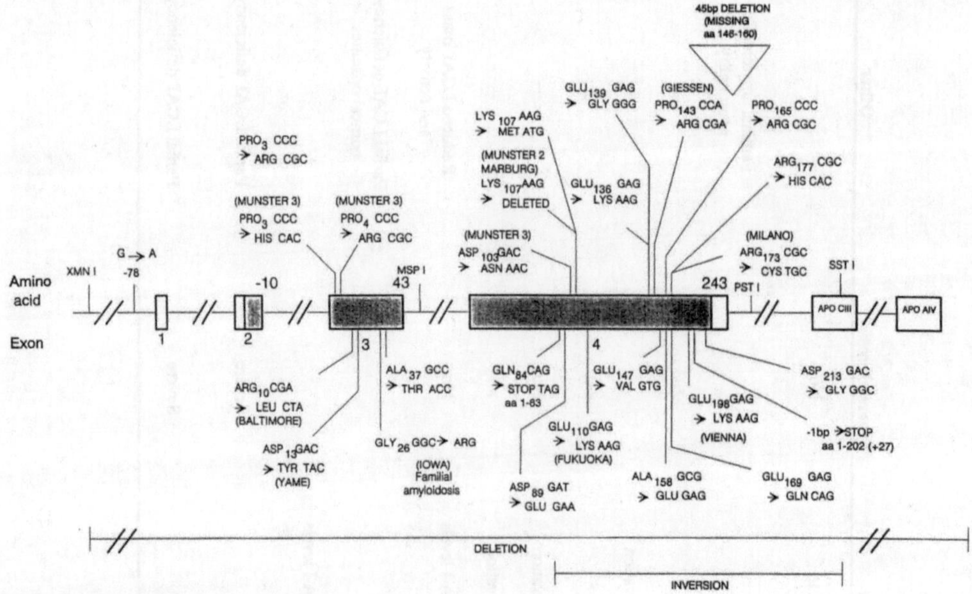

Figure 10. Schematic representation of the human apoA-I gene showing the mutations which have been encountered in humans (68,107-130).

particles by LCAT will generate spherical particles where the hinged apoA-I region is associated with the lipoprotein surface [59] (Figure 9). Despite the remaining controversies, it is generally accepted that apoA-I as a structural component of HDL plays an important role in

Factors Affecting ApoA-I and HDL Levels

Epidemiological and genetic data have shown convincingly that low [13-15] or high [13,15,20-22] levels of HDL or apoA-I are associated with increased or decreased risk of developing atherosclerosis, respectively, indicating the importance of apoA-I in cellular cholesterol homeostasis. ApoA-I and HDL levels can be affected by a variety of genetic and environmental factors. The genetic factors may affect the apoA-I gene or other structural or regulatory genes involved in the cholesterol transport system. It has been observed that genetic mutations which decrease the activity of the hepatic lipase [103] and the cholesteryl ester transfer protein [104,105] increase the HDL levels and may be beneficial. In contrast, mutations which decrease the lipoprotein lipase activity increase the plasma triglyceride and decrease the plasma HDL levels; thus increasing the risk of cardiovascular disease [106]. The environmental factors include nutrition, exercise, cigarette smoking, and moderate consumption of alcohol [15].

ApoA-I Mutations and Their Physiological Importance

The first structural mutation of apoA-I described was found in Italy and has been called apoA-I Milano [107,108]. Subjects with apoA-I Milano have low HDL cholesterol and apoA-I levels, but do not appear to suffer from premature atherosclerotic disease [108,109]. Further analysis indicated that the mutant apoA-I Milano allele results from a structural mutation that causes a Arg173 Cys substitution [110]. Affected individuals are heterozygotes and carry one normal and one variant apoA-I allele [110]. Subsequently, other apoA-I variants were described and named apoA-I Giessen and apoA-I Marburg or Munster-2 [68,111,112]. The subjects with apoA-I Marburg had hypertriglyceridemia and reduced HDL cholesterol levels [112]. Subjects with apoA-I Marburg and apoA-I Giessen also represent heterozygosity for one normal and one variant allele. The apoA-I Giessen results from a Pro-143→Arg substitution [68] and the apoA-I Marburg

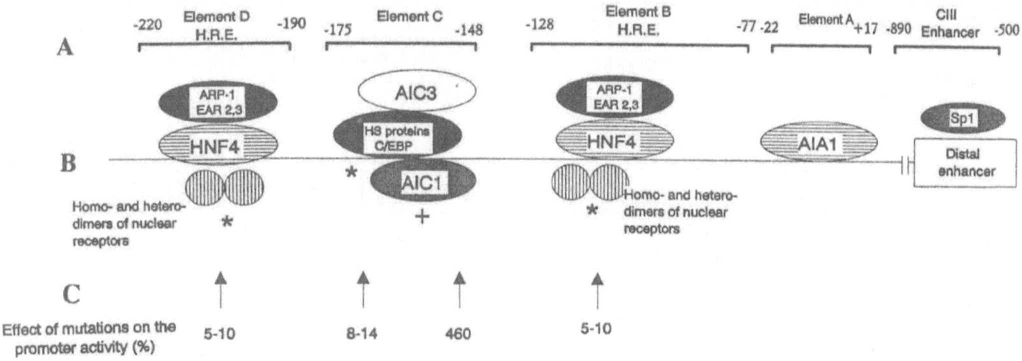

Figure 11. Schematic representation of the regulatory elements and the transcription factors which participate in the regulation of the human apoA-I gene. Panel A: proximal and long distal regulatory elements. Panel B: Factors bound to the proximal and distal elements. Panel C: Effects of mutations which prevent the binding of the factors to their binding sites (indicated by arrows) on hepatic transcription. The symbol * indicates decrease in transcription 5-15% of control; + indicates increase in transcription 4.6-fold.

from a deletion of Lys-107 (111). Both mutant apoA-I forms have 40-70% normal ability to activate LCAT as compared to normal apoA-I (68,111). A third apoA-I mutation which is associated with HDL deficiency, partial LCAT deficiency and corneal opacities results from a frameshift starting at residue 202 which causes incorporation of 27 random carboxy terminal residues including three cysteines (113). A patient with deletion of residues 146-160 of apoA-I had 15% of normal apoA-I levels due to rapid catabolism of the mutant protein (114). Several other mostly heterozygote variants of apoA-I have been described and are shown in Figure 10 and Table VII. The majority of the mutations are the result of amino acid substitutions (115-122). Most of the subjects with apoA-I mutations are asymptomatic and have normal lipids and lipoprotein profiles (123). However, in addition to apoA-I Milano and Giessen, the Pro-165→Arg was associated with low apoA-I and HDL levels (124) and displayed 50-60% of normal ability to activate LCAT (125). The Pro-3 Ar g or His substitutions were associated with increased plasma concentration of proapoA-I suggesting a low proapoA-I to apoA-I conversion (124). The Gly-26 to Arg substitution is associated with familial amyloidotic polyneuropathy III and is inherited as an autosomal dominant trait (118,119). The amyloid fibrils formed in this condition consist of the 83 amino terminal residues of apoA-I. Two human conditions characterized by severe deficiencies of plasma apoCIII and HDL are associated with premature atherosclerosis. One condition has resulted from deletion of apoA-I, apoCIII, apoA-IV locus (126,127) and the other from an inversion of the apoA-I, apoCIII locus. The break points of this inversion are in the fourth exon of the apoA-I gene and the first intron of the apoCIII gene (128-131) (Figure 10). Development of premature atherosclerosis has been manifested in a patient deficient in apoA-I synthesis. This condition results from a homozygous nonsense mutation at the code specifying residue 84 of apoA-I (121). The rather moderate severity of this condition and the absence of atherosclerosis in the HDL deficient patient with the partial LCAT deficiency (113) suggest that some of the functions of apoA-I may be assumed by other apolipoproteins, perhaps apoA-IV.

The Importance of ApoA-I Deduced from Epidemiological and Genetic Data in Humans and Transgenic Mouse Experiments. Regulation of ApoA-I Synthesis.

The transgenic animal experiments indicate that an increased level of apoA-I is protective against atherosclerosis. One potential mechanism which may lead to increased apoA-I synthesis is the activation of transcription of the human apoA-I gene. Recently we have made progress towards the understanding of the molecular mechanisms which are responsible for the transcriptional regulation of the human apoA-I gene. The hepatic transcription is controlled by

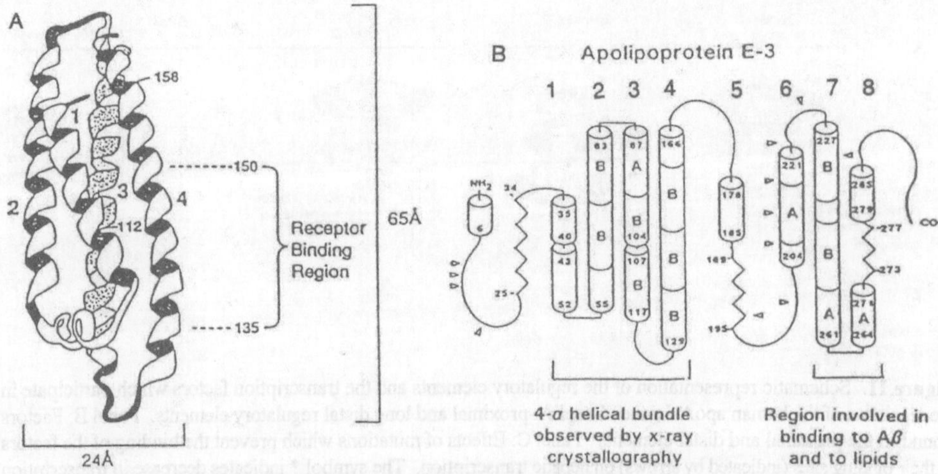

Figure 12. Panel A: Antiparallel arrangement of the sequences comprising the aminoterminal 4 helix-bundle of apoE as seen by X-ray crystallography. From ref 34, with permission. Panel B: Antiparallel arrangement of the carboxy terminal region 241 to 277 as determined by protein modeling. From ref 37, with permission. This region is required for binding of apoE to amyloid peptide β (Aβ) [23]

three regulatory elements, B (-128 to -77), C (-175 to -148) and D (-220 to -190) (Figure 11A) [132,133]. The hepatic transcription is enhanced approximately 10-fold by an enhancer region localized -890 to -500 nucleotides in the 5′ position of the apoCIII gene [134]. Elements B and D bind the hepatic nuclear factor 4 (HNF-4) as well as the orphan receptors EAR-2, EAR-3, ARP-1. In addition the enhancer binds homodimers of 9-cis retinoic acid receptor (RXR) and heterodimers of RXR with all-trans retinoic acid receptor (RAR) and thyroid hormone receptor (THR) (Figure 11B) [135]. The binding of the nuclear receptors to the regulatory elements D and B of apoA-I appears to be important. Mutations which abolished the binding of these activities to their cognate sites on elements D or B decreased hepatic transcription to very low levels (Figure 11C) [132,133]. This indicates that the hepatic factor(s) which bind to these elements are positive activators of transcription.

The regulatory region C is recognized by both positive and negative regulators which bind in overlapping domains. The region -148 to -168 is recognized by two activities designated AIC1 and AIC3. Competition experiments suggest that this latter activity may be related to HNF-3. Mutations which affected the binding of AIC1 increased transcription 4.6-fold indicating that this protein acts as a negative regulator (Figure 11C). Element C is also recognized by the heat-stable activities which bind to several elements of the apoB and apoCIII promoter as well as by C/EBP. Mutations which affected the binding of these activities reduced transcription to 8-14% of control (Figure 11C) [132,133]

Finally, the enhancer region is recognized mainly by activities related to SP1. This region, in addition to enhancing the hepatic transcription ten-fold, appears to be necessary for intestinal transcription [136]. This analysis suggests potential mechanisms for increasing apoA-I gene transcription. For instance, increased transcription could be achieved by a) prevention of binding to the regulatory region C of the negative regulatory factor; b) recruitment of stimulatory homo- or heterodimers of nuclear receptors to the regulatory region D; c) potentiation of the interactions of the proteins which bind to the enhancer with those which bind to the proximal regulatory region; d) combination of the above processes.

CONCLUSIONS ON APOA-I

1. ApoA-I is the major protein component of HDL. It activates the enzyme LCAT and participates in the reverse transport of cholesterol.

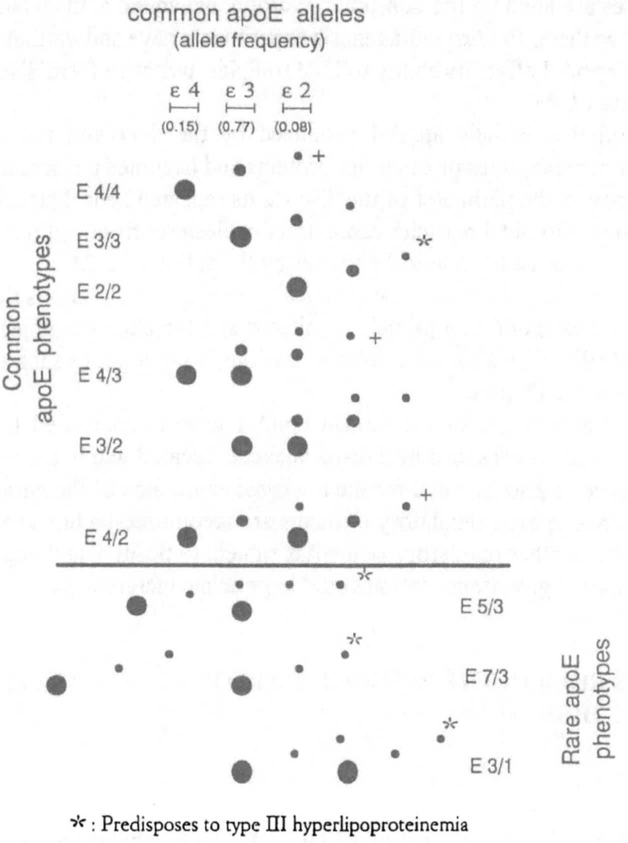

Figure 13. Schematic representation of the common and rare apoE alleles and apoE phenotypes. Large and small spheres represent the unmodified and the O-glycosylated sialyated forms of apoE, respectively. The symbols * and + indicate phenotype associated with type III hyperlipoproteinemia and Alzheimer's disease, respectively.

2. Epidemiological and genetic data have established that a decrease in HDL or apoA-I levels represents an important risk factor for CHD. In contrast, increased apoA-I and HDL levels protect against CHD and may confer longevity to humans. In human subjects, there is a very strong positive correlation between HDL and apoA-I levels. A large percentage (~5%) of human subjects have low HDL levels and are at risk for CHD.
3. The plasma HDL levels are influenced by other genetic factors and correlate positively with the activity of the lipoprotein lipase and negatively with the activity of CETP.
4. Experiments in transgenic mice have shown that increased apoA-I synthesis (which is achieved by introduction of the human apoA-I gene in mice) is associated with an increase in plasma HDL levels. These transgenic animals do not develop atherosclerosis in response to atherogenic diets whereas the non-transgenic animals develop atherosclerosis in response to the same diet. Thus high HDL and apoA-I levels confer protection against atherogenic substances of the circulation in this experimental animal model.
5. ApoA-I has a unique structure which consists of repeated units composed mostly of 22 amino acids. These repeated units can form amphipathic α-helices. On one side of the α-helices

are lined up the polar hydrophilic amino acids which are exposed to the water. On the other side of the helices are lined up the non-polar hydrophobic amino acids which can bind to the lipids and solubilize them. *In vitro* mutagenesis experiments have shown that alterations in the repeated units of apoA-I affect its ability to bind to lipids and/or to form discoidal particles as well as to activate LCAT.

6. It is believed that initially apoA-I produced by the liver and the intestine extracts phospholipids from membranes or other lipoproteins and becomes a discoidal particle. In this particle apoA-I covers the perimeter of the disc via its repeated helical structures. It has been proposed that these discoidal particles can extract cholesterol from cell membranes, and can be subsequently converted to spherical particles by the action of LCAT.

7. When apoA-I is changed critically by genetically transmitted mutations then the formation or the stability of the lipoprotein particles is altered and the affected subjects may have low HDL levels. Low HDL may also result from alterations in other genes such as the lipoprotein lipase gene and the CETP gene.

8. The hepatic transcription of the human apoA-I gene is controlled by three proximal regulatory elements and is enhanced by a distal enhancer located within the apoCIII promoter. The apoCIII enhancer is also essential for the intestinal expression of the human apoA-I gene. Two of the proximal apoA-I regulatory elements are recognized by ligand activated nuclear hormone receptors. Another regulatory element is subject to positive and negative regulation. Upregulation of apoA-I gene transcription could in principle increase apoA-I and HDL levels.

APOE AND ITS ROLE IN THE PATHOGENESIS OF ATHEROSCLEROSIS AND IN ALZHEIMER'S DISEASE

ApoE Structure

Apolipoprotein E is a component of VLDL, IDL, HDL, chylomicrons and chylomicron remnants. The primary sequence of the human apoE was obtained by conventional protein sequencing methods [137] and was verified in the 1983 to 1985 period by cloning cDNAs and the gene encoding this protein [138-141]. Biochemical studies combined with secondary structure predictions indicate that apoE contains two independently folded domains which are separated by a region of random coil [142,143]. Similar to the other apolipoproteins, both the amino acid [137] and the nucleotide sequence of apoE [138-141] showed that the protein, as well as the gene, contains repeated units which can form amphipathic α-helices [31,32]. Such amphipathic α helices have been observed by X-ray crystallography of the 22 KDa amino terminal portion of the human apoE which forms a four helix bundle [33-35]. The four defined antiparallel helical bundles extend between residues 24 to 42, 54 to 81, 87 to 122, and 130 to 164. Adjacent helices are stabilized by hydrophobic interactions and salt bridges. The relative orientation of these α-helical segments of apoE as determined by x-ray crystallography [33-35] as well as the arrangement of the amino acids in the carboxy terminal 241 to 277 region of apoE3, as suggested by protein modeling [37], is shown in Figure 12A and B. The receptor binding activity of apoE is localized in the amino terminal fragment whereas the lipid binding is mediated by the carboxy terminal fragment [144,145].

ApoE Biosynthesis

The early work indicated that the major site of apoE synthesis was the liver [1]. Subsequent work involving protein and mRNA analysis showed that apoE is a ubiquitous protein which is synthesized by a variety of tissues and cells [16,146,147]. The induction of apoE mRNA and protein synthesis in human monocyte-macrophage cultures correlates with the monocyte-macrophage differentiation [16]. Several studies also indicate that a large portion of apoE synthesized by cells

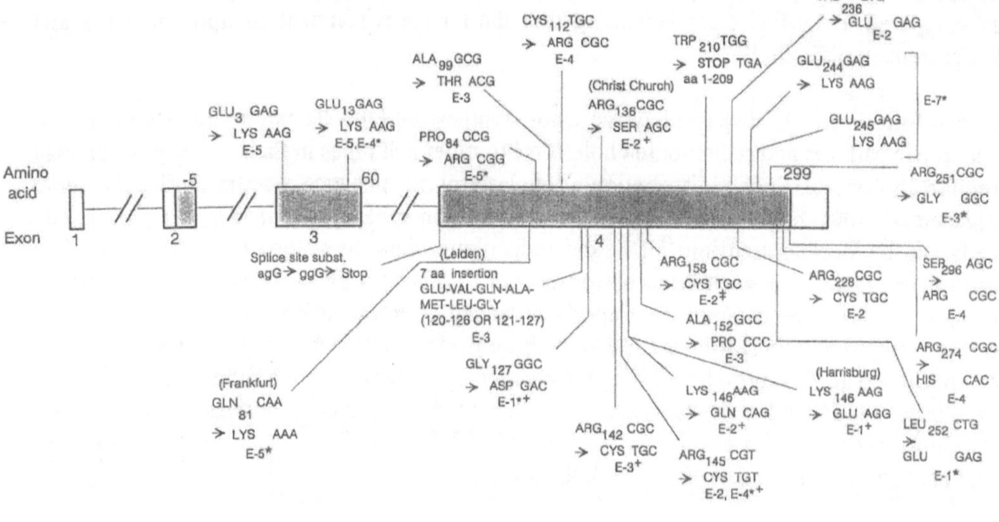

Figure 14. Schematic representation of the human apoE gene, the various apoE mutations, and the phenotypes they generate. The Arg-158→Cys substitution is associated with the recessive form of type III HLP. ApoE forms associated with dominant forms of type III HLP are indicated by +; recessive inheritance is indicated by ‡. The symbols E1*, E4*, E5*, and E7* indicate double mutations as follows: E1*=Gly-127→Asp and Arg-158→Cys or E1*=Arg-158→Cys and Leu-252→Glu; E2*= Val-236→Gly, and Cys-112→Arg; E4*=Glu-13→Lys and Arg-145→Cys; E5*=Pro-84→Arg and Cys-112→Arg or E5*=Gln-81→Lys and Cys-112→Lys; E7*=Glu-244→Lys and Glu-245→Lys. The apoE mutations are described in refs. 190-207.

and liver perfusion systems is secreted in a lipid-poor form and is subsequently incorporated into preexisting lipoprotein particles such as IDL, HDL and chylomicron remnants [(1)]. Alternatively, lipid-free apoE may function proximal to the site of synthesis, for instance within the brain tissue. ApoE can be isolated from plasma [(3)] and culture media of HepG-2 cells [(4,17)] as a component of lipoprotein particles (LP) with defined lipid and apoprotein composition designated LpB:E, LpB:C:E, LpA-II:B:C:D:E. The physiological significance as well as the sites of assembly of these particles are not known.

ApoE Functions

Receptor and Lipid Binding. ApoE is the ligand which promotes the recognition and catabolism of apoE-containing lipoproteins such as chylomicron remnants, βVLDL, hypertriglyceridemic VLDL, IDL and HDL with apoE by the LDL receptor [(148)] and the E/α2M macroglobulin receptor [(6)]. The receptor binding domain of apoE was identified by binding studies of normal variant and truncated apoE forms as well as by the use of receptor blocking monoclonal antibodies [(144,145)]. The combined data indicated that the receptor binding domain of apoE is in the vicinity of residues 130-160. Site-directed *in vitro* mutagenesis also verified that residues in the region 130-160 are important for receptor binding [(149,150)]. The apoE4 form has normal receptor binding. The sequence of apoE between residues 142-149 has 63% homology with the region between residues 3359 and 3366 of apoB-100 [(1,151)]. The heparin binding domains of apoE were localized in the second half of the molecule, between amino acids 142-147, 243-272, and 211-218 [(152,153)]. The amino terminal 22 KDa apoE fragment does not bind lipids indicating that the lipid binding domain of apoE is in the carboxyterminal domain. Nevertheless a single amino acid substitution of Cys-112 by Arg in apoE4 results in preferential distribution of apoE to the triglyceride-rich lipoproteins [(154,155)]. ApoE conformation is profoundly affected by the lipid environment. In discoidal particles, infrared spectroscopy showed that the axis of the helices is parallel to the acyl chains of the phospholipids [156]. In addition, ^{14}C NMR spectroscopy showed different pKa values for Lys residues in lipoproteins as compared to DMPC discs indicating that the microenvironment of apoE in discs and lipoproteins is different [157].

Other functions. As discussed above, apoE synthesized by cells can be incorporated into lipoprotein particles and redistribute cholesterol to other cell types in their vicinity which may require cholesterol. Such redistribution of cholesterol, for instance, occurs during the nerve regeneration following injury [158-162]. In a similar fashion apoE synthesized by the peripheral cells can be incorporated into HDL and redistribute cholesterol directly to the liver thus contributing to the reverse cholesterol transport or to other peripheral cells [163-165]. The ability of apoE to direct the catabolism of apoE-containing lipoproteins by the LDL and the LDL receptor related protein (LRP) or E/α2M receptor is inhibited by apoCI and to a lesser extent by apoCII and apoCIII proteins [166,167]. Introduction by transfection of a functional apoE gene in mouse Y1 adrenal cells decreases dramatically steroidogenesis suggesting a potential regulatory role of apoE in adrenal steroidogenesis [168]. Finally, *in vitro* experiments have shown that apoE promotes the conversion of IDL to LDL [169].

Common Genetic Polymorphism in Human apoE

Based on biochemical and genetic data, we have proposed that there are three common apoE alleles in humans. ϵ4, ϵ3, and ϵ2 which give rise to three homozygous (E4/4, E3/3, E2/2), and three heterozygous (E4/3, E3/2, E4/2) phenotypes (Figure 13) [170,172]. The average frequency of the apoE alleles in Caucasians is ϵ4=0.15, ϵ3=0.77, and ϵ2=0.08. This corresponds to a calculated phenotype frequency of E4/4=3%, E3/3=60%, E2/2=1%, E4/2=22%, E3/2=12%, and E4/2=2% [173] (Figure 13). The isolation of cDNAs and the gene encoding human apoE provided the molecular explanation for the common polymorphism of apoE [170-173]. The three different apoE isoproteins, E4, E3 and E2, observed in humans are accounted for by mutations in amino acids 112 and 158. The E4 isoprotein contains Arg-112, Arg-158, the E3 contains Cys-112, Arg-158, and the E2 contains Cys-112, Cys-158. The substitution of Cys for Arg at one or two positions explains the observed charge differences between apoE isoproteins E4, E3, and E2 [171-173].

Common apoE Alleles Affect Plasma Lipid and Lipoprotein Levels in the General Population

A variety of studies have shown uniformly that the effect of apoE genotypes on plasma and LDL cholesterol follows the order ϵ4/ϵ4=ϵ4/ϵ3=ϵ4/ϵ2>ϵ3/ϵ3>ϵ3/ϵ2>ϵ2/ϵ2 [173]. Homozygosity or heterozygosity for the ϵ2 allele is also associated with increases in VLDL cholesterol and triglycerides [174,175] compared to ϵ3/ϵ3 genotypes. Homozygosity or heterozygosity for ϵ2 or ϵ4 alleles is also associated with increased and decreased plasma apoE, respectively [176,177]. Furthermore, the ϵ2/ϵ2 genotype is associated with decreased plasma apoB levels as compared to ϵ4/ϵ3 and ϵ4/ϵ4 genotypes [176,177]. A variety of recent studies with hyperlipidemic populations have shown that homozygosity and/or heterozygosity [1,178-181] for the ϵ2 allele is associated with hyperlipidemia and hypertriglyceridemia. In addition, homozygosity and/or heterozygosity for the ϵ4 allele was associated in some studies with elevated plasma and LDL cholesterol levels [173,178-183]. In plasma, apoE4 is distributed preferentially in TG-rich lipoproteins [154,155].

Certain apoE Phenotypes and Genotypes are Associated with Type III Hyperlipoproteinemia (type III HLP)

Familial type III HLP, also called familial dysbetalipoproteinemia, or broad β, or floating β disease, is characterized by premature atherosclerosis, xanthomas, elevated plasma

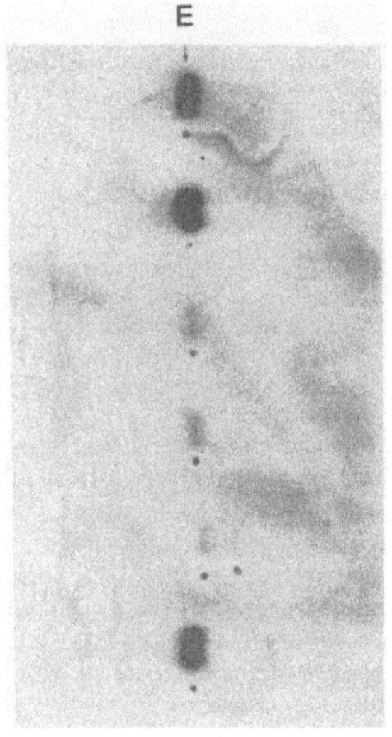

Figure 15. Blotting analysis of RNA isolated from human tissues and cultured human monocyte macrophages obtained from normal subjects and an apoE-deficient patient. Lanes a&b: 5μg total RNA from HepG2 cells and human liver, respectively. Lanes c&d: total RNA (20μg) obtained from human intestine and muscle, respectively. Lanes e&f: total mRNA obtained from three 100mm diameter petri dishes of the apoE-deficient patient and normal subjects, respectively. Note that apoE-deficient cultures have two mRNA species and greatly reduced apoE mRNA concentrations as compared to control cultures. Hybridization of the same blot with chicken α-actin cDNA probe gave similar intensities of α-actin mRNA for both the normal and deficient monocyte-macrophage cultures. From ref 216 with permission.

cholesterol and triglyceride levels, cholesterol-enriched βVLDL and IDL particles, increased plasma apoE levels, premature coronary and peripheral atherosclerosis [184]. The frequency of the disease was estimated to be 0.1-0.01% in the population [184]. The great majority of the patients with type III HLP have the E2/2 phenotype [174] which results from the substitution of Cys for Arg-158. This mutation combined with other genetic or environmental factors affects the catabolism of apoE containing lipoproteins and causes type III HLP [174,184-186]. Another feature of type III HLP is that it results in the accumulation in plasma of remnants of lipoprotein metabolism enriched in cholesteryl esters and apoE [187,188].

Dominant and Recessive Forms of Type III Hyperlipoproteinemia

The Arg-158→Cys mutation in the homozygote state depending on other genetic or environmental factors may result in type III hyperlipoproteinemia. This form of type III HLP is inherited in an autosomal recessive mode [189]. A variety of rare apoE mutations have also been described (Figure 14) [190-207]. Some of these mutations are associated with a dominant mode of inheritance of type III HLP which is expressed at an early age. These include substitutions Arg-136→Glu, Arg-142→Cys, Arg-145→Cys, Lys-146→Gln,Lys-146→Glu and an insertion of seven amino acids (duplication of residues 121-127) [192-197]. These apoE mutations which are associated with dominant forms of type III HLP are between residues 136 to 152. The importance of the 136 to 152 region of apoE for receptor binding was also assessed by *in vitro*

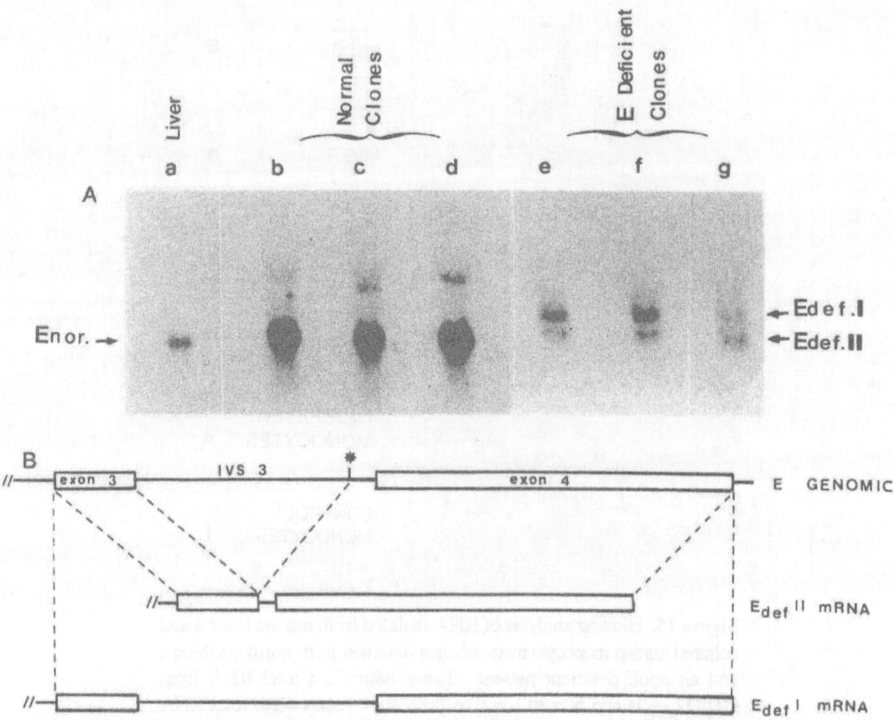

Figure 16. Panel A: Autoradiogram of blotting analysis of RNA isolated from human liver and transformed mouse C127 cells expressing either the normal or apoE-deficient gene. Lane a: mRNA (5μg) obtained from fetal human liver. Lanes b-g: RNA isolated from one 50mm diameter petri dish of clones of C127 cells expressing the normal or deficient apoE gene. Lanes b-d: RNA isolated from clones transfected with the normal apoE gene (E_{nor}). Lanes e-g: RNA isolated from clones transfected with the deficient apoE gene (E_{def}). Note the similarity of the RNA produced by the apoE-deficient clones with the RNA obtained from the monocyte-macrophage cultures of the patient with apoE deficiency (Figure 15). Panel B: Schematic representation of the mRNA species that are generated by aberrant splicing in the apoE-deficient (E_{def}) gene. *: The position of the activated cryptic splice site, as determined by S1 nuclease mapping and sequence analysis of the deficient apoE gene.

mutagenesis. Mutations within this region reduced the receptor binding activity 10 to 50% of control [149,150]. New insights into the structure/function relationship of the human apoE have been obtained with X-ray crystallography of the 22 kD amino terminal thrombolytic fragment of apoE [33-35]. The electrostatic potential map of the surface of the 22 K amino terminal apoE region shows a large positive potential around residues 135-150 [33-35] and has been proposed that these residues make direct ionic interactions with the negatively charged residues present at the ligand binding domain of the LDL receptor [33-35]. The X-ray crystallographic data show that Arg-158 is within the fourth α-helical bundle at a distance greater than 15Å from Asp-150 which in apoE3 is exposed to solvent. Arg-158 has the potential to form salt bridges with the adjacent Asp-154 of bundle 4 or Glu-96 of bundle 3 [33-35] (Figure 12). In apoE2, Arg-150 becomes salt-bridged with Asp-154. This reduces the electrostatic surface potential of the receptor binding region of residues 135-150. This change combined with the disruption of an extensive salt bridge between helices 3 and 4 in apoE3 accounts for the decreased receptor binding activity of apoE2 [33-35]. The disruption of the salt bridge network in apoE2 causes movement of the polypeptide backbone of helix 3 at Gly-105 as compared to apoE3 [34]. Thus the alteration of a single charged residue in apoE3 is sufficient to alter substantially the structure and thus affect the functions of apoE.

Additional information on the mechanisms which may be responsible for the expression of type III hyperlipoproteinemia came from the work of Innerarity and colleagues [207]. These

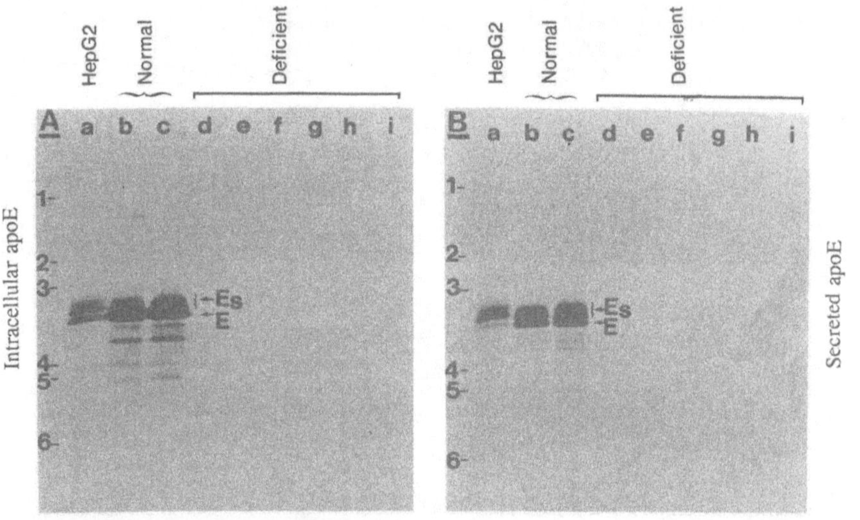

Figure 17. Analysis of proteins produced by HepG2 cells and C127 cells transfected with either the normal or apoE-deficient gene following 1 hour of labeling with [^{35}S]methionine. The figure shows an autoradiogram of a sodium dodecyl sulfate-polyacrylamide gel electrophoresis of apoE secreted by HepG2 cells and transformed C127 cells expressing either the normal or apoE-deficient gene. Numbers 1-6 indicate the position of protein molecular weight markers as follows: 1, bovine serum albumin (67,200); 2, ovalbumin (43,000); 3, aldolase (40,000); 4, human apoA-I (28,000); 5, bovine chymotrypsinogen (25,700); and 6, lysozyme (14,700). Lane a: apoE secreted by HepG2 cells. Lanes b and c: apoE secreted by two stable C127 clones transformed with the normal apoE gene. Lanes d-l: lack of apoE synthesis by six C127 clones transformed with the apoE-deficient gene. Panel A: intracellular; Panel B: secreted apoE. From ref. 217 with permission.

investigators have shown that the receptor binding properties of apoE containing lipoproteins obtained from type III patients with the Cys-158 mutation depend on the lipid composition of the particles. Particles obtained from untreated patients with high values of cholesterol and triglycerides bind poorly to cultured skin fibroblasts and their binding could be improved by treatment of the particles with cysteamine. In contrast, particles obtained following weight loss could bind efficiently and the binding was not affected by cysteamine treatment of the lipoprotein particles [(189,207)]. Some of the apoE variants are the result of double mutations in the apoE gene [(189,141,201,202)]. A variant has been described which has E4/4 phenotype, carries a Gln-13→Lys and an Arg-145→Cys substitution, and is associated with severe dominant form of type III HLP [(203)]. A single mutation Arg-145→Cys was shown previously to have 45% normal receptor binding and less severe clinical symptoms [(189)]. In contrast, a subject with a Glu-13→Cys substitution has normal lipid profile [(204)]. It is possible that changes in the receptor binding domain of apoE combined with distal changes may affect more severely apoE functions.

Factors Affecting Phenotypic Expression of Type III HLP

Numerous studies suggest that the E2/2 phenotype alone which results from an Arg→Cys substitution is not sufficient to cause type III HLP and that other genetic factors may be required for the phenotypic expression of the disease [(174,208-213)]. In some cases type III HLP is expressed in families with a tendency toward hypertriglyceridemia based on environmental or other genetic factors [(208-213)].

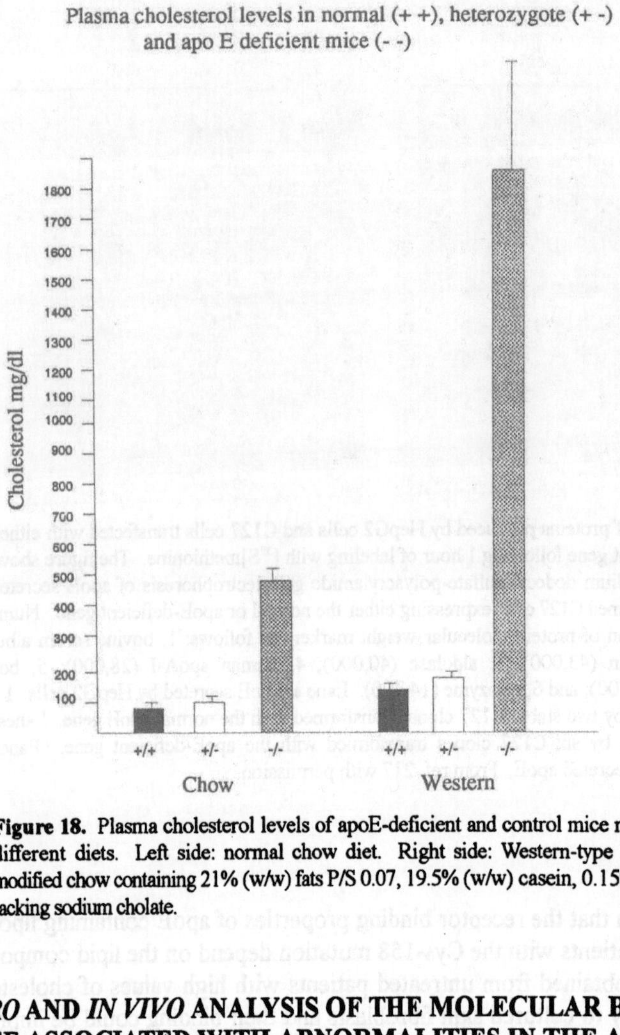

Figure 18. Plasma cholesterol levels of apoE-deficient and control mice maintained on two different diets. Left side: normal chow diet. Right side: Western-type diet consisting of modified chow containing 21% (w/w) fats P/S 0.07, 19.5% (w/w) casein, 0.15% cholesterol and lacking sodium cholate.

IN VITRO AND *IN VIVO* ANALYSIS OF THE MOLECULAR BASIS OF A HUMAN DISEASE ASSOCIATED WITH ABNORMALITIES IN THE APOE GENE

In Vitro Molecular Analysis of a Familial apoE Deficiency

Familial apoE deficiency in humans is associated with increased plasma cholesterol, accumulation of remnants in the VLDL and IDL region and premature atherosclerosis[184]. A rare form of type III HLP associated with familial apoE deficiency was first described in blood relatives in 1981 [214]. The biochemical and clinical features of this and other patients with apoE deficiency were a) lack of plasma apoE, b) Chol: 529±74; TG 221±62; VLDL Chol:243±9; IDL Chol: 230±41, c) Cholesterol-rich VLDL and IDL remnants which could be lowered by diets, d) The patients presented the clinical picture of type III hyperlipoproteinemia: appearance of tubo-eruptive and palmar xanthomas, and e) development of premature atherosclerosis [214,215].

Initial studies using cultures of peripheral blood human monocyte-macrophages obtained from an apoE-deficient patient and normal controls showed that the apoE deficient cultures synthesize low amounts of two aberrant forms of apoE mRNA [216](Figure 15) and do not produce any immunoprecipitable forms of apoE [217]. To investigate the molecular basis of this condition, we cloned the deficient apoE gene and performed comparative expression studies of the normal and deficient genes. For this purpose, the apoE genes were placed under the

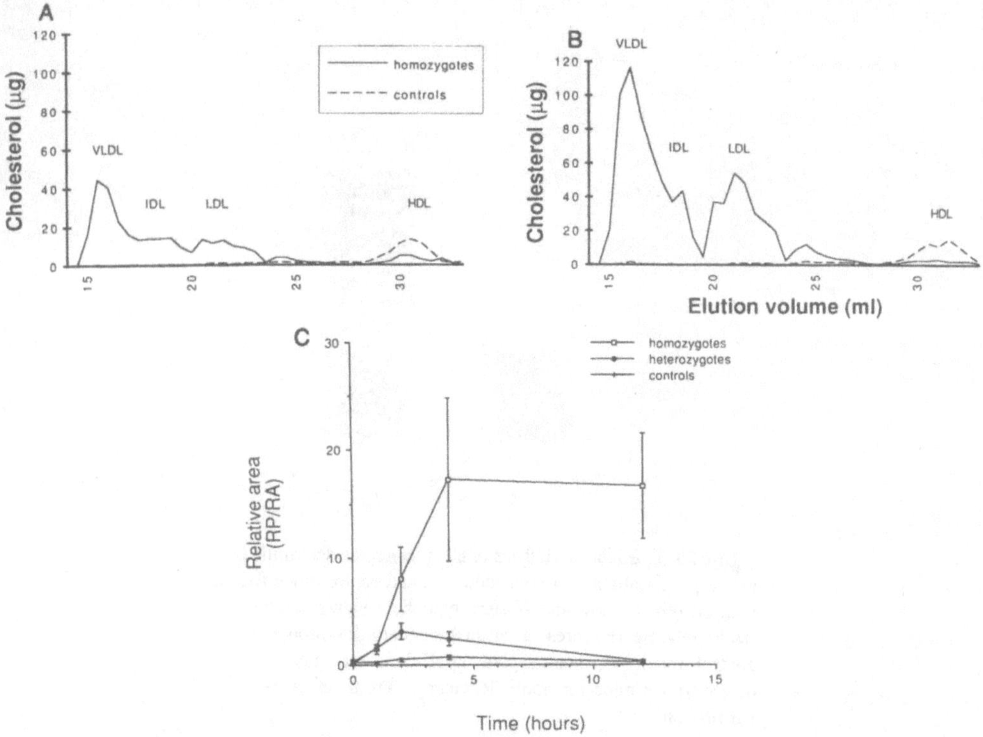

Figure 19. FPLC fractionation of 100 μl plasma from apoE-deficient and control mice. The approximate elution peaks of the major lipoprotein fractions are shown. Each profile represents the average of two independent pools of plasma. Panel A: mice on a chow diet. Panel B: Mice on the Western-type diet. Panel C: Vitamin A-fat tolerance test in apoE-deficient and control mice after administration of an intragastric bolus of 3000 U of retinol of 100 μl of corn oil in homozygotes, heterozygotes, and controls after an overnight fast. Approximately 60 μl of blood was drawn from each animal at 0, 1, 2, 4, and 12hr and analyzed for the presence of retinol palmitate which associates with lipoprotein remnants. From ref. 8, with permission.

control of the mouse metallothionein I promoter in a bovine papilloma virus vector and the resulting plasmids were used to generate permanent cell lines in C127 cells. Analysis of these cell lines showed that in the normal gene the mouse metallothionein I promoter directs the synthesis of normal apoE mRNA and protein. In contrast, in the deficient apoE gene the same promoter directs the synthesis of two abnormal apoE mRNA forms, which are similar to those observed in the peripheral blood monocyte-macrophage cultures of the patient (Figures 15 and 16A,B).

DNA sequence analysis has shown an A→G substitution at the 3′ splice site of the gene derived from the apoE deficient patients. This substitution creates two abnormally spliced mRNA forms. The smaller form contains 53 nucleotides, and the larger form contains the entire third intron of the apoE gene. Both of these mRNA species contain chain termination codons within the intronic sequence and code for short apoE peptides that are not detectable by gel electrophoretic techniques [217]. This analysis established that mutations in the apoE gene which preclude the synthesis of apoE can lead to type III hyperlipoproteinemia which is characterized by increased plasma cholesterol and triglyceride levels and the accumulation in plasma of lipoprotein remnants in the VLDL and IDL region and development of premature atherosclerosis.

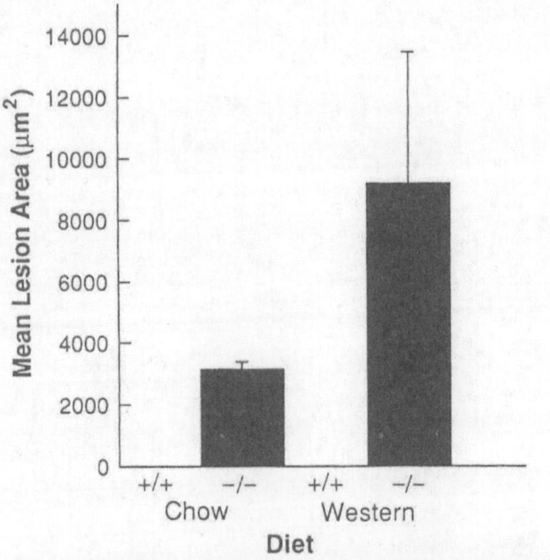

Figure 20. Quantitative analysis of atherosclerosis determined in the proximal aorta in apoE-deficient and control mice fed both the mouse chow and Western-type diet. Values summarized in the figure represent: mean ± standard deviation; +/+, control mice; -/-, homozygous apoE-deficient mice. +/-, heterozygous mice for apoE deficiency. From ref. 8, with permission.

Generation of Animal Models of ApoE Deficiency and Type III Hyperlipoproteinemia

The importance of apoE for the clearance of lipoprotein remnants has been established both by transgenic as well as gene inactivation experiments. Thus overexpression of the rat apoE gene in mice [218-220] increased clearance of VLDL and LDL, decreased VLDL and LDL levels and conferred resistance to diet-induced hypercholesterolemia. In other studies, apoE deficient mice have been generated by homologous recombination [8,221]. Comparison of the lipid and lipoprotein profiles of the deficient and control mice on chow diet showed that there is a dramatic increase in total cholesterol in the deficient mice maintained on chow diet. All of this increase could be accounted for by increases in the VLDL and LDL cholesterol (Figure 18,19A,B) [8]. More dramatic increases in total VLDL and IDL cholesterol occur when the apoE-deficient mice are fed western-type diets rich in cholesterol and saturated fats (Figure 18,19,A,B). A vitamin A fat tolerance test performed on the apoE-deficient and normal mice showed that administration of intragastric bolus of retinol in corn oil impaired the clearance of retinol palmitate esters (which accumulate in the chylomicron remnant fractions) in the deficient but not in the control mice (Figure 19C). These observations establish that apoE is required for the clearance of the remnants of lipoprotein metabolism which are rich in cholesteryl esters. The human patient information as well as the deficient mouse studies establish that formation of these remnants occurs independently of apoE synthesis. However clearance of these remnants which float in the VLDL and IDL region is critically dependent on the presence of apoE. The deficient mice develop coronary and pulmonary atherosclerotic lesions within 10 weeks on chow diet which is accelerated further when the animals are placed on western-type diets (Figure 20). Thus the biochemical and clinical features of apoE deficiency in mice resemble closely the picture observed previously in apoE-deficient humans.

The contribution of apoE mutations in the development of atherosclerosis has also been demonstrated using transgenic mice expressing a mutant apoE form in which Arg-142 was mutated into Cys as well as mice expressing the 121 to 127 deletion mutation (apo-Leiden) [222-225]. These mutations in humans are associated with a dominant form of type III hyperlipoproteinemia [1,189]. Compared to control littermates, the transgenic mice have increased cholesterol and triglyceride levels distributed in the VLDL and LDL region and develop atherosclerotic lesions within the first 4 months of life on chow diets. Diets enriched in cholesterol and saturated fats resulted in the appearance of βVLDL in plasma and the development of more extensive complex lesions in the transgenic mice as compared to the control mice which developed small lesions. The data confirm the cause-effect relationship between structural apoE mutations and the lipid abnormalities and resulting atherosclerosis which occurs in patients with type III hyperlipoproteinemia [184,187-189].

ApoE and Alzheimer's Disease

ApoE synthesis is increased following injury and is implicated in the growth and repair of the nervous system during development or after injury. In the peripheral nervous system, the synthesis of apoE increases 250-350 fold within 3 weeks of sciatic nerve crush, and apoE constitutes 5% of the total soluble extracellular protein [163]. Recently, it has been shown that apoE is found in the brain lesions of AD patients and that the apoE4 phenotype predominates in patients with late onset familial AD [23]. It has been suggested that apoE is a risk factor and contributes to the pathogenesis of the disease. Two hypotheses have been advanced to explain the involvement of apoE in AD. The first hypothesis implies extracellular interaction of apoE with the 42 amino acid long amyloid peptide β (Aβ). Initial observations indicated that apoE from cerebrospinal fluid binds with high affinity to Aβ [23,226]. *In vitro* experiments have shown that apoE can form a complex with Aβ peptide which is stable following boiling for 5 min at 5% SDS. Formation of stable complexes of apoE4 with Aβ is fast whereas formation of complexes of apoE3 with Aβ is slow (requires hours) [23]. First, apoE from cerebrospinal fluid binds with high avidity to Aβ *in vitro* [23]. Electron microscopy studies showed that a combination of apoE4 with Aβ resulted in the formation of monofibrils 7 nm in diameter. The matrix of monofibrils formed with apoE4 was much denser than that with apoE3 [228] (Figure 21). The second hypothesis implies lack of interaction of apoE4 with the microtubule associated protein tau [228]. It has been suggested that apoE3 binds to tau, and protects it from phosphorylation in the microtubule-binding domain, while apoE4 does not bind to tau. As a result, tau gets hyperphosphorylated and can no longer bind efficiently and stabilize microtubules, thus leading to the formation of neurofibrillary tangles inside neurons. These events may cause neuronal dysfunction or death [228]. Recent studies have shown that apoE3 in the presence of βVLDL increased neurite outgrowth and decreased branching in cultures of dorsal root ganglion neurons and apoE4 decreased neurite outgrowth and branching [229] (Figure 22). The requirement for βVLDL indicates that apoE is taken up by the neurons either by the LDL receptor or the LRP, or both. Upon entry into the cell, it is possible that apoE3 may interact and stabilize the cytoskeleton, thus supporting neurite extension. As discussed above a single amino acid substitution Arg-112 to Cys, causes significant changes in the tertiary structure of apoE. ApoE4 differs from apoE3 by substitution of Cys-112 by Arg. X-ray crystallography showed that in apoE3 Glu-109 is not salt bridged and Arg-61 lies over Cys-112 between helices 2 and 3. In apoE4, Glu-109 forms a salt bridge with Arg-112 [34]. This causes Arg-61 of helix 2 to rotate towards the opposite direction. This, in turn, may affect the interaction of the apoE segment with the carboxy terminal region which is required for the binding of apoE to Aβ [23]. Protein modeling by Nolte & Atkinson [37] indicates that similar to the amino terminal region the carboxy terminal apoE region may also have an up and down helical bundle arrangement (Figure 12). The Cys-112 to Arg change in E4 does not cause significant distortions of the polypeptide backbone similar to that seen when Arg-158 changes to Cys in apoE2 [34].

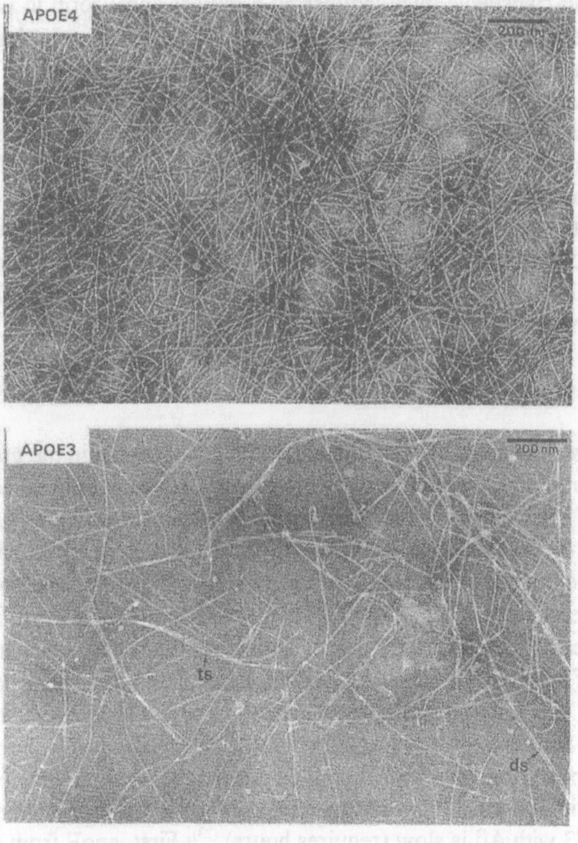

Figure 21. Comparison of monofibrils generated by co-incubations of apoE4 or apoE3 with Aβ peptide. Panel A: Very dense apoE4/Aβ monofibril matrix devoid of double- or triple-stranded Aβ ribbons. Panel B: Less dense apoE3/Aβ matrix consisting mainly of monofilaments but with residual double-stranded (ds) and triple-stranded (ts) ribbons, the strands of which are slightly separated. From ref. 227, with permission.

Parallel studies have also established that mutations in the amyloid precursor protein (APP) are associated with early onset of AD and amyloidogenesis. The APP is a transmembrane protein of yet unidentified function. The APP gene can generate transcripts encoding for proteins of 695 to 751 or 770 amino acid long (the 751 residue form contains a region homologous to the Kunitz family of serine inhibitors named KPI) [(230-232)]. Following synthesis in cell cultures the APP undergoes constitutive cleavage resulting in the release of a 110 Kd fragment into the medium and the retention of 10 KDa fragment into the membrane (233). The amyloid peptide β (Aβ) which forms the precipitates found in brains corresponds to an APP fragment containing 11 to 13 amino acids of the transmembrane and 29 to 31 amino acids of the extracellular domain [(234,235)]. The existing evidence is consistent with the hypothesis that variations in APP and apoE as well as the interaction of apoE variants with Aβ may play an important but still unidentified role in the pathogenesis of AD. Linkage studies indicate that the gene(s) responsible for some forms of familial AD are localized on chromosome 21. Several mutations linked to familial AD have been found in APP [(236-239)].

Accelerated AD is observed in patients with Down's syndrome (trisomy 21) and large amounts of Aβ are deposited in the blood vessels in hereditary cerebral hemorrhage with amyloidosis (Dutch type) which results from a substitution of Gln-693 by Glu in APP770 [(240)]. Familial AD is also linked to markers on chromosome 14 [(241)]. Thus the existing data support

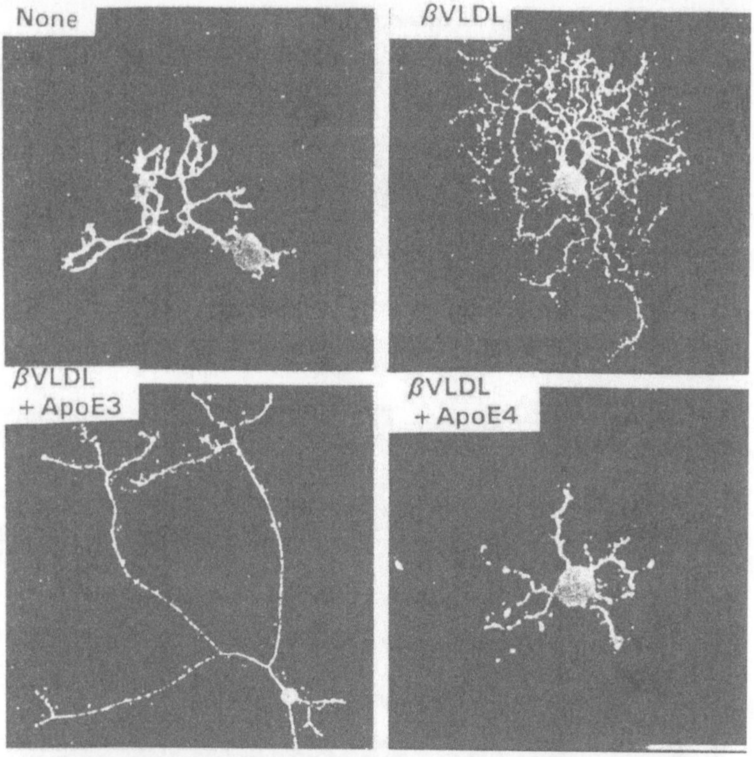

Figure 22. Photomicrographs of representative DRG neurons grown. Panel A: in N2 medium alone, Panel B: with βVLDL (40μg of cholesterol per milliliter), Panel C: with apoE3 (30μg/ml) and βVLDL (40μg/ml) or Panel D: with apoE4 (30μg/ml) and βVLDL (40μg/ml). Cells were nonspecifically stained with oil. Scale bar, 100μm. Reprinted with permission from Nathan et al., *Science* 264:850-852. Copyright 1994 American Association for the Advancement of Science.

the hypothesis that familial AD results from the interaction of genetic and environmental factors which result in the extracellular deposition of Aβ amyloid.

CONCLUSIONS ON APOE

1. Apolipoprotein E is a ubiquitous 34.2 KDa protein found in VLDL, IDL, HDL chylomicrons and chylomicron remnants.
2. X-ray crystallography has shown that the amino terminal domain of apoE forms a four-helix bundle. Computer modeling suggests that the carboxy terminal region may form similar antiparallel helical bundle structure.
3. ApoE-containing lipoproteins are recognized by the LDL receptor, the VLDL receptor, and the apoE/α_2 macroglobulin receptor.
4. There is a common genetic polymorphism of apoE in humans which generates six common apoE phenotypes: E4/4, E3/3, E2/2, E4/3, E3/2 and E4/2.
5. The E2/2 phenotype which results from a substitution of Cys for Arg-158 is associated with a recessive form of the type three hyperlipoproteinemia (type III HLP).
6. Other mutations in apoE within the receptor-binding domain of apoE (residues 136 to 152) are associated with dominant forms of type III HLP.

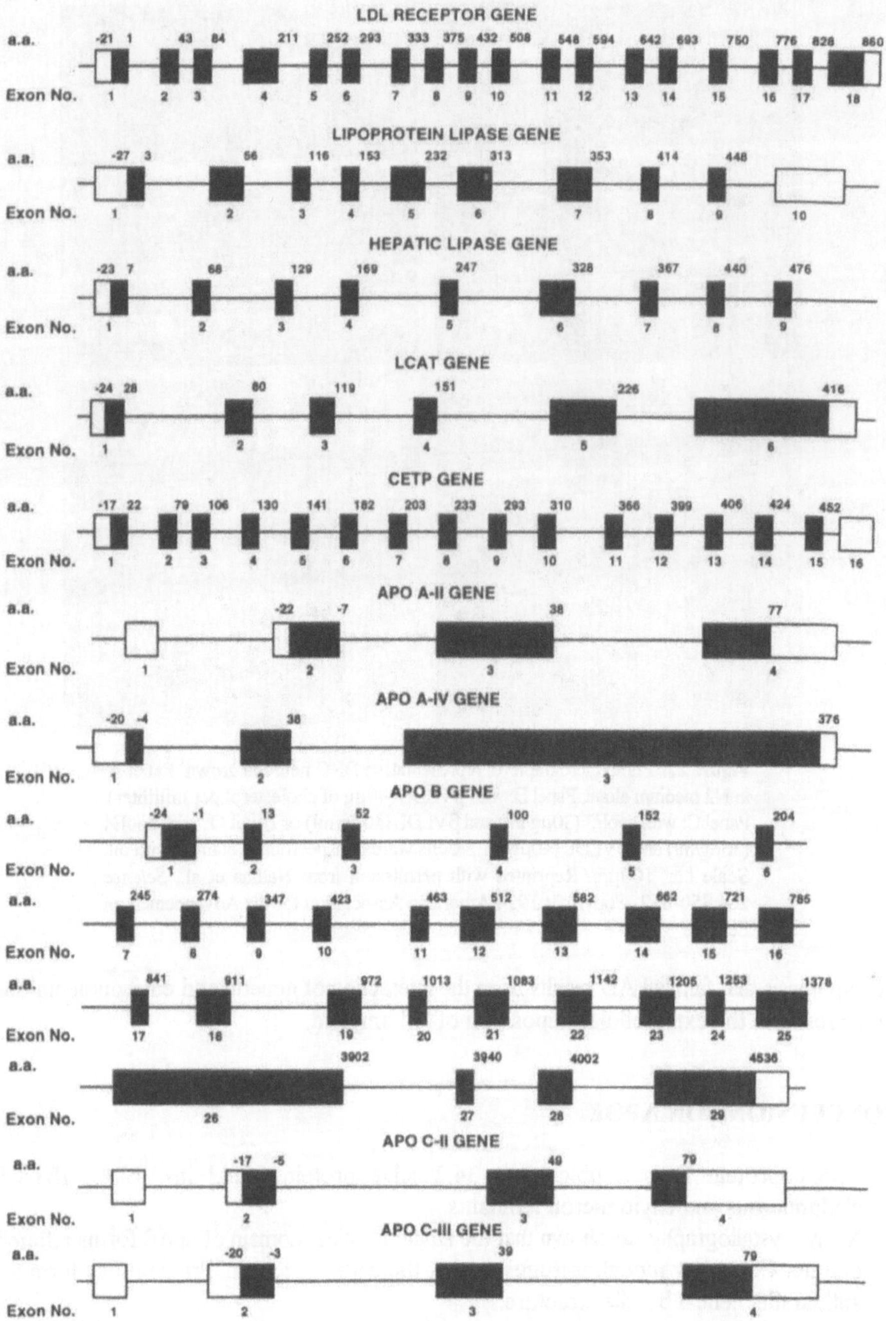

Figure 23. Intron exon organization of the genes referred to in Tables VIII to XIV. White and black boxes indicate untranslated and coding exonic sequences. Lines indicate the introns connecting them. The exons of the different genes are drawn to scale; introns are not to scale. The numbers on top of each gene indicate the amino acid sequence. The sequence of the mature protein is indicated by positive numbers at the exon/intron junction. The sequence of the signal peptide is indicated by negative numbers. The numbers at the bottom of the figure identify the exons. This figure allows the localization of the mutations in Tables VIII to XIV to the corresponding gene sequence.

7. Human subjects with a deficiency in apoE as well as apoE-deficient mice generated by inactivation of the apoE gene have high plasma cholesterol and develop premature atherosclerosis. The increase in cholesterol is due to the accumulation in plasma, of cholesteryl-ester-rich lipoprotein remnants which float in the VLDL and IDL region.
8. Homozygosity or heterozygosity for apoE4 is a risk factor for late onset of Alzheimer's disease (AD). Although the mechanism which leads to the pathogenesis of AD is not known, the following two hypotheses have been advanced currently to explain the association of apoE4 with AD. The first hypothesis implies strong extracellular interaction of apoE4 with the 42-residue long amyloid peptide β (Aβ) which contributes to the formation of amyloid plaques. The second hypothesis implies lack of interactions of apoE4 with tau which result in hyperphosphorylation of tau leading to the formation of neurofibrillary tangles, neuronal dysfunction, and death.
9. This chapter illustrates how current advances in biological sciences allow analysis of the molecular basis of a human disease. This analysis entails: a) cloning and characterization of the mutant genes; b) generation of cell lines which express the mutant genes; c) analysis of alterations in the mutant mRNA and protein; d) introduction of mutant apoE genes into mice to generate animal models of a dominant form of type III hyperlipoproteinemia; e) inactivation of the apoE gene in mice to generate an animal model of apoE deficiency. The apoE-deficient mice are particularly valuable for studying the conditions which lead to the progression or regression of atherosclerotic lesions.

Genetic Variations in Other Proteins of the Lipoprotein Transport System

As shown in Figure 1B, lipoprotein metabolism can be viewed as a complex biological pathway which requires the balanced functions of several proteins. Thus it is reasonable to expect that changes in the structure of any of these proteins which modify their function(s) or changes in the regulation of their synthesis will affect the plasma lipid and lipoprotein profiles and may in some instances contribute to the risk of CHD. The functions of the proteins which participate in this pathway and their association with human diseases is shown in Tables II and III. Normal and mutant forms of several of these proteins have been studied intensively by molecular approaches similar to those described for apoA-I and apoE. An updated list of the genetic mutations which have been described for the lipoprotein lipase and the hepatic lipase, the LCAT, the CETP, the LDL receptor, and the various apolipoprotein genes is shown in Tables VIII to XIV [1,143-276]. The intron/exon organization of the genes refered to in these tables is shown in Figure 23. A detailed review of these proteins has been provided recently [1] and is beyond the scope of this chapter.

Animal Models with Deficiencies or Overexpression of Genes Involved in Lipoprotein Transport

The importance of apolipoproteins, plasma enzymes, lipid transfer proteins and lipoprotein receptors for the maintenance of cholesterol homeostasis has been further studied by transgenic and gene inactivation experiments in mice and rabbits [277]. One of the most striking findings which emerged from these studies was the demonstration that inactivation of a single gene (apoE) in transgenic mice is sufficient to change an atherosclerosis-resistant into an atherosclerosis-prone mouse species [8,221,278-281]. Other transgenic experiments established an *in vivo* function of apoCIII as well as apoCII and apoCI genes in the development of hypertriglyceridemia [9,282-287]. Finally, generation of mice expressing human apoA-I, apoCIII and CETP resulted in low HDL high triglyceride levels a phenotype associated with a risk for CHD [288]. This and other interesting information derived from the transgenic mice as well as mice deficient in different components of the lipoprotein transport system is shown in Table XV [8-12,277-325].

TABLE VIII: MUTATIONS IN THE LDL RECEPTOR GENE ASSOCIATED WITH FAMILIAL HYPERCHOLESTEROLEMIA

VARIANT FORM	GENETIC ALTERATION	REF
Deletion of Asp^{26}-Gly^{27}	Δ6bp	1*
Trp^{66}→Gly	TGG→GGG	243
Asp^{69}→Tyr	GAT→TAT	1
Glu^{119}→Lys	GAG→AAG	1
Asp^{147}→His	GAC→CAC	244
Ser^{156}→Leu	TCG→TTG	1
Deletion of Gly^{197}	Δ3bp	1
Asp^{206}→Glu	GAC→GAG	1
Glu^{207}→Lys	GAG→AAG	1
Asp^{283}→Asn	GAC→AAC	1
5bp insertion at nucleotide 395 leading to Glu^{351}-Gly^{352}-Gly^{353}-Tyr^{354}→Ala-Leu-Asn	AGGGTGGCT→CACTGA	245
Val^{408}→Met	GTG→ATG	1
Ala^{410}→Thr	GCT→ACT	1
Asp^{412}→His	GAC→CAC	246
Val^{502}→Met	GTG→ATG	1
1bp deletion at 531→Stop (aa1-530)		247
Gly^{525}→Asp	GGC→GAC	1
Gly^{528}→Asp	GGT→GAT	1
Gly^{544}→Val	GGC→GTC	1
Trp^{660}→Stop (aa1-659)	TGC→TGA	1
Pro^{664}→Leu	CCT→CTT	1
Trp^{792}→Stop (aa1-791)	TGC→TGA	1
4bp insertion→Stop (aa1-795)		1
Tyr^{807}→Cys	TAT→TGT	1
Deletion of 15 Kb	involving exon 1+promoter	1
Deletion of 5 Kb	involving exon 1+promoter	1
Deletion of 5 Kb	involving exons 2+3	1
Deletion of 10 Kb	involving exons 2+3	1
Deletion of 1 Kb	involving exon 5	1
Deletion of 0.8 Kb	involving exon 5	1
Deletion of 2.5 Kb	involving exons 7+8	1
Deletion of 4 Kb	involving exons 7+8	1
Deletion of 12 Kb	involving exons 7-14	1

VARIANT FORM	GENETIC ALTERATION	REF
Deletion of 4 Kb	involving exons 13+14	1
Deletion of 5 Kb	involving exons 13-15	1
Deletion of 5.5 Kb	involving exons 16-18	1
Deletion of 7.8 Kb	involving exons 16-18	1
Deletion of exon 15		248
Deletion of 7bp	involving exon 6	249
Duplications:		1
+14 Kb	Duplication of exons 2-8	1
+>5 Kb	Duplication of exons 2-5	1

* Primary citation is provided in ref. 1.

TABLE IXA: MUTATIONS IN THE LIPOPROTEIN LIPASE GENE ASSOCIATED WITH FAMILIAL TYPE I HYPERLIPOPROTEINEMIA

VARIANT FORM	GENETIC ALTERATION	REF
Exon 1 3′ Splice site substitution	Cgt→Cct→Stop	1*
Asp^{9}→Asn	GAC→AAC	1
Exon 2 3′ Splice site substitution	Ggt→Gat→Stop	1
Exon 3 5′ splice site substitution	agG→aaG→Stop	1
Asp^{21}→Val	GAT→GTC	1
His^{44}→Tyr	CAC→TAC	1
Tyr^{61}→Stop	TAT→TAA	1
Trp^{64}→Stop	TGG→TAG	1
Val^{69}→Leu	GTG→GTC	250
Tyr^{73}→Stop	TAC→TAA	251
Arg^{75}→Ser	AGA→AGC	251
Trp^{86}→Stop	TGG→TGA	1
Trp^{86}→Arg	TGG→CGG	1
Gln^{106}→Stop	CAG→TAG	1
His^{136}→Arg	CAT→CGT	1
Gly^{139}→Ser	GGCAGC	252
Gly^{142}→Flu	GGA→GAA	1
Gly^{154}→Ser	GGC→AGC	253
Asp^{156}→Gly	GAT→GGT	1
Pro^{157}→Arg	CCA→CGA	1
Ser^{172}→Cys	TCT→TGT	254
Ala^{176}→Thr	GCA→ACA	1
Asp^{180}→Glu	GAC→GAG	255

VARIANT FORM	GENETIC ALTERATION	REF
$His^{183} \rightarrow Gln$	CAC→CAG	256
$Gly^{188} \rightarrow Glu$	GGG→GAG	1
$Ile^{194} \rightarrow Thr$	ATT→ACT	1
$Gly^{195} \rightarrow Glu$	GGA→GAA	1
$Ile^{205} \rightarrow Ser$	ATT→AGT	257
$Pro^{207} \rightarrow Leu$	CCG→CTG	1
$Cys^{216} \rightarrow Ser$	TGT→AGT	1
1bp deletion → Stop (aa1-220)	$ATT^{219} \rightarrow TGA^{220}$	1
$Arg^{243} \rightarrow His$	CGC→CAC	1
$Arg^{243} \rightarrow Cys$	CGC→TGC	258
$Ser^{244} \rightarrow Thr$	TCC→ACC	1
$Asp^{250} \rightarrow Asn$	GAC→AAC	1
$Leu^{252} \rightarrow Arg$	CTG→CGG	259
$Ala^{261} \rightarrow Thr$	GCC→ACC	259
$Tyr^{262} \rightarrow Stop$	TAC→TAG	1
$Asn^{291} \rightarrow Ser$	AAT→AGT	259
$Ala^{334} \rightarrow Thr$	GCC→ACC	260
$Leu^{365} \rightarrow Val$	CTA→GTA	261
$Trp^{382} \rightarrow Stop$	TGG→TAG	262
Addition of 2 Kb	duplication between exons 6&7	1
$Ser^{447} \rightarrow Stop$	TCA→TGA	1

*Primary citation is provided in ref. 1.

TABLE IXB: MUTATIONS IN THE HEPATIC LIPASE GENE ASSOCIATED WITH HEPATIC LIPASE DEFICIENCY

VARIANT FORM	GENETIC ALTERATION	REF
$Phe^{267} \rightarrow Ser$	TTC→TCC	263
$Met^{383} \rightarrow Thr$	ATG→ACG	1*

* Primary citation is provided in ref. 1.

TABLE X: MUTATIONS IN THE LCAT GENE ASSOCIATED WITH FAMILIAL LCAT DEFICIENCY

VARIANT FORM	GENETIC ALTERATION	REF
$Pro^{10} \rightarrow Leu$	CCG→CTG	1*
$Pro^{10} \rightarrow Frameshift$		1
$Leu^{32} \rightarrow Pro$	CTG→CCG	1
$Tyr^{83} \rightarrow Stop$	TAC→AAT	1
$Ala^{93} \rightarrow Thr$	GCC→ACC	1

VARIANT FORM	GENETIC ALTERATION	REF
$Gly^{119}\rightarrow$Frameshift		264
+$Thr^{123}\rightarrow$Ile	ACA→ATA	1
$Arg^{135}\rightarrow$Trp	CGG→TGG	1
Insertion of Gly between residues 140 &141	Addition of a new codon, GGC	1
$Tyr^{144}\rightarrow$Cys	TAT→TGT	1
$Arg^{147}\rightarrow$Trp	CGG→TGG	1
$Tyr^{156}\rightarrow$Asn	TAC→CAA	1
$Arg^{158}\rightarrow$Cys	CGC→TGC	1
$Gly^{183}\rightarrow$Ser	GGC→AGC	1
$Leu^{209}\rightarrow$Pro	CTT→CCT	1
$Asn^{228}\rightarrow$Lys	AAC→AAA	1
$Met^{252}\rightarrow$Lys	ATG→AAG	265
$Met^{293}\rightarrow$Ile	ATG→ATA	1
Deletion of Leu^{300}	CTC→delete	266
$Thr^{321}\rightarrow$Ile	ACG→ATA	1
$Thr^{347}\rightarrow$Met	ACG→ATG	1
$Ile^{375}\rightarrow$Frameshift		1
$Asn^{391}\rightarrow$Ser	AAT→AGT	267

Fish-eye disease
** Primary citation is provided in ref. 1.

TABLE XIA: MUTATIONS IN THE CETP GENE ASSOCIATED WITH CETP DEFICIENCY

VARIANT FORM	GENETIC ALTERATION	REF
$Asp^{442}\rightarrow$Gly	GAC→GGC	268
$Gln^{309}\rightarrow$Stop	CAA→TAA	269
Intron 15 3′ splice site substitution	Cgt→Cat	1*

TABLE XIB: MUTATIONS IN THE MICROSOMAL TRIGLYCERIDE TRANSFER PROTEIN ASSOCIATED WITH ABETALIPOPROTEINEMIA

VARIANT FORM	GENETIC ALTERATION	REF
1bp deletion →frameshift	$C^{215}\rightarrow$delete	270
$Arg^{595}\rightarrow$Stop	CGA→TGA	1*

TABLE XIIA: MUTATIONS IN THE APOLIPOPROTEIN AII GENE

VARIANT FORM	GENETIC ALTERATION	REF
Splice site substitution in exon 3	Agt→Aat→Stop	1*

TABLE XIIB: MUTATIONS IN THE APOLIPOPROTEIN AIV GENE

VARIANT FORM	GENETIC ALTERATION	REF
$Glu^{230} \rightarrow Lys$ (AIV-3)	GAG→AAG	1*
$Thr^{347} \rightarrow Ser$ (AIV-1)	ACT→TCT	1
$Glu^{165} \rightarrow Lys$ (AIV-3)	GAG→AAG	1
$Lys^{167} \rightarrow Glu$ (AIV-0)	AAG→GAG	1
4aa insertion ^{361}-Glu-Gln-Gln-Gln-^{362} (AIV0)	Insertion of 12 nucleotides	1
$Gln^{360} \rightarrow His$ (AIV-2)	CAG→CAT	1

*Primary citation is provided in ref. 1.

TABLE XIIIA: TRUNCATED FORMS IN THE APOB GENE ASSOCIATED WITH HYPOBETALIPOPROTEINEMIA

VARIANT FORM	GENETIC ALTERATION	REF
5 bp deletion frameshift at Val^{2357} and termination at residue 2362	$^{bp\ 7276}$ AAGTTAAG ^{7283} ↓ AAG	271
ApoB-82 $Cys^{3734} \rightarrow Stop$ (aa 1-3733)	TGC→TGA	272
ApoB-32.5 $Tyr^{1474} \rightarrow Stop$ (aa 1-1473)	TAT→TAG	272
ApoB-52.8 (aa 1-2395)	Deletion of 1 nucleotide at position 7286 frameshift, addition of 33 residues, Stop	272
ApoB-52.8 (aa 1-2395)	Deletion of 1 nucleotide at position 7359 frameshift, addition of 12 residues, Stop	272
ApoB-27.6 5′ splice substitution of intron 24	gt→gc	273
ApoB-25 (aa 1-1012)	Deletion in Exon 21→Stop	1*
ApoB-29 $Ser^{1305} \rightarrow Stop$ (aa 1-1304)	TCT→TAA	1
$Arg^{1306} \rightarrow Stop$ (aa 1-1305)	CGA→TGA	1
ApoB-31 (aa 1-1423)	Δ of 1 bp→Stop	1
ApoB-37 (aa 1-1727)	Δ of 4 bp→Stop	1
ApoB-39 (aa 1-1795)	Δ of 1 bp→Stop	1
ApoB-40 (aa 1-1828)	Δ of 2bp→Stop	1
ApoB-46 $Arg^{2058} \rightarrow Stop$ (aa 1-2057)	CGA→TGA	1
ApoB-48 $Gln^{2153} \rightarrow Stop$ (aa 1-2152)	CAA→TAA	1
ApoB-50 $Gln^{2522} \rightarrow Stop$ (aa 1-2251)	CAG→TAG	1
ApoB-54.8 $Arg^{2486} \rightarrow Stop$ (aa 1-2485)	CGA→TGA	1
ApoB-61 (aa 1-2772)	Δ of 37bp→Stop	1
ApoB-67 (aa 1-3039)	Δ of 1 bp→Stop	1
ApoB-83 $Ser^{3750} \rightarrow Stop$ (aa 1-3749)	TCA→TGA	1
ApoB-86 $Phe^{3896} \rightarrow Stop$ (aa 1-3895)	TTC→TAA	1
ApoB-87 (aa 1-3941)	Δ of 1 bp→Stop	1

VARIANT FORM	GENETIC ALTERATION	REF
ApoB-90 (aa 1-4033)	Δof 1 bp→Stop	1

TABLE XIIIB: MUTATIONS IN THE APOB GENE RESPONSIBLE FOR THE IMMUNOCHEMICAL APOB POLYMORPHISMS

VARIANT FORM	GENETIC ALTERATION	REF
Thr^{71}→Ile	ACC→ATC	1*
Arg^{3611}→Gln	CGG→CAG	1
Lys^{4154}→Glu	AAA→GAA	1

*Primary citation if provided in ref. 1.

TABLE XIIIC: MUTATIONS IN THE APOB GENE

VARIANT FORM	GENETIC ALTERATION	REF
Arg^{3500}→Gln	CGG→CAG	1*
Thr^{3705}→Ile	ACT→ATT	1
Phe^{3937}→Leu	TTT→CTT	1
Arg^{4019}→Tyr	CGG→TAC	1

TABLE XIVA: MUTATIONS IN THE APOLIPOPROTEIN CII GENE ASSOCIATED WITH FAMILIAL TYPE I HYPERLIPOPROTEINEMIA

VARIANT FORM	GENETIC ALTERATION	REF
Met^{1}→Val (Paris)	ATG→GTG	1*
Arg^{19}→Stop	CGA→TGA	274
Trp^{26}→Arg	TGG→CGC	275
Tyr^{37}→Stop (Padova)	TAC→TAA	1
Glu^{38}→Lys	GAG→AAG	276
Lys^{55}→Gln (African variant)	AAA→CAA	1
Splice site substitution (Hamburg) in donor splicing site of exon 2	Cgt→Gct→Stop	1
-1bp→Stop (Nijmegen) (aa 1-17)		1
-1 bp→Stop (Toronto) (aa 1-68 (+6))		1
+1bp→Stop (St. Michael) (aa 1-69(+27))		1

TABLE XIVB: MUTATIONS IN THE APOLIPOPROTEIN CIII GENE

VARIANT FORM	GENETIC ALTERATION	REF
Lys^{58}→Gly	AAG→GAG	1
Thr^{74}→Ala	ACT→GCT	1

*Primary citation is provided in ref. 1.

TABLE XV: IN VIVO PHYSIOLOGICAL FUNCTIONS OF THE PROTEINS INVOLVED IN LIPOPROTEIN TRANSPORT

Protein Name	Gene Alteration	Physiological Consequences	Ref
ApoA-I	ApoA-I transgenic mice (overexpression of the human apoA-I gene)	Increased HDL and apoA-I levels; decreased mouse apoA-I levels; generation of 10.3 and 8.8 nm size particles. Mice with high (2-fold increased) HDL levels are better protected against endotoxin than controls. In an atherosclerosis-susceptible strain A-I overexpression was protective against the formation of sinus foam cell lesions.	10,11 289,290 12
	ApoA-I gene knockout (apoA-I deficiency)	ApoA-I deficiency in an atherosclerosis resistant mouse strain reduced HDL levels by 75% as compared to the control, and reduced cholesterol ester flux to 12% of control. The deficient mice did not develop atherosclerosis on chow diet or on atherogenic diets.	291-293
ApoA-II	ApoA-II transgenic mice (overexpression of human or mouse apoA-II gene)	Overexpression of human apoA-II did not increase plasma HDL levels. Overexpression of mouse apoA-II increased by two-fold HDL cholesterol and by 2-3-fold non-HDL cholesterol and triglyceride levels. Overexpression of apoA-II increased synthesis and decreased metabolism of HDL and plasma HDL fractions and reduced cholesterol efflux and LCAT activity in vitro.	294,295
	Human apoA-I X apoA-II transgenic mice	Mice overexpressing both apoA-I and apoA-II are less protected against atherosclerosis in response to atherogenic diets as compared with the apoA-I expressing mice.	296
ApoA-IV	ApoA-IV transgenic mice (overexpression of human apoA-I gene in intes-tine or liver)	Intestinal expression is associated with normal phenotype. Hepatic expression is associated with an increase in total and VLDL triglycerides, and a decrease in VLDL catabolism.	297,298
ApoB	ApoB transgenic mice (overexpression of human apoB gene)	High plasma and LDL cholesterol.	299-301
	Expression of a truncated human apoB gene	Reduction of total VLDL, IDL and LDL cholesterol. The lipid profile resembles hypobetalipoproteinemia syndromes. Increase in apoB-100 and modest increase in apoB-48 levels.	302
ApoCI	ApoCI transgenic mice (overexpression of human apoCI gene)	Mild hypertriglyceridemia. Diet-induced hypercholesterolemia.	281,287

Protein Name	Gene Alteration	Physiological Consequences	Ref
ApoCII	ApoCII transgenic mice (overexpression of human apoCII gene)	Severe hypertriglyceridemia which is exacerbated by high fat diet, normal VLDL with an increase in apoC:E ratio. Accumulation of total and VLDL triglycerides. Delayed clearance of VLDL.	284
ApoCIII	ApoCIII transgenic mice (overexpression of human apoCIII gene)	Severe hypertriglyceridemia. Accumulation of large VLDL with an increase in apoCIII:E ratios. Delayed clearance of VLDL. Accumulation of apoB-48 containing remnants.	9,285
	ApoCIII gene knockout (ApoCIII deficiency)	Homozygous apoCIII deficiency reduced triglyceride levels to 70% of control, increased the clearance of chylomicrons, and conferred resistance to post-prandial hypertriglyceridemia.	286
ApoE	ApoE transgenic mice (overexpression of human or rat apoE gene)	Increased clearance of VLDL and LDL. Decrease in plasma VLDL and LDL cholesterol levels. Resistance to diet-induced hypercholesterolemia. No effect on bile acid secretion and metabolism.	218-220
	ApoE transgenic mice (overexpression of dominant mutations of human apoE gene encoding an apoE variant defective in receptor binding)	Increase in total cholesterol and triglycerides, increase in apoB-48. Increase in cholesterol is due to the accumulation of cholesteryl ester-rich lipoprotein remnants which float in the VLDL and IDL region.	278-281
	ApoE knockout mice (ApoE deficiency)	8-fold increase in total cholesterol on chow diet and 25-fold increase on Western-type diets, due to accumulation of cholesteryl ester-rich remnants of intestinal origin in the VLDL and IDL region which resemble βVLDL. Decrease in postprandial lipoprotein clearance. ApoE-deficient mice develop fibroproliferative atherosclerotic lesions when fed a chow diet. This condition is exacerbated by Western-type diets. The plasma of deficient mice contains auto-antibodies to oxidized lipoproteins. The apoE-deficient mice are an excellent model to study the progression and regression of atherosclerosis. The deficient mice have normal nerve regeneration.	8,221 302-305
	Human apoA-I transgenic X apoE-deficient mice	Three-fold increase in HDL and six-fold reduction in the development of atherosclerosis as compared to the apoE deficient mice.	306
Apo(a)	Apo(a) transgenic mice (overexpression of human apo(a) cDNA)	Lp(a) is not formed. Apo(a) found in the lipoprotein-free fraction, forms Lp(a) upon infusion of human LDL. Development of atherosclerosis.	307,308
	Apo(a) X apoB transgenic mice	Lp(a) is formed.	300,301 309

Protein Name	Gene Alteration	Physiological Consequences	Ref
Cholesteryl ester transfer protein	CETP transgenic mice (overexpression of human or monkey CETP gene)	Moderate reduction in HDL cholesterol and apoA-I levels. Reduction in HDL size. Downregulation of the LDL receptor gene.	320
	Human CETP X apoA-I mice	Reduction by 66% of HDL cholesterol and by 42% of apoA-I and HDL size. Inverse correlation exists between CETP expression and HDL and apoA-I levels and HDL size and positive correlation between CETP levels and apoB levels and LDL size.	288,321,322
	Human CETP X apoA-II mice	Production of TG-enriched HDL. No effects on HDL size.	323
	Human CETP X apoA-I X apoCIII mice	Severe reduction in HDL and apoA-I levels and HDL size. The observed high triglyceride and low HDL levels resemble the human profile which is associated with risk for CHD.	277,288
	Human CETP X apoB transgenic mice	Increase in total cholesterol and triglycerides. Decrease in HDL and apoA-I levels.	324
	Human CETP X mutant apoE (Arg-142→Cys) mice	Reduction by 50% of HDL cholesterol and apoA-I levels. 2.5-fold increase in the cholesteryl ester content of VLDL. Accumulation of βVLDL in lines overexpressing the mutant apoE.	325

Protein Name	Gene Alteration	Physiological Consequences	Ref
LDL receptor	LDL receptor transgenic mice (overexpression of the human LDL receptor cDNA)	10-fold higher clearance of LDL. Decrease by 90% of ligand (apoB and apoE) levels. Transgenic mice are resistant to increases in LDL and IDL in response to atherogenic diets.	310,311,312
	LDL receptor knockout mice (LDL receptor deficiency)	Increase in IDL and LDL cholesterol by 8-fold and in total cholesterol by 2-fold. Further increases in total LDL and IDL cholesterol in response to atherogenic diets.	313
	Gene replacement in the LDL deficient mice by administration of adenoviral vectors carrying the LDL receptor gene	Reduction of IDL and LDL cholesterol to normal levels. Increased clearance of apoB-containing lipoproteins.	313
	LDL receptor deficient X apoE deficient mice	Similar profiles to those of LDL and apoE deficient mice with marked elevations in both apoB-100 and apoB-48.	314
ApoE/α_2M receptor (LRP)	LRP knockout mice (LRP deficiency)	Homozygous deficiency is lethal. Heterozygous has normal phenotype.	315
Lipoprotein lipase	Lipoprotein lipase transgenic mice (overexpression of the human lipoprotein lipase gene)	5-fold increase in LPL activity in adipose tissue and 1.7-fold in post-heparin plasma. 75% reduction in plasma triglycerides and 1.4-fold increase in HDL_2 cholesterol. Rapid clearance of VLDL. Protection from hypercholesterolemia in response to atherogenic diets.	316
Hepatic lipase	Hepatic lipase transgenic rabbits (overexpression of human hepatic lipase gene)	80-fold increase in HL activity in post-heparin plasma. 5-fold decrease in plasma cholesterol. Marked decrease in HDL and IDL cholesterol. Suppression of hypercholesterolemia in response to a diet containing 0.3% cholesterol.	317
LCAT	LCAT transgenic mice (overexpression of human LCAT gene)	11-fold increase in LCAT activity. Decreased catabolism of apoA-I and apoA-II. Increase in total & HDL cholesterol on normal diets. Further increase in HDL cholesterol on high fat diets.	318
	Human LCAT X human apoA-I mice	Alteration in HDL size.	319
	Human LCAT X apoA-I X apoA-II mice	Increase in HDL size.	319

In conclusion, this chapter provides an updated account of the structure, function, and genetic variation of the proteins involved in lipoprotein transport. In addition, it summarizes the lipid and lipoprotein profiles of animal models either overexpressing or lacking one of the above proteins. From this analysis it becomes apparent that the different lipid and lipoprotein profiles may affect differently the functions of the endothelium in normal state as well as following injury[326].

The advances in biological sciences have provided the molecular basis of several diseases of lipoprotein metabolism. This information provides unequivocal molecular markers for the early diagnosis of the disease. Animal models have been generated with precise molecular defects which mimic a specific human disease and can be utilized to study the progression and potential treatment of the disease. This new knowledge may allow the design of new pharmacological agents for the treatment of specific diseases including the potential for gene therapy in the future.

ACKNOWLEDGEMENTS

This work was supported by grants from the National Institutes of Health (HL-48739) and a grant from the EEC (BIOT-CT93-0473).

REFERENCES IN ORDER OF APPEARANCE

1. Zannis, V.I., Kardassis, D., and Zanni, E.E. 1993. Genetic mutations affecting human lipoproteins, their receptors and their enzymes. *Adv. Hum. Genetics* (H. Harris and K. Hirschorn, eds.) Plenum Press, NY. Vol. 21, pp. 145-319.
2. Herbert, P.N., Assmann, G., Gotto, Jr., A.M., and Fredrickson, D.S. 1982. Familial lipoprotein deficiency: Abetalipoproteinemia, hypobetalipoproteinemia, and Tangier disease, in: "The Metabolic Basis of Inherited Disease" (J.B. Stanbury, J.B. Wyngaarden, D.S. Fredrickson, J.L. Goldstein, M.D. Brown, eds.), McGraw-Hill, New York, pp. 589-651.
3. Cheung, M.C., and Alberts, J.J. 1984. Characterization of lipoprotein particles isolated by immuno-affinity chromatography. *J. Biol. Chem.* 259:12201-12209.
4. Dashti, N., Alaupovic, P., Knight-Gibson, C., and Koren, E. 1987. Identification and partial characterization of discrete apolipoprotein B containing particles human hepatoma cell line HepG2. *Biochemistry* 26:4837-4846.
5. Alaupovic, P. 1971. Conceptual development of the classification systems of plasma lipoproteins. In: "Protides of the Biological Fluids" (H. Peeters, ed.) Pergammon Press, Oxford, pp.9-20.
6. Brown, M.S., Herz, J., Kowal, R.C., and Goldstein, J.L. 1991. The low-density lipoprotein receptor-related protein: Double agent or decoy? *Curr. Opin. Lipidol.* 2:65-72.
7. Brown, M.S., and Goldstein, J.L. 1983. Lipoprotein metabolism in the macrophage: Implications for cholesterol deposition in atherosclerosis. *Ann. Rev. Biochem.* 52:223-261.
8. Plump, A.S., Smith, J.D., Hayek, T., Aalto-Setala, K., Walsh, A., Verstuyft, J.D., Rubin, E.M., Breslow, J.L. 1992. Severe hypercholesterolemia and atherosclerosis in apolipoprotein E-deficient mice created by homologous recombination in ES cells. *Cell.* 71:343-353.
9. Ito, Y., Azrolan, N., O'Connel, A., Walsh, A., Breslow, J.L. 1990. Hypertriglyceridemia as a result of apoCIII gene expression in transgenic mice. *Science*. 249:790-793.
10. Walsh, A., Ito, Y., Breslow, J.L. 1989. High levels of the human apolipoprotein A-I in transgenic mice result in increased plasma levels of small high density lipoprotein (HDL) particles comparable to human HDL3. *J. Biol. Chem.* 264:6488-6494.
11. Rubin, E.M., Ishida, B.Y., Clift, S.M,. Krauss, R.M. 1991. Expression of human apolipoprotein A-I in transgenic mice results in reduced plasma levels of murine apolipoprotein A-I and the appearance of two new high density lipoprotein size subclasses. *Proc. Natl. Acad. Sci. USA* 88:434-438.
12. Rubin, E.M., Krauss, R.M., Spangler, E.A., Verstuyft, J.G., Clift, S.M. 1991. Inhibition of early atherogenesis in transgenic mice by human apolipoprotein A-I. *Nature* 353:265-26
13. Castelli, W.P., J.T. Doyle, T. Gordon, C.G. Hames, M.C. Hjortland, S.B. Hulley, A. Kagan, W.J. Zukel. 1977. HDL cholesterol and other lipids in coronary heart disease: The cooperative lipoprotein phenotyping study. *Circulation* 55:767-772.
14. Heiss, G., H. Tyroler. 1982. In: Proceedings of the Workshop on Apolipoprotein Quantification, NIH Publ. No. 83-1266. U.S. Dept. Health and Human Services, NIH, Bethesda, Maryland. 7-24.

15. Anderson, K.M., Wilson, P.W.F., Odell, P.M., and Kannel, W.B. 1991. An updated coronary risk profile: A statement for health professionals. *Circulation* 8:356-362.

16. Zannis, V.I., S.F. Cole, C. Jackson, D.M. Kurnit and S.K. Karathanasis. 1985. Distribution of apoA-I, apoC-II, apoC-III and apoE mRNA in human tissues, time dependent induction of apoE mRNA by cultures of human monocyte-macrophages. *Biochemistry* 24:4450-4455.

17. Fruchart, J.C., and Bard, J.M. 1991. Lipoprotein particle measurement. An alternative approach to classification of lipid disorders. *Curr. Opin. Lipidol.* 2:362-366.

18. Lagrost, L., L. Persegol, C. and P. Gambert. 1994. Influence of apolipoprotein composition of high density lipoprotein particles on cholesteryl ester transfer protein activity. Particles containing various proportions of apolipoproteins A-I and A-II. *J. Biol. Chem.* 269:3189-3197.

19. Vanloo, B., Taveirne, J., Baert, J., Lorent, G., Lins, L., Ruyschaert, J.M., and Rosseneu, M. 1992. LCAT activation properties of apoA-I CNBr fragments and conversion of discoidal complexes into spherical particles. *Biochim. Biophys. Acta* 1128:258-266.

20. Glueck, C.J., P. Gartside, R.W. Fallat, J. Sielski, P.M. Steiner. 1976. Longevity syndromes: Familial hypobeta- and hyperalphalipoproteinemia. *J. Lab. Clin. Med.* 88:941-957.

21. Glueck, C.J., R.W. Fallat, F. Millet, and M.F. Steiner. 1975. Familial hyperalphalipoproteinemia. *Arch. Intern. Med.* 135:1025-1028.

22. Patsch, W., J. Kuisk, C. Glueck, and G. Schonfeld. 1981. Lipoproteins in familial hyperalphalipo proteinemia. *Arterioscler.* 1:156-161.

23. Strittmatter, W.J., Saunders, A.M., Schmechel, D., Pericak-Vance, M., Enghild, J., Salvesen, G.S., Roses, A.D. 1993. Apolipoprotein E: High avidity binding to β-amyloid and increased frequency of type 4 allele in late onset familial Alzheimer's disease. *Proc. Natl. Acad. Sci. USA* 90:1977-1981.

24. Brewer, H.B. Jr., T. Fairwell, A. Larue, R. Ronan, A. Houser, T. Bronzert. 1978. The amino acid sequence of human apoA-I, an apoprotein isolated from high density lipoproteins. *Biochem. Biophys. Res. Commun.* 80:623-630.

25. Zannis, V.I., Karathanasis, S.K., Keutmann, H.T., Goldberger, G., and Breslow, J.L. 1983. Intracellular and extracellular processing of human apoA-I. Secreted apoA-I isoprotein 2 is a propeptide. *Proc. Natl. Acad. Sci. USA* 80:2574-2578.

26. Gordon, J.I., Sims, H.F., Lentz, S.R., Edelstein, C., Scanu, A.M., and Strauss, A.W. 1983. Proteolytic processing of human preproapolipoprotein A-I. A proposed defect in the conversion of pro A-I to A-I in Tangier's disease. *J. Biol. Chem.* 258:4037-4044.

27. Barker, W.C., M.O. Dayhoff. 1977. Evolution of lipoproteins deduced from protein sequence data. *Comp. Biochem. Physiol.* 57:309-315.

28. McLachlin, A.D. 1977. Repeated helical pattern in apolipoprotein A-I. *Nature* 267:465-466.

29. Fitch, W.M. 1977. Phylogenies constrained by the crossover process as illustrated by hemoglobins and a thirteen-cycle, eleven-amino-acid repeat in human apolipoprotein A-I. *Genetics* 86:623-644.

30. Karathanasssis S.K., V.I. Zannis, J.L. Breslow. 1983. Isolation and characterization of the human apolipoprotein A-I gene. *Proc. Natl. Acad. Sci. USA* 80:6147-6151.

31. Li, W.H., M. Tanimura, C.C. Luo, S. Datta, L. Chan. 1988. The apolipoprotein multigene family: biosynthesis, structure, structure-function relationships and evolution. *J. Lipid Res.* 29:245-271.

32. Segrest, J.P., R.L. Jackson, J.D. Morrisett, A.M. Gotto Jr. 1974. A molecular theory of lipid-protein interactions in the plasma lipoproteins. *FEBS Letters* 38:247-253.

33. Wilson, C., M.R. Wardell, K.H. Weisgraber, R.W. Mahley, D.A. Agard. 1991. Three-dimensional structure of the LDL receptor-binding domain of human apolipoprotein E. *Science* 252:1817-1822.

34. Wardell, M.R., C. Wilson, A. Agard, R.W. Mahley, and K.H. Weisgraber. 1993. Crystal structures of the common apolipoprotein E variants: Insights into functional mechanisms. "NATO ASI Series". Vol. II, 73:81-96.

35. Wilson C., T. Mau, K.H. Weisgraber, M.R. Wardell, R.W. Mahley, D.A. Agard. 1994. Salt bridge relay triggers defective LDL receptor binding by a mutant apolipoprotein. *Structure* 2:713-718.

36. Breiter, D.R., M.R. Kanost, M.M. Benning, G. Wesenberg, J.H. Law, M.A. Wells, I. Rayment, H.M. Holden. 1990. Molecular structure of an apolipoprotein determined at 2.5Å resolution. *Biochemistry* 30:603-608.

37. Nolte, R.T., and D. Atkinson. 1992. Conformational analysis of apolipoprotein A-I and E-3 based on primary sequence and circular dichroism. *Biophys. J.* 63:1221.

38. Marcel, Y.L., P.R. Provost, H. Kos, E. Rattai, N.V. Dac, J.C. Fruchart, E. Rassart. 1991. The epitopes of apolipoprotein A-I define distinct structural domains including a mobile middle region. *J. Biol. Chem.* 266:3644-3653.

39. Segrest, J.P., M.K. Jones, H. De Loof, C.G. Brouillette, Y.V. Venkatachalapathi, and G.M. Ananth aramaiah. 1992. The amphipathic helix in the exchangeable apolipoproteins: A review of secondary structure and function. *J. Lipid Res.* 33:141-146.

40. Srinivas, R.V., Y.V. Venkatachalapathi, R. Zheng, R.J. Owens, K.B. Gupta, S.K. Srinivas, G.M. Anantharamaiah, J.P. Segrest, and R.W. Compans. 1991. Inhibition of virus-induced cell fusion by apolipoprotein A-I and its amphipathic peptide analogs. *J. Cell. Biochem.* 45:224-237.
41. Kroon, D.J., J.P. Kupferberg, E. Kaiser, and F.J. Kezdy. 1978. Mechanism of lipid-protein interaction in lipoproteins: A synthetic peptide:lecithin vesicle model. *J. Am. Chem. Soc.* 100:5975-5977.
42. Sparrow, J.T., and A.M. Gotto Jr. 1980. Phospholipid binding studies with synthetic apolipoprotein fragments. *Ann. N.Y. Acad. Sci.* 348:187-211.
43. Skipski, V. 1972. In: "Blood Lipids and Lipoproteins": Quantitation, Composition and Metabolism. G. J. Nelson, editor. Wiley-Interscience, New York. 471-583.
44. Scanu, A.M., C. Edelstein, P. Keim. 1975. In: "The Plasma Proteins", vol. 1, 2nd ed. F. W. Putnam, editor. Academic Press, New York. 317-334.
45. Atkinson, D., D.M. Small. 1986. Recombinant lipoproteins: implications for structure and assembly of native lipoproteins. *Ann. Rev. Biophys. Chem.* 15:403-456.
46. Atkinson, D., G.G. Shipley. 1984. Structural studies of plasma lipoproteins. Brookhaven Symposium in Biology, 32 Neutrons in Biology. Neutron Scattering Analysis for Biological Studies. (P.B. Schoenborn, ed.) Plenum Publ. pp. 211-226.
47. Atkinson, D., H.M. Smith, J. Dickson, J.P. Austin. 1976. Interaction of apoprotein from porcine high-density lipoprotein with dimyristoyl lecithin. *Eur. J. Biochem.* 64:541-547.
48. Atkinson, D., D.M. Small, G.G. Shipley. 1980. X-ray and neutron scattering studies of plasma lipoproteins. *Ann. N.Y. Acad. Sci.* 348:284-288.
49. Pownall, H.J., J.B. Massey, S.K. Kusscrow, A.M. Gotto Jr. 1979. Kinetics of lipid protein interactions: Effect of cholesterol on the association of human plasma high density lipoprotein A-I with L-a-dimyristoylphosphatidylcholine. *Biochemistry* 18:574-579.
50. Jonas, A., K.E. Kezdy, and J.H. Wald. 1989. Defined apolipoprotein A-I conformations in reconstituted high density lipoprotein discs. *J. Biol. Chem.* 264:4818-4824.
51. Brasseur, R., J. De Meutter, B. Vanloo, E. Goormaghtigh, J.M. Ruysschaert, and M. Rosseneu. 1990. Mode of assembly of amphipathic helical segments in model high-density lipoproteins. *Biochem. Biophys. Acta* 1043:245-252.
52. Anantharamaiah, G.M., C.G. Brouillette, J.A. Engler, H. De Loof, Y.V. Venkatachalaphathi, J. Boogaerts, and J.P. Segrest. 1990. Role of amphiphatic helices in HDL structure/function. In: "Hypercholesterolemia, Hypocholesterolemia, Hypertriglyceridemia". (C.L. Malmendier, ed.) Plenum Press, NY.
53. Jonas, A., J.H. Wald, K.L. Toohill, E.S. Krul, K.E. Kězdy. 1990. Apolipoprotein A-I structure and lipid properties in homogeneous, reconstituted spherical and discoidal high density lipoproteins. *J. Biol. Chem.* 265:22123-22129.
54. Sparks, D.L., M.C. Phillips, and S. Lund-Katz. 1992. The conformation of apolipoprotein A-I in discoidal and spherical recombinant high density lipoprotein particles: ^{13}C NMR studies of lysine ionization behavior. *J. Biol. Chem.* 267:25830-25838.
55. Sparks, D.L., S. Lund-Katz, M.C. Phillips. 1992. The charge and structural stability of apolipoprotein A-I in discoidal and spherical recombinant high density lipoprotein particles. *J. Biol. Chem.* 267:25839-25847.
56. Curtiss, L.K. and R.S. Smith. 1988. Localization of two epitopes of apolipoprotein A-I that are exposed on human high density lipoproteins using monoclonal antibodies and synthetic peptides. *J. Biol. Chem.* 263:13779-13785.
57. Silverman, S.R., F. Bernini, J.T. Sparrow, A.M. Gotto Jr., and L.C. Smith. 1987. Monoclonal antibodies as probes of high density lipoprotein structure: Identification and localization of a lipid-dependent epitope. *Biochemistry* 26:5833-5843.
58. Collet, X., B. Perret, G. Simard, E. Raffai, and Y.L. Marcel. 1991. Differential effects of lecithin and cholesterol on the immunoreactivity and conformation of apolipoprotein A-I in high density lipoproteins. *J. Biol. Chem.* 266:9145-9152.
59. Rothblat, G.H., F.H. Mahlberg, W.J. Johnson, and M.C. Phillips. 1992. Apolipoproteins, membrane cholesterol domains, and the regulation of cholesterol efflux. *J. Lipid Res.* 33:1091-1097.
60. Fielding, C.J. 1990. Lecithin:cholesterol acyltransferase. In: "Advances in Cholesterol Research". M. Esfahani, J. B. Swaney, editors. Telford Press, Philadelphia. 271-314.
61. Fielding, C.J., V.G. Shore, P.D. Fielding. 1972. A protein cofactor of lecithin:cholesterol acyltransferase. *Biochem. Biophys. Res. Commun.* 46:1943-1949.
62. Soutar, A.K., C.W. Garner, H.N. Baker, J.T. Sparrow, R.L. Jackson, A.M. Gotto, L.C. Smith. 1975. Effect of the human plasma apolipoproteins and phosphatidylcholine acyl donor on the activity of lecithin: cholesterol acyltransferase. *Biochemistry* 14:3057-3064.
63. Fielding, C.J., P.E. Fielding. 1981. Evidence for a lipoprotein carrier in human plasma catalyzing sterol efflux from cultured fibroblasts and its relationship to lecithin:cholesterol acyltransferase. *Proc. Natl. Acad. Sci USA* 78:3911-3914.

64. Dory, L., C.H. Sloop, L.M. Boquet, R.L. Hamilton, P.S. Roheim. 1983. Lecithin:cholesterol acyltransferase-mediated modification of discoidal peripheral lymph high density lipoproteins: possible mechanism of formation of cholesterol-induced high density lipoproteins (HDLc) in cholesterol-fed dogs. *Proc. Natl. Acad. Sci USA* 80:3489-3493.

65. Nichols, A. 1990. Conversions in the origins and metabolism of human plasma. In: "Advances in Cholesterol Research". M. Esfahani, J.M. Swaney, editors. Telford Press, New Jersey. 315-365.

66. Segrest, J.P., H. DeLoof, J.G. Dohlman, C.G. Brouillette, G.M. Anantharamaiah. 1990. Amphipathic helix motif: classes and properties. *Proteins* 8:103-117.

67. Fukushima, D., S. Yokoyama, D.J. Kroon, F.J. Kėzdy, E.T. Kaiser. 1980. Chain length-function correlation of amphiphilic peptides. Synthesis and surface properties of a tetratetracontapeptide segment of apolipoprotein A-I. *J. Biol. Chem.* 255:10651-10657.

68. Utermann, G., J. Haas, A. Steinmetz, R. Paetzold, S.C. Rall Jr., K.H. Weisgraber, R.W. Mahley. 1984. Apolipoprotein A-I and Giessen (Pro143→Arg). A mutant that is defective in activating lecithin:cholesterol acyltransferase. *Eur. J. Biochem.* 144:325-331.

69. Roghani, A., and Zannis, V.I. 1988. Alterations of the Gln residues of human proapoAI by in vitro mutagenesis. Characterization of the normal and mutant protein forms. *Biochemistry* 27:7428-7435.

70. Minnich, A., Collet, X., Roghani, A., Cladaras, C., Hamilton, R.L., Fielding, C.J., and Zannis, V.I. 1992. Site-directed mutagenesis and structure-function analysis of the human apolipoprotein A-I. *J. Biol. Chem.* 267:16553-16560.

71. Sorci-Thomas, M., M.W. Kearns, and J.P. Lee. 1993. Apolipoprotein A-I domains involved in lecithin-cholesterol acyltransferase activation. *J. Biol. Chem.* 268:21403-21409.

72. Redgrave, T.G., D.C.K. Roberts, C.F. West. 1975. Separation of plasma lipoproteins by density gradient ultracentrifugation. *Anal. Biochem.* 65:42-49.

73. Rall, S.C. Jr., K.H. Weisgraber, R.W. Mahley, Y. Ogawa, C.J. Fielding, G. Utermann, J. Haas, A. Steinmetz, H.J. Menzel, G. Assman. 1984. Abnormal lecithin:cholesterol acyltransferase activation by a human apolipoprotein A-I variant in which a single lysine residue is deleted. *J. Biol. Chem.* 259: 10063-10070.

74. Banka, C.L., R.S. Smith, D.J. Bonnet, L.K. Curtiss. 1990. Localization of an apolipoprotein A-I epitope critical for LCAT activation. *J. Biol. Chem.* 266:23776-23892.

75. Stein, O., Y. Stein. 1973. The removal of cholesterol from Landschutz ascites cells by high-density apolipoprotein. *Biochim. Biophys. Acta* 326:232-244.

76. Nakai, T., P.S. Otto, D.L. Kennedy, T.F. Whayne Jr. 1976. Rat high density lipoprotein subfraction (HDL3) uptake and catabolism by isolated rat liver parenchymal cells. *J. Biol. Chem.* 251:4914-4921.

77. Drevon, C.A., T. Berg, K.R. Norum. 1977. Uptake and degradation of cholesteryl ester-labeled rat plasma lipoproteins in purified rat hepatocytes and non-parenchymal liver cells. *Biochim. Biophys. Acta* 487:122-136.

78. Ose, L., I. Roken, K.R. Norum, C.A. Drecon, T. Berg. 1981. The binding of high density lipoproteins to isolated rat hepatocytes. *Scand. J. Clin. Invest.* 41:63-73.

79. Miller, N.E., D.B. Weinstein, D. Steinberg. 1977. Binding, internalization and degradation of high density lipoprotein by culture normal human fibroblasts. *J. Lipid Res.* 18:438-450.

80. Kovanen, P.T., W.J. Schneider, G.M. Hillman, J.L. Goldstein, M.S. Brown. 1979. Separate mechanism for the uptake of high and low density lipoprotein by mouse adrenal gland in vivo. *J. Biol. Chem.* 2 54:5498-5505.

81. Kovanen, P.T., M.S. Brown, J.L. Goldstein. 1979. Increased binding of low density lipoprotein to liver membranes from rats treated with i-a-ethinyl estradiol. *J. Biol. Chem.* 254:11367-11373.

82. Hui, D.Y., T.L. Innerarity, R.W. Mahley. 1981. Lipoprotein binding to canine hepatic membranes. Metabolically distinct apo-E and apo-B,E receptors. *J. Biol. Chem.* 256:5646-5654.

83. Chen, Y.D., F.B. Kroemer, G.M. Reaven. 1980. Identification of specific high density lipoprotein--binding sites in rat testis and regulation of binding by human chorionic gonadotropin. *J. Biol. Chem.* 255:9162-9167.

84. Rifici, V.A., H.A. Eder. 1984. A hepatocyte receptor for high-density lipoproteins specific for apolipoprotein A-I. *J. Biol. Chem.* 259:13814-13818.

85. Fidge, N.H., P.J. Nestel. 1985. Identification of apolipoprotein involved in the interaction of human high density lipoprotein 3 with receptors on cultured cells. *J. Biol. Chem.* 260:3570-3575.

86. Hwang, J., K.M.J. Memon. 1985. Binding of apolipoprotein A-I and A-II after recombination with phospholipid vesicles to the high density lipoprotein receptor of luteinized rat ovary. *J. Biol. Chem.* 60:5660-5668.

87. Steinmetz, A., R. Barbaras, N. Ghalim, V. Clavey, J.C. Fruchart, G. Ailhaud. 1990. Human apoli poprotein A-IV binds to apolipoprotein A-I/A-II receptor sites and promotes cholesterol efflux from adipose cells. *J. Biol. Chem.* 265:7859-7863.

88. Tabas, I., A.R. Tall. 1984. Mechanism of the association of HDL3 with endothelial cells, smooth muscle cells and fibroblasts. *J. Biol. Chem.* 259:13897-13905.

89. Biesbroeck, R., M.F. Oram, J.J. Albers, E.L. Bierman. 1983. Specific high-affinity binding of high density lipoproteins to cultured human skin fibroblasts and arterial smooth muscle cells. *J. Clin. Invest.* 71:525-539.

90. Slotte, J.P., J. F. Oram, E.L. Bierman. 1987. Binding of high density lipoproteins to cell receptors promotes translocation of cholesterol from intracellular membranes to the cell surface. *J. Biol. Chem.* 262:12904-12907.

91. Graham, D.L., J.F. Oram. 1987. Identification and characterization of a high density lipoprotein-binding protein in cell membranes by ligand blotting. *J. Biol. Chem.* 262:7439-7442.

92. Schmitz, G., G. Assmann, H. Robenek, B. Brennhausen. 1985. Tangier disease: A disorder of intracellular membrane traffic. *Proc. Natl. Acad. Sci. USA* 82:6305-6309.

93. Glomset, J.A. 1980. High-density lipoproteins in human health and disease. *Adv. Intern. Med.* 25:91-116.

94. Barbaras, R., X. Collett, H. Chap, and B. Perret. 1994. Specific binding of free apolipoprotein A-I to a high-affinity binding site on HepG2 cells: Characterization of two high-density lipoprotein sites. *J. Am. Chem. Soc.* 33:2335-2340.

95. Leblond, L.Y., L. Marcel. 1991. The amphipathic alpha-helical repeats of apolipoprotein A-I are responsible for binding of high density lipoproteins to HepG2 cells. *J. Biol. Chem.* 266:6058-6067.

96. Eisenberg, S. 1984. High density lipoprotein metabolism. *J. Lipid Res.* 25:1017-1058.

97. Johnson, W.J., M.J. Bamberger, R.A. Latta, P.E. Rapp, M.C. Phillips, G.H. Rothblat. 1986. The bidirectional flux of cholesterol between cells and lipoproteins: Effects of phospholipid depletion of high density lipoprotein. *J. Biol. Chem.* 261:5766-5776.

98. Phillips, M.C., W.J. Johnson, G.H. Rothblat. 1987. Mechanisms and consequences of cellular cholesterol exchange and transfer. *Biochim. Biophys. Acta* 906:223-276.

99. Johnson, W.J., G.K. Chacko, M.C. Phillips and G.H. Rothblat. 1990. The efflux of lysosomal cholesterol from cells. *J. Biol. Chem.* 265:5546-5553.

100. Bielicki, J.K., W.J. Johnson, R.B. Weinberg, J.M. Glick, G.H. Rothblat. 1992. Efflux of lipid from fibroblasts to apolipoproteins: Dependence on elevated levels of cellular unesterified cholesterol. *J. Lipid Res.* 33:1699-1709.

101. Monaco, L., H.M. Bond, K.E. Howell, R. Cortese. 1987. A recombinant apoA-1-protein A hybrid reproduces the binding parameters of HDL to its receptor. *EMBO J.* 6:3253-3260.

102. Mendel, C.M., S.T. Kunitake, J.P. Kane, E. Kempner. 1988. Radiation inactivation of binding sites for high density lipoproteins in human fibroblast membranes. *J. Biol. Chem.* 263:1314-1319.

103. Breckenridge, W.C., J. A. Little, P. Alaupovic, C. S. Wang, A. Kuksis, G. Kakis, F. Lindgren, and G. Gardiner. 1982. Lipoprotein abnormalities associated with a familial deficiency of hepatic lipase. *Atherosclerosis* 45:161-179.

104. Koizumi, J., H. Mabuchi, and A. Yoshimura. 1985. Deficiency of serum cholesteryl-ester transfer activity in patients with familial hyperalphalipoproteinaemia. *Atherosclerosis* 58:175-186.

105. Imazu, A., M.L. Brown, C.B. Hesler, L.B. Agellon, J. Koizumi, K. Takata, Y. Maruhama, H. Mabuchi, and A.R. Tall. 1990. Increased high-density lipoprotein levels caused by a common cholesteryl-ester transfer protein gene mutation. *New Engl. J. Med.* 323:1234-1238.

106. Babirak, S.P., P. H. Iverius, W. Y. Fujimoto, and J. D. Brunzell. 1989. The detection of the heterozy gote state for lipoprotein lipase deficiency. *Arterioscler.* 9:326-334.

107. Weisgraber, K.H., T. P. Bersot, R. W. Mahley, G. Franceschini, and C. R. Sirtori. 1980. A-I Milano apoprotein. Isolation and characterization of a cysteine-containing variant of the A-I apoprotein from human high density lipoproteins. *J. Clin. Invest.* 66:901-907.

108. Franceschini, G., C. R. Sirtori, A. Capurso, K. H. Weisgraber, and R. W. Mahley. 1980. Decreased high density lipoprotein cholesterol levels with significant lipoprotein modification and without clinical atherosclerosis in an Italian family. *J. Clin. Invest.* 66:892-900.

109. Franceschini, G., M. Baio, L. Calabresi, C. R. Sirtori, and M. C. Cheung. 1990. Apolipoprotein AI Milano. Partial lecithin:cholesterol acyltransferase deficiency due to low levels of a functional enzyme. *Biochim. Biophys. Acta* 1043:1-6.

110. Weisgraber, K.H., S. C. Rall,Jr., T. P. Bersot, R. W. Mahley, G. Franceschini, and C. R. Sirtori. 1983. Apolipoprotein A-IMilano. Detection of normal A-I in affected subjects and evidence for a cysteine for arginine substitution in the variant A-I. *J. Biol. Chem.* 258:2508-2513.

111. Rall, S.C. Jr., K. H. Weisgraber, R. W. Mahley, Y. Ogawa, C. J. Fielding, G. Utermann, J. Haas, A. Steinmetz, H. J. Menzel, and G. Assmann. 1984. Abnormal lecithin:cholesterol acyltransferase activation by a human apolipoprotein A-I variant in which a single lysine residue is deleted. *J. Biol. Chem.* 259: 10063-10070.

112. Utermann, G., G. Feussner, G. Franceschini, J. Haas, and A. Steinmetz. 1982. Genetic variants of group A apolipoproteins. Rapid methods for screening and characterization without ultracentrifugation. *J. Biol. Chem.* 257:501-507.558.

113. Funke, H., A. von Eckardstein, P. H. Pritchard, M. Karas, J. J. Albers, and G. Assmann. 1991. A frameshift mutation in the human apolipoprotein A-I gene causes high density lipoprotein deficiency, partial lecithin: cholesterol-acyltransferase deficiency, and corneal opacities. *J. Clin. Invest.* 87: 371-376.

114. Deeb, S.S., M.C. Cheung, R.L. Peng, A.C. Wolf, R. Stern, J.J. Albers, and R.H. Knopp. 1991. A mutation in the human apolipoprotein A-I gene. Dominant effect on the level and characteristics of plasma high density lipoproteins. *J. Biol. Chem.* 266:13654-13660.

115. Menzel, H.J., G. Assmann, S. C. Rall,Jr., K. H. Weisgraber, and R. W. Mahley. 1984. Human apolipoprotein A-I polymorphism. Identification of amino acid substitutions in three electrophoretic variants of the Munster-3 type. *J. Biol. Chem.* 259:3070-3076.

116. Rall, S.C., K. H. Weisgraber, R. W. Mahley, O. Ehnholm, D. Schamaun, B. Olaisen, J. P. Blomhoff, and P. Teisberg. 1986. Identification of homozygosity for a human apolipoprotein A-I variant. *J. Lipid Res.* 27:436-441.

117. Strobl, W., H. U. Jabs, M. Hayde, T. Holzinger, G. Assmann, and K. Widhalm. 1988. Apolipoprotein A-I (Glu 198----Lys): a mutant of the major apolipoprotein of high-density lipoproteins occurring in a family with dyslipoproteinemia. *Pediatr. Res.* 24:222-228.

118. Nichols, W.C., F. E. Dwulet, J. Liepnieks, and M. D. Benson. 1988. Variant apolipoprotein AI as a major constituent of a human hereditary amyloid. *Biochem. Biophys. Res. Commun.* 156:762-768.

119. Nichols, W.C., R. E. Gregg, H. B. Brewer,Jr., and M. D. Benson. 1990. A mutation in apolipoprotein A-I in the Iowa type of familial amyloidotic polyneuropathy. *Genomics* 8:318-323.

120. Jeenah, M., A. Kessling, N. Miller, and S. Humphries. 1990. G to A substitution in the promoter region of the apolipoprotein AI gene is associated with elevated serum apolipoprotein AI and high density lipoprotein cholesterol concentrations. *Molec. Biol. Med.* 7:233-241.

121. Matsunaga, T., Y. Hiasa, H. Yanagi, T. Maeda, N. Hattori, K. Yamakawa, Y. Yamanouchi, I. Tanaka, T. Obara, and H. Hamaguchi. 1991. Apolipoprotein A-I deficiency due to a codon 84 nonsense mutation of the apolipoprotein A-I gene. *Proc. Natl. Acad. Sci USA* 88:2793-2797.

122. Breslow, J.L. 1989. Familial disorders of high density lipoprotein metabolism. In "The Metabolic Basis of Inherited Disease". C. R. Scriver, A. I. Beaudet, W. S. Sly, and D. Valle, editors. McGraw-Hill, New York. 1251-1266.

123. Schonfeld, G. 1990. The genetic dyslipoproteinemias-nosology update 1990. *Atherosclerosis* 81:81-93.

124. von Eckardstein, A., H. Funke, A. Henke, K. Altland, A. Benninghoven, and G. Assmann. 1989. Apolipoprotein A-I variants..Naturally occurring substitutions of proline residues affect plasma concentration of apolipoprotein A-I. *J. Clin. Invest.* 84:1722-1730.

125. Jonas, A., A. von Eckardstein, K. E. Kězdy, A. Steinmetz, and G. Assmann. 1991. Structural and functional properties of reconstituted high density lipoprotein discs prepared with six apolipoprotein A-I variants. *J. Lipid Res.* 32:97-106.

126. Schaefer, E.J., J.M. Ordovas, S.W. Law, G. Ghiselli, M.L. Kashyap, L.S. Srivastava, W.H. Heaton, J.J. Albers, W.E. Connor, F.T. Lindgren, Y. Lemeshev, J.P. Segrest, and H.B. Brewer Jr. 1985. Familial apolipoprotein A-I and C-III deficiency, variant II. *J. Lipid Res.* 26:1089-1101.

127. Ordovas, J.M., D.K. Cassidy, F. Civeira, C.L. Bisgaier, and E.J. Schaefer. 1989. Familial apoli poprotein A-I, C-III, and A-IV deficiency and premature atherosclerosis due to deletion of a gene complex on chromosome 11. *J. Biol. Chem.* 264:16339-16342.

128. Norum R.A. 1982. Familial deficiency of apolipoproteins AI and CIII and precocious coronary artery disease. *New Engl. J. Med.* 306:1513-1519.

129. Karathanasis, S.K., E. Ferris, and I. A. Haddad. 1987. DNA inversion within the apolipoproteins AI/CIII/AIV-encoding gene cluster of certain patients with premature atherosclerosis. *Proc. Natl. Acad. Sci USA* 84:7198-7202.

130. Karathanasis S.K., V.I. Zannis, J.L. Breslow. 1983. A DNA insertion has occurred in the apoli poprotein A-I gene of patients with premature atherosclerosis. *Nature* 305:823-825.

131. Karathanasis, S.K., McPherson, J., Zannis, V.I., Breslow, J.L. 1983. Linkage of human apolipoprotein A-I and C-III genes. *Nature* 304:371-373.

132. Papazafiri, P., Ogami, K., Ramji, D.P., Nicosia, A., Monaci, P., Cladaras, C., and Zannis, V.I. 1991. Promoter elements and factors involved in hepatic transcription of the human apoA-I gene positive and negative regulators bind to overlapping sites. *J. Biol. Chem.* 266:5790.

133. Zannis, V.I., D. Kardassis, P. Cardot, M. Hadzopoulou-Cladaras, E.E. Zanni and C. Cladaras. 1992. Molecular biology of the human apolipoprotein genes: Gene regulation and structure-function relationship. *Curr. Opin. Lipid.* 3:96-113.

134. Zannis, V.I., C. Cladaras, and I. Talianidis. 1994. The role of the hormone response elements (HRE) and the apoCIII enhancer on the transcriptional regulation of genes involved in lipid transport. Xth International Symposium on Atherosclerosis. Montreal 1994. (in press).

135. Tzameli, E., C. Cladaras, and V.I. Zannis. 1993. Effect of mutations in the nuclear receptor response element of apoA-I on transcription and the binding of homo- and heterodimeric complexes of nuclear receptors. *Circulation* 88:422.

136. Walsh, A., N. Azrolan, K. Wang, A. Marcigliano, A. O'Connel, and J.L. Breslow. 1993. Intestinal expression of the human apoA-I gene is controlled by a DNA region, 3′ to the gene in the promoter of the adjacent convergently transcribed apoCIII gene. *J. Lipid Res.* 34:617.

137. Rall, S.C., K. H. Weisgraber and R. W. Mahley. 1981. Human apolipoprotein E: The complete amino acid sequence. *J. Biol. Chem.* 257:4171-4178.

138. Zannis, V.I., J. McPherson, G. Goldberger, S.K. Karathanasis and J.L. Breslow. 1984. Synthesis, intracellular processing and signal peptide of human apoE. *J. Biol. Chem.* 259:5495-5499.

139. Paik, Y.K., D. J. Chang, C. A. Reardon, G. E. Davies, R. W. Mahley and J. M. Taylor. 1985. Nucleo tide sequence and structure of the human apolipoprotein E gene. *Proc. Natl. Acad. Sci USA* 82:3445-3451.

140. Das, H.K., J. McPherson, G.A.P. Bruns, S.K. Karathanasis, and J.L. Breslow. 1985. Isolation, characterization and mapping to chromosome 19 of the human apolipoprotein E gene. *J. Biol. Chem.* 260:6240-6246.

141. McLean, J.W., N.A. Elshourbagy, D.J. Chang, R.W. Mahley, and J.M. Taylor. 1984. Human apolipoprotein E mRNA. cDNA cloning and nucleotide sequencing of a new variant. *J. Biol. Chem.* 259:6498-7504.

142. Wetterau, J.R., L.P. Aggerbeck, S.C. Rall, Jr. and K.H. Weisgraber. 1988. Human apolipoprotein E3 in aqueous solution. I. Evidence for two structural domains. *J. Biol. Chem.* 263:6240-6248.

143. Aggerbeck, L.P., J.R. Wetterau, K.H. Weisgraber, C.S.C. Wu and F.T. Lindgren. 1988. Human apolipoprotein E3 in aqueous solution. II. Properties of the amino- and carboxyl-terminal domains. *J. Biol. Chem.* 263:6249-6258.

144. Innerarity, T.L., E.J. Friedlander, S.C. Rall, Jr., K.H. Weisgraber and R.W. Mahley. 1983. The receptor-binding domain of human apolipoprotein E. Binding of apolipoprotein E fragments. *J. Biol. Chem.* 258:12341-12347.

145. Weisgraber, K.H., T.L. Innerarity, K.J. Harder, R.W Mahley, R.W. Milne, Y.L. Marcel and J.T. Sparrow. 1983. The receptor-binding domain of human apolipoprotein *E. J. Biol. Chem.* 258:12348-12354.

146. Zannis, V.I., C. Cladaras and E.E. Zanni. 1991. Apolipoprotein and lipoprotein synthesis and modifica tions. *Current Opin. Lipid.* 2:149-155.

147. Newman, T.C., P.A. Dawson, L.L. Rudel., and D.L. Williams. 1985. Quantitation of apolipoprotein E mRNA in the liver and peripheral tissues of nonhuman primates. *J. Biol. Chem.* 260:2452-2457.

148. Pitas, R.E., T.L. Innerarity and R.W. Mahley. 1980. Cell surface receptor binding of phospholipid-pro tein complexes containing different ratios of receptor-active and inactive E apoprotein. *J. Biol. Chem.* 255:5454-5460.

149. Lalazar, A., K.H. Weisgraber, S.C. Rall Jr., H. Giladi, T.L. Innerarity, A.Z. Levanon, J.K. Boyles, B. Amit, M. Gorecki and R.W. Mahley. 1988. Site-specific mutagenesis of human apolipoprotein E. Receptor binding activity of variants with single amino acid substitutions. *J. Biol. Chem.* 263:3542-3545.

150. Lalazar, A. and R.W. Mahley. 1989. Human apolipoprotein E. Receptor binding activity of truncated variants with carboxyl-terminal deletions. *J. Biol. Chem.* 264:8447-8450.

151. Cladaras, C., M. Hadzopoulou-Cladaras, R.T. Nolte, D. Atkinson and V.I. Zannis (1986) The complete sequence and structural analysis of human apolipoprotein B-100: Relationship between apoB-100 and apoB-48 forms. *EMBO J.* 5:3495-3508.

152. Weisgraber, K.H., S.C. Rall, Jr., R.W. Mahley, R.W. Milne, Y.L. Marcel, and J.T. Sparrow. 1986. Human apolipoprotein E. *J. Biol. Chem.* 261:2068-2076.

153. Cardin, A.D., N. Hirose, D.T. Blankenship, R.L. Jackson, and J.A.K. Harmony. 1986. Binding of a high reactive heparin to human apolipoprotein E: Identification of two heparin-binding domains. *Biochem. Biophys. Res. Commun.* 134:783-789.

154. Weisgraber K.H. 1990. Apolipoprotein E distribution among human plasma lipoproteins: role of the cysteine-arginine interchange at residue 112. *J. Lipid Res.* 31:1503-1511

155. Gregg R.E., L.A. Zech, E.J. Schaefer, D. Stark, D. Wilson, H.B. Brewer Jr. 1986. Abnormal in vivo metabolism of apolipoprotein E4 in humans. *J. Clin. Invest.* 78:815-821.

156. Hefele Wald J., Goormaghtigh E., De Meutter J., Ruysschaert J.M. and Jonas A. 1990. Investigation of the lipid domains and apolipoprotein orientation in reconstituted high density lipoproteins by fluorescence and I.R. methods. *J. Biol. Chem.* 265: 20044-20050

157. Lund-Katz S., Weisgraber K.H., Mahley R.W. and Phillips M.C. 1993. Conformation of apolipoprotein E. *J. Biol. Chem.* 268:23008-23015.

158. Pitas, R.E., J.K. Boyles, S.H. Lee, D. Foss and R.W. Mahley. 1987. Astrocytes synthesize apolipoprotein E and metabolize apolipoprotein E-containing lipoproteins. *Biochim. Biophys. Acta* 917:148-161.

159. Mahley, R.W. 1988. Apolipoprotein E: Cholesterol transport protein with expanding role in cell biology. *Science* 240:622-630.

160. Muller, H.W., M.J. Ignatius, D.H. Hanagen and E.M. Shooter. 1986. Expression of specific sheath cell proteins during peripheral nerve growth and regeneration in mammals. *J. Cell. Biol.* 102:393-402.

161. Ignatius, M.J., P.J. Gebicke-Harter, J.H.P. Skene, J.W. Schilling, K.H. Weisgraber, R.W. Mahley and E.M. Shooter. 1986. Expression of apolipoprotein E during nerve degeneration and regeneration. *Proc. Natl. Acad. Sci USA* 83:1125-1129.

162. Pitas, R.E., J.K. Boyles, S.H. Lee, D. Hui and K.H. Weisgraber. 1987. Lipoproteins and their receptors in the central nervous system. *J. Biol. Chem.* 262:14352-14360.

163. Boyles, J., C.D. Zoellner, L.J. Anderson, L.M. Kosik, R.E. Pitas, K.H. Weisgraber, D.Y. Hui, R.W. Mahley, P.J. Gebicke-Harter, M.J. Ignatius and E.M. Shooter. 1989. A role of apolipoprotein E, apolipoprotein A-I, and low density lipoprotein receptors in cholesterol transport during regeneration and remyelination of the rat sciatic nerve. *J. Clin. Invest.* 83:1015-1031.

164. Gwynne, J.T. and B. Hess. 1980. The role of high density lipoproteins in rat adrenal cholesterol metabolism and steroidogenesis. *J. Biol. Chem.* 255:10875-10883.

165. Gordon, V., T.L. Innerarity and R.W. Mahley. 1983. Formation of cholesterol- and apoprotein E-enriched high density lipoproteins in vitro. *J. Biol. Chem.* 258:6202-6212.

166. Kowal, R.C., J. Herz, K.H. Weisgraber, R.W. Mahley, M.S. Brown, J.L. Goldstein. 1990. Opposing effects of apo-E and C on lipoprotein binding to low density lipoprotein receptor-related protein. *J. Biol. Chem.* 265:10771-10779.

167. Weisgraber, K.H., R.W. Mahley, R.C. Kowal, J. Herz, J.L. Goldstein and M.S. Brown. 1990. Apolipoprotein C-I modulates the interaction of apolipoprotein E with beta-migrating very low density lipoprotein (beta-VLDL) and inhibits binding of beta-VLDL to low density lipoprotein receptor-related protein. *J. Biol. Chem.* 265:22453-22459.

168. Reyland, M.E., J.T. Gwynne, P. Forgez, M.M. Prack and D.L. Williams. 1991. Expression of the human apolipoprotein E gene suppresses steroidogenesis in mouse Y1 adrenal cells. *Proc. Natl. Acad. Sci USA* 88:2375-2379.

169. Ehnholm, C., R.W. Mahley, D.A. Chappell, K.H. Weisgraber, E. Ludwig, and J.L. Witztum. 1984. Role of apolipoprotein E in the lipolytic conversion of beta-very low density lipoproteins to low density lipoproteins in type III hyperlipoproteinemia. *Proc. Natl. Acad. Sci. USA* 81:5566-5570.

170. Zannis, V.I., P. W. Just and J. L. Breslow. 1981. Human apolipoprotein E isoprotein subclasses are genetically determined. *Am. J. Hum. Genet.* 33:11-24.

171. Zannis, V.I., J. L. Breslow, G. Utermann, R. W. Mahley, K. H. Weisgraber, R. J. Havel, J. L. Goldstein, M. S. Brown, G. Schonfeld, W. R. Hazzard and C. B. Blum. 1982. Proposed nomenclature of apoE isoprotein genotypes and phenotypes. *J. Lipid Res.* 23:911-914.

172. Zannis, V.I. and J. L. Breslow. 1981. Human very low density lipoprotein apolipoprotein E isoprotein polymorphism is explained by genetic variation and post-translational modification. *Biochemistry* 20:1033-1041.

173. Davignon, J., R.E. Gregg and C.F. Sing. 1988. Apolipoprotein E polymorphism and atherosclerosis. *Arterioscler.* 8:1-21.

174. Breslow, J.L. and V. I. Zannis. 1986. Genetic variation in apolipoprotein E and type III hyperlipoproteinemia. *Arterioscler. Rev.* 14:119-141.

175. Zannis, V.I. 1989. Molecular biology of human apolipoproteins B and E and associated diseases of lipoprotein metabolism. In: "Advances in Lipid Research". Vol. 23, pp. 1-64, Academic Press, San Diego, CA.

176. Utermann, G. 1985. In "Diabetes, Obesity and Hyperlipidemias". G. Crepaldi, A. Tiengo and G. Baggio, editors. Excerpta Medica, Amsterdam. 1-28.

177. Utermann, G. 1985. ApoE mutants, hyperlipidemia and arteriosclerosis. *Adv. Exp. Med. Biol.* 183:173-188.

178. Sing, C.F. and J. Davignon. 1985. Role of the apolipoprotein E polymorphism in determining normal plasma lipid and lipoprotein variation. *Am. J. Hum. Genet.* 37:268-285.

179. Assmann, G., G. Schmitz, H. J. Menzel, and H. Schultz. 1984. Apolipoprotein E polymorphism and hyperlipidemia. *Clin. Chem.* 30:641-643.

180. Lussier-Cacan, S., D. Bouthillier, and J. Davignon. 1985. ApoE allele frequency in primary endogenous hypertriglyceridemia (type IV) with and without hyperapobetalipoproteinemia. Arterioscler. 5:639-643.

181. Eto, M., K. Watanabe, Y. Iwashima, A. Morikawa, E. Oshima, M. Sekiguchi, and K. Ishii. 1986. Apolipoprotein E polymorphism and hyperlipemia in type II diabetics. *Diabetes* 35:1374-1382.

182. Leren, T.P., A. L. Borresen, K. Berg, P. Hjermann, and P. Leren. 1985. Increased frequency of the apolipoprotein E-4 isoform in male subjects with multifactorial hypercholesterolemia. *Clin. Genet.* 27:458-462.

183. Utermann, G., A. Hardewig, and F. Zimmer. 1984. Apolipoprotein E phenotypes in patients with myocardial infarction. *Human. Genetics* 65:237-241.

184. Goldstein, J.L. and M.S. Brown. 1983. Familial hypercholesterolemia. In: "The Metabolic Basis of Inherited Disease". (J.B. Stanbury, J. Wyngaarden and D.S. Fredrickson, eds.) McGraw-Hill, New York. pp. 672-712.

185. Mahley, R.W. and S.C. Rall, Jr. 1989. Type III hyperlipoproteinemia (Dysbetalipoproteinemia): The role of apolipoprotein E in normal and abnormal lipoprotein metabolism. In: "The Metabolic Basis of Inherited Disease". (C.R. Scriver, A.I. Beaudet, W.S. Sly and D. Valle, eds.) McGraw-Hill, New York. pp. 1195-1211.

186. Breslow, J.L., V.I. Zannis, R.R. SanGiacomo, J.L.H.C. Third, T. Tracy and C.J. Glueck. 1982. Studies of familial type III hyperlipoproteinemia using as a genetic marker the apoE phenotype E2/2. *J. Lipid Res.* 23:1224-1235.

187. Havel, R.J., L. Kotite, J.L. Vigne, J.P. Kane, P. Tun, N. Phillips and G.C. Chen. 1980. Radioimmunoassay of human arginine-rich apolipoprotein, apoprotein E. Concentration in blood plasma and lipoproteins as affected by apoprotein E-3 deficiency. *J. Clin. Invest.* 66:1351-1362.

188. Havel, R.J. and J.L. Kane. 1973. Primary dysbetalipoproteinemia: Predominance of a specific apoprotein species in triglyceride-rich lipoproteins. *Proc. Natl. Acad. Sci USA* 70:2015-2019.

189. Mahley, R.W., T.L. Innerarity, S.C. Rall, Jr., K.H. Weisgraber and J.M. Taylor. 1990. Apolipoprotein E: genetic variants provide insights into its structure and function. *Curr. Opin. Lipid* 1:87-95.

190. Bersot, T.P., T.L. Innerarity, R.W. Mahley and R.J. Havel. 1983. Cholesteryl ester accumulation in mouse peritoneal macrophages induced by beta-migrating very low density lipoproteins from patients with atypical dysbetalipoproteinemia. *J. Clin. Invest.* 72(3):1024-1033.

191. Havel, R.J., L. Kotite, J.P. Kane, P. Tun and T. Bersot. 1983. Atypical familial dysbetalipoproteinemia associated with apolipoprotein phenotype E3/3. *J. Clin. Invest.* 72:379-387.

192. Wardell, M.R., S.O. Brennan, E.D. Janus, R. Fraser, and R.W. Carrell. 1987. Apolipoprotein E2-Christchurch (136Arg→Ser). New variant of human apolipoprotein E in a patient with type III hyperlipoproteinemia. *J. Clin. Invest.* 80:483-490.

193. Rall, S.C. Jr., Y.M. Newhouse, H.R.G. Clarke, K.H. Weisgraber, B.J. McCarthy, R.W. Mahley and T.P. Bersot. 1989. Type III hyperlipoproteinemia associated with apolipoprotein E phenotype E3/3: structure and genetics of an apolipoprotein E3 variant. *J. Clin. Invest.* 83:1095-1101.

194. Mann, W.A., R.E. Gregg, D.L. Sprecher, and H.B. Brewer, Jr. 1989. Apolipoprotein E-1Harrisburg: a new variant of apolipoprotein E dominantly associated with type III hyperlipoproteinemia. *Biochim. Biophys. Acta* 1005:239-244.

195. Wardell, M.R., K.H. Weisgraber, L.M. Havekes, and S.C. Rall, Jr. 1989. Apolipoprotein E3-Leiden contains a seven-amino acid insertion that is a tandem repeat of residues 121-127. *J. Biol. Chem.* 264:21205-21210.

196. van den Maagdenberg, A.M., P. de Knijff, A.F. Stalenhoef, J.A. Gevers Leuven, L.M. Havekes and R.R. Frants. 1989. Apolipoprotein E*3-Leiden allele results from a partial gene duplication in exon 4. *Biochem. Biophys. Res. Commun.* 165:851-857.

197. Smit, M., P. de Knijff, E. van der Kooij-Meijs, C. Groenendijk, A.M. van den Maagdenberg, J.A. Gevers Leuven, A.F. Stalenhoef, P.M. Stuyt, R.R. Frants and L.M. Havekes. 1990. Genetic heterogeneity in familial dysbetalipoproteinemia. The E2(lys146----gln) variant results in a dominant mode of inheritance. *J. Lipid Res.* 31:45-53.

198. Weisgraber, K.H., S.C. Rall Jr., T.L. Innerarity and R.W. Mahley. 1984. A novel electrophoretic variant of human apolipoprotein E: Identification and characterization of apolipoprotein E1. *J. Clin. Invest.* 73:1024-1033.

199. Steinmetz, A., N. Assefbarkhi, C. Eltze, K. Ehlenz, H. Funke, A. Pies, G. Assmann, and H. Kaffarnik. 1990. Normolipemic dysbetalipoproteinemia and hyperlipoproteinemia type III in subjects homozygous for a rare genetic apolipoprotein E variant (apoE1). *J. Lipid Res.* 31:1005-1013.

200. Wardell, M.R., S.C. Rall Jr., E.J. Schaefer, J.P. Kane and K.H. Weisgraber. 1991. Two apolipoprotein E5 variants illustrate the importance of the position of additional positive charge on receptor-binding activity. *J. Lipid Res.* 32:521-528.

201. Maeda, H., H. Nakamura, S. Kobori, M. Okada, H. Mori, H. Niki, T. Ogura, and S. Hiraga. 1989. Identification of human apolipoprotein E variant gene: apolipoprotein E7 (Glu244,245→Lys 244,245). *J. Biochem.* 105:51-54.

202. Tajima S., T. Yamamura, M. Menju, A. Yamamoto. 1989. Analysis of apolipoprotein E7 (apolipoprotein E-Suita) gene from a patient with hyperlipoproteinemia. *J. Biochem.* 105:249-253.

203. Lohse P., W.A. Mann, E.A. Stein, H.B. Brewer Jr. 1991. Apolipoprotein E-4 Philadelphia (Glu13-Lys, Arg145-Cys). Homozygosity for two rare point mutations in the apolipoprotein E gene combined with severe type III hyperlipoproteinemia. *J. Biol. Chem.* 266:10479-10484.

204. Mailly, F., C. F. Xu, M. Xhignesse, S. Lussier-Cacan, P. J. Talmud, J. Davignon, S. E. Humphries, A. C. Nestruck. 1991. Characterization of a new apolipoprotein E5 variant detected in two French-Canadian subjects. *J. Lipid Res.* 32:613-620.

205. van den Maagdenberg, A.M., W. Weng, I.H. de Bruijn, P. de Knijff, H. Funke, A.H. Smelt, J.A. Gebers Leuven, F.M. van't Hooft, G. Assmann, M.H. Hofker. 1993. Characterization of five new mutants in the carboxyl-terminal domain of human apolipoprotein E: No cosegregation with severe hyperlipidemia. *Am. J. Hum. Genet.* 52:937-946.

206. Ruzicka, V., W. Marz, A. Russ, E. Fisher, W. Mondorf, W. Gross. 1993. Characterization of the gene for apolipoprotein E5-Frankfurt (Gln71→Lys, Cys112→Arg) by polymerase chain reaction, restriction isotyping, and temperature gradient gel electrophoresis. *Electrophoresis* 14:1032-1037.

207. Innerarity T.L., D.Y. Hui, T.P. Bersot, R.W. Mahley. 1986. Type III hyperlipoproteinemia: A focus on lipoprotein receptor-apolipoprotein E2 interactions. *Adv. Exp. Med. Biol.* 201:273-288.

208. Fredrickson, D.S. and R.I. Levy. 1972. Familial hyperlipoproteinemia in: "The Metabolic Basis of Inherited Disease". (J.B. Stanbury, J.B. Wyngaarden, and D.S. Fredrickson, eds.), McGraw-Hill, New York, pp. 545-614.

209. Utermann, G., K. H. Vogelberg, A. Steinmetz, W. Schoenborn, N. Pruin, M. Jaeschke, M. Hees and H. Canzler. 1979. Polymorphism of apolipoprotein E. II. Genetics of hyperlipoproteinemia type III. *Clin. Genet.* 15:37-62.

210. Hazzard W.R., G.R. Warnick, G. Utermann, J.J. Albers. 1981. Genetic transmission of isoapolipoprotein E phenotypes in a large kindred: Relationship to dysbetalipoproteinemias and hyperlipidemia. Metabolism: *Clinical and Experimental* 30:79-89.

211. Marien J.K., H.A. Hulsmans, C.M. vanGent. 1974. On a family with coexistence of phenotypes II and III hyperlipoproteinemia. *Acta Med. Scand.* 196:149-153.

212. Morganroth, J., Levy, R.I., and Fredrickson, D.S. 1975. The Biochemical, Clinical, and Genetic Features of Type III Hyperlipoproteinemia. *Ann. Intern. Med.* 82:158-174.

213. Hazzard, W.R., and Bierman, E.L. 1972. Aggravation of broad-B disease (type III hyperlipoproteinemia) by hypothyroidism. *Arch. Intern. Med.* 130:822-828.

214. Ghiselli, G., E.J. Schaefer, P. Gascon, and H.B. Breser Jr. 1981. Type III hyperlipoproteinemia associated with apolipoprotein E deficiency. *Science* 214:1239-1241.

215. Schaefer, E.J., R.E. Gregg, G. Ghiselli, T.M. Forte, J.M. Ordovas, L.A. Zech, and H.B. Brewer. 1986. Familial apolipoprotein E deficiency. *J. Clin. Invest.* 78:1206-1219.

216. Zannis, V.I., J.M. Ordovas, C. Cladaras, F.S. Cole, G. Forbes and E.J. Schaefer. 1985. mRNA and apolipoprotein E synthesis abnormalities in peripheral blood monocyte macrophages in familial apolipoprotein E deficiency. *J. Biol. Chem.* 24:12891-12894.

217. Cladaras, C., M. Hadzopoulou-Cladaras, B.K. Felber, G. Pavlakis, and V.I. Zannis. 1987. The molecular basis of a familial apoE deficinecy. An acceptor splice site mutation in the third intron of the deficient apoE gene. *J. Biol. Chem.* 262:2310-2315.

218. Smith, J.D., Plump, A.S., Hayek, T., Breslow, J.L. 1990. Accumulation of human apolipoprotein E in the plasma of transgenic mice. *J. Biol. Chem*; 265:14709-14712.

219. Shimano H., N. Yamada, M. Katsuki, M. Shimada, T. Gotoda, K. Harada, T. Murase, C. Fukazawa, F. Takaku, Y. Yazaki. 1992. Overexpression of apolipoprotein E in transgenic mice: Marked reduction in plasma lipoprotein except high density lipoprotein and resistance against diet-induced hypercholesterolemia. *Proc. Natl. Acad. Sci. USA* 89:1750-1754.

220. Shimano H., N. Yamada, M. Katsuki, K. Yamamoto, T. Gotoda, K. Harada, M. Shimada, Y. Yazaki. 1992. Plasma lipoprotein metabolism in transgenic mice overexpressing apolipoprotein *E. J. Clin. Invest.* 90:2084-2091.

221. Piedrahita J.A., S.H. Zhang, J.R. Hagaman, P.M. Oliver, N. Maeda. 1992. Generation of mice carrying a mutant apolipoprotein E gene inactivated by gene targeting in embryonic stem cells. *Proc. Natl. Acad. Sci. USA* 89:4471-4475.

222. Van den Maagdenberg A.M.J., M.H. Hofker, P.J.A. Krimpenfort, I. De Bruijn, B. Van Vlijmen, H. van der Boom, L.M. Havekes, R.R. Frants. 1993. Transgenic mice carrying the apolipoprotein E3-Leiden gene exhibit hyperlipoproteinemia. *J. Biol. Chem.* 268:10540-10545.

223. Fazio S., Y. Lee, X. Sheng, S.C. Rall Jr. 1993. Type III hyperlipoproteinemic phenotype in transgenic mice expressing dysfunctional apolipoprotein. *J. Clin. Invest.* 92:1497-1503.

224. Fazio S., Y. Horie, W.S. Simonet, K.H. Weisgraber, J.M. Taylor, S.C. Rall Jr. 1994. Altered lipoprotein metabolism in transgenic mice expressing low levels of a human receptor-binding-defective apolipoprotein E variant. *J. Lipid Res.* 35:408-416.

225. Fazio S., D.A. Sanan, Y-L. Lee, Z-S. Ji, R.W. Mahley, S.C. Rall Jr. 1994. Susceptibility to diet-induced atherosclerosis in transgenic mice expressing a dysfunctional human apolipoprotein E (Arg112, Cys142). *Arterioscler. Thromb.* 14:1873-1879.

226. Wisniewski, T., Golabek, A., Matsubara, E., Ghiso, J. and Frangione, B. (1993) Apolipoprotein E: binding to soluble Alzheimer's β-amyloid. *Biochem. Biophys. Res. Commun.* 192:359-365.

227. Sanan, D.A., K.H. Weisgraber, S.J. Russell, R.W. Mahley, D. Hunng, A. Saunders, D. Schmochol, T. Wianlowski, B. Frangiono, A.D. Roses and W.J. Strittmatter. 1994. Apolipoprotein E associates with β amyloid peptide of Alzheimer's disease to form novel monofibrils. Isoform apoE4 associates more efficiently than apoE3. *J. Clin. Invest.* 94:860-869.

228. Strittmatter, W.J., K.H. Weisgraber, M. Goedert, A.M.Saunders, D. Huang, E.H. Corder, L.-M. Dong, R. Jakes, M.J. Alberts, J.R. Gilbert, S.-H. Han, C. Hulette, G. Einstein, D.E. Schmechel, M.A. Pericak-Vance and A.D. Roses. 1994. Hypothesis: Microtubule instability and paired helical filament formation in the Alzheimer disease brain are related to apolipoprotein E genotype. *Exper. Neurol.* 125:163-171.

229. Nathan, B.P., S. Bellosta, D.A. Sanan, K.H. Weisgraber, R.W. Mahley and R.E. Pitas. 1994. Differential effects of apolipoprotein E3 and E4 on neuronal growth in vitro. *Science* 264:850.

230. Ponte, P., P. Gonzalez-DeWhitt, J. Schilling, J. Miller, D. Hus, B. Greenberg, K. Davis, W. Wallace, I. Lieberburg, F. Fuller and B. Cordell. 1988. A new A4 amyloid mRNA contains a domain homologous to serine proteinase inhibitors. *Nature* 331:525-527.

231. Tanzi, R.E., A.I. McClatchey, E.D. Lamperti, L.L. Villa-Komaroff, J.F. Gueslla, R.L. Neve. 1988. Protease inhibitor domain encoded by an amyloid protein precursor mRNA associated with Alzheimer's disease. *Nature* 331:528-530.

232. Kitaguchi N., Y. Takahashi, Y. Tolkushima, S. Shiojiri, H. Ito. 1988. Novel precursor of Alzheimer's disease amyloid protein shows protease inhibitory activity. *Nature* 331:530-532.

233. Selkoe D.J., M.B. Podisny, C.L. Joachim, E.A. Vickers, G. Lee, T. Oltersdorf. 1988. β-Amyloid precursor protein of Alzheimer disease occurs as 110- to 135-kilodalton membrane-associated proteins in neural and non-neural tissues. *Proc. Natl. Acad. Sci. USA* 85:7341-7345.

234. Esch F.S., P.S. Keim, E.C. Beattie, R.W. Blacher, A.R. Culwell, T. Oltersdorf, D. McClure, P.J. Ward. 1990. Cleavage of amyloid β peptide during constitutive processing of its precursor. *Science* 248:1122-1124.

235. Sisodia S.S., E.H. Koo, K. Beyreuther, A. Unterbeck, D.L. Price. 1990. Evidence that beta-amyloid protein in Alzheimer's disease is not derived by normal processing. *Science* 248:492-495.

236. Goate A., M.-C. Chartier-Harlin, M. Mullan. 1991. Segregation of a missense mutation in the amyloid precursor protein with familial Alzheimer's disease. *Nature* 349:704-706.

237. Chartier-Harlin M.-C., F. Crawford, H. Houlden. 1991. Early-onset Alzheimer's disease caused by mutations at codon 717 of the β amyloid precursor protein gene. *Nature* 353:844-846.

238. Murrell J., M. Farlow, B. Ghetti, M.D. Benson. 1991. A mutation in the amyloid precursor protein associated with hereditary Alzheimer's disease. *Science* 254:97-99.

239. Mullan M., F. Crawford, K. Axelman, H. Houlden, L. Lilius, B. Winblad, L. Lannfeldt. 1993. A pathogenic mutation for probable Alzheimer's disease in the APP gene at the N-terminus of bamyloid. *Nature Genet.* 1:345-347.

240. Levy E., M.D. Carman, I.J. Fernandez-Madrid, M.D. Power, I. Lieberburg, S.G. van Duinen, G.T.A.M. Bots, W. Luyendijk and B. Frangione. 1990. Mutation of the Alzheimer's disease amyloid gene in hereditary cerebral hemorrhage, Dutch-type. *Science* 248:1124-1126.

241. Schellenberg, G.D., T.D. Bird, E.M. Wijsman, H.T. Orr, L. Anderson. E. Nemens, White, Genetic linkage evidence for a familial Alzheimer's disease locus on chromosome 14. *Science* 258:668-671.

242. Pericak-Vance M.A., J.L. Bebout, P.C. Gaskell. 1991. Linkage studies in familial Alzheimer's disease: Evidence for chromosome 19 linkage. *Am. J. Hum. Genet.* 48:1034-1050.

243. Moorjani, S., M. Roy, A. Torres, C. Betard, C. Gagne, M. Lambert, D. Brun, J. Davignon, P. Lupien. 1993. Mutations of low-density-lipoprotein-receptor gene, variation in plasma cholesterol, and expression of coronary heart disease in homozygous familial hypercholesterolemia. *Lancet* 341:1303-1306.

244. Leitersdorf, E., A. Reshef, V. Meiner, E.J. Dann, Y. Beigel, F.G. Roggen, D.R. van der Westhuyzen, G.A. Coetzee. 1993. A missense mutation in the low density lipoprotein receptor gene causes familial hypercholesterolemia in Sephardic Jews. *Hum. Genet.* 91:141-147.

245. Yamakawa-Kobayashi, K., T. Kobayashi, K. Saku, K. Arakawa, H. Hamaguchi. 1993. Two novel frameshift mutations associated with the presence of direct repeats of hte LDL receptor gene in familial hypercholesterolemia. *Hum. Genet.* 92:331-335.

246. Miyake, Y., S. Tajima, T. Funahashi, T. Yamamura, A. Yamamoto. 1992. A point mutation of low-density-lipoprotein receptor causing rapid degradation of the receptor. *Eur. J. Biochem.* 210:1-7.

247. Yamakawa-Kobayashi, K., T. Kobayashi, H. Yanagi, Y. Shimakura, J. Satoh, H. Hamaguchi. 1994. A novel complex mutation in the LDL receptor gene probably caused by the simultaneous occurrence of deletion and insertion in the same region. *Human Genet.* 93:625-628.

248. Koivisto, P.V., U.M. Koivisto, P.T. Kovanen, H. Gylling, T.A. Miettinen, K. Kontula. 1993. Deletion of exon 15 of the LDL receptor gene is associated with a mild form of familial hypercholesterolemia. *Arterioscler. Thromb.* 13:1680-1688.

249. Koivisto, U-M., L. Hamalainen, M-R. Taskinen, K. Kettunen, and K. Kontula. 1993. Prevalence of familial hypercholesterolemia among young North Karelian patients with coronary heart dissease: A study based on diagnosis by polymerase chain reaction. *J. Lipid Res.* 34:269.

250. Bruin, T., S. Tuzgol, W.J. Mulder, A.E. van den Ende, H. Jansen, M.R. Hayden, and J.J. Kastelein. 1994. A compound heterozygote for lipoprotein lipase deficiency, Val69→Leu and Gly188→Glu: Correlation between in vitro LPL activity and clinical expression. *J. Lipid Res.* 35:438-445.

251. Wilson, D.E., A. Hata, L.K. Kwong, A. Lingam, J. Shuhua, D.N. Ridinger, C. Yeager, K.C. Kaltenborn, P.H. Iverius, J.M. Lalouel. 1993. Mutations in exon 3 of the lipoprotein lipase gene segregating in a family with hypertriglyceridemia, pancreatitis, and non-insulin-dependent diabetes. *J. Clin. Invest.* 92:203-211.

252. Bijvoet, S.M., T. Bruin, S. Tuzgol, H.D. Bakker, M.R. Hayden, and J.J. Kastelein. 1994. Homozygosity for a mutation in the lipoprotein lipase gene (Gly139→Ser) causes chylomicronemia in a boy of Spanish descent. *Human Genet.* 93:339-343.

253. Bruin, T., S. Tuzgol, D.E. van Diermen, N. Hoogerbrugge-van der Linden, J.D. Brunzell, M.R. Hayden, J.J. Kastelein. 1993. Recurrent pancreatitis and chylomicronemia in an extended Dutch kindred is caused by a Gly154→Ser substitution in lipoprotein lipase. *J. Lipid Res.* 34:2109-2119.

254. Ma, Y., M.S. Liu, D. Ginzinger, J. Frohlich, J.D. Brunzell, and M.R. Hayden. 1993. Gene-environment interaction in the conversion of a mild-to-severe phenotype in a patient homozygous for a Ser^{172}→Cys mutation in the lipoprotein lipase gene. *J. Clin. Invest.* 91:1953-1958.

255. Haubenwallner, S., G. Horl, N.S. Shachter, E. Presta, S.K. Fried, G. Hofler, G.M. Kostner, J.L. Breslow, and R. Zechner. 1993. A novel missense mutation in the gene for lipoprotein lipase resulting in a highly conservative amino acid substitution (Asp180→Glu) causes familial chylomicronemia (type I hyperlipoproteinemia). *Genomics* 18:392-396.

256. Tenkanen, H., M-R. Taskinen, M. Antikainen, I. Ulmanen, K. Kontula, and C. Ehnholm. 1994. A novel amino acid substitution (His183→Gln) in exon 5 of the lipoprotein lipase gene results in loss of catalytic activity: Phenotypic expressin of the mutant gene in a heterozygous state. *J. Lipid Res.* 35:220-2283

257. Reina, M., J.D. Brunzell, and S.S. Deeb. 1992. Molecular basis of familial chylomicronemia: Mutations in the lipoprotein lipase and apolipoprotein C-II genes. *J. Lipid Res.* 33:1823-1832.

258. Ma, Y., M.S. Leu, D. Chitayat, T. Bruin, U. Beisiegel, P. Benlian, L. Foubert, J.L. DeGennes, H. Funke, I. Forsythe. 1994. Recurrent mis-sense mutations at the first and second base of codon Arg243 in human lipoprotein lipase in patients of different ancestries. *Human Mutation* 3:52-58.

259. Ma, Y., T.C. Ooi, M.S. Liu, H. Zhang, R. McPherson, A.L. Edwards, I.J. Forsythe, J. Frohlich, J.D. Brunzell, and M.R. Hayden. 1994. High frequency of mutations in the human lipoprotein lipase gene in pregnancy-induced chylomicronemia: Possible association with apolipoprotein E2 isoform. *J. Lipid Res.* 35:1066-1075.

260. Kobayashi, J., N. Sasaki, J. Tashiro, H. Inadera, Y. Saito, and S. Yoshida. 1993. A missense mutation (Ala →Thr) in exon 7 of the lipoprotein lipase gene in a case iwth type I hyperlipidemia. *Biochem. Biophys. Res. Commun.* 191:1046-1054.

261. Pepe, G., G. Chimienti, F. Resa, V. DiPerma, C. Tarricone, M. Lovecchio, A.M. Colacicco, and A. Capurso. 1994. A new Italian case of lipoprotein lipase deficiency: A Leu^{365}→Val change resulting in loss of enzyme activity. *Biochem. Biophys. Res. Commun.* 199:570-576.

262. Kozaki, K., T. Gotoda, M. Kawamura, H. Shimano, Y. Yazaki, Y. Ouchi, H. Orimo, and N. Yamada. 1993. Mutational analysis of human lipoprotein lipase by carboxy-terminal truncation. *J. Lipid Res.* 34:1765-1772.

263. Durstenfeld, A., O. Ben-Zeev, K. Reue, G. Stahnke, and M.H. Doolittle. 1994. Molecular characterization of human hepatic lipase deficinecy. In vitro expression of two naturally occurring mutations. *Arteriosclerosis & Thrombosis* 14:381-385.

264. Gotoda, T., N. Yamada, T. Murase, M. Sakuma, N. Murayam, H. Shimano, K. Kozaki, J.J. Albers, Y. Yazaki, and Y. Akanuma. 1991. Differential phenotypic expression by three mutant alleles in familial lecithin:cholesterol acyltransferase deficiency. *Lancet* 338:778-781.

265. Skretting, G., J.P. Blomhoff, J. Solheim, and H. Prydz. 1992. The genetic defect of the original Norwegian lecithin:choelsterol acyltransferase deficiency families. FEBS Lett. 309:307-310.

266. Klein, H.G., S. Santamarina-Fojo, N. Duverger, M. Clerc, M.F. Dumon, J.J. Albers, S. Marcovina, H.B. Brewer Jr. 1993. Fish eye syndrome: A molecular defect in the lecithin-cholesterol acyltransferase (LCAT) gene associated with normal alpha-LCAT-specific activity. Implications for classification and prognosis. *J Clin. Invest.*92:479-485.

267. Funke, H., A. von Eckardstein, P.H. Pritchard, A.E. Hornby, H. Wiebusch, C. Motti, M.R. Hayden, C. Dachet, B. Jacotot, U. Gerdes. 1993. Genetic and phenotypic heterogeneity in familial lecithin: Cholesterol acyltransferase (LCAT) deficiency. Six newly identified defective alleles further contribute to the structural heterogeneity in this disease. *J. Clin. Invest.* 91:677-683.

268. Takahashi, K., X.C. Jiang, N. Sakai, S. Yamashita, K. Hirano, H. Bujo, H. Yamazaki, J. Kusunoki, T. Miura, P. Kussie. 1993. A missense mutation in the cholesteryl ester transfer protein gene with possible dominant effects on plasma high density lipoproteins. *J. Clin. Invest.* 92:2060-2064.

269. Gotoda, T., M. Kinoshita, H. Shimano, K. Harada, M. Shimada, J-I. Ohsuga, T. Teramoto, Y. Yazaki, and N. Yamada. 1993. Cholesteryl ester transfer protein deficiency caused by a nonsense mutation detected in the patient's macrophage mRNA. *Biochem. Biophys. Res. Commun.* 194:519-524.

270. Sharp, D., L. Blinderman, K.A. Combs, B. Klenzie, B. Ricci, K. Wager-Smith, C.M. Gil, C.W. Turck, M.-E. Bouma, D.J. Rader, L.P. Aggerbeck, R.E. Gregg, D.A. Gordon, and J.R. Wetterau. 1993. Cloning and gene defects in microsomal triglyceride transfer protein associated with abetalipoproteinemia. *Nature* 365:65-69.

271. Groenewegen, W.A., E.S. Krul, and G. Schonfeld. 1993. Apolipoprotein B-52 mutation associated with hypobetalipoproteinemia is compatible with a misaligned pairing deletion mechanism. *J. Lipid Res.* 34:971-981.

272. Young, S.G., C.R. Pullinger, B.R. Zysow, H. Hofmann-Radvani, M.F. Linton, R.V. Farese, Jr., J.F. Terdiman, S.M. Snyder, S.M. Grundy, G.L. Vega. 1993. Four new mutations in the apolipoprotein B gene causing hypobetalipoproteinemia, including two different frameshift mutations that yield truncated apolipoprotein B proteins of identical length. *J. Lipid Res.* 34:501-507.

273. Talmud, P.J., E.S. Krul, M. Pessah, G. Gay, G. Schonfeld, S.E. Humphries, and R. Infante. 1994. Donor splice mutation generates a lipid-associated apolipoprotein B-27.6 in a patient with homozygous hypobetalipoproteinemia. *J. Lipid Res.* 35:468-477.

274. Reina, M., J.D. Brunzell, and S.S. Deeb. 1992. Molecular basis of familial chylomicronemia: Mutations in the lipoprotein lipase and apolipoprotein CII genes. *J. Lipid Res.* 33:1823-1832.

275. Inadera, H., A. Hibino, J. Kobayashi, T. Kanzaki, K. Shirai, S. Yukawa, Y. Saito, and S. Yushida. 1993. A missense mutation (Trp26→Arg) in exon 3 of the apolipoprotein CII gene in a patient with apolipoprotein CII deficiency (apoCII-Wakayama). *Biochem. Biophys. Res. Commun.* 193:1174-1183.

276. Pullinger, C.R., B.R. Zysow, L.L. Hennessy, P.H. Frost, M.J. Malloy, and J.P. Kane. 1993. Molecular cloning and characteristics of a new apolipoprotein CII mutant identified in three unrelated individuals with hypercholesterolemia and hypertriglyceridemia. *Hum. Mol. Genet.* 2:69-74.

277. Breslow, J.L. 1994. Lipoprotein metabolism and atherosclerosis susceptibility in transgenic mice. *Curr.Opin.Lipid* 5:175-184.

278. Zhang S.H., R.L. Reddick, J.A. Piedrahita, N. Maeda. 1992. Spontaneous hypercholesterolemia and arterial lesions in mice lacking apolipoprotein *E. Science* 258:468-471.

279. Reddick R.L., S.H. Zhang, N. Maeda. 1994. Atherosclerosis in mice lacking apoE. Evaluation of lesional development and progression. *Arterioscler. Thromb.* 14:141-147.

280. Zhang, S.H. R.L. Reddick, B. Burkey. N. Maeda. 1994. Diet-induced atherosclerosis in mice heterozygous and homozygous for apolipoprotein E gene disruption. *J. Clin. Invest.* 94:937-945.

281. Nakashima Y., A.S. Plump, E.W. Raines, J.L. Breslow, R. Ross. 1994. ApoE-deficient mice develop lesions of all phases of atherosclerosis throughout the arterial tree. *Arterioscler. Thromb.* 14:133-140.

282. Simonet W.S., N. Bucay, S.J. Lauer, J.M. Taylor. 1993. Multiple tissue-specific elements control the apolipoprotein E/C-I gene locus in transgenic mice. *J. Biol. Chem.* 265:8651-8654.

283. Aalto-Setala K., E.A. Fisher, X. Chen, T. Chajek-Shaul, T. Hayek, R. Zechner, A. Walsh, R. Ramakrishnan, H.N. Ginsberg, J.L. Breslow. 1992. Mechanism of hypertriglyceridemia in human apoCIII transgenic mice: Diminished VLDL fractional catabolic rate associated with increased apoCIII and reduced apoE on the particles. *J. Clin. Invest.* 90:1889-1900.

284. Schacter N.S., T. Hayek, T. Leff, J.D. Smith, D.W. Rosenberg, A. Walsh, R. Ramakrishnan, I.J. Goldberg, H.N. Ginsberg, J.L. Breslow. 1994. Overexpression of apolipoprotein C-II causes hypertriglyceridemia in transgenic mice. *J. Clin. Invest.* 93:1683-1690.

285. de Silva H.V., S.J. Lauer, J. Wang, W.S. Simonet,K.H. Weisgraber, R.W. Mahley, J.M. Taylor. 1994. Overexpression of human apolipoprotein C-III in transgenic mice results in an accumulation of apolipoprotein B48 remnants that is corrected by excess apolipoprotein E*. *J. Biol. Chem.* 269:2324.

286. Maeda N., H. Li, D. Lee, P. Oliver, S.H. Quarfordt, J. Osada. 1994. Targeted disruption of the apolipoprotein C-III gene in mice results in hypotriglyceridemia and protection from postprandial hypertriglyceridemia. *J. Biol. Chem.* 269:23610-23616.

287. van Ree J.H., W.J.A.A. van den Broek, R.R. Frants, M.H. Hofker, L.M. Havekes. 1994. Diet induced hypercholesterolemia in apolipoprotein C1-deficient mice. *Circulation* 90:I-135.

288. Hayek T., N. Azrolan, R.B. Verdery, A. Walsh, T. Chajek-Shaul, L.B. Agellon, A.R. Tall, J.L. Breslow. 1993. Hypertriglyceridemia and cholesteryl ester transfer protein interact to dramatically alter high density lipoprotein levels, particle sizes, and metabolism. *J. Clin. Invest.* 92:1143-1152.

289. Levine, D.M., Parker, T.S., Donnelly, T.M., Walsh, A., Rubin, A.L. 1994. In vivo protein against endotoxin by plasma high density lipoprotein. *Proc. Natl. Acad. Sci. USA* 90:12040-12044.

290. Chajek-Shaul T., T. Hayek, A. Walsh, J.L. Breslow. 1991. Expression of the human apolipoprotein A-I gene in transgenic mice alters high density lipoprotein (HDL) particle size distribution and diminishes selective uptake of HDL cholesteryl esters. *Proc. Natl. Acad. Sci. USA* 88:6731-6735.

291. Williamson R., D. Lee, J. Hagaman, N. Maeda. 1992. Marked reduction of high density lipoprotein cholesterol in mice genetically modified to lack apolipoprotein A-I. *Proc. Natl. Acad. Sci. USA* 89:7134-7138.

292. Plump, A.S., T. Hayek, A. Walsh, J.L. Breslow. 1993. Diminished HDL cholesterol ester flux in apoA-I deficient mice. *Circulation* 88:2266a.

293. Li, H., R.L. Reddick, and N. Maeda. 1993. Lack of apoA-I is not associated with increased susceptibility to atherosclerosis in mice. *Arteriosclerosis & Thrombosis* 13:1814-1821.

294. Schultz J.R., E.L. Gong, M.R. McCall, A.V. Nichols, S.M. Clift, E.M. Rubin. 1993. Expression of human apolipoprotein A-II and its effect on high density lipoproteins in transgenic mice. *J. Biol. Chem.* 267:21630-21636.

295. Hedrick C.C., W.L. Castellani, C.H. Warden, D.L. Puppione, A.J. Lusis. 1993. Influence of mouse apolipoprotein A-II on plasma lipoproteins in transgenic mice. *J. Biol. Chem.* 268:20676-20682.

296. Schultz J.R., J.G. Verstuyft, E.L. Gong, A.V. Nichols, E.M. Rubin. 1993. Protein composition determines the anti-atherogenic properties of HDL in transgenic mice. *Nature* 365:762-764.

297. Aalto-Setala K., C.L. Bisgaier, A. Ho, K. Kieft, M.G. Traber, H.J. Kayden, R. Ramakrishnan, A. Walsh, A.D. Essenburg, J.L. Breslow. 1994. Intestinal expression of human apolipoprotein A-IV in transgenic mice fails to influence dietary lipid absorption or feeding behavior. *J. Clin. Invest.* 93:1776-1786.

298. Duverger N., G. Tremp, B. Schombert, F. Emmanuel, S. Cune, F. Attenot, N. Touchet, P. Denefle. 1994. Hepatic overexpression of human apolipoprotein A-IV causes hypertriglyceridemia in transgenic mice. *Circulation* 90:I-135.

299. Chiesa G., D.F. Johnson, Z. Yao, T.L. Innerarity, R.W. Mahley, S.G. Young, R.H. Hammer, H.H. Hobbs. 1993. Expression of human apolipoprotein B-100 in transgenic mice. *J. Biol. Chem.* 268:23747-23750.

300. Linton M.F., R.V. Farese Jr., G. Chiesa, D.S. Grass, P. Chin. R.E. Hammer, H.H. Hobbs, S.G. Young. 1993. Transgenic mice expressing high plasma concentrations of human apolipoprotein B100 and lipoprotein (a). *J. Clin. Invest.* 92:3029-3037.

301. Callow M.J., L.J. Stoltzfus, R.M. Lawn, E.M. Rubin. 1994. Expression of human apolipoprotein B and assembly of lipoprotein (a) in transgenic mice. *Proc. Natl. Acad. Sci. USA* (in press)

302. Homanics G.E., T.J. Smith, S.H. Zhang, D. Lee, S.G. Young, N. Maeda. 1993. Targeted modification of the apolipoprotein B gene results in hypobetalipoproteinemia and developmental abnormalities in mice. *Proc. Natl. Acad. Sci. USA* 90:2389-2393.

303. Plump A.S., T.M. Forte, S. Eisenberg, J.L. Breslow. 1993. Atherogenic V-VLDL in the apoE-deficient mouse: Composition, origin, and fate. [Abstract] *Circulation* 88:A9.

304. Palinski W., V.A. Ord, A.S. Plump, J.L. Breslow, D. Steinberg, J.L. Witzum. 1994. Apoprotein E-deficient mice are a model of lipoprotein oxidation in atherogenesis: Demonstration of oxidation-specific epitopes in lesions and high titers of autoantibodies to malondialdehyde-lysine in serum. *Arterioscler. Thromb.* 14:606-616.

305. Popko B., J.F. Goodrum, T.W. Bouldin, S.H. Zhang, N. Maeda. 1993. Nerve regeneration occurs in the absence of apolipoprotein E in mice. *J. Neurochem.* 60:1155-1158.

306. Paszty C., N. Maeda, J. Verstuyft, E.M. Rubin. 1994. Apolipoprotein A-I transgene corrects apolipoprotein E deficiency-induced atherosclerosis in mice. *J. Clin. Invest.* 94:899-903.

307. Chiesa G., H.H. Hobbs, M.L. Koschinsky, R.M. Lawn, S.D. Maika, R.E. Hammer. 1992. Reconstitution of lipoprotein (a) by infusion of human low density lipoprotein into transgenic mice expressing human apolipoprotein (a). *J. Biol. Chem.* 267:24369-24374.

308. Lawn R.M., D.P. Wade, R.E. Hammer, G. Chiesa, J.G. Verstuyft, E.M. Rubin. 1992. Atherogenesis in transgenic mice expressing human apolipoprotein (a). *Nature* 360:670-671.

309. Frazer K.A., G. Narla, J. Zhang, E.M. Rubin. 1994. A 270 Kb yeast artificial chromosome containing apolipoprotein (a) appropriately expresses in transgenic mice. *Circulation* 90:I-624.

310. Hofmann S.L., D.W. Russell, M.S. Brown, J.L. Goldstein, R.E. Hammer. 1988. Overexpression of low density lipoprotein (LDL) receptor eliminates LDL from plasma in transgenic mice. *Science* 239:1277-1281.

311. Hofmann S.L., D.L. Eaton, M.S. Brown, W.J. McConathy, J.L. Goldstein, R.E. Hammer. 1990. Overexpression of human low density lipoprotein receptors leads to accelerated catabolism of Lp(a) lipoprotein in transgenic mice. *J. Cli. Invest.*85:1542-1547.

312. Yokode M., R.E. Hammer, S. Ishibashi, M.S. Brown, J.L. Goldstein. 1990. Diet-induced hypercholesterolemia in mice: Prevention by overexpression of LDL receptors. *Science* 250:1273-1275.

313. Ishibashi S., M.S. Brown, J.L. Goldstein, R.D. Gerard, R.E. Hammer, J. Herz. 1993. Hypercholesterolemia in low density lipoprotein receptor knockout mice in its reversal by adenovirus-mediated gene delivery. *J. Clin. Invest.* 92:883-893.

314. Ishibashi S., Herz J., Maeda N., Golstein J.L., Brown M.S. 1994. The two receptor model of lipoprotein clearance: Tests of the hypothesis in "knockout" mice lacking the low density lipoprotein receptor, apolipoprotein E, or both proteins. *Proc. Natl. Acad. Sci. USA* 91:4431-4435.

315. Herz J., D.E. Clouthier, R.E. Hammer. 1992. LDL receptor-related protein internalizes and degrades uPA-PA-1 complexes and is essential for embryo implantation. *Cell.*71(3):411-421.

316. Shimada M., H. Shimano, T. Gotoda, K. Yamomoto, M. Kawamura, T. Inaba, Y. Yazaki, N. Yamada. 1993. Overexpression of human lipoprotein lipase in transgenic mice. *J. Biol. Chem.* 268:17924-17929.

317. Fan J., J. Wang, Z-S. Ji, D. Sanan, Q. Dang, A. Bensadoun, R.W. Mahley, J.M. Taylor. 1994. Transgenic rabbits overexpressing human hepatic lipase have reduced levels of high density lipoproteins and intermediate density lipoproteins and diminished responses to dietary cholesterol. *Circulation* 90:I-289.

318. Berard A., R.D. Shamburek, B. Vaisman, S. Meyn, M. Kindt, J. Albers, S. Santamarina-Fojo, H.B. Brewer Jr. 1994. Differential dietary response of mice overexpressing hLCAT. *Circulation* 90:I-82.

319. Francone O.L., E.L. Gong, P. Cooper, C.J. Fielding, E.M. Rubin. 1994. Expression of human lecithin:cholesterol acyltransferase in transgenic mice: Studies on substrate specificity. *Circulation* 90:I-82.

320. Agellon L.B., A. Walsh, T. Hayek, P. Moulin, X.C. Jiang, S.A. Shelanski, J.L. Breslow, A.R. Tall. 1991. Reduced high density lipoprotein cholesterol in human cholesterol ester transfer protein transgenic mice. *J. Biol. Chem.* 266:10796-10801.

321. Marotti K.R., C.K. Castle, R.W. Murray, E.F. Rehberg, H.G. Polites, G.W. Melchior. 1992. The role of cholesteryl ester transfer protein in primate apolipoprotein A-I metabolism. Insights from studies with transgenic mice. *Arterioscler. Thromb.* 12:736-744

322. Jiang X.C., L. Masucci-Magoulas, J. Mar, M. Lin, A. Walsh, J.L. Breslow, A. Tall. 1993. Down-regulation of mRNA for the low density lipoprotein receptor in transgenic mice containing the gene for human cholesteryl ester transfer protein. Mechanism to explain accumulation of lipoprotein B particles. *J. Biol. Chem.* 268:27406-27412.

323. Zhong S., I.J. Goldberg, C. Bruce, E. Rubin, J.L. Breslow, A. Tall. 1994. Human apoA-II inhibits the hydrolysis of HDL triglyceride and the decrease of HDL size induced by hypertriglyceridemia and CETP in transgenic mice. *Circulation* 90:I-81.

324. Grass D.S., U. Saini, R.H. Felkner, S.G. Young, R.E. Wallace, T.D. Yeck, M.E. Swanson. 1994. Transgenic mice expressing human apoB and human cholesteryl ester transfer protein have a lipoprotein-cholesterol profile similar to that of normolipidemic humans. *Circulation* 90:I-134

325. Fazio S., K.R., Marotti, Y-L. Lee, C.K. Castle, G.W. Melchior, S.C. Rall, Jr. 1994. Co-expression of cholesteryl ester transfer protein and defective apolipoprotein E in transgenic mice alaters plasma cholesterol distribution. *J. Biol. Chem.* 269:32368-32372.

326. Bockxmeer M.V., C.D. Mamotte, F.R. Gibbons and R.R. Taylor. 1993. Apolipoprotein E4 homozygosity is an important determinant of restenosis after coronary angioplasty. *Circulation* 88:519.

III. INFLAMMATION

ACUTE INFLAMMATORY LUNG INJURY: MECHANISMS AND INTERVENTIONS

Peter A. Ward, Andreas Seekamp, and Michael S. Mulligan

Department of Pathology
The University of Michigan Medical School
Ann Arbor, Michigan USA

ABSTRACT

Acute inflammatory responses involving the lung occur rather commonly during adult respiratory distress syndrome (ARDS), during intrapulmonary presence of infectious agents (bacteria, viruses), after ischemia reperfusion events, and in a variety of other situations. The key issues to be addressed are: identifying the triggers for these inflammatory responses, defining the role of complement and the participation of leukocytes, and determining the requirements for cytokines and the role of adhesion molecules in these events. Experimental rat models have provided important insights into mechanisms of acute lung injury. The first model is produced by systemic activation of complement following intravenous infusion of purified cobra venom factor (CVF). Infusion of CVF results in intravascular activation of neutrophils and upregulation of neutrophil CD11b/CD18, resulting in intrapulmonary sequestration and oxidant-mediated damage of vascular endothelial and alveolar epithelial cells. In this model the morphological features in rat lungs are similar to those found in human ARDS. Injury in this lung model is neutrophil- and complement-dependent, cytokine-independent, and requires β_2 integrins (LFA-1, Mac-1), ICAM-1 and L- and P-selectins. Blocking of any of the relevant adhesion molecules reduces neutrophil accumulation and tissue injury. In the second model, injury follows intrapulmonary deposition of IgG immune complexes and extensive intra-alveolar accumulation of neutrophils. The morphological features of this inflammatory reaction would be akin to inflammatory responses to infectious bacterial agents in lung. In this model complement and neutrophils are both required for induction of injury. Neutrophil recruitment requires β_2 integrins (CD11a/CD18), ICAM-1, E- and L-selectins, and cytokines (IL-1, TNFα). Injury is due to combined release of oxidants and proteases from lung macrophages and recruited neutrophils. The third model of acute lung injury follows in vivo ischemia, (followed by reperfusion) of distal extremities. Pulmonary vascular damage occurs, is complement- and neutrophil-dependent, and requires participation of cytokines (TNFα and IL-1) and adhesion molecules (CD11a/CD18, CD11b/CD18, L- and E-selectin). Thus, inflammatory lung injury varies with respect to the causes of injury, the cytokines involved, and the requirement for adhesion molecules. In all cases lung injury can be reduced by appropriate blocking of adhesion molecules and cytokines.

Inflammatory responses are assumed to play major roles in events that lead to acute tissue injury and, ultimately, multiorgan failure. The triggers that drive these adverse responses may be extrinsic (such as bacteria, bacterial lipopolysaccharide, viruses, etc.) or may be the result of activation of proinflammatory cascades (e.g., an immune response, complement activation, cytokine expression, etc.). Identifying the factors responsible for initiation of inflammatory cascades and understanding the mechanisms responsible for ensuing tissue damage are critical

Vascular Endothelium: Responses to Injury
Edited by John D. Catravas *et al.*, Plenum Press, New York, 1996

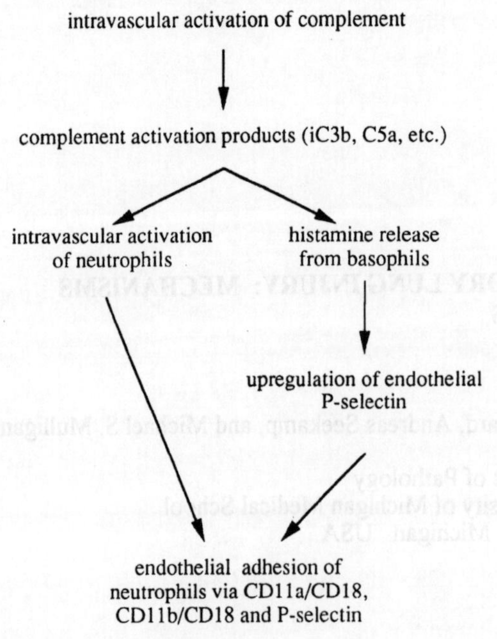

Figure 1. Pathophysiological events related to lung injury following systemic activation of complement.

to development of more effective therapeutic interventions. The following studies involving the use of animal models of lung injury provide insights into mechanisms of inflammatory tissue damage and how the injury can be effectively averted.

LUNG INJURY FOLLOWING SYSTEMIC ACTIVATION OF COMPLEMENT

It was postulated many years ago that systemic activation of complement would lead to neutrophil activation, their sequestration in lung, and accompanying lung injury *(Craddock et al., 1977)*. Using a potent activator of the alternative complement pathway, namely, cobra venom factor (CVF), intravenous infusion of CVF resulted in prompt activation of plasma complement, appearance in plasma of C5a, profound neutropenia due to intrapulmonary sequestration of neutrophils, and pulmonary vascular endothelial damage in capillaries with accompanying morphological damage of endothelial cells together with interstitial edema and intra-alveolar hemorrhage *(Till et al., 1982)*. Pathophysiological events are described in Figure 1. The injury, which peaks very early (within 30 min), was found to be neutrophil-dependent and oxygen radical mediated and was shown to be dependent on availability of complement, since complement blockade achieved by the use of recombinant human soluble complement receptor-1 (sCR1) was substantially protective *(Mulligan et al., 1992)*. Blocking of either TNFα or IL-1 did not provide protection from CVF-induced lung injury *(Mulligan et al., 1992)*. This is to be expected, given the brief duration of the model and the rapid onset of injury (see above). The pathophysiological events that seem to be involved in this type of lung injury are outlined in Figure 1. One of the key events resulting from complement activation appears to be upregulation of P-selectin on the pulmonary vascular endothelium *(Mulligan et al., 1992)*. Whether this upregulation is due to C5a-induced release of histamine from basophils and mast cells or due to some other factor remains to be determined. The requirements for selectins (P, E and L) in this model of injury have been determined by the use of blocking antibodies, selectin-Ig chimeric proteins, or by the use of oligosaccharides such as

Table 1. Selectin Requirements for Lung Injury

Lung Injury	Requirements		
	P-selectin	L-Selectin	E-Selectin
CVF(systemic activation of complement)	+	+	-
IgG imune complex	-	+	+
ischemia-reperfusion (of hind limb)	±	+	+

Table 2. Cytokine Requirements for Lung Injury

Lung Injury	Requirements	
	TFNα	IL-1
CVF	-	-
IgG immune complexes	+	+
ischemia-reperfusion	+	+

Table 3. β2 Integrin Requirements for Lung Injury

Lung Injury	Requirements	
	LFA-1 (CD11a/CD18)	Mac-1 (CD11b/CD18)
CVF	+	+
IgG immune comples	+	-
ischemia-reperfusion	+	+

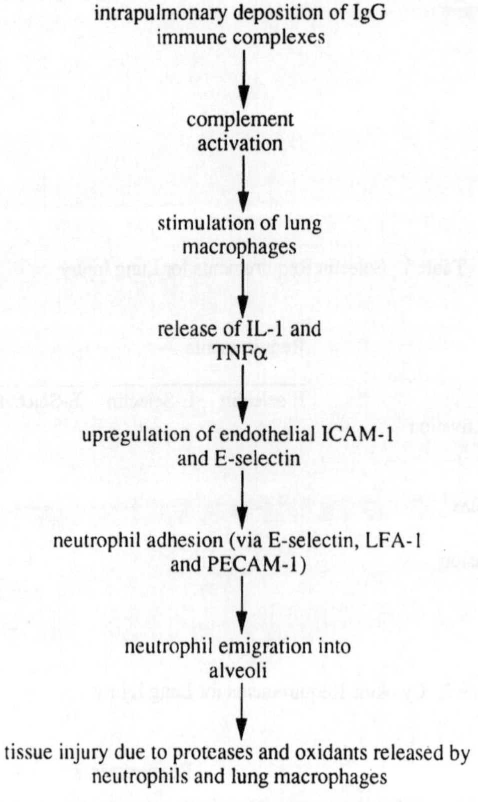

Figure 2. Pathophysiological events related to lung injury following intrapulmonary deposition of IgG immune complexes.

sialyl Lewis^x, which are known to be reactive with selectins *(Mulligan et al., 1993; Mulligan et al., 1994; Mulligan et al., 1993)*. On the basis of these approaches, the CVF model of lung injury is P- and L-selectin-dependent but independent of a role for E-selectin. Upregulation of endothelial P-selectin following complement activity initiates the sequence of engagement of adhesion molecules: selectin upregulation followed by engagement of endothelial ICAM-1, resulting in intermittent adhesion of neutrophils to the endothelium, which is s-electin-dependent, followed by firm adhesion of neutrophils to the endothelium, which is ICAM-1 dependent. Because of the rapidity of the inflammatory response and development of injury in the CVF model, cytokine (TNFa, IL-1) involvement has been ruled out *(Mulligan et al., 1992)*. The ultimate efficient "capture" by the vascular endothelium of neutrophils in this model of injury also requires involvement of two of the neutrophil β_2 integrins, namely CD11a/CD18 (LFA-1) and CD11b/CD18 (Mac-1) *(Mulligan et al., 1992)*. Blocking of any of the β_2 integrin subunits substantially reduces neutrophil accumulation in lung capillaries and, correspondingly, reduces injury. One "counter-receptor" for neutrophil LFA-1 and Mac-1 is ICAM-1. In fact, ICAM-1 is required for the full recruitment of neutrophils in the CVF model of lung injury, as revealed by the use of blocking antibodies *(Mulligan et al., 1992)*. Since the time course of events leading to injury in this model is so brief, upregulation of ICAM-1 would not be expected. The requirement for ICAM-1 in this model presumably relates to constitutively expressed ICAM-1 and is probably also linked to the fact that complement activation would be expected to cause significant upregulation of neutrophil Mac-1, which would enhance Mac-1/ICAM-1 adhesive interactions. Requirements for selectins, cytokines and β2 integrins in this model of injury are described in Tables 1-3.

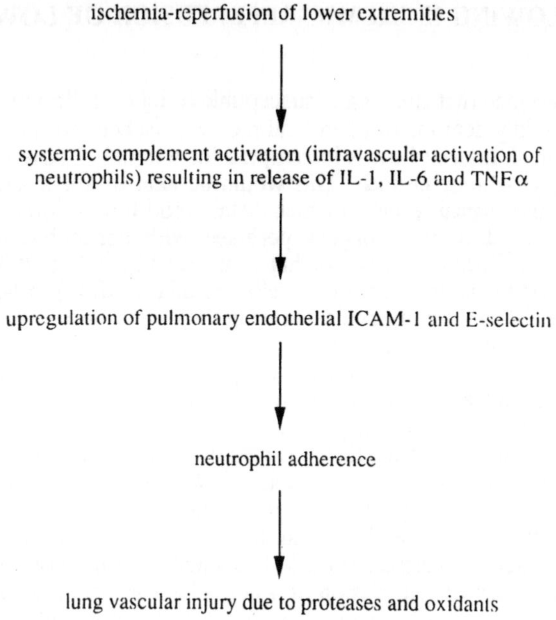

Figure 3. Pathophysiological events in lung injury following hind-limb ischemia and reperfusion.

LUNG INJURY FOLLOWING DEPOSITION OF IgG IMMUNE COMPLEXES

A rather different series of events leads to acute lung injury in rats following intrapulmonary deposition of IgG immune complexes. In this model, injury develops over a four hour period and is associated with the intra-alveolar influx of large numbers of neutrophils *(Johnson and Ward, 1974)*. Both vascular endothelial cells of lung as well as alveolar epithelial cells are damaged by toxic products of neutrophils, including oxidants and proteases. Activation of residential lung macrophages (interstitial and intra-alveolar) seems to be the key early response that results in elaboration by of TNFα and IL-1 (*Warren et al., 1989, Warren, 1991)*. The proposed sequence of pathophysiological events is shown in Figure 2. The release of these cytokines and their access to nearby endothelial cells results in upregulation of endothelial E-selectin and ICAM-1, which sets the stage for the sequence of adhesion events required for recruitment of blood neutrophils into the alveolar compartment *(Mulligan et al., 1993, Mulligan et al., 1991)*. P-selectin appears to play no role in the adhesion-promoting events. Somewhat surprisingly, LFA-1, but not Mac-1, is required for neutrophil recruitment. Reasons for the divergency in β2 integrin requirements are not understood. As would be expected, a requirement for LFA-1 is also coupled with a requirement for its endothelial counter-receptor, ICAM-1. Injury which follows accumulation of neutrophils within the alveolar compartment appears to be linked both to toxic oxygen products ($O2^{\cdot}$, H2O2, HO·) as well as toxic products from L-arginine (including ·NO, ONOO· and HO·), together with released leukocytic metalloproteinases *(Johnson and Ward, 1981; Mulligan et al., 1991; Mulligan et al., 1993)*. It seems probable that both neutrophils and lung macrophages contribute to this potpourri of noxious substances. See Tables 1-3 for a summary of requirements for selectins, cytokines and β2 integrins in this model of injury.

LUNG INJURY FOLLOWING ISCHEMIC-REPERFUSION OF LOWER EXTREMITIES

There is abundant evidence that the lung is susceptible to injury following adverse events in other organs. This is vividly demonstrated in a rat model of ischemia-reperfusion involving the lower extremities in which arterial flow is interrupted for 4 hours, followed by reperfusion which occurs over another 4 hour period. This would be akin to aortic cross-clamping in humans in cases of aortic aneurysmal repair, in which distal blood flow is interrupted, followed by re-establishment of blood flow to organs perfused with aortic blood flow. Using measurements of leakage of ^{125}I-albumin, hemorrhage (assessed by ^{51}Cr-RBC) and neutrophil accumulation (as determined by tissue content of myeloperoxidase, MPO), it has been possible to measure changes in the distal extremities of rats (using the crural muscle mass) as well as in the lungs. During the reperfusion process it is evident that a systemic inflammatory response is occurring *(Seekamp et al., 1993)*. Proposed pathophysiological events in this model are shown in Figure 3. For instance, during reperfusion there is evidence of progressive complement depletion, as assessed by whole plasma complement measurements *(Seekamp, A. and Ward, P.A., unpublished)*. Evidence that this results in functional outcomes is found in analysis of blood neutrophils for β2 integrin content *(Seekamp et al., 1993)*. In contrast to neutrophils obtained at the beginning of the reperfusion process, at 4 hours these cells demonstrate approximately a 1.6 fold and 2.0 fold increase in surface content of CD11b and CD18, respectively, while there is no increase in CD11a content. Furthermore, if serum/plasma obtained at the commencement of reperfusion or 4 hours later is incubated with neutrophils obtained from normal rat blood, virtually identical changes in β_2 integrin content can be induced in these cells. While it remains to be demonstrated that the factor(s) in serum is (are) in fact derived from the complement system, the data do show that during the reperfusion process systemic activation of the inflammatory cascade is occurring, leading to intravascular activation of leukocytes.

During the reperfusion process progressive injury occurs both locally (in the crural muscle mass) and remotely (in the lungs), as reflected by leakage of plasma albumin and hemorrhage. In parallel, tissue MPO also increases *(Seekamp et al., 1993)*. Hemorrhage and albumin leakage depend on the availability of both complement and neutrophils. In addition, TNFα and IL-1, which appear in the plasma and reach peak concentration approximately 2 hours during reperfusion, are both involved in local and remote vascular injury, since their blockade by antibody, IL-1ra or sTNFaR-1 reduces evidence of vascular injury *(Seekamp et al., 1993)*. In the lung, E-selectin upregulation has been found in the pulmonary vasculature 4 hours after commencement of reperfusion. Not unexpectedly, both CD11a/CD18 and CD11b/CD18 integrins are required for neutrophil recruitment and for the full development of vascular injury. Selectin requirements for the development of local and remote injury have also been determined. Lung, injury occurring 4 hours after reperfusion is both E- and L--selectin-dependent but independent of P-selectin. However, if ischemia and reperfusion times were limited to 1 and 1.5 hour, respectively, a clear requirement for P-selectin is also found *(Seekamp et al., 1994)*. Thus, it seems that in ischemia/reperfusion events leading to injury, P-selectin is involved in the early events whereas E- and L-selectins are involved in the later events. Using similar strategies to define requirements in the vasculature of the crural muscle mass, both E- and L-selectins are required, but not P-selectins. In neither local or remote injury following ischemia and reperfusion has a requirement for VLA-4 (the leukocyte but non-neutrophil "counter-receptor" for VCAM-1) been identified. These studies reveal the multiplicities of requirements for cytokines and adhesion molecules in events related to ischemia/reperfusion injury. Summaries of the requirements for selectins, cytokines and β_2 integrins are found in Tables 1-3.

It is apparent that inflammatory injury can be blocked by a variety of approaches that are linked to inhibition of neutrophils, complement, cytokine or adhesion molecules (both leukocytic and endothelial). Whether in an experimental model or in human inflammatory injury, it is vital to have information on what type of inflammatory reaction is under consideration. For instance, the cytokine profile required for a cell-mediated immune type of response will probably be quite different from that related to a neutrophil-enriched type of inflammatory injury. The ability to estimate type of inflammatory cell profile will permit logical decisions related to interventional approaches.

REFERENCES

Craddock, P.R., Fahr, J., Dallas, A.P., Brigham, K.L. and Jacob, H.S., 1977, Pulmonary vascular leukostasis resulting from complement (C1) activation by dialyzer cellophane membranes. *J. Clin. Invest.*, 59:879.

Johnson, K.J. and Ward, P.A., 1974, Acute immunologic pulmonary alveolitis. *J. Clin. Invest.*, 54:349.

Johnson, K.J. and Ward, P.A., 1981, Role of oxygen metabolites in immune complex injury of lung. *J. Immunol.*, 126:2365.

Mulligan, M.S., Desrochers, P.E., Chinnaiyan, A.M., Gibbs, D.F., Varani, J., Johnson, K.J., and Weiss, S.J., 1993, In vivo suppression of immune complex-induced alveolitis by secretory leukoproteinase inhibitor and tissue inhibitor of metalloproteinases 2 *Proc. Natl. Acad. Sci. U.S.A.*, 90:11523..

Mulligan, M.S., Hevel, J.M., Marletta, M.A., and Ward, P.A., 1991, Tissue injury caused by deposition of immune complexes is L-arginine dependent. *Proc. Natl. Acad. Sci. U. S. A.*, 88:6338.

Mulligan, M.S., Miyasaka, M., Tamatani, T., Jones, M.L., and Ward, P.A., 1994, Requirements for L-selectin in neutrophil-mediated lung injury in rats. *J. Immunol.*, 152:832.

Mulligan, M.S., Paulson, J.C., De Frees, S., Zheng, Z-L., Lowe, J.B., and Ward, P.A., 1993, Protective effects of oligosaccharides in P-selectin-dependent lung injury. *Nature*, 364:149.

Mulligan, M.S., Polley, M.J., Bayer, R.J., Nunn, M.F., Paulson, J.C., and Ward, P.A., 1992, Neutrophil-dependent acute lung injury. Requirement for P-selectin (GMP-140). *J. Clin. Invest.*, 90:1600.

Mulligan, M.S., Vaporciyan, A.A., Miyasaka, M., Tamatani, T., and Ward, P.A., 1993, Tumor necrosis factor alpha regulates in vivo intrapulmonary expression of ICAM-1. *Am. J. Pathol.*, 142:1739.

Mulligan, M.S., Varani, J., Dame, M.K., Lane, C.L., Smith, C.W., Anderson, D.C., and Ward, P.A., 1991, Role of endothelial-leukocyte adhesion molecule 1 (ELAM-1) in neutrophil-mediated lung injury in rats. *J. Clin. Invest.*, 88:1396.

Mulligan, M.S., Varani, J., Warren, J.S., Till, G.O., Smith, C.W., Anderson, D.C., Todd III, R.F., and Ward, P.A., 1992, Roles of β_2 integrins of rat neutrophils in complement- and oxygen radical-mediated acute inflammatory injury. *J. Immunol.*, 148:1847.

Mulligan, M.S., Watson, S.R., Fennie, C., and Ward, P.A., 1993, Protective effects of selectin chimeras in neutrophil-mediated lung injury. *J. Immunol.*, 151:6410.

Mulligan, M.S., Yeh, C.G., Rudolph, A.R. and Ward, P. A., 1992, Protective effects of soluble CR1 in complement- and neutrophil- mediated tissue injury. *J. Immunol.*, 148:1479.

Seekamp, A., Mulligan, M.S., Till, G.O., and Ward, P.A., 1993, Requirements for neutrophil products and L-arginine in ischemia- reperfusion injury. *Am. J Pathol.*, 142:1217.

Seekamp, A., Till, G.O., Paulson, J.C., Mulligan, M.S., Anderson, D.C., Miyasaka, M., and Ward, P.A., 1994, Role of selectins in local and remote tissue injury following ischemia and reperfusion. *Amer. J. Pathol.*, 144:592.

Seekamp, A., Warren, J.S., Remick, D.G., Till, G.O., and Ward, P.A., 1993, Requirements for tumor necrosis factor-a and interleukin-1 in limb ischemia/reperfusion injury and associated lung injury. *Amer. J. Pathol.*, 143:453.

Till, G.O., Johnson, K.J., Kunkel, R., and Ward, P.A., 1982, Intravascular activation of complement and acute lung injury. Dependency on neutrophils and toxic oxygen metabolites. *J. Clin. Invest.*, 69:1126.

Warren, J.S., 1991, Intrapulmonary interleukin-1 mediates acute immune complex alveolitis in the rat. *Biochem. Biophys. Res. Commun.*, 175:604.

Warren, J.S., Yabroff, K.R., Remick, D.G., Kunkel, S.L., Chensue, S.W., Kunkel, R.G., Johnson, K.J. and Ward, P.A., 1989, Tumor necrosis factor participates in the pathogenesis of acute immune complex alveolitis in the rat. *J. Clin. Invest.*, 84:1873.

CYTOKINE ACTIVATION OF ENDOTHELIUM: CLONING AND CHARACTERIZATION OF A NEW IL-1 INDUCIBLE GENE

M. Introna, V. Vidal Alles, G. Picardi, A. Basile, C. Matteucci, B. Bottazzi, G. Peri, E. D'Aniello, F. Breviario, and A. Mantovani

Laboratory of Immunology
Istituto di Ricerche Farmacologiche "Mario Negri"
via Eritrea 62
20157 Milano
ITALY

INTRODUCTION

Endothelial cells play an active role in several different inflammatory and immunological reactions which take place in the tissues.

Many observations have been produced to document the responsiveness of endothelial cells both to many soluble mediators and to several populations of circulating and tissue-resident leukocytes *(Mantovani ét al., 1992; Mantovani and Dejana, 1989)*. We have in particular focused on the metabolic response of the human umbilical vein endothelial cells (HUVEC) to the Interleukin-1 (IL-1) cytokine. IL-1 has multiple effects on endothelial cells, which can be schematically synthetized in the concept that the cells, upon exposure to this cytokine, become oriented in a prothrombotic, proinflammatory sense *(Mantovani et al., 1992)*. This metabolic program involves the induction or the repression of several genes, which have been studied at several time points following the exposure to the cytokine *(Mantovani et al., 1992)*. Nevertheless, we think that the entire "program" may be rapidly and specifically induced by IL-1. In order to conduct a systematic study on the genes involved in the "immediate-early" response of the HUVEC to IL-1, we set out several years ago to construct a cDNA library from HUVEC stimulated for 1 hour with IL-1 in the presence of Cycloheximide in order to inhibit the synthesis of possible intermediate proteins, and then performed a "differential screening" using ^{32}P-labelled cDNA probes obtained from the same cells and from unstimulated HUVEC *(Introna et al., 1993)*.

THE ISOLATION AND CHARACTERIZATION OF PTX3 cDNA

Over 4000 clones were individually screened and 40 were identified (1%) which gave a stronger signal with the cDNA from IL-l treated cells. Partial sequencing of these clones revealed that 38 of them corresponded to genes already cloned *(Introna et al., 1993; D'Aniello et al., 1993)* that is interleukin-8 (IL-8) (16 clones) *(Strieter et al., 1989)*, endothelial leukocyte adhesion molecule 1 (ELAM-1) (7 clones) *(Bevilacqua et al., 1989)*, Gro-α (6 clones) *(Wen et al., 1989)*, Gro-β (5 clones) *(Haskill et al., 1990)*, plasminogen activator inhibitor (pA-I) (2 clones) *(Gramse et al., 1986)* and the inhibitor of the transcription factor NF-κB (I-κB) (2 clones) *(Haskill et al., 1991)*. It is interesting to observe that, with the exception of the I-κB gene, all other cloned genes had been previously reported to be induced in HUVEC by IL-1

Vascular Endothelium: Responses to Injury
Edited by John D. Catravas *et al.*, Plenum Press, New York, 1996

```
MAN      CRP     MEK---------LL-CFLVLTSLSHAFGQTDMSRKAFVFPKESDTSYVSL      40
RABBIT   CRP     MEK---------LLWCFLTLVSFSNMSDQAGMHKKAFVFPKESDNSYVSL      41
MOUSE    CRP     MEK---------LLWCLLIMISFSRTFGHEDMFKKAFVFPKESDTSYVSL      41
HAMSTER  FP      MDK--------MLL-LLGVSILLSEVFAQTDLTGKVFVFPRESESDYVKL      41
MOUSE    SAP     MDK--------LLLWMFVFTSLLSEAFCQTDLKRKVFVFPRESETDHVKL      42
MAN      SAP     MNK--------PLLWISVLTSLL-EAFAHTDLSGKVFVFPRESVTDHVNL      41
CRAB     CRP 1.1 MKTFHGPTCGTAVSLCLLLF--LTSALEEGEITSKV-KFPPSSSPSFPRL      47
CRAB     CRP 3.3 MKTFHGPTCGTAVSLCLLLF--LTSALEEGEITSKV-KFPPSSSPSFPRL      47
CRAB     CRP 1.4 MKTFHGPTFGTAVFLYLLLF--LPSALEEGEITSKV-KFPPSSSPSFPRL      47
MAN      PTX3    ---------------------------------ETAILFPMRSKKIFGSV      17
                                                  ..  **  *      ..

                        △
MAN      CRP     KAPLTKP-LKAFTVCLHFYTELSSTRGYSIFSYATKRQDNEILIFWSKDI      89
RABBIT   CRP     NAQLKKP-LKAFTVCLYFYTDLSMTRGYSIFSYATRRQFNEILLFWSKDI      90
MOUSE    CRP     EAESKKP-LNTFTVCLHFYTALSTVRSFSVFSYATKKNSNDILIFWNKDK      90
HAMSTER  FP      IPRLEKP-LENFTLCFRTYTDLS--RPHSLFSYNTKNKDNELLIYKERMG      88
MOUSE    SAP     IPHLEKP-LQNFTLCFRTYSDLS--RSQSLFSYSVKGRDNELLIYKEKVG      89
MAN      SAP     ITPLEKP-LQNFTLCFRAYSDLS--RAYSLFSYNTQGRDNELLVYKERVG      88
CRAB     CRP 1.1 VMVGTLPDLQEITLCYW-FKVNRLKGTLHMFSYATAKKDNELLTLIDEQG      96
CRAB     CRP 3.3 VMVGTLPDLQEITLCYW-FKVNCLKGTLHMFSYATAKKDNELLTLLDEQG      96
CRAB     CRP 1.4 VMVGTLPDLQEITLCYW-FKVNHLKSTLTIFSYNTAKNDNELLTSLEKQG      96
MAN      PTX3    HPVRPMR-LESFSACIWVKATDVLNKTI-LFSYGTKRNPYEIQLYLSYQS      65
                         . *.... *                .***..  .  ..    .

                                              △
MAN      CRP     GYSFTV-GGSEILFEVPEVTVAPV--HICTSW-----ESASGIVEFWVDG     131
RABBIT   CRP     GYSFSV-GGDEIIFKVSDIPVDPT--HLCASW-----ESSTGIAELWVDG     132
MOUSE    CRP     QYTFGV-GGAEVRFMVSEIPEAPT--HICASW-----ESATGIVEFWIDG     132
HAMSTER  FP      EYGLYI-ENLGAIVRGVEEFASPV--HFCTSW-----ESSSGIAEFWVNG     130
MOUSE    SAP     EYSLYI-GQSKVTVRGMEEYLSPV--HLCTTW-----ESSSGIVEFWVNG     131
MAN      SAP     EYSLYI-GRHKVTPKVIEKFPAPV--HICVSW-----ESSSGIAEFWING     130
CRAB     CRP 1.1 DFLFNVHGAPQLKVQCPNKIHIGKWHHVCHTWSSWEGEATIAVDGFHCKG     146
CRAB     CRP 3.3 DFLFNVHGAPQLKVQCPNKIHIGKWHHVCHTWSSWEGEATIAVDGFHCKG     146
CRAB     CRP 1.4 AFHMNVHGAPQLKVQCPNKIHIGKWHHVCHTWSSWEGEATIGVDGFHCKG     146
MAN      PTX3    --IVFVVGGEENKLVAEAMVSLGRWTHLCGTWNS-----EEGLTSLWVNG     108
                     . .      .              * * .*      . .. ..  .*
                                             |_____|

MAN      CRP     K-PRVRKSLKKGYTVGAEASIILGQEQDS--FGGNFEGSQSLVGDIGNVN     178
RABBIT   CRP     K-PMVRKSLKKGYILGPEASIILGQDQDS--FGGSFEKQQSLVGDIGNVN     179
MOUSE    CRP     K-AKVRKSLHKGYTVGPDASIILGQEQDS--YGGDFDAKQSLVGDIGDVN     179
HAMSTER  FP      I-PWVKKGLKKGYTVKTQPSIILGQEQDN--YGGGFDKSQSFVGEMGDLN     177
MOUSE    SAP     K-PWVKKSLQREYTVKAPPSIVLGQEQDN--YGGGFQRSQSFVGEFSDLY     178
MAN      SAP     T-PLVKKGLRQGYFVEAQPKIVLGQEQDS--YGGKFDRSQSFVGEIGDLY     177
CRAB     CRP 1.1 N-A---TGIAVGRTLSQGGLVVLGQDQDS--VGGKFDATQSLEGELSELN     190
CRAB     CRP 3.3 N-A---TGIAVGRTLSQGGLVVLGQDQDS--VGGKFDATQSLEGELSELN     190
CRAB     CRP 1.4 N-A---TGIAMGVTLSQGGLVVLGQEQDS--VGGEYDAEQSLEGELSELN     190
MAN      PTX3    ELAATTVEMATGHIVPEGGILQIGQEKNGCCVGGGFDETLAFSGRLTGFN     158
                 .   ..  .  .     . .**....   ** ..   .. * ...

MAN      CRP     MWDFVLSPDEINTIYL-----GGPFSPNVLNWRALKYEVQGEVFTKPQL--------WP     224
RABBIT   CRP     MWDYALSPEEINTIYA-----GGTFSPNVLDWRELTYQVRGEVHVKPQL--------WP     225
MOUSE    CRP     MWDFVLSPEQINTVYV-----GGTLSPNVLNWRALNYKAQGDVFIKPQL--------WS     225
HAMSTER  FP      MWDSVLTPEEIKSVYE-----GSWLEANILDWRTLNYEMSGYAVIRPRCVALSSYNKIS     231
MOUSE    SAP     MWDYVLTPQDILFVYR-----DSPVNPNILNWQALNYEINGYVVIRPRV--------WD     224
MAN      SAP     MWDSVLPPENILSAYQ-----GTPLPANILDWQALNYEIRGYVIIKPLV--------WV     223
CRAB     CRP 1.1 LWNTVLNHEQIKYLSKCAHPSERHIYGNIIQWDKTQFKAYDGVVLSPNEIC-------A     242
CRAB     CRP 3.3 LWNTVLNHEQIKYLSKCAHPSERHIYGNIIQWDKTQFKAYDGVVLSPNEIC-------A     242
CRAB     CRP 1.4 LWNTVLNHEQIKHLSKCAHPSERHIHGNIIQWDKTQFQAYDGVVLSPNEIC-------A     242
MAN      PTX3    IWDSVLSNEEIRET---GGAESCHIRGNIVGWGVTEIQPHGGA----------QYV--S     202
                 .*. .* ..*        .       *.. *        . . .
```

Figure 1. Computer alignment between all known members of pentraxins. Numbers on right indicate aminoacid position. The asterisks indicate identical aminoacids; the dots indicate conserved residues. The triangles indicate the conserved cysteines. The horizontal bar indicates the pentraxin family consensus signature.

(Sironi et al., 1989; Rajavashisth et al., 1990; Bussolino et al., 1989; Kojima et al., 1989; Anagnostou et al., 1990) and in fact belong to the prothrombotic/proinflammatory program *(Mantovani et al., 1992)*. The induction of I-κB, a negative regulator of a transcriptional factor may provide a genetic background for the observed inhibition of several functions in HUVEC upon exposure to IL-1 *(D'Aniello et al., 1993)*. One such candidate is the ELAM-1 gene itself which is rapidly induced and then repressed *(Bevilacqua et al., 1989)* and contains a functional NFκB binding site in its promoter region *(Whelan et al., 1991)*.

The last two clones contained an insert of 900 base pairs belonging to a previously unknown gene, as determined by screening of the EMBL Nucleotide Sequence Database. Since it contained an incomplete open reading frame, it was necessary to screen other HUVEC cDNA libraries. The full length cDNA was isolated and called PTX3. It contains a 68 base pair of 5' untranslated region, an open reading frame of 1143 base pairs coding for 381 amino acids, and a TAA stop codon followed by a polyadenylation signal *(Breviario et al., 1992)*.

The 3' terminal half of the predicted protein can be significantly aligned with the nine cloned members of the pentraxin gene family (Fig. 1) including the two human members C-reactive protein (CRP) and serum amyloid P component (SAP) *(Pepys and Baltz, 1992)*. There is an overall identity of 22 amino acids (indicated with asterisks in the figure) over 259 consensus length (9%) and a similarity of 55 amino acids (indicated with dots in the figure) (21%) *(Breviario et al., 1992)*.

In particular, PTX3 contains 8 amino acids (H-L-C-G-T-W-N-S) (underlined in the figure) which fit perfectly the pentraxin family signature (H-X-C-X-S/T-W-X-S) as identified by the PROSITE protein pattern database *(Bairoch, 1991)*. Finally, two cysteines at position 210 and 271 of PTX3 are conserved in all members of the family from horseshoe crab to man (Fig. 1, indicated with triangles).

Although studied for several years, the precise biological functions of the pentraxin still remain elusive *(Pepys and Beltz, 1983)* in spite of conflicting reports suggesting that they may play a role within the natural immune system, for example, as opsonins for macrophages and neutrophils *(Mortensen et al., 1976)*, as inhibitors of neutrophil chemotaxis *(Kew et al., 1990)*, by increasing tumor-cytotoxicity mediated by macrophages *(Zahedi and Mortensen, 1986)*, by inducing the release of several cytokines by macrophages *(Sarlo and Mortensen, 1985)*, and by inducing the synthesis of tissue factor by monocytes *(Galve-de Rochemonteix et al., 1993)*. More recently, the three-dimensional structure of SAP has been derived which shows a remarkable similarity with that of the legume lectins and which is consistent with its described calcium dependent protein and sugar binding function *(Emsley et al., 1994)*.

The 5' end portion of the protein contains a 17 amino acids signal peptide, indicating that the protein product may be secreted. No other motif or homology could be identified in this region, which has a very hydrophilic profile *(Breviario et al., 1992)*.

THE ISOLATION OF THE HUMAN GENOMIC CLONE

The screening of a human placenta library with the full length cDNA probe allowed us to isolate two partially overlapping phage inserts containing the entire genomic information for human PTX3 *(Breviario et al., 1992)*. The gene is organized into three exons, the third one coding exactly for the 3' half of PTX3 protein which shows homology with the other pentraxins *(Breviario et al., 1992)*.

The gene is localized on the human chromosome 3q2.5 as determined by Southern blots analysis of somatic cell hybrids and by in situ hybridization *(Breviario et al., 1992)*. In the light of the similarities and the differences between PTX3 and the classical pentraxins, it is interesting to note that the other two pentraxins are organized in two exons (the first one coding for the signal peptide), and the human CRP and SAP are localized on chromosome 1 *(Breviario et al., 1992)*.

STUDIES ON THE EXPRESSION OF PTX3

Northern analysis confirmed that PTX3 is an IL-1 responsive gene in human endothelial cells with a transient peak of expression between 4 and 6 hours *(Breviario et al., 1992)*. In addition we found that PTX3 is inducible upon stimulation with αTNF but not with IL-6. This same pattern of expression and responsiveness has been found also in hepatocytes and fibroblasts *(Breviario et al., 1992)*. This analysis clearly shows major differences with CRP and SAP genes whose expression is mainly hepatic and dependent upon IL-1 and IL-6 stimulation.

We were intrigued, on the one hand, by its promiscuous responsiveness in several different cell types and, on the other hand, by one report of very low levels of expression of CRP (detectable only after Polymerase Chain Reaction amplification) in a lymphocyte subpopulation *(Kuta and Baum, 1986)*. Therefore we wondered whether PTX3 expression could also be observed in human circulating leukocytes. Indeed we have observed that LPS, IL-1 and TNF are all able to induce PTX3 expression in human mononuclear cells with the usual rapid and transient kinetic profile *(Vidal Alles et al.,1994)*. In order to test the sensitivity of PTX3 mRNA response in human circulating cells to LPS, a dose response experiment was performed. Even very low doses of LPS (2ng/ml) are able to induce mRNA expression in human mononuclear cells (Fig. 2). Subsequent studies have clarified that the monocytes are the unique responder population among leukocytes, and that at least part of the monocytic differentiation lineage has retained this capacity from immature blasts to *in vitro* derived macrophages *(Vidal Alles et al.,1994)*. Interestingly enough, the cytokine-induced PTX3 expression is protein

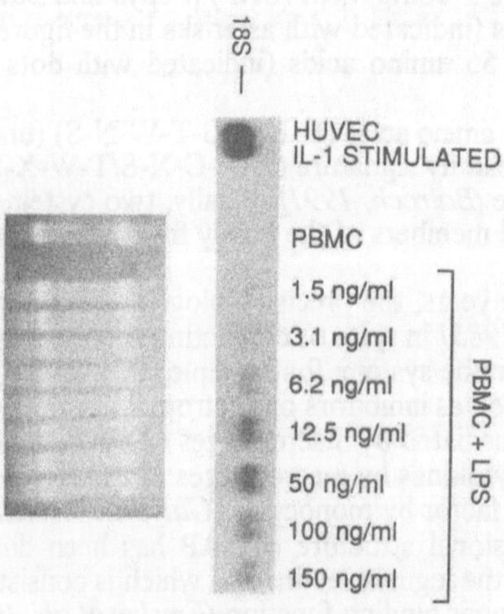

Figure 2. Dose response curve to LPS by human peripheral blood mononuclear cells (PBMC). The PBMC were exposed to the indicated doses of LPS for 6 hours and then total RNA was isolated. 18s indicate the position of the small ribosomal RNA.

synthesis independent in the monocytic cell line U937 as it is in the HUVEC (Breviario et al., 1991) but seems to require some protein intermediate in the peripheral monocytes *(Vidal Alles et al.,1994)*.

Finally, we have observed the production and the release of PTX3 protein in the supernatants of cytokine stimulated endothelial, fibroblastic and monocytic cells *(Breviario et al., 1992; Vidal Alles et al.,1994)* which take place in the 48-72 hours following the exposure to the signal, and therefore much later respect to the transient appearance of the mRNA (Breviario et al., 1992). One interesting peculiarity of PTX3 expression is its restricted responsiveness to IL-1 and TNF in all different cell types tested so far.

To test for the specificity of PTX3 induction in endothelial cells, we decided to expose identical cultures of HUVEC to two other well defined cytokines, i.e., interferon γ (IFNγ) and Fibroblast Growth Factor (FGF), because are representative of the inducers of the "immunological" and the "proliferative" metabolic programs, respectively, which can schematically be differentiated from the "prothrombotic/proinflammatory" program induced by the IL-1 in endothelial cells *(Mantovani et al., 1992; Mantovani and Dejana, 1989)*.

Interestingly enough, while IL-1 and TNF are able to induce PTX3 transcription, both FGF and IFNγ are unable to do so above background levels (Fig. 3).

CONCLUSION AND SPECULATIONS

All in all, these and other data (data not shown) favor the hypothesis that PTX3 may well represent a selective marker for proinflammatory activation of the endothelial cells *(Mantovani and Dejana, 1989)*.

Due to its belonging to the classical pentraxin family *(Breviario et al., 1992)* and to the observation that PTX3 is indeed produced and released by normal cell types, PTX3 may well represent a novel marker of the acute phase.

More interestingly, PTX3 can be produced also by endothelial cells directly lining the blood vessels, and, finally, its induction seems to be restricted to few cytokines representative of the prototypic tissue inflammation *(Mantovani and Dejana, 1989)*.

In the near future, with the aid of monoclonal antibodies directed against the recombinant protein, we will investigate whether PTX3 behaves as an acute phase reactant in animal species with particular interest for the clinical situations of *"in vivo"* endothelial cell activation.

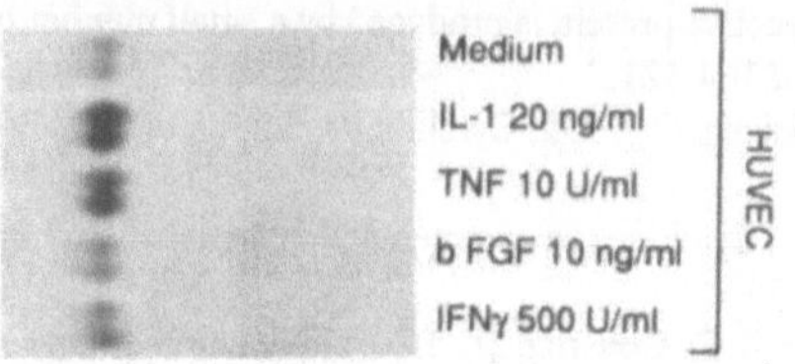

Figure 3. PTX3 expression in HUVEC exposed to different cytokines. HUVEC have been exposed to the indicated cytokines for 4 hours and then total RNA was isolated.

ACKNOWLEDGEMENTS

This work was supported in part by Ministero Sanità - Istituto Superiore Sanità (National AIDS Project), by Consiglio Nazionale delle Ricerche, Roma (Progetto Finalizzato A.C.R.0.) and by Italian Association for Cancer Research (AIRC). The financial support of telethon-Italy (project n.371) is gratefully acknowledged.

REFERENCES

Anagnostou, A., Lee, E.S., Kessimian, N., Levinson, R. and Steiner, M., 1990, Erythropoietin has a mitogenic and positive chemotactic effect on endothelial cells. *Proc. Nat. Acad. Sci. U.S.A.* 87:5978-5982.

Bairoch, A., 1991, PROSITE: a dictionary of sites and patterns in proteins. *Nucleic Acids Res.* 19 Suppl :2241-2245.

Bevilacqua, M.P., Stengelin, S., Gimbrone, M.A.J. and Seed, B., 1989, Endothelial leukocyte adhesion molecule 1: An inducible receptor for neutrophils related to complement regulatory proteins and lectins. *Science* 243:1160-1165.

Breviario, F., D'Aniello, E.M., Golay, J., Peri, G., Bottazzi, B., Bairoch, A., Saccone, S., Marzella, R., Predazzi, V., Rocchi, M., Della Valle, G., Dejana, E., Mantovani, A., Introna, M, 1992, Interleukin-1-inducible genes in endothelial cells. Cloning of a new gene related to C-reactive protein and serum amyloid P component. *J. Biom. Chem.* 267:22190-22197.

Bussolino, F., Wang, J.M., Defilippi, P., Turrini, F., Sanavio, F., Edgell, C.J., Aglietta, M., 1989, Granulocyte- and granulocyte-macrophage-colony stimulating factors induce human endothelial cells to migrate and proliferate. *Nature* 337:471-473.

D'Aniello, E.M., Breviario, F., Martin Padura, I., Lampugnani, M.G., Dejana, E., Mantovani, A., Introna, M., 1993, Interleukin-1 and Tumor Necrosis Factor induce transient expression of an inhibitor of nuclear factor κB in endothelial cells. *Endothelium* 1:161-165.

Emsley, J:, White, H.E., O'Hara, B.P., Oliva, G., Srinivasan, N., Tickle, I.J., Blunbell, T.L., Pepys, M.B., Wood, S.P., 1994, Structure of pentameric human serum amyloid P component. *Nature* 367:338-345.

Galve-de Rochemonteix, B., Wiktorowicz, K., Kushner, I., Dayer, J.M., 1993, C-Reactive protein increases production of IL-lα, IL-1 and TNF-a and expression of mRNA by human alveolar macrophages. *J. Leukocyte Biol.* 53:439.

Gramse, M., Beviario, F., Pintucci, G., Millet, I., Dejana, E., van Damme, J., Donati M.B., Mussoni, L., 1986, Enhancement by Interleukin-1 (IL-1) of plasminogen activator inhibitor (PA-I) activity in cultured human endothelial cells. *Biochem. Biophys. Res. Comun.* 139:720-727.

Haskill, S., Beg, A.A., Tompkins, S.M., Morris, J.S., Yurochko, A.D., Sampson-Johannes, A., Monbal, K., Ralph, P., Baldwin Jr., A.S., 1991, Characterization of an immediate-early gene induced in adherent monocytes that encodes I kappa B-like activity. *Cell* 65:1281-1289.

Introna, M., Breviario, F., D'Aniello, E.M., Golay, J., Dejana, E., Mantovani, A.. 1993, IL-1 inducible genes in human umbilical vein endothelial cells. *European Heart J.* 14:78-81.

Kew, R.R., Hyers, T.M., Webster, R.0., 1990, Human C-reactive protein inhibits neutrophil chemotaxis *in vitro*: Possible implications for the adult respiratory distress syndrome. *J. Lab. Clin. Med.* 115:339.

Kojima, S., Tadenuma, H., Inada, Y. and Saito, Y., 1989, Enhancement of plasminogen activator activity in cultured endothelial cells by granulocyte colony-stimulating factor. *J. Cell Physiol.* 138:192-196.

Kuta, A.E., Baum, L.L., 1986, C-Reactive protein is produced by a small number of normal human peripheral blood lymphocytes. *J. Exp. Med.*164:321.

Mantovani, A., Bussolino, F. and Dejana, E., 1992, Cytokine regulation of endothelial cell function. *FASEB J.* 6:2591-2599.

Mantovani, A. and Dejana, E., 1989, Cytokines as communication signals between leukocytes and endothelial cells. *J. Immunology* 10:370-375.

Mortensen, R.F., Osmand, A.P., Lint,T.F., Gewurtz, H., 1976, Interaction of Creactive protein with lymphocytes and monocytes: Complement-dependent adherence and phagocytosis. *J. Immunol.* 117:774.

Pepys, M.B. and Baltz, M.L., 1983, Acute phase proteins with special reference to C-reactive protein and related proteins (pentaxis) and serum amyloid A protein. *Adv. Immunol.* 34:141-160.

Rajavashisth, T.B., Andalibi, A., Territo, M.C., Berliner, J.A., Navab, M., Fogelman, A.M., Lusis, A.J., 1990, Induction of endothelial cell expression of granulocyte and macrophage colony-stimulating factors by modified low density lipoproteins. *Nature* 344:254-257.

Sarlo, K. T. and Mortensen, R.F., 1985, Enhance interleukin-1 (IL-1) production mediated by mouse serum amyloid P component. *Cell Immunol.* 93:398.

Sironi, M., Breviario, F., Proserpio, P., Biondi, A., Vecchi, A., Van Damme, J., Dejana, E., Mantovani, A., 1989, IL-1 stimulates IL-6 production in endothelial cells. *J. Immunol.* 142:549-553.

Strieter, R.M., Kunkel, S.L., Showell, H.J., Remick, D.G., Phan, S.H., Ward, P. A., Marks, R.M., 1989, Endothelial cell gene expression of a neutrophil chemotactic factor by TNF-alpha, LPS, and IL-1 beta. *Science* 243:1467- 1469.

Wen, D.Z., Rowland, A. and Derynck, R, 1989, Expression and secretion of gro/MGSA by stimulated human endothelial cells. *EMBO J.* 8:1761-1766.

Whelan, J., Ghersa, P., Hooft Van Huijsduijnen, R., Gray, J., Chandra, G., Talabot, F., Delamarter,J. F., 1991, An NF kappa B-like factor is essential but not sufficient for cytokine induction of endothelial leukocyte adhesion molecule 1 (ELAM-1) gene transcription. *Nucleic Acids Res.* 19:2645- 2653.

Zahedi, K., Mortensen, R.F., 1986, Macrophage tumoricidal activity induced by human C-reactive protein. *Cancer Res.* 46:5077.

PRO-INFLAMMATORY ADHESINS

Dorian O. Haskard

Department of Medicine
Royal Postgraduate Medical School
Hammersmith Hospital
Du Cane Road, London W12 ONN

INTRODUCTION

Because endothelium forms the first barrier between circulating leucocytes and the tissues, adhesion interactions between endothelial cells and leukocytes are central to the control of leucocyte recruitment into inflammatory sites and are also critical for leukocyte-mediated vascular injury. Over the past ten years there has been an enormous increase in understanding of the cellular and molecular nature of adhesion interactions between leukocytes and EC.

CLASSIFICATION OF PRO-INFLAMMATORY ADHESINS

The molecules that have been most fully characterized as being involved in leukocyte interactions with EC can be classified both in terms of their molecular structure and in terms of their specialized adhesive functions (Table 1). The nature of these molecules has been well reviewed in a number of recent articles (*eg., Bevilacqua, 1993; Springer, 1994*).

Leukocyte rolling

Direct observation of the microvasculature *in vivo* shows that the first event in leukocyte emigration into inflammatory lesions is leukocyte tethering, following which leukocytes roll on the endothelial lining of the vessel wall (*Atherton and Born, 1972*). There is now a large body of evidence implicating the involvement of the three selectin molecules (L-selectin (CD62L), P-selectin (CD62P) and E-selectin (CD62E) in this process, both *in vivo* (*Ley et al., 1991; Von Andrian et al., 1992; Doré et al, 1993; Mayadas et al., 1993*) and *in vitro* (*Lawrence and Springer, 1991; Smith et al., 1991; Lawrence and Springer, 1993; Abbassi et al., 1993*). The ability of the selectins to mediate rolling, but not firm adhesion (see below), is thought to be related to the fast association and dissociation constants of the adhesive bonds between selectins and their ligands under conditions of flow (*Williams, 1991*).

Selectins are single-chain glycoproteins, each consisting of an N-terminal domain homologous to C-type animal lectins, an epidermal growth factor-like domain and a series of short consensus repeat units homologous to those found in complement regulatory proteins. As might be expected from the lectin-like N-terminal domains, each of the three molecules has been shown to bind specific carbohydrate structures. There is considerable overlap in the abilities of the selectins to recognize isolated oligosaccharides, and they are each able to bind sialyl-Lewis x (CD15s) and sialyl-Lewis a (*Berg et al., 1991; Polley et al., 1991; Berg et al., 1992; Tsan et al., 1992; Foxall et al., 1992*). Furthermore both P- and E-selectin are able to recognize a novel cell surface mucin designated PSGL-1 (*Sako et al., 1994*). On the other hand

Table 1. Pro-Inflammatory adhesins involved in leukocyte-endothelial cell interactions

Molecule	Distribution	Ligands
Rolling		
L-selectin	Neutrophils Eosinophils Monocytes Lymphocyte subpopulation	Sialyated, sulphated carbohydrates decorating GlyCAM-1, CD34 MadCAM-1 and others
P-selectin	Activated platelets Activated EC	Sialyated Le^x decorating PSGL-1 and others
E-selectin	Cytokine-activated EC	Sialyated Le^x decorating ESL-1 and ? others
Firm adhesions		
LFA-1 (αLβ2)	All leukocytes	ICAM-1, ICAM-2, ICAM-3
Mac-1 (αMβ2)	Myeloid cells Lymphocyte subpopulation	ICAM-1, iC3b, fibrinogen Factor X
p150,95 (αXβ2)	Myeloid cells Lymphocyte subpopulation	iC3b, fibrinogen
VLA-4 (α4β1)	Memory lymphocytes, Monocytes	VCAM-1, fibronectin
LPAM-1 (α4β7)	Lymphocyte subpopulation	MadCAM-1, VCAM-1, fibronectin
Transmigration		
CD31	EC, Neutrophils, Monocytes, Lymphocyte subpopulation	CD31 and other(s)

it is becoming clear that it is possible to define experimental conditions in which particular selectins show preferential adhesion (*Aruffo et al., 1992; Larsen et al., 1992; Steininger et al., 1992; Vestweber et al., 1994; Kansas et al., 1994*) suggesting that the affinity of selectins for carbohydrate residues is influenced by the context in which the carbohydrate is presented. Recently Kansas and colleagues have reported that the epidermal growth factor domain of P-selectin may be critical in determining the binding specificity of P-selectin, and that a protein-protein interaction may occur between this region of P-selectin and the P-selectin ligand (*Kansas et al., 1994*). It is also possible that the specificity of the carbohydrate for individual selectins may be influenced by sulphation or other modification (*Imai et al., 1993*).

Leukocyte activation

The process of rolling slows the passage of leukocytes along the vessel by up to several hundred fold and facilitates their exposure to activating factors arising from the local inflammatory microenvironment. In addition to classical chemoattractants such as FMLP, platelet activating factor and activated C5, there is a rapidly expanding list of factors that potentially may influence leukocyte activation on the blood-tissue interface. Of emerging importance is the family of 70-80 amino acid peptides known as chemokines or intercrines, which includes factors such as interleukin-8, monocyte chemotattractant protein-1, and macrophage inflammatory proteins 1α and 1ß (*Oppenheim et al., 1991*). These and other chemokines act on leukocytes through specific G-protein-coupled surface molecules known as serpentine receptors, which are variably expressed on leukocytes of different lineage. Heterogeneity between leukocytes in responsiveness to the large number of potential chemoattractants is likely to be at least as important as heterogeneity of adhesion molecule expression in determining the selectivity of leukocyte traffic into inflamed tissues (see below). A notable feature of many chemokines is that they are heparin-binding proteins and may be presented to the rolling leukocyte bound to proteoglycans and other negatively charged structures on the endothelial cell surface (*Rot, 1993; Tanaka et al., 1993a; Tanaka et al., 1993b*).

Firm adhesion

One of the main effects of chemoattractants on leukocytes is to alter the conformation of surface integrins to a high affinity state capable of binding their ligands (*Hynes, 1992*). The mechanisms whereby the affinity of integrins becomes enhanced are complex but appear to include both signalling via the cytoplasmic domains ("inside-out signalling") (*Dustin and Springer, 1989*) and conformational adjustments secondary to initial contact with ligand ("ligand-induced binding") (*Cabanas and Hogg, 1993*). The upregulated affinity of leukocyte integrins occurs very rapidly, thereby promoting firm adhesion of rolling leukocytes to endothelial cells. Importantly, the process is also rapidly reversible, thus permitting deadhesion and migration into the tissues (see below).

The integrins of particular importance for leukocyte adhesion to EC are the three ß2 integrins LFA-1 (αLß2), Mac-1 (αMß2), p150,95 (αXß2) (*Larson and Springer, 1990*) and the two α4 integrins VLA-4 (α4ß1) (*Hemler et al., 1990*) and LPAM-1 (α4ß7) (*Holzmann and Weissman, 1989*). Of the ß2 integrins, LFA-1 is expressed by virtually all leukocytes, whilst Mac-1 and p150,95 are predominantly expressed by cells of the myeloid series and are only found on relatively small subsets of lymphocytes. On the other hand α4 integrins are not expressed by neutrophils and may be particularly important for the selective accumulation of mononuclear cells and eosinophils in allergic and chronic inflammatory lesions.

The best characterized ligands for integrins on EC are intercellular adhesion molecules-1 and -2 (ICAM-1, CD54 and ICAM-2, CD102), vascular cell adhesion molecule-1 (VCAM-1, CD106) and the mucosal addressin MadCAM-1. These molecules are members of the immunoglobulin (Ig) superfamily, each being single chain glycoproteins with Ig domains arranged in tandem. ICAM-1 has five Ig domains with binding sites for LFA-1 on the first and possibly the second domains nearest the N-terminal (*Berendt et al., 1992*), and for Mac-1 on the central third domain (*Diamond et al., 1991*). ICAM-2 is a smaller molecule with just two Ig domains which bind LFA-1 but not Mac-1 (*Staunton et al., 1989*).

VCAM-1, which in humans has isoforms with six or seven Ig domains, binds VLA-4 (*Elices et al., 1990*) and also LPAM-1, although this latter interaction is of lower affinity (*Chan et al., 1992; Rüegg et al., 1992*). Evidence has recently been provided that the N-terminal domains one and two support VLA-4 binding on both six and seven domain forms of VCAM-1 (*Osborn et al., 1994*). In addition, the seven domain form has another VLA-4 binding site provided by the inserted fourth domain (*Osborn et al., 1992, Vonderheide et al., 1994*).

MadCAM-1, which is the most recently described ligand for a leukocyte integrin on EC, has a novel structure composed of two N-terminal Ig domains with homology to VCAM-1 and ICAM-1 (domain 1) and VCAM-1 (domain 2), together with a third proximal Ig domain with homology to IgA1 (*Briskin et al., 1993)*. This molecule, which has a distribution which is restricted to the mucosal lymphoid system, acts as a ligand for LPAM-1 (*Berlin et al., 1993*). Interestingly, the second and third Ig domains are separated by a mucin-like region which is now thought to be capable of binding L-selectin under shear, providing a mechanism whereby rolling and firm integrin-mediated adhesion can be mediated by the same endothelial ligand (*Berg et al., 1993*).

Apart from these integral membrane proteins, it is possible that luminally expressed matrix molecules may play a role in binding leukocyte integrins to EC. Thus, Elices et al., have recently reported that alternatively spliced fibronectin containing the CS1 peptide sequence can be detected on the luminal surface of vascular endothelium in rheumatoid synovium and may act as a ligand for VLA-4 (*Elices et al., 1994*).

Transmigration

Following firm adhesion to the luminal surface of EC, leukocytes transmigrate through endothelium into the tissues, guided by gradients of chemoattractants (chemotaxis) and adhesion ligands (haptotaxis). The process of transmigration is not yet well understood but in many experimental situations has been found to depend on the function of ß2 integrins (*Smith et al., 1989; Van Epps et al., 1989; Hakkert et al., 1991; Kavanaugh et al., 1991*). Recent attention has focused on a possible role of CD31 (PECAM-1) in leukocyte transmigration, as monoclonal antibodies against this molecule inhibit neutrophil and monocyte passage across cultured EC monolayers *in vitro* (*Muller et al., 1993*) and also inhibit the accumulation of leukocytes in inflamed skin (*Vaporciyan et al., 1993*) or peritoneum (*Bogen et al., 1994*) *in vivo*. The precise mechanism whereby CD31 participates in transmigration is unclear, but may include signalling and the upregulation of integrin affinities (*Tanaka et al., 1992*).

DETERMINANTS OF SELECTIVITY

Although the processes of rolling, activation, firm adhesion and transmigration have been best described for neutrophils, it is likely that the same principles apply also to lymphocytes (*Bargatze and Butcher, 1993*) and leukocytes of other lineage. The large degree of diversity of adhesion molecules and chemoattractants involved in these steps makes it difficult to explain the selectivity of leukocyte traffic into inflamed tissues on the basis of a single "lock and key" homing interaction between leukocytes and EC. It is more likely that whether or not a particular leukocyte migrates into a lesion depends upon its being able to participate in each step of the adhesion cascade described above (*Butcher, 1991*). This model, which has been referred to as the "area code model", now best explains how endothelium can act as a filter for different leukocyte subsets (*Springer, 1994*). With this model in mind, the endothelial adhesion molecules which might influence selective leukocyte traffic into extra-lymphoid tissue in inflammation are P- and E-selectin and the ligands for $\alpha 4$ integrins (VCAM-1, MadCAM-1 and the CS1 peptide of fibronectin). Thus, only a relatively small subpopulation of lymphocytes are able to bind the two endothelial selectins, whilst neutrophils do not express $\alpha 4$ integrins.

THE CONTROL OF ENDOTHELIAL CELL ADHESION MOLECULE EXPRESSION

A large number of studies have explored how EC adhesion molecule expression is controlled *in vitro*. There are two principal mechanisms whereby the expression of EC adhesion molecules is regulated.

Rapid protein-synthesis independent activation

Stimulation of EC with agonist such as histamine, thrombin or C5b-9 complexes cause P-selectin to be rapidly expressed through a mechanism involving the translocation of presynthesized P-selectin from Weibel-Palade bodies to the cell surface (*Hattori et al., 1989a; Hattori et al., 1989b; Geng et al., 1990*). Although the expression of P-selectin by this mechanism is transient, there is evidence that oxidants can cause a more delayed and sustained expression of P-selectin (*Patel et al., 1991*). Furthermore, P-selectin expression has been found to be cytokine inducible (see below) in mouse and bovine EC (*Weller et al., 1992; Sanders et al., 1992*).

Protein synthesis-dependent activation

Stimulation of cultured EC with interleukin-1 (IL-1), tumor necrosis factor (TNF) or lipopolysaccharide (LPS) results in the induction of gene transcription and de novo protein synthesis of a number of pro-inflammatory factors including E-selectin, ICAM-1 and VCAM-1 (*Pober and Cotran, 1990*). Although IL-1 and TNF are the best characterized cytokines in this regard, there are a number of other factors that have been shown to be capable of altering expression of these molecules. Examples include the selective upregulation of ICAM-1 expression by interferon gamma (IFNγ) (*Pober et al., 1986b*), the selective induction of VCAM-1 by IL-4 (*Thornhill and Haskard, 1990*), the upregulation of E-selectin expression by thrombin (*Shankar et al., 1992*) and IL-3 (*Brizzi et al., 1993*), the induction of VCAM-1 by IL-13 (*Peri et al., 1994*), the upregulation of ICAM-1 expression by hydrogen peroxide (*Lo et al., 1993; Bradley et al., 1993*), the upregulation of VCAM-1 and ICAM-1 by l-ysophosphatidylcholine, a component of atherogenic lipoproteins (*Kume et al., 1992*) and the upregulation of E-selectin, ICAM-1 and VCAM-1 by direct contact with T cells (*Damle et al., 1991; Yarwood et al., 1994*). Further investigation is needed to establish the relative importance of these mechanisms for different pathological situations. In many inflammatory tissues *in vivo* endothelial cells are likely to be exposed to a number of pro-inflammatory agents. It is therefore of interest to explore the effects on adhesion molecule expression of stimulating EC with combinations of cytokines. Perhaps the best studied combinations are those of TNF + IFNγ and TNF + IL-4, which tend to have additive or synergistic effects that may be of relevance to the mode and duration of endothelial cell activation in immune-mediated inflammatory responses (*Dustin and Springer, 1988; Thornhill and Haskard, 1990; Doukas and Pober, 1990; Leeuwenberg et al., 1990; Thornhill et al., 1991; Paleolog et al., 1992*).

The second messenger systems which initiate the induction of the cytokine responsive adhesion molecules are poorly understood. Although activating protein kinase C (PKC) by stimulating human umbilical endothelial cells with phorbol esters leads to increased expression of E-selectin, ICAM-1 and VCAM-1, experiments with PKC inhibitors indicate that PKC activation is not critical for expression of these molecules in response to TNF and other cytokines (*Ritchie et al., 1991*). With cultured dermal microvascular EC this conclusion is even more definite, since phorbol esters are less potent in stimulating expression of the three adhesion molecules (*Mason et al., 1994*). Although the immediate second messenger systems are poorly understood, there is increasing evidence for a role of the transcription factor NFkB in the induction of gene transcription in IL-1 and TNF regulated adhesion molecules (*Bradley et al., 1993, Read et al., 1994*).

The mechanisms whereby adhesion molecule expression is downregulated are as important as the mechanisms of upregulation. Both E-selectin and ICAM-1 have relatively fast turn-over times in the cell membrane, with proteins being removed both by internalization and cleavage (*von Asmuth et al., 1992*). It is therefore possible that cleavage from the cell surface accounts for the soluble forms of E-selectin, VCAM-1 and ICAM-1 that are detectable in plasma (*Gearing and Newman, 1993; Wellicome et al., 1993*). Importantly, EC have an intrinsic capacity to downregulate their responsiveness to cytokines, as shown particularly by the kinetics of E-selectin expression *in vitro* (*Pober et al., 1986a*). Recently it has become evident that amongst the genes induced by IL-1 and TNF are those of proteins of the IkB family which function to inhibit NFkB activity and which may downregulate endothelial activation at later stages of the inflammatory response (*De Martin et al., 1993, Read et al., 1994*).

HETEROGENEITY OF ENDOTHELIAL CELL RESPONSES TO CYTOKINES

Since a large proportion of our understanding of the responsiveness of EC to activation comes from studies utilizing EC derived from human umbilical veins, there has been much debate as to what degree conclusions can be generalized to other types of EC. It is becoming increasingly apparent that although the general principles of rapid and cytokine-mediated activation probably apply to most EC, there are clear differences to be found when comparing different forms of EC *in vitro* or *in vivo*. For example, although IL-1 can induce both E-selectin and VCAM-1 expression by HUVEC, it induces E-selectin but not VCAM-1 expression by dermal microvascular EC (DMEC). In contrast TNF leads to expression of both molecules by both HUVEC and DMEC (*Swerlick et al., 1992*). There is also evidence from the immunocytochemical analysis of vessels *in vivo* that EC are heterogeneous in the capacity to express cytokine-inducible adhesion molecules (*Fries et al., 1993, Keelan et al., 1994a*).

THE KINETICS OF E-SELECTIN EXPRESSION *IN VIVO*

Although experiments with cultured EC have shown that expression of adhesion molecules is tightly regulated, it has been less clear to what degree the time-course of expression of individual adhesion molecules relates to the time-course of emigration of different types of leukocyte into the tissues during inflammatory responses *in vivo*. To an extent, this question can be addressed by comparing the immunocytochemical staining of vessels by specific monoclonal antibodies with the presence in the tissues of leukocytes of different lineage at different stages of developing lesions (*Norris et al., 1991*). However, this approach has a number of technical limitations including difficulties quantifying antigen expression by immunocytochemistry, the definition of luminal rather than cytoplasmic antigen expression and the separation of migrating from sessile tissue leukocytes.

We have found that it is possible to use a radiolabelled monoclonal antibody (mAb 1.2B6) to quantify E-selectin expression *in vivo* with a precision that has not previously been possible using immunohistochemical techniques. This antibody was generated against E-selectin on human EC (*Wellicome et al., 1990*) but also reacts with porcine E-selectin (*Tsang Y et al: manuscript in preparation*). By studying the uptake of radiolabelled mAb 1.2B6 in the tissues following its intravenous injection we have obtained quantitative data on E-selectin expression in both systemic and localized inflammatory responses (*Keelan et al., 1993a*).

The ability to quantify E-selectin expression in the pig has allowed the comparison of the time-course of expression of this molecule in relation to leukocyte traffic, using a model of cutaneous inflammation previously characterized for lymphocyte migration (*Binns et al., 1990; Binns et al., 1992*). We found that the injection of four separate agents IL-1, TNF, phytohaemagglutinin (PHA) and phorbol myristate acetate (PMA), into pig skin led to a stereotyped time-course of neutrophil accumulation with maximal localization at 2 hr followed by a sharp drop in the ability of neutrophils to localize over the ensuing hours. In contrast, the time-course of lymphocyte accumulation was found to be stimulus-dependent, peaking at about 3 hours with IL-1, 6 hours with TNF, and 8-9 hours with PHA. Whilst the onset of E-selectin expression parallelled the onset of neutrophil rather than that of lymphocyte accumulation, the duration of E-selectin expression with each of the mediators was more long-lasting than that of neutrophil localization and showed mediator specific time-courses similar to those seen for lymphocytes (*Binns et al., 1994*). One can conclude from these experiments that E-selectin may be involved in the accumulation in inflamed skin of both neutrophils and lymphocytes, but that the kinetics of expression of the molecule cannot fully account for the kinetics of accumulation of either cell type. Further work in this area is directed at characterizing the roles of other adhesion molecules and chemokines in determining the kinetics of leukocyte traffic into these lesions.

IMAGING E-SELECTIN

The ability to specifically localize E-selectin expression *in vivo* using radiolabelled an anti-E-selectin monoclonal antibody offers the opportunity to non-invasively image expression of this molecule. We have performed preliminary studies to this end using a porcine model of arthritis in which inflammation is stimulated by intraarticular injection of phytohaemagglutinin (*Binns et al., 1990*). Using mAb labelled with 111-In, we have shown that it is possible to image E-selectin expression not only in the inflamed joint but also in the activated draining regional lymph nodes (*Keelan et al., 1993b*). More recently we have found that a better signal to noise ratio is obtained using F(ab')2 fragments (*Jamar et al.*, 1994). Interestingly, imaging the inflamed joint with mAb 1.2B6 F(ab')2 fragments gave superior images to those obtained with radiolabelled neutrophils (*Jamar et al., 1994*), suggesting that this technique may be useful for the localization and quantification of activated inflammation in the clinical setting.

ADHESION MOLECULES AS THERAPEUTIC TARGETS

There is now a substantial literature on the effects of inhibiting individual adhesion molecules *in vivo* using monoclonal antibodies (*Keelan and Haskard, 1992*). This body of work has established that targeting any one of a number of adhesion molecules can have anti-inflammatory effects, consistent with a model in which adhesion molecules act sequentially rather than in parallel during leukocyte emigration. However, the majority of studies have tended to focus on relatively acute models, and there is less information on the effects of blocking adhesion molecules in chronic inflammation. It is likely that the various genetically-engineered strains of mice with adhesion molecule deletions will go some way to addressing this issue (*Wilson et al., 1993; Sligh et al., 1993; Mayadas et al., 1993*). These mice will also be useful for testing the long-term safety of inhibiting adhesion molecule function.

ACKNOWLEDGEMENTS

I would like to acknowledge the important contributions of Drs. Richard Binns, Peter Chapman, Andrew Harrison, Francois Jamar, Edward Keelan, Justin Mason, Mike Peters, Martyn Robinson, and Helen Yarwood to the recent work described in this article.

REFERENCES

Abbassi, O., Kishimoto, T.K., McIntire, L.V., Anderson, D.C. and Smith, C.W., 1993, E-selectin supports neutrophil rolling under conditions of flow. *J. Clin. Invest.*, 92:2719.

Aruffo, A., Dietsch, M.T., Wan, H., Hellström, K.E. and Hellström, I., 1992, Granule membrane protein 140 (GMP140) binds to carcinomas and carcinoma-derived cell lines. *Proc. Natl. Acad. Sci. USA*, 89:2292.

Atherton, A. and Born, G.V.R., 1972, Quantitative investigations of the adhesiveness of circulating polymorphonuclear leukocytes to blood vessel walls. *J. Physiol.*, 222:447.

Bargatze, R.F. and Butcher, E.C., 1993, Rapid G protein-regulated activation event involved in lymphocyte binding to high endothelial venules. *J. Exp. Med.*, 178:367.

Berendt, A.R., McDowall, A., Craig, A.G.,1992, et al. The binding site on ICAM-1 for plasmodium falciparum-infected erythrocytes overlaps, but is distinct from, the LFA-1-binding site. *Cell*, 68:71.

Berg, E.L., Robinson, M.K., Mansson, O., Butcher, E.C. and Magnani, J.L., 1991, A carbohydrate domain common to both sialyl Le^a and sialyl Le^X is recognized by the endothelial cell leukocyte adhesion molecule ELAM-1. *J. Biol. Chem.*, 266:14869.

Berg, E.L., Magnani, J., Warnock, R.A., Robinson, M.K. and Butcher, E.C., 1992, Comparison of L-selectin and E-selectin ligand specificities: The L-selectin can bind the E-selectin ligands sialyl Le^x and sialyl Le^a. *Biochem. Biophys. Res. Commun.*, 184:1048.

Berg, E.L., McEvoy, L.M., Berlin, C., Bargatze, R.F. and Butcher, E.C., 1993, L-selectin-mediated lymphocyte rolling on MadCAM-1. *Nature*, 366:695.

Berlin, C., Berg, E.L., Briskin, M.J.,1993, et al. α4β7 integrin mediates lymphocyte binding to the mucosal vascular addressin MAdCAM-1. *Cell*, 74:185.

Bevilacqua, M.P., 1993, Endothelial-leukocyte adhesion molecules. *Annu. Rev. Immunol.*, 11:767.

Binns, R.M., Licence, S.T. and Wooding, S.T., 1990, Phytohemagglutinin induces major short-term protease--sensitive lymphocyte traffic involving high endothelium venule-like blood vessels in acute delayed-type hypersensitivity-like reactions in skin and other tissues. *Eur. J. Immunol.*, 20:1067.

Binns, R.M., Licence, S.T., Wooding, F.B.P. and Duffus, W.P.H. , 1992, Active lymphocyte traffic induced in the periphery by cytokines and phytohemagglutinin: three different mechanisms? *Eur. J. Immunol.*, 22:2195.

Binns, R.M., Keelan, E.T.D., Licence, S.T., Harrison, A.A., Robinson, M. and Haskard, D.O., 1994, Neutrophil and lymphocyte accumulation in inflammation: lack of correlation with E-selectin expression and dependence on β2 integrins. *British Journal of Rheumatology*, 33 (suppl 1):15.

Bogen, S., Pak, J., Garifallou, M., Deng, X. and Muller, W.A. , 1994, Monoclonal antibody to murine PECAM-1 (CD31) blocks acute inflammation in vivo. *J Exp Med*, 179:1059.

Bradley, J.R., Johnson, D.R. and Pober, J.S., 1993, Endothelial activation by hydrogen peroxide. *Am. J. Pathol.*, 142:1598.

Briskin, M.J., McEvoy, L.M. and Butcher, E.C., 1993, MAdCAM-1 has homology to immunoglobulin and mucin-like adhesion receptors and to IgA1. *Nature*, 363:461.

Brizzi, M.F., Garbarino, G., Rossi, P.R.,1993, et al. Interleukin 3 stimulates proliferation and triggers endothelial-leukocyte adhesion molecule 1 gene activation of human endothelial cells. *J. Clin. Invest.*, 91:2887.

Butcher, E.C., 1991, Leukocyte-endothelial cell recognition: Three (or more) steps to specificity and diversity. *Cell*, 67:1033.

Cabanas, C. and Hogg, N., 1993, ICAM-1 is required for activation of LFA-1. *Proc Natl Acad Sci USA*, 90:5838.

Chan, B.M., Elices, M.J., Murphy, E. and Hemler, M.E., 1992, Adhesion to vascular cell adhesion molecule 1 and fibronectin. Comparison of alpha 4 beta 1 (VLA-4) and alpha 4 beta 7 on the human B cell line JY. *J Biol Chem.*, 267:8366.

Damle, N.K., Eberhardt, C. and Van der Vieren, M., 1991, Direct interaction with primed CD4⁺ CD45R0⁺ memory T lymphocytes induces expression of endothelial leukocyte adhesion molecule-1 and vascular cell adhesion molecule-1 on the surface of vascular endothelial cells. *Eur. J. Immunol.*, 21:2915.

De Martin, R., Vanhove, B., Cheng, Q., 1993, et al. Cytokine-inducible expression in endothelial cells of an IkappaBα-like gene is regulated by NFkappaB. *EMBO J.*, 12:2773.

Diamond, M.S., Staunton, D.E., Marlin, S.D. and Springer, T.A., 1993, Binding of the integrin Mac-1 (CD11b/CD18) to the third immunoglobulin-like domain of ICAM-1 (CD54) and its regulation by glycosylation. *Cell*, 65:961.

Doré, M., Korthuis, R.J., Granger, D.N., Entman, M.L. and Smith, C.W., 1993, P-selectin mediates spontaneous leukocyte rolling *in vivo*. *Blood*, 82:1308.

Doukas, J. and Pober, J.S., 1990, IFNγ enhances endothelial activation induced by tumor necrosis factor but not IL-1. *J. Immunol.*, 145:1727.

Dustin, M.L. and Springer, T.A., 1988, Lymphocyte function-associated antigen-1 (LFA-1) interaction with inter-cellular adhesion molecule-1 (ICAM-1) is one of at least three mechanisms for lymphocyte adhesion to cultured endothelial cells. *J. Cell. Biol.*, 107:321.

Dustin, M.L. and Springer, T.A., 1989, T-cell receptor cross-linking transiently stimulates adhesiveness through LFA-1. *Nature*, 341:619.

Elices, M.J., Osborn, L., Takada, Y., 1990, et al. VCAM-1 on activated endothelium interacts with the leukocyte integrin VLA-4 at a site distinct from the VLA-4/fibronectin binding site. *Cell*, 60:577.

Elices, M.J., Tsai, V., Strahl, D.,1994, et al. Expression and functional significance of alternatively spliced CS1 fibronectin in rheumatoid arthritis microvasculature. *J. Clin. Invest.*, 93:405.

Foxall, C., Watson, S.R., Dowbenko, D., 1992, et al. The three members of the selectin receptor family recognize a common carbohydrate epitope, the sialyl Lewis^x oligosaccharide. *J. Cell Biol.*, 117:895.

Fries, J.W.U., Williams, A.J., Atkins, R.C., Newman, W., Lipscomb, M.F. and Collins, T., 1993, Expression of VCAM-1 and E-selectin in an *in vivo* model of endothelial activation. *Am. J. Pathol.*, 143:725.

Gearing, A.J.H. and Newman, W., 1993, Circulating adhesion molecules in disease. *Immunology Today*, 14:506.

Geng, J.-G., Bevilacqua, M.P., Moore, K.L., 1990,et al. Rapid neutrophil adhesion to activated endothelium mediated by GMP-140. *Nature*, 343:757.

Hakkert, B.C., Kuijpers, T.W., Leeuwenberg, J.F.M., van Mourik, J.A. and Roos, D. , 1991, Neutrophil and monocyte adherence to and migration across monolayers of cytokine-activated endothelial cells: The contribution of CD18, ELAM-1, and VLA-4. *Blood*, 78:2721.

Hattori, R., Hamilton, K.K., Fugate, R.D., McEver, R.D. and Sims, P.J. , 1989a,Stimulated secretion of endothelial von Willebrand factor is accompanied by rapid redistribution to the cell surface of the intracellular granule membrane protein GMP-140. *J. Biol. Chem.*, 264:7768.

Hattori, R., Hamilton, K.K., McEver, R.P. and Sims, P.J., 1989b, Complement proteins C5b-9 induce secretion of high molecular weight multimers of endothelial von Willebrand factor and translocation of granule membrane protein GMP-140 to the cell surface. *J. Biol. Chem.*, 264:9053.

Hemler, M.E., Elices, M.J., Parker, C. and Takada, Y., 1990, Structure of the integrin VLA-4 and its cell-cell and cell-matrix adhesion functions. *Immunological reviews*, 114:45.

Holzmann, B. and Weissman, I.L., 1989, Integrin molecules involved in lymphocyte homing to Peyer's Patches. *Immunological reviews*, 108:45.

Hynes, R.O., 1992, Integrins: Versatility, modulation, and signaling in cell adhesion. *Cell*, 69:11.

Imai, Y., Lasky, L.A. and Rosen, S.D., 1993, Sulphation requirement for GlyCAM-1, an endothelial ligand for L-selectin. *Nature*, 361:555.

Jamar, F., Chapman, P.T., Harrison, A.A., Binns, R.M., Haskard, D.O., Peters, A.M., 1995, Inflammatory arthritis: imaging of endothelial activation with an indium-111-F(ab')2 fragment of anti-E-selectin monoclonal antibody. *Radiology*, 194:843.

Kansas, G.S., Saunders, K.B., Ley, K.,1994, et al. A role for the epidermal growth factor-like domain of P-selectin in ligand recognition and cell adhesion. *J. Cell. Biol.*, 124:609.

Kavanaugh, A.F., Lightfoot, E., Lipsky, P.E. and Oppenheimer-Marks, N. , 1991, Role of CD11/CD18 in adhesion and transendothelial migration of T cells: Analysis utilizing CD18-deficient T cell clones. *J. Immunol.*, 146:4149.

Keelan, E. and Haskard, D.O., 1992, CAMs and anti-CAMs: The clinical potential of leukocyte adhesion molecules. *Journal of the Royal College of Physicians*, 26:17.

Keelan, E.T.M., Licence, S.T., Peters, A.M., Binns, R.M. and Haskard, D.O., 1994a, Characterization of E-selectin expression *in vivo* using a radiolabelled monoclonal antibody. *Am. J. Physiol. Heart Circ. Physiol.*, 266:H279.

Keelan, E.T.M., Harrison, A.A., Chapman, P.T., Binns, R.M., Peters, A.M. and Haskard, D.O., 1994b, Imaging vascular endothelial activation: a novel approach using radiolabelled monoclonal antibody against the endothelial cell adhesion molecule E-selectin. *J. Nucl. Med.*, 35:*276*.

Kume, N., Cybulsky, M.I. and Gimbrone, M.A.,Jr., 1992, Lysophosphatidylcholine, a component of atherogenic lipoproteins, induces mononuclear leukocyte adhesion molecules in cultured human and rabbit arterial endothelial cells. *J. Clin. Invest.*, 90:1138.

Larsen, G.R., Sako, D., Ahern, T.J.,1992, et al. P-selectin and E-selectin. Distinct but overlapping leukocyte ligand specificities. *J. Biol. Chem.*, 267:11104.

Larson, R.S. and Springer, T.A., 1990, Structure and function of leukocyte integrins. *Immunological reviews*, 114:181.

Lawrence, M.B. and Springer, T.A., 1991, Leukocytes roll on an a selectin at physiologic flow rates: distinction from and prerequisite for adhesion through integrins. *Cell*, 65:859.

Lawrence, M.B. and Springer, T.A.,1993, Neutrophils roll on E-selectin. *J Immunol*, 151:6338.

Leeuwenberg, J.F.M., von Asmuth, E.J.U., Jeunhomme, T.M.A.A. and Buurman, W.A., 1990, Interferon-gamma regulates the expression of an adhesion molecule ELAM-1 and IL-6 by human endothelial cells *in vitro*. *J. Immunol.*, 145:2110.

Ley, K., Gaehtgens, P., Fennie, C., Singer, M.S., Lasky, L.A. and Rosen, S.D., 1991, Lectin-like cell adhesion molecule 1 mediates leukocyte rolling in mesenteric venules in vivo. *Blood*, 77:2553.

Lo, S.K., Janakidevi, K., Lai, L. and Malik, A.B., 1993, Hydrogen peroxide-induced increase in endothelial adhesiveness is dependent on ICAM-1 activation. *Am. J. Physiol. Lung Cell. Mol. Physiol.*, 264:L406.

Mason, J.C., Yarwood, H., Mahiouz, D. and Haskard, D.O. , 1994, Activation of protein kinase C leas to markedly different effects on expression of surface antigens by dermal microvascular endothelial cells (DMEC) and human microvascular endothelial cells (HUVEC). *FASEB J*, 8(4):A132.

Mayadas, T.N., Johnson, R.C., Rayburn, H., Hynes, R.O. and Wagner, D.D. , 1993, Leukocyte rolling and extravasation are severely compromised in P selectin-deficient mice. *Cell*, 74:541.

Muller, W.A., Weigl, S.A., Deng, X. and Phillips, D.M., 1993, PECAM-1 is required for transendothelial migration of leukocytes. *J. Exp. Med.*, 178:449.

Norris, P., Poston, R.N., Thomas, D.S., Thornhill, M., Hawk, J. and Haskard, D.O., 1993, The expression of endothelial leukocyte adhesion molecule-1 (ELAM-1), Intercellular adhesion molecule-1 (ICAM-1) and vascular cell adhesion molecule -1 (VCAM-1) in experimental cutaneous inflammation: a comparison of ultraviolet- B erythema and delayed hypersensitivity. *J. Invest. Dermatol.*, 96:763.

Oppenheim, J.J., Zachariae, C.O.C., Mukaida, N. and Matsushima, K., 1991, Properties of the novel proinflammatory supergene "intercrine" cytokine family. *Annu. Rev. Immunol.*, 9:617.

Osborn, L., Vassallo, C. and Benjamin, C.D. , 1992, Activated endothelium binds lymphocytes through a novel binding site in the alternately spliced domain of vascular cell adhesion molecule-1. *J. Exp. Med.*, 176:99.

Osborn, L., Vassallo, C., Browning, B.G., 1994,et al. Arrangement of domains, and amino acid residues required for binding of vascular adhesion molecule-1 to its counter-receptor VLA-4 ($\alpha4\beta1$). *J. Cell Biol.*, 124:601.

Paleolog, E.M., Aluri, G.R. and Feldmann, M., 1992, Contrasting effects of interferon y and interleukin 4 on responses of human vascular endothelial cells to tumour necrosis factor α. *Cytokine*, 4:470.

Patel, K.D., Zimmerman, G.A., Prescott, S.M., McEver, R.P. and McIntyre, T.M., 1991, Oxygen radicals induce human endothelial cells to express GMP-140 and bind neutrophils. *J. Cell Biol.*, 112:749.

Peri, G., Bernasconi, S., Sciacca, F.L., 1994, et al. Effects of Il-13 on endothelial and mesothelial cells: selective induction of VCAM-1 and stimulation of Il-6 production. *J. Cell. Biochem.*, 18A (suppl):287.

Pober, J.S., Bevilacqua, M.P., Mendrick, D.L., Lapierre, L.A., Fiers, W. and Gimbrone, M.A., 1986a, Two distinct monokines, interleukin 1 and tumor necrosis factor, each independently induce the biosynthesis and transient expression of the same antigen on the surface of cultured human vascular endothelial cells. *J. Immunol.*, 136:1680.

Pober, J.S., Gimbrone, M.A., Lapierre, L.A., 1986b, et al. Overlapping patterns of activation of human endothelial cells by interleukin 1, tumor necrosis factor, and immune interferon. *J. Immunol.*, 137:1893.

Pober, J.S. and Cotran, R.S., 1990, Cytokines and endothelial cell biology. *Physiological Reviews*, 70:427.

Polley, M.J., Phillips, M.L., Wayner, E., 1991, et al. CD62 and endothelial cell-leukocyte adhesion molecule 1 (ELAM-1) recognize the same carbohydrate ligand, sialyl-Lewis x. *Proc. Natl. Acad. Sci. USA*, 88:6224.

Read, M.A., Whitley, M.Z., Williams, A.J. and Collins, T., 1994, NF-kB and IkBα: an inducible regulatory system in endothelial activation. *J Exp Med*, 179:503.

Ritchie, A.J., Johnson, D.R., Ewenstein, B.M. and Pober, J.S., 1991, Tumor necrosis factor induction of endothelial cell surface antigens is independent of protein kinase C activation or inactivation: Studies with phorbol myristate acetate and staurosporine. *Cell. Immunol.*, 146:3056.

Rot, A.,1993, Neutrophil attractant/activation protein-1 (interleukin-8) induces *in vitro* neutrophil migration by haptotactic mechanism. *Eur. J. Immunol.*, 23:303.

Rüegg, C., Postigo, A.A., Sikorski, E.E., Butcher, E.C., Pytela, R. and Erle, D.J. , 1992, Role of integrin $\alpha4\beta7/\alpha4\beta P$ in lymphocyte adherence to fibronectin and VCAM-1 and in homotypic cell clustering. *J. Cell Biol.*, 117:179.

Sako, D., Chang, X.-J., Barone, K.M., 1994,et al. Expression cloning of a functional glycoprotein ligand for P-selectin. *Cell*, 75:1179.

Sanders, W.E., Wilson, R.W., Ballantyne, C.M. and Beaudet, A.L., 1992, Molecular cloning and analysis of *in vivo* expression of murine P-selectin. *Blood*, 80:795.

Shankar, R., de la Motte, C.A. and DiCorleto, P.E., 1992, 3-Deazaadenosine inhibits thrombin-stimulated platelet-derived growth factor production and endothelial-leukocyte adhesion molecule-1-mediated monocytic cell adhesion in human aortic endothelial cells. *J. Biol. Chem.*, 267:9376.

Sligh, J.E.,Jr., Ballantyne, C.M., Rich, S.S., et al. Inflammatory and immune responses are impaired in mice deficient in intercellular adhesion molecule 1. *Proc. Natl. Acad. Sci. USA*, 90:8529, *1993*.

Smith, C.W., Marlin, S.D., Rothlein, R., Toman, C. and Anderson, D.C. Cooperative interactions of LFA-1 and Mac-1 with intercellular adhesion molecule-1 in facilitating adherence and transendothelial migration of human neutrophils in vitro. *J. Clin. Invest.*, 83:2008, *1989*.

Smith, C.W., Kishimoto, T.K., Abbass, O., et al. Chemotactic factors regulate lectin adhesion molecule 1 (LECAM-1)-dependent neutrophil adhesion to cytokine-stimulated endothelial cells in vitro. *J. Clin. Invest.*, 87:609, *1991*.

Springer, T.A. Traffic signals for lymphocyte recirculation and leukocyte emigration: the multistep paradigm. *Cell*, 76:301, *1994*.

Staunton, D.E., Dustin, M.L. and Springer, T.A. Functional cloning of ICAM-2, a cell adhesion ligand for LFA-1 homologous to ICAM-1. *Nature*, 339:61, *1989*.

Steininger, C.N., Eddy, C.A., Leimgruber, R.M., Mellors, A. and Welply, J.K. The glycoprotease of *Pasteurella haemolytica* A1 eliminates binding of myeloid cells to P-selectin but not to E-selectin. *Biochem. Biophys. Res. Commun.*, 188:760, *1992*.

Swerlick, R.A., Lee, K.H., Li, L.-J., Sepp, N.T., Caughman, S.W. and Lawley, T.J. Regulation of vascular cell adhesion molecule 1 on human dermal microvascular endothelial cells. *J. Immunol.*, 149:698, *1992*.

Tanaka, Y., Albelda, S.M., Horgan, K.J., et al. CD31 expressed on distinctive T cell subsets is a preferential amplifier of β1 integrin-mediated adhesion. *J. Exp. Med.*, 176:245, *1992*.

Tanaka, Y., Adams, D.H., Hubscher, S., Hirano, H., Siebenlist, U. and Shaw, S. T-cell adhesion induced by proteoglycan-immobilized cytokine MIP-1β. *Nature*, 361:79, *1993a*.

Tanaka, Y., Adams, D.H. and Shaw, S. Proteoglycans on endothelial cells present adhesion-inducing cytokines to leukocytes. *Immunology Today*, 14:111, *1993b*.

Thornhill, M.H. and Haskard, D.O. IL-4 regulates endothelial activation by IL-1, tumor necrosis factor or IFNγ. *J. Immunol.*, 145:865, *1990*.

Thornhill, M., Wellicome, S.M., Mahiouz, D.L., Lanchbury, J.S.S., Kyan-Aung, U. and Haskard, D.O. Tumor necrosis factor combines with IL-4 or IFN-γ to selectively enhance endothelial cell adhesiveness for T cells: the contribution of VCAM-1 dependent and independent binding mechanisms. *J. Immunol.*, 146:592, *1991*.

Van Epps, D.E., Potter, J., Vachula, M., Smith, C.W. and Anderson, D.C. Suppression of human lymphocyte chemotaxis and transendothelial migration by anti-LFA-1 antibody. *J. Immunol.*, 143:3207, *1989*.

Vaporciyan, A.A., DeLisser, H.M., Yan, H.-C., et al. Involvement of platelet-endothelial cell adhesion molecule-1 in neutrophil recruitment in vivo. *Science*, 262:1580, *1993*.

Vestweber, D., Levinovitz, A., Lenter, M. Monospecific and common glycoprotein ligands for E- and P-selectin on myeloid cells. J. Cell. Biochem. suppl 18A:263.

Von Andrian, U.H., Hansell, P., Chambers, J.D., et al. L-selectin function is required for β_2-integrin-mediated neutrophil adhesion at physiological shear rates in vivo. *Am. J. Physiol. Heart Circ. Physiol.*, 263:H1034, *1992*.

von Asmuth, E.J.U., Smeets, E.F., Ginsel, L.A., Onderwater, J.J.M., Leeuwenberg, J.F.M. and Buurman, W.A. Evidence for endocytosis of E-selectin in human endothelial cells. *Eur. J. Immunol.*, 22:2519, *1992*.

Vonderheide, R.H., Tedder, T.F., Springer, T.A. and Staunton, D.E. Residues within a conserved amino acid motif of domains 1 and 4 of VCAM-1 are required for binding to VLA-4. *J. Cell. Biol.*, 125:215, *1994*.

Weller, A., Isenmann, S. and Vestweber, D. Cloning of the mouse endothelial selectins. Expression of both E- and P-selectin is inducible by tumor necrosis factor α. *J. Biol. Chem.*, 267:15176, *1992*.

Wellicome, S.M., Thornhill, M.H., Pitzalis, C., et al. A monoclonal antibody that detects a novel antigen on endothelial cells that is induced by tumor necrosis factor, IL-1 or lipopolysaccharide. *J. Immunol.*, 144:2558, *1990*.

Wellicome, S., Kapahi, P., Mason, J.C., Lebranchu, Y., Yarwood, H. and Haskard, D.O. Detection of a circulating form of vascular cell adhesion molecule-1: raised levels in rheumatoid arthritis and systemic lupus erythematosus. *Clin Exp Immunol*, 92:412, *1993*.

Williams, A.F. Out of equilibrium. *Nature*, 352:473, *1991*.

Wilson, R.W., Ballantyne, C.M., Smith, C.W., et al. Gene targeting yields a CD18-mutant mouse for study of inflammation. *J. Immunol.*, 151:1571, *1993*.

Yarwood, H., Mason, J.C., Mahiouz, D., Sugars, K. and Haskard, D.O. Characterisation of E-selectin and vascular cell adhesion molecule-1 (VCAM-1) induction on human umbilical vein endothelial cells and dermal microvascular endothelial cells during co-culture with T-lymphocytes. *FASEB J*, 8(4):A133, *1994*.

IV. ANGIOGENESIS/VASCULOGENESIS

ANGIOGENESIS: AN OVERVIEW OF REGULATION AND POTENTIAL CLINICAL APPLICATION

Michael E. Maragoudakis

Department of Pharmacology
Medical School
University of Patras
Patras 26 500
GREECE

INTRODUCTION

The development of the adult vasculature proceeds via two distinct mechanisms: Vasculogenesis and angiogenesis. Vasculogenesis involves *de novo* development of blood vessels and occurs only during embryonic development. *(Dieterlen-Liévre and Pardanaud, 1992)*. The term angiogenesis describes collectively the formation of new blood vessels by spouting from pre-existing vessels and involves a complex cascade of events *(D'Amore and Thomson, 1987)*. Angiogenesis occurs during later development of the organism and sporadically throughout adult life. In the adult organism under physiological conditions angiogenesis is an extremely slow process, restricted to situations such as ovulation and wound healing. This type of "physiological angiogenesis" proceeds for a short time only in an orderly and highly regulated fashion and is self-limited *(Folkman and Shing, 1992)*. On the contrary in many pathological states, such as a tumor growth and metastasis, diabetic retinopathy, retrolental fibroplasia, rheumatoid arthritis, psoriasis, hemangiomas, rejection of organ transplants etc., angiogenesis is uncontrolled, no longer self limited and may contribute to the pathology of the disease. For example, in diabetic retinopathy the proliferating capillaries, which are leaky cause hemorrhage resulting in blindness. Practically every medical specialty deals with disease states where some derangement of angiogenesis is involved *(Folkman, 1985)*.

The lack of angiogenesis in most normal tissues probably results from the interaction of complex multifactorial regulatory system which maintains a balance between stimulation and suppression of angiogenesis *(Maragoudakis, 1993)*. The ability for the ordered formation of new blood vessels, however, is ever present in all tissues and can be initiated and controlled by synergistic effects, not only of soluble factors but also extracellular matrix proteins, on the normally quiescent vascular endothelium *(Folkman and Shing 1992; Grant et al., 1992)*. In addition other accessory cell types such as pericytes, leucocytes, macrophages and smooth muscle cells play a pivotal role in the regulation of physiological and pathological angiogenesis *(Polverini and DiPietro, 1992; D'Amore and Thomson, 1987)*.

Understanding the mechanisms involved in the regulation of angiogenesis is of paramount importance both for locating the site of derangement in disease states and also for developing promoters or inhibitors of angiogenesis as potential therapeutic agents.

Vascular Endothelium: Responses to Injury
Edited by John D. Catravas *et al.*, Plenum Press, New York, 1996

METHODOLOGY

A major difficulty in the field of angiogenesis research is the existing methodology. Most of the methods used, such as the chick chorioallantoic membrane (CAM), the rabbit cornea, intradermal assays etc., are morphological (for review see Auerback et al., 1991). They are based on application of the test material either directly or in slow release form on the model system and evaluation of the results either visually or by computer assisted image analysis. All these methods are cumbersome, expensive, time consuming and at best give only semi-quantitative results. To overcome these difficulties many methods of endothelial cell cultures in extracellular matrix components have been developed *(Williams, 1993)*. The endothelial cells under these conditions acquire the angiogenic phenotype and form tube-like structures. Although these *in vitro* methods are very convenient and represent great simplification over the *in vivo* methods they may be conceptually wrong, since they are focusing on only one aspect of the angiogenic cascade. Furthermore, the important involvement of other accessory cell types, such as pericytes, inflammatory and smooth muscle cells is not taken under consideration. In addition the interaction and sequestering of the many soluble angiogenic and anti-angiogenic factors with specific extracellular matrix proteins as well as the direct effects of these extracellular matrix proteins can not be evaluated in these endothelial cell culture systems.

A major improvement was reported by Maragoudakis et al., (1988), which is based on measuring the rate of basement membrane synthesis as a biochemical index of angiogenesis. The rationale for this methodology is based on the fact that new capillaries at least at some point of their development are bare endothelial cell tubes with their basement membrane produced by the proliferating endothelial cells, thus forming a sleave supporting the abluminal side of the new vessel. Only later the supporting structures of the blood vessels are formed by the surrounding mesenchyme. The finding, therefore, that a quantitative relationship exists between the rate of basement membrane synthesis and angiogenesis in the CAM system *in vivo* and the Matrigel System *in vitro (Haralabopoulos et al.,)* was reasonable to expect. Since basement membrane synthesis, like endothelial cell proliferation, is an extremely slow process under normal physiological conditions, the background activity of basement membrane synthesis in the absence of angiogenesis is extremely slow, thus increasing the sensitivity of the method. We have used the rate of basement membrane collagen biosynthesis, which is the major component of basement membrane as a measure of basement membrane biosynthesis. This is easily measured as collagenous protein digestible by collagenase. We are using, therefore newly-synthesized collagenase digestible proteins labeled from UC^{14}-proline as a quantitative and sensitive index of angiogenesis.

We have validated this technique with the most rigorous method of assessing angiogenesis which is direct counting of blood vessels and also more recently by computer assisted image analysis *(Harris-Hooker et al., 1983)*. Using this technique we are able to evaluate numerous promoters and inhibitors of angiogenesis *(Maragoudakis et al., 1993)* and to study many aspects of the regulation of angiogenesis *(Maragoudakis, 1993)*.

Individual Steps in the Angiogenesis Cascade

Activation of the normally quiescent endothelial cell by angiogenic stimuli initiates a series of well characterized events. Although there are fundamental differences between controlled (physiological) and uncontrolled (pathological) angiogenesis in the same basic sequential steps are involved.

The first of these events is the local dissolution of the surrounding extracellular matrix, consisting of collagen type IV, laminin, proteoglycans etc. This local degradation of basement membrane of the parent vessel, usually a venule, is mediated by specific proteolytic enzymes, including plasminogen activators and collagenases, which are produced by the activated endothelial cells and leucocytes. This leads to release of growth factors sequestered in the extracellular matrix and formation of angiogenic peptides *(Ausprunk and Folkman, 1977; Folkman and Klagshrun, 1987, 1990)*.

Another well characterized step is the migration of endothelial cells *(Sholley et al., 1984)*. The endothelial cells protrude through the holes of the basement membrane and assume bipolar configuration forming a sprout off the parent vessel. The cells at the tip of the sprout migrate and are responsible for the elongation of the sprout. As the endothelial cells are released from

Table I. Endogenous Modulators of Angiogenesis

Promoters	**Inhibitors**
b-FGF	TGF-β
VEGF/VPF	α-Interferon
TNF-α	TNF-γ
TNF-β	$TIMP_1$
PD-ECGF	$TIMP_2$
PDGF	Platelet factor 4
IL-8	Thrombospondin
LTC_4	Heparin + Hydrocortisone
LTD_4	Cartillage extract
ESAF	Retinoids
PGE_1, PGE_2	PKC Inhibitors
Angiotropin	Nitric Oxide (NO)
Angiogenin	2-methoxyestradiol
Adenosine	
Histamine	
Lactic acid	
Gangliosides	
Thrombin	
PMA	
Plasminogen activator	
Hyaluronan fragments	
Laminin peptides	

their albuminal basement membranes they are exposed to interstitial collagens I and III, fibrorectin and vinculin, which are factors associated with migration. Chemotactic substances released by leucocytes and growth factors such as b-FGF are also involved in endothelial cell migration.

Near the tip of the capillary sprout endothelial cells proliferate. The proliferation step follows migration by 24h and helps in retaining the vascular continuity of the forming vessel. The proliferating endothelial cells canalize to form the lumen of the new vessels *(Folkman and Klagsbrum, 1987)*. Anastomosis of these newly formed channels and synthesis of extracellular matrix results in patent microvessel with the onset of blood flow. The most extensively studied step in the angiogenic cascade is the proliferation of endothelial cells. Regulation of endothelial cells turn over is a result of the coordinated action of a number of factors and matrix components. It should be emphasized, however, that despite the intensive study of this step, endothelial cell proliferation is not the most critical step of angiogenesis. As reported by Sholley et al., 1984, angiogenesis can take place without endothelial cell proliferation. In fact, in the various *in vitro* models of angiogenesis, such as the Matrigel assay *(Grant et al., 1992)*, the emphasis is on migration and organization of endothelial cells to form tube-like structures.

We have shown recently that although endothelial cells are plated in dishes coated with a gel of basement membrane proteins, newly synthesized basement membrane collagen is essential for the expression of the angiogenic phenotype, *(Haralabopoulos et al., 1994)*.

A complete understanding of the individual steps involved in angiogenesis and the mechanisms by which the plethora of the reported promoters or inhibitors of angiogenesis regulate these steps, will provide an insight to numerous developmental and physiological or pathological processes.

Endogenous Promoters and Inhibitors of Angiogenesis

An attempt has been made to summarize the endogenous modulators of angiogenesis reported thus far that either promote or suppress angiogenesis (see Table 1). For reviews see Klagsburn and Folkman, 1990 and Maragoudakis, 1993.

The criteria for considering a substance as promoter or suppressor of angiogenesis are not rigorous for all the agents listed in Table I. In most cases a mitogen in endothelial cell cultures and a promotion of angiogenesis in an *in vivo* model system such as CAM or rabbit cornea, is considered an angiogenic factor. There are many reasons, however, that many of these agents are not likely to play a role in physiological or pathological angiogenesis. Even the best studied angiogenic factor, b-FGF, has questionable role in angiogenesis. b-FGF is a potent

Table II. Criteria for demonstrating that a factor is a modulator of angiogenesis *in vivo*

- Specificity of mitogenic activity of the factor on endothelial cells only
- Promotion by the factor of endothelial cell invasion into extracellular matrix in culture and formation of tube like structures
- Promotion of angiogenesis by the factor *in vivo* in experimental model system (CAM, rabbit cornea, etc.)
- Demonstration of the temporal and spatial formation of the alleged factor in embryonic angiogenesis, during female reproductive cycle and in cancer growth and metastasis. This can be done for peptide factors by *in situ* analysis of m-RNA and protein synthesis
- Demonstration of the expression of m-RNA for the receptor of the alleged factor and binding of the factor to the receptor during angiogenesis
- Demonstration of the specificity of the expression of both the alleged factor and its receptors in endothelial cell of proliferation blood vessels
- Demonstration of the inhibition of angiogenesis antitumor growth by inhibiting the production or the activity of the alleged factor. This can be done by antisense techniques, specific antibodies or specific inhibitors of the angiogenic factor
- Demonstration that preventing the expression of the receptors or interfering with receptor antagonists leads to inhibition of angiogenesis and tumor growth

mitogen in endothelial cell cultures, cause angiogenesis when administered *in vivo* and is expressed in tumor cells. However the mitogenic activity of b-FGF is not restricted to endothelial cells but also to other cell types and most important it is not a secreted factor. Furthermore, there is no evidence for a temporal and spatial correlation between expression of b-FGF and its receptors and angiogenesis *in vivo*.

In contrast Klagsbrun and Soker, 1993 has reviewed the evidence for the angiogenic role of VEGF/VPF, which are more compelling than that of b-FGF, and considered VEGF/VPF the best candidate for the principal promoter of physiological and pathological angiogenesis. VEGF/VPF is mitogenic specifically for endothelial cells and is a secreted growth factor. This factor is angiogenic *in vivo* and its expression as well as the expression of its receptors strongly correlates with periods of vasculogenesis and angiogenesis during embryonic development and physiological angiogenesis in the female reproductive cycle. In addition VEGF/VPF is expressed in tumor cells and in small vessels, and its receptors are expressed only in the adjacent small blood vessels *(Klagsbrun and Soker, 1993)*.

Table II summarizes some of the experimental approaches, which can be used to demonstrate that a given agent is angiogenic.

Many endogenous suppressors of angiogenesis have been reported, which may play a role in the *in vivo* regulation of angiogenesis (see Table I). Of special interest is the inhibition of angiogenesis reported recently by the endothelium-derived relaxing factor (NO) *(Pipili-Synetos et al., 1994)*. This may be nature's way of saying "no" to angiogenesis.

Regulation of Angiogenesis

The multiplicity of endogenous factors that have been reported thus far to either promote or suppress angiogenesis in various *in vitro* and *in vivo* models is analogous in many ways to the situation that existed in the blood clotting cascade before the interrelation and interaction of the various clotting factors was elucidated. Similarly in the study of angiogenesis we are dealing with a complex cascade of events leading from activation of the normally quiescent endothelial cells to the formation of new blood vessels at a moments notice with many factors involved. This process similarly to blood coagulation is tightly regulated and involves the interaction of endothelial cells with other cell types, such as pericytes, inflammatory cells, and smooth muscle cells, with extracellular matrix components and with endogenous promoters or inhibitors of angiogenesis. Each of these modulators of angiogenesis is connected to a particular transduction mechanism after interaction with specific receptors. The complexity of such a system requires that overall controls of the multiple transduction mechanisms must exist, so that a sensitive and effective regulation of both the neovascularization and the disappearance of blood capillaries is maintained under physiological conditions.

We have studied the overall regulation of angiogenesis through two major signaling pathways, the protein kinase C (PKC) and c-AMP-dependent protein kinase (PKA) *(Tsopanoglou et al., 1993, 1994)*. These pathways mediate the effects of growth factors hormones, neurotransmitters and oncogenes on cell proliferation and cellular responses. We have demonstrated that they play a key role in the signal transduction mechanisms that regulate angiogenesis.

In the CAM system of angiogenesis we have shown that activators of PKC such as 4β-phorbol-12-myristate-13-acetate (4-β-PMA) or 1,2-dioctanoyl-sn-glycerol (DIC_8) cause a dose-dependent promotion of angiogenesis. The effect is specific since the inactive analogs of 4-α-PMA and diolein, which do not activate PKC, are not promoters of angiogenesis. In addition, the specific inhibitor of PKC, ro 318220, cause a dose-dependent suppression of basal angiogenesis and also reverses the angiogenesis-promoting effect of 4-β-PMA and DIC_8 to levels below control. This suggests that both basal and stimulated angiogenesis are controlled by PKC activity. Similar conclusions were reached from experiments with the *in vitro* system of angiogenesis using endothelial cells on Matrigel *(Tsopanoglou et al., 1994)*. In the Matrigel system endothelial cells from human umbilical cord (HUVECS) form tube-like structures. The PKC inhibitor Ro-318220 prevents tube formation.

Contrary to PKC activation, when activators of PKA were used in the CAM or the Matrigel system of angiogenesis resulted in suppression of angiogenesis. Thus forskolin or Sp-diastereoisomer of adenosine-3', 5'-cyclic monophosphotionate (Sp-c-AMPS) caused inhibition on basal angiogenesis. The inactive analog Rp-c-AMPS, which is a competitive inhibitor of c-AMP, was without effect. when angiogenesis was stimulated by 4-β-PMA both Forskolin and Sp-c-AMPS could reverse this stimulatory effect. We conclude from these studies that activators of PKC and PKA have opposing effects on angiogenesis *(Tsopanoglou et al., 1994)*. In the regulation of angiogenesis, therefore, we are dealing with a cross-talk phenomenon between the two major signaling pathways.

We have recently reported that thrombin may play a major regulatory role in angiogenesis *(Tsopanoglou et al., 1993)*. A relationship of hypercoagulability and cancer has been recognized for more than 125 years. However, the molecular basis for this association between blood coagulation and cancer growth and metastasis has not been completely understood. We have established that thrombin is powerful promoter of angiogenesis and this action is independent of fibrin formation. Both α- Thrombin and γ-Thrombin cause a dose-dependent promotion of angiogenesis in the CAM system. γ-Thrombin lacks the catalytic site for fibrin formation, yet it promotes angiogenesis to the same extent as α-thrombin. Hirudin, which binds both the catalytic and anion binding exocites of α-thrombin abolishes the angiogenesis-promoting effects of both a α & γ-thrombin. Similarly heparin, which binds thrombin and accelerate thrombin inactivation by antithrombin III abolishes the effects of α & γ-thrombin on angiogenesis *(Tsopanoglou et al., 1993)*.

The mechanism by which thrombin promotes angiogenesis has not been established. We have shown that both PKC inhibitors and PKA activators reverse the angiogenic action of thrombin suggesting that the angiogenesis promoting effect of thrombin is also mediated through these signaling pathways. Thrombin in addition to its effects on the blood coagulation has diverse actions on many cell types including platelets, endothelial cells, cancer cells, leucocytes, smooth muscle cells and neurones (for review, see Tapparelli et al., 1993). It is possible that some of these effects of thrombin may activate the angiogenic cascade. In any case there is an increasing body of evidence that patients with cancer and certain inflammatory diseases (e.g. sepsis) are associated with an elevated procoagulant activity. Present day understanding of the cellular events of thrombin action suggests that thrombin is directly involved and provides a rational for treatment of both inflammatory diseases and cancer with anti-coagulant drugs *(Tapparelli et al., 1993)*. The angiogenesis-promoting effect of thrombin that we have demonstrated provides further support for the potential therapeutic application of anti-coagulant therapy for cancer and inflammatory diseases, since in both cases angiogenesis is a prominent event in the pathology of these states. Because modified thrombins (e.g. γ-thrombin) which devoid of coagulant activity can mimic the actions of α-thrombin on angiogenesis it is likely that the angiogenic effect of thrombin is related to specific thrombin receptors unrelated to coagulation. This provides the possibility for pharmacological interventions on angiogenesis with specific inhibitors without interfering with the role of thrombin in hemostasis. Similarly agonists to the receptors of thrombin-related to angiogenesis which are devoid of coagulant activity have the potential for therapeutic applications in wound healing and other situation where promotion of angiogenesis is desirable.

As already mentioned the endothelium relaxing factor (NO) is a potent inhibitor of angiogenesis *(Pipili-Synetos et al., 1993)*. It suppresses both the basal angiogenesis and the stimulated angiogenesis by 4-β-PMA or thrombin, *(Maragoudakis et al., 1994)*. The findings suggest that NO may be acting as inhibitor of angiogenesis via inhibition of PKC.

Indeed S-nitrosylation of protein thiols is one of the recognized cellular regulatory mechanisms induced by nitric oxide (NO) and nitric oxide generating agents. PKC has critical thiol residues, which influence its Kinase activity. It has been shown that NO induces a reversible inactivation of PKC activity and phorbol ester binding. The inactivation by NO of PKC is blocked by the NO scavenger oxyhemoglobin and is reversed by dithiotheitol *(Gopalakrishna et al., 1993)*.

The aforementioned studies on the regulation of angiogenesis could contribute in the understanding of the derangement of this process in disease states. This could provide this basis for developing promoters or inhibitors of angiogenesis for potential therapeutic applications.

Diseases States in Which Suppressors of Angiogenesis Have Therapeutic Potential: Tumor Growth and Metastasis

Uncontrolled angiogenesis is particularly evident in solid tumors *(Folkman and Shing, 1992)*. The original hypothesis of Folkman that tumor growth and metastasis are angiogenesis-dependent events has been born out experimentally by many investigators in a variety of systems. Furthermore Gullino (1985) has shown that angiogenesis precedes neoplastic transformation and as such angiogenesis could provide a valuable diagnostic tool for early detection of tumors and/or evaluation of antitumor therapy. Tumors cell secrete angiogenic factors and it has been shown that they are over expressed in variety of tumors *(Folkman, 1992)*. Quantitation of elevated angiogenic factors in the blood or the urine of cancer patients is being evaluated for prognosis and guide to therapy *(Folkman, 1992)*.

The importance of angiogenesis in tumor growth and metastasis has also been established from clinical studies. For example, in human breast carcinoma the extent of angiogenesis correlates with the occurrence of metastasis *(Weidner et al., 1991)*. The same holds true for cutaneous melanoma *(Folkman, 1987)*. It has been established that quantitation of angiogenesis in biopsy specimens of breast cancer provides an independent marker of future metastatic risk and as such a guide for the selection of cases for aggressive therapy *(Weidner, 1991)*.

The effectiveness of a variety of inhibitors as anti-tumor agents has been tested in a number of animal tumor models *(Maragoudakis et al., 1990; Folkman and Ingber, 1992)*. Clinical trials are under way to evaluate the effectiveness of such inhibitors in human cancers. For example, fumagillin (AGM 1470), an antibiotic which is a potent inhibitor of endothelial cell proliferation and neovascularization is in clinical trials *(Folkman and Ingber, 1992)*. Also a hydroxamate derivative (BB-94), which is an inhibitor of metalloproteinases is being evaluated in the clinic for cancer therapy *(Brown and Phil, 1994)*.

The first successful treatment of a life-threatening angiogenic disease, a child with pulmonary hemangioma, has been accomplished using interferon alpha-2a *(White et al., 1989)*. Interferon alpha-2a inhibits angiogenesis by blocking endothelial cell motility and proliferation.

This success provides an optimistic outlook for the potential of anti-angiogenic agents for cancer therapy. It is anticipated that such therapy will be combined with existing anti-cancer treatments and it will be long term. However, the possibility exists that the derangement of angiogenesis in cancer will be located and thus it will become possible to interfere with the specific angiogenic stimulus.

Chronic Inflammation

Inflammation resulting from a variety of causes leads to pathological angiogenesis. The destruction of articular cartilage in rheumatoid arthritis may be due at least in part to uncontrolled angiogenesis. Drugs used for treatment of rheumatoid arthritis, such as penicillamine, are inhibitors of angiogenesis and have been used also for brain tumors *(Brem et al., 1990)*. The presence of inflammatory cells such as leucocytes and macrophages in cancer may be a major contributing factor to tumor angiogenesis. Polverini and DiPietro, (1992) have reviewed the role of macrophages in both physiological and pathological angiogenesis.

Ocular Diseases

The pathology from ocular injuries and many ocular diseases such as diabetic retinopathy, glaucoma and retrolental fibroplasia, is related to uncontrolled angiogenesis *(BenEzra, 1987)*.

Psoriasis

In psoriasis there is a chronic inflammation and epidermal proliferation associated with neovascularization of the superficial vascular plexus of the dermis *(Elder et al., 1989)*.

All the aforementioned disease states could be controlled by specific non-toxic inhibitors of angiogenesis.

Therapeutic Potential for Promoters of Angiogenesis

Peptic Ulcer

In a very elegant study Folkman et al., 1991 have shown that promotion of angiogenesis may be a new approach to accelerating peptic ulcer healing. The presented evidence that even the conventional therapy of peptic ulcers with H_2 blockers, antacids or sucralfate may be a result of protection of the endogenous angiogenic molecule b-FGF from acid degradation. This opens up a new mechanism for peptide ulcer healing by providing locally acid-stable angiogenic factors.

Non-healing Wounds

In diabetic or old people stimulation of angiogenesis may benefit non-healing wounds and fractures. Chronic application of an angiogenic peptide has been reported to accelerate wound healing in the clinic *(Folkman et al., 1991)*. Our recent finding that thrombin by a mechanism independent of fibrin information promotes angiogenesis provides a new approach to wound healing by thrombin-peptides or modified thrombins *(Tsopanoglou et al., 1994)*.

Scleroderma

The pathogenesis of scleroderma is unknown. In this condition it is desirable either to prevent capillaries from disappearing or to stimulate the formation of new capillaries. Polverini et al., (1992) have shown that macrophages from patients with scleroderma fail to stimulate angiogenesis after treatment with lipopolysaccharide as do macrophages from normal people.

Myocardial Infarction

Experiments with pigs have shown that coronary stenosis leads to progressive enhancement of the expression of endogenous angiogenic factors such as b-FGF, VEGF etc. *(Sharma et al., 1992)*. This suggests that these factors play an important role in the ischaemia-induced angiogenesis. Therefore, local delivery of angiogenic factors may further promote angiogenesis in the infarcted myocardium.

Transplants

The engraftment of skin or other organ transplants could benefit from local availability of angiogenic factors, since angiogenesis is required for successful transplantation. On the contrary, in cases of organ transplant rejection control of the uncontrolled inflammatory angiogenesis may be desirable.

Hair growth

The growth of hair follicles requires active angiogenesis and many investigators have suggested that stimulation of angiogenesis may improve hair growth and prevent hair loss.

In conclusion it has become evident that new blood vessel formation is essential to numerous developmental and physiological processes. Understanding the molecular and cellular aspects of angiogenesis and the role of the many factors involved both in physiological and pathological angiogenesis will have immense therapeutic applications. This knowledge will shed light into the derangement of this complex cascade in many disease states and will provide the basis for developing novel therapeutic agents by modulating angiogenesis.

REFERENCES

Auerbach, R., Auerbach, W. and Polakowski, I., 1991, Assays for Angiogenesis: A review. *Pharmac. Ther.* 51:1-11.

Ausprunk, D.H. and Folkman, J., 1977, Migration and proliferation of endothelial cells in preformed and newly formed blood vessels during tumor angiogenesis. *Microvascular Res.* 14:53-68.

BenEzra, D., 1987, ed: Ocular circulation and Neovascularization, Amsterdam: Martinus Nijhoft/junk.

Brem, S., Sagzag, D., Tsanaclis, A.M., Gately, S., Elkouby, M.P. and Brien, S.E., 1990, Inhibition of angiogene is and tumor growth in the brain. *Am. J. Pathol.* 137:1121-1141.

Brown, P., and Phil, D., 1994 (in press), Clinical trials of a low MW inhibitor in cancer, in Inhibition of Matrix Metalloproteinases: Therapeutic potential. Edited by Greenwald, R.A. and Golub, L.M., N.Y. Acad Sci.

Haralabopoulos, G., Grant, D., Kleinman, H., Lelkes, P., Papaioannou, S. and Maragoudakis, M., 1994 (in press), *Lab Invest.*

D'Amore, P. and Thomson, R.W., 1987, Mechanisms of angiogenesis. *Ann. Rev. Physiol.* 49:453-464.

Dieterlen-Liévre, F. and Pardanau, L., 1992, Embryonic development of blood vessels, *In:* Angiogenesis in Health and Disease. Maragoudakis, M.E., Gullino, P., and Lelkes, P. (eds), Plenum Press, New York, p.p. 1-16.

Elder, J.T., Fisher, H.J., Lindquist, P.B., Bennett, G.L., Helkow, M.R., Coffey, R.J., Ellingsworth, L., Derynck, R., and Voorhees, J.I., 1989, Over-expression of transforming growth factor a in psoriatic epidermis. *Science* 243:811-814.

Folkman, J., 1985, Toward an understanding of angiogenesis: Search and Discovery. *Perspect. Biol. Med.* 29:10-35.

Folkman, J., 1987, What is the role of angiogenesis in metastasis from cutaneus melanoma. *Eur. J. Clin. Onc.* 23:361-363.

Folkman, J., 1992, The role of angiogenesis in tumor growth. *Semin. Cancer Biol.* 3:65-67.

Folkman, J. and Ingber, D., 1992, Inhibition of Angiogenesis. *Semin. Cancer Biol.* 3:89-96.

Folkman, J. and Klagsbun, M., 1987, Angiogenic factors. *Science* 235:442-447.

Folkman, J. and Shing, Y., 1992, Angiogenesis: Minireview *J. Biol. Chem.* 267:10931-10933.

Gopalakrishna, R., Chen, Z.H. and Gundimeda, U., 1993, Nitric oxide and nitric oxide generating agents induce a reversible inactivation of protein Kinase C and phorbol ester binding. *J. Biol. Chem.* 268:27180-27185.

Grant, D.S., Kinsella, J., Cid, M.C. and Kleinman, H., 1992, Specific laminin domains mediate endothelial cell adhesion alignment and angiogenesis, *In:* Angiogenesis in Health Disease. Maragoudakis, M.E., Gullino, P. and Lelkes, P. (eds), p.p. 99-110, Plenum Press, New York.

Gullino, P.M., 1992, Angiogenesis and neoplastic transformation. *In:* Angiogenesis in Health and Disease. p.p. 149-155, Maragoudakis, M.E., Gullino, P.M., and Lelkes, P. (eds), Plenum Press, New York.

Haralabopoulos, G., Grant, D., Kleinman, H., Lelkes, P., Papaioannou, S. and Maragoudakis, M.E., 1994 (in press), *Lab. Invest.*

Harris-Hooker, S.A., Gajdusek, C.M., Wight, T.N. and Schwartz, S.M., 1983, Neovascular response induced by cultured aortic endothelial cells. *J. Cell Phys.* 114:302-310.

Klagsbrun, M. and Folkman,J., 1990, Angiogenesis. *In:* Peptide Growth Factors and their Receptors II, Handbook of Experimental Pharmacology. Vol. 95, p.p. 544-546, Spom, M.B. and Roberts, A.B. (eds), Berlin-Spingel-Verlag.

Klagsbrun, M. and Soker, S., 1993, VEGF/VPF: The angiogenic factor found? *Current Biology* 3:699-702.

Maragoudakis, M.E., 1993, Angiogenesis. *In:* Annual of Cardiac Surgery, M. Yacoub and J. Pepper (eds), p.p. 13-19, 6th Ed. Current Science Ltd ISBM.

Maragoudakis, M.E., Missirlis, Karakiulakis, G., Sarmonika, M. and Tsopanoglou, N. Basement membrane biosynthesis as a target for developing inhibitors of angiogenesis with anti-tumor properties. *Kidney Intern. Nephrology* 43:147-150.

Maragoudakis, M.E., Missirlis, E., Sarmonika, M., Panoutsakopoulou, M. and Karakioulakis, G., 1990, 252:753-757. Basement membrane synthesis as a target for tumor therapy. *J. Pharmacol. Exp. Ther.* 252:753-757.

Maragoudakis, M.E., Sarmonika, M. and Panoutsakopoulou, 1988, Rate of basement membrane synthesis as an index of angiogenesis. *Tissue and Cell* 20:531-539.

Maragoudakis, M.E., Pipili-Synetos, E., Tsopanoglou, N., Haralabopoulos, G. and Peristeris, P., 1994 (in press), On the regulation of angiogenesis. *In:* Angiogenesis: Molecular and Clinical Aspects. Maragoudakis, M.E., Gullino, P., Lelkes, P. (eds), Plenum Press, New York.

Pipili-Synetos, E., Sakkoula, E. and Maragoudakis, M.E., 1993, Nitric oxide is involved in the regulation of angiogenesis. *Br. J. Pharmacol.* 108:855-857.

Pipili-Synetos, E., Sakkoula, E., Haralabopoulos, G., Andriopoulou, P., Peristeris, P. and Maragoudakis, M.E., 1994, Evidence that nitric oxide is an endogenous mediator of angiogenesis. *Br. J. Pharmacol.* 111:894-902.

Polverini, P.J. and DiPietro, L.A., 1992, Role of macrophage in the regulation of physiological and pathological angiogenesis. *In:* Angiogenesis in Health, Diseases. Maragoudakis, M.E., Gullino, P. and Lelkes, P. (eds) p.p. 43-53, Plenum Press, New York.

Sharma, H.S., Wünsch, M., Sack, S. and Schapper, W., 1992, Vascular growth factor in the intremittently ischemic heart: A study on Growth factor expression. *In:* Angiogenesis in Health and Disease 1991, p.p. 195-204, Maragoudakis, M.E., Gullino, P.M. and Lelkes, P. (eds), Plenum Press, New York, London.

Sholley, M.M., Ferguson, G.P., Seibel, H.R., Montour, J.L. and Wilson, J.D., 1984, Mechanisms of neo vascularization vascular sprouting can occur without proliferation of endothelial cells. *Lab. Invest.* 51:624-634.

Tapparelli, C., Metternich, R., Ehrhardt, C. and Cook, N.S., 1993, Synthetic low-molecular weight thrombin inhibitors: molecular design and pharmacological profile. *TIPS* 14:366-375.

Tsopanoglou, N., Haralabopoulos, G. and Maragoudakis, M.E., 1994, Opposing effects on the modulation of angiogenesis by protein Kinase C and cyclic AMP mediated pathways. *J. Vasc. Res.* 723:

Tsopanoglou, N., Pipili-Synetos, E. and Maragoudakis, M.E., 1993, Protein Kinase C involvement in the regulation of angiogenesis. *J. Vasc. Res.* 30:202-208.

Tsopanoglou, N., Pipili-Synetos, E. and Maragoudakis, M.E., 1993, Thrombin promotes angiogenesis by a mechanism independent of fibrin formation. *Am. J. Physiol.* 264:C1302-C1307.

Weidner, N., 1992, The regulation of tumor angiogenesis and metastasis with emphasis on invasive breast carcinoma. *In:* Advances in Pathology and Laboratory Medicine-Morsby Year Book, Weinstein (ed), p.p. 101-122, Chicago III., V.5.

Weidner, N., Semple, J.P., Welch, W.R. and Folkman, J., 1991, Tumor angiogenesis and metastasis - Correlation in invasive breast carcinoma. *N. Engl. J. Med.* 324:1-8.

While, C.W., Sondheimer, H.M., Crouch, E.C., Wilson, H. and Fan, L.L., 1989, Treatment of pulmonary heamangiomatosis with recombinant interferon alpha-2a. *N. Engl. J. Med.* 320:1197-1200.

Williams, S., 1993, Angiogenesis in three dimentional cultures. *Lab. Invest.* 60:491-493.

MOLECULAR ORGANIZATION OF ENDOTHELIAL CELL TO CELL JUNCTIONS

Elisabetta Dejana and Jean-Luc Plantier

CEA Departement de Biologie Moleculaire et Structurale
Laboratoire d'Hematologie, INSERM U217
CEN-G, Grenoble, France

INTRODUCTION

The endothelial monolayer forms the main barrier to macromolecular passage from blood to tissues and regulates the extravasation of circulating cells such as leukocytes. When endothelial permeability is altered such as in inflammatory conditions, edema takes place. Inflammatory edema is a significant component of many diseases e.g. adult respiratory syndrome, septicemia and hypoxia/reperfusion syndrome. This is frequently the result of both pressure-induced filtration of plasma proteins and changes in the permeability properties of the endothelial lining of the vascular wall *(Majno et al., 1961)*.
Many evidences indicate that intercellular junctions are the major structural determinant of endothelial permeability. Interendothelial junctions are complex structures formed by transmembrane adhesive molecules linked to a network of cytoskeletal proteins inside of the cells. *(Franke et al., 1988)*.

Endothelial cells are able to rapidly change the architecture of their junctions to allow the passage of plasma proteins and cells. This effect is quickly reversible and in many cases the endothelium is able to reorganize cell to cell contacts within few minutes. Despite their functional relevance little is known about the molecular components of endothelial junctions and the mechanisms that regulate the organization/disorganization of these structures. Cytoskeletal protein assembly is regulated through signals transmitted via receptors and second messengers. A mechanistic hypothesis is that changes in endothelial cell permeability might occur through a chain of events that includes activation of a specific receptor, generation of an intracellular signal via a second messenger, cytoskeletal changes and subsequent alteration of the geometry of the interendothelial junctions *(Haselton et al., 1992)*. A similar sequence of events could also take place considering cell extravasation. In this case leukocyte interaction with specific endothelial adhesive receptors and/or the release of active components might constitute the trigger for endothelial cell cytoskeletal modifications and opening of intercellular junctions. Structural characteristics of endothelial junctions At ultrastructural level, the type and the number of endothelial cell junctions varies along the vascular tree, they are well organized and numerous in the large vessels where the control of permeability is strict while they almost disappear in the postcapillary venules where cell extravasation and exchange of plasma constituents need to be particularly efficient *(Simionescu and Simionescu, 1991)*.

In general, cell junctions have been divided on the basis of morphological and functional characteristics in: i) communicating (gap) junctions, ii) occluding (tight) junctions and iii) adherens junctions (figure 1, for review see *Simionescu and Simionescu, 1991)*. Communicating junctions mediate the passage of chemical or electrical signals from one endothelial cell to another; tight junctions seal cell together in a way that inhibits even small molecules from leaking from the luminal to the abluminal side of the endothelial monolayer; adherence junctions

Vascular Endothelium: Responses to Injury
Edited by John D. Catravas *et al.*, Plenum Press, New York, 1996

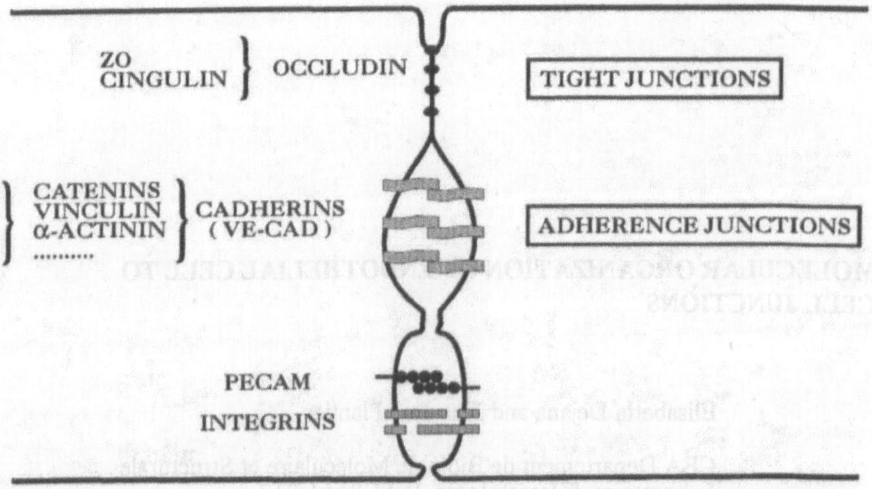

Figure 1. *Schematic representation of junctional adhesion structures in endothelial cells.* The different types of intercellular junctions are localized at the lateral sides of the cells. Tight junctions occupy the most apical position to seal the cell layer, adherence junctions are found more basally. The junctions are formed by adhesive molecules such as occluding and cadherins (in particular VE-cadherin) that are responsible of the physical attachment of the two cell membranes. Inside of the cells occluding and VE-cadherins codistribute with specific cytoplasmic proteins like ZO proteins, cingulin (occluding) or catenins, vinculin, α-actinin, etc. (VE-cadherin). Some kinases (raf, ras, yes, src) have been also found to localize with adherence junction. Other adhesive molecules such as PECAM or integrins (see TExt) localize at cell to cell junctions in endothelial cells but their relationship with cytoplasmic proteins or with other molecules at junctions remains to be established.

leave a detectable space (about 20 nm) between the membranes of two adherent endothelial cells (Figure 1). No data have been reported on the presence of desmosomes (another cell to cell adherence structure present in epithelial cells of different origin, *(Franke et al., 1988)* in endothelial cells. However, the presence of desmosomal markers (such as desmoplakin) has been observed in lymphatic endothelium *(Schmelz and Franke, 1993)*.

Distinct molecular components are organized in the different types of cell junctions. Gap junctions are formed by clusters of transmembrane proteins called connexins; one transmembrane component of tight junctions has been recently identified and denominated occludin *(Furusue et al., 1993)* in addition specific cytoplasmic proteins have been localized in tight junctions such as ZO-1 and cingulin; adherence junctions are formed by Ca2+ dependent transmembrane adhesion proteins such as cadherins (see below) and a complex series of cytoplasmic proteins such as catenins, plakoglobin, vinculin, a-actinin, zyxin and actin.(for review see *Tsukita et al., 1992)*.

In relation to its functional properties, endothelium expresses different types of junctions. The endothelium of large arteries presents complex junctions consisting of occluding junctions intercalated with communicating junctions. The occluding junctions are expected to seal off the intercellular spaces conferring a low degree of permeability on the endothelial layer. The gap junctions are required to maintain direct cell to cell communication. In veins the occluding junctions are less elaborate than in arteries and the communicating junctions occur at less frequency than in the arterial endothelium. In the microvasculature characteristic variations exist in the extent and organization of intercellular junctions from one microvessel to the other. The junctional system (similar to large arteries) is maximally elaborated in arterioles while in postcapillary venules (where exchange of cells and plasma components most frequently occurs) is loosely organized, communicating junctions are absent and about 30% of the junctions are open *(Simionescu and Simionescu, 1991)*.

MOLECULAR COMPONENTS OF ENDOTHELIAL CELL JUNCTIONS

The morphological structure of cell to cell junctions in the endothelium presents many similarities with that of other types of epithelia. Many cytoskeletal molecules present in the epithelial junctions (for instance vinculin, plakoglobin, catenins, actin...) have been also found in endothelial cells. However, among the transmembrane components, endothelial cells express specific molecules that are not found in other epithelia and, in some cases, in any other cell type.

Three different types of adhesive molecules have been found to be organized at cell to cell contacts in the endothelium: integrins, immunoglobulins and cadherins (figure 1).

Integrins

In general, the integrin family consists of a large number of heterodimers that regulate a variety of cell adhesive functions (for review see *Hynes, 1992)*. Almost all cell types express one or various members of this family of proteins and integrins are well preserved along the phylogenetic tree. All integrins are formed by two non covalently linked subunits: the larger termed α chain (120-180 Kd) and the smaller β chain (90-110 Kd). Both subunits are integral membrane proteins that present a small C-terminal cytoplasmic domain and a large N-terminal extracellular domain and a transmembrane segment.

In endothelial cells integrins promote cell attachment to extracellular matrix proteins. However, besides their location at the basal aspect of the membrane, in contact with the subendothelium, integrins (in particular $\alpha_2\beta_1$ and $\alpha_5\beta_1$) have been also found at cell-cell junctions in endothelial cells *(Lampugnani et al., 1991)*. This type of distribution seems to be specific of polarized epithelial cells. For instance in keratinocytes, two β_1 integrins are located at cell-cell contacts ($\alpha_2\beta_1$ and $\alpha_3\beta_1$) *(Carter et al., 1990; De Luca et al., 1990)* while no such distribution was found in smooth muscle cells or skin fibroblasts.

Immunoglobulins

Many adhesive molecules belong to the immunoglobulin superfamily *(Albelda et al., 1991)*. These transmembrane proteins sustain a variety of homophilic and heterophilic intercellular contacts. In endothelial cells it has been recently discovered a new immunoglobulin located at cell to cell contacts. This molecule has been called PECAM-1: platelet-endothelial-adhesion molecule 1 *(Muller et al., 1989)* but also endoCAM *(Albelda et al., 1990)* or CD31 *(Newman et al., 1990; Simmons et al., 1990)*.

This protein is present not only on the surface of continuous (but not sinusoidal) endothelium in situ *(Muller et al., 1989)*, but also in platelets, in cells of myeloid lineages *(Goyert et al., 1986 and Stockinger et al., 1990)* and in several lymphocyte subsets *(Tanaka et al., 1992)*. PECAM-1 is a 120-130 Kd transmembrane protein that has been fully sequenced *(Newman et al, 199O; Albelda et al., 1991)*. It presents at the N-terminus, exposed to the extracellular environment, six immunoglobulin homology units. These domains, in different number, are typical of all the members of the family.

The aggregation mediated by PECAM-1 could be both hetero and homophilic *(Albelda et al., 1991)*. Data reported are consistent both with a PECAM-1 to PECAM-1 recognition and interaction of this molecule with another, still unknown, ligand. This ligand might be either an integrin, since this aggegation is divalent cation and temperature dependent, or a glycosaminoglycan moiety, since the second Ig loop of PECAM-1 contains a typical (LKREKN) glycosaminoglycan recognition sequence. For other members of the immunoglobulin family both homophilic and heterophilic recognition mechanisms have been reported *(Albelda et al., 1991 and references therein)*. PECAM may function in endothelial cell to cell recognition and adhesion. One of the most striking activity of PECAM is that of directing leukocyte migration through endothelial junctions as demonstrated in vitro and in vivo *(Muller et al., 1993; Vaporcyan et al., 1993)*.

Cadherins

Cadherins are Ca2+-dependent and protease-sensitive adhesive proteins *(Takeichi, 1991)* that mediate homotypic cell to cell interaction *(Miyatami, 1989; Nose, 1988)*. Structurally, cadherins are single chain (120-140 Kd), transmembrane proteins localized in many epithelia at adherence junctions. Through their cytoplasmic domain they are linked to three proteins called catenins that mediate their anchorage to actin microfilaments of the cytoskeleton *(Tsukita et al., 1992; Ozawa et al., 1989; Takeichi, 1991)*. Several members of the cadherin family have been characterized in different tissues *(Takeichi, 1991)*.

Evidence has been reported that endothelial cells express two cadherins: N-(neural) and P (placental)-cadherins that have been also found in other tissues *(Liaw et al., 1990; Salomon et al., 1992)*. Interestingly N-cadherin is not localized at cell to cell junctions but remains diffuse on the cell membrane *(Salomon et al., 1992)*.

We have recently discovered a new endothelial specific cadherin (also called 7B4 from the name of the monoclonal antibody able to bind to it or cadherin-5 or VE(vascular endothelial)-cadherin *(Lampugnani et al., 1992)*. VE-cadherin is strictly located at endothelial junctions both in vitro and in situ *(Lampugnani et al., 1992)*. VE-cadherin is bound at appositional surfaces of cultured endothelial cells only as they become confluent and is stably expressed at intercellular boundaries of confluent monolayers. This molecule is present in all types of endothelium (veins and arteries, capillary and large vessels). In vitro, addition of antibodies directed to this molecule increased permeation of macromolecules across monolayers even without any obvious change of cell morphology. When endothelial cell permeability was increased by agents such as thrombin, or elastase its distribution pattern at intercellular contact rims was severely altered.

The interrelationship between PECAM-1, VE-cadherin and integrins at endothelial cell-cell contacts is unclear. By electron microscopy analysis, only VE-cadherin has been found to be localized in specific structures such as adherence junctions. PECAM-1 does not codistribute with VE-cadherin suggesting that it is probably organized in another type of structures *(Leach et al., 1993)*.

Overall these data indicate that cell to cell junctions in endothelial cells are very complex structures where different molecules can be organized and play a coordinated role in maintaining the permeability of the monolayer. Most probably other components will be identified in the future. Further studies are required to understand how these molecules interplay and which is their reciprocal relevance in modulating endothelial permeability.

CONCLUDING REMARKS

Endothelial cell junctions are very important structures for the regulation of vascular permeability, leukocyte extravasation and vascular remodelling. Despite their functional relevance little is known about their molecular components and the mechanisms that control their assembly/disassembly.

Junction type and number vary along the vascular tree in relation to the different functional properties of the endothelium. For instance, in postcapillary venules, where the passage of blood components and cells needs to be very rapid and efficient, junctions are lower in number and less organized than in large arteries where permeability must be limited.

Agents that increase vascular permeability and inflammatory cytokines disrupt junctional organization leading to increased permeability and edema. This effect might be mediated by regulated processes that include intracellular second messengers and reorganization of the cytoskeletal and transmembrane junction proteins. Similar processes might also regulate leukocyte extravasation where leukocyte binding to endothelial adhesive molecules might be the trigger for induction of junction opening. When more information on the molecular mechanisms that regulate the local and selective opening of endothelial junctions will be available, specific inhibitors might be developed. These agents will act as selective modulators of endothelial permeability and leukocyte extravasation and therefore as potent inhibitors of inflammatory reactions.

REFERENCES

Albelda, S.M., Oliver, P., Romer, L., and Buck, C.A., 1990, EndoCAM: a novel endothelial cell-cell adhesion molecule. *J. Cell. Biol.* 110:1227-1237.

Albelda, S.M., Muller, W.A., Buck, C.A., and Newman, P.J., 1991, Molecular and cellular properties of PECAM-1 (endoCAM/CD31): A novel vascular cell-cell adhesion molecule. *J. Cell Biol.* 114:1059-1068.

Carter, W.G., Wayner, E.A., Bouchard, T.S., and Kaur, P., 1990, The role of integrins $\alpha3\beta1$ and $\alpha3\beta1$ in cell-cell and cell-substrate adhesion of human epidermal cells. *J. Cell. Biol.* 110:1378-1404.

De Luca, M., Tamura, R.N., Kajiji, S., Bondanza, S., Rossino, P., Cancedda, R., Marchisio, P.C., and Quaranta, V., 1990, Polarized integrin mediates human keratynocyte adhesion to basal lamina. *Proc. Natl. Acad. Sci. U.S.A.* 87:6888-6892.

Franke, W.W., Cowin, P., Grund, C., Kuhn, C., and Kepprell, H.P., 1988, The endothelial junction: the plaque and its components.In Simionescu N and Simionescu M (eds.) *Endothelial Cell Biology in Health and Disease* 147-166.

Furuse, M., Hirase, T., Itoh, M., Nagafuchi, A., Yonemura, S., Tsukita, S., and Tsukita, S., 1993, Occludin: a novel integral membrane protein localizing at tight junctions. *J. Cell. Biol.* 123:1777-1788.

Goyert, S.M., Ferrero, E.M., Seremetis, S.V., Winchester, R.J., Silver, J., and Mattison , A.C., 1986, Biochemistry and expression of myelomonocytic antigens. *J. Immunol.* 137:3909-3914.

Haselton, F.R., Alexander, J.S., Mueller, S.N., and Fishman, A.P., 1992, Modulation of endothelial paracellular permeability. A mechanistic approach.In Simionescu N and Simionescu M (eds.) *Endothelial cell disfunctions* 103-126.

Hynes, R.O., 1992, Integrins: Versatility, modulation and signaling in cell adhesion. *Cell* 69:11-25.

Joris, I., Cuénoud, H.F., Doern, G.V., Underwood, J.M., and Majno, G., 1992, Endothelial permeability in Inflammation. The role of capillaries versus venules. In Simionescu N and Simionescu M (eds.) *Endothelial cell disfunctions* 233-242.

Lampugnani, M.G., Resnati, M., Dejana, E., and Marchisio, P.C., 1991, The role of integrins in the maintenance of endothelial monolayer integrity. *J. Cell. Biol.* 112:479-490.

Lampugnani, M.G., Resnati, M., Raiteri, M., Pigott, R., Pisacane, A., Houen, G., Ruco, L.P., and Dejana, E., 1992, A novel endothelial specific memebrane protein is a marker of cell-cell contacts. *J. Cell. Biol.* 119:239-246.

Leach, L., Clark, P., Lampugnani, M.G., Arroyo, A.G., Dejana, E., and Firth, J.A., 1993, Immunoelectron characterization of the inter-endothelial junctions of human term placenta. *J. Cell. Sci.*, in press.

Liaw, C.W., Cannon, C., Power, M.D., Kiboneka ,P.K., and Rubin ,L.L., 1990, Identification and cloning of two species of cadherins in bovine endothelial cells. *EMBO J.* 9:2701-2708.

Majno, G., Palade, G.E., and Schoefl, G.I., 1961, Studies on inflammation.II.The site of action of histamine and serotonin along the vascular tree: A topographic study. *J. Biophys. Biochem. Cytol.* 11:607-626.

Miyatami, S., Shimamura, K., and Hatta, M. et al., 1989, Neural cadherin: role in selective cell-cell adhesion. *Science* 245:631-635.

Muller, W.A., Ratti, C.M., McDonnell, S.L., and Cohn, Z.A., 1989, A human endothelial cell-restricted, externally disposed plasmalemmal protein enriched in intercellular juntions. *J. Exp. Med.* 170:399-414.

Muller, W.A., Weigl, S.A., Deng, X., and Philipps, D.M., 1993, PECAM-1 is required for transendothelial migration of leucocytes. *J. Exp. Med.* 178:449-460.

Newman, P.J., Berndt, M.C., Gorski, J., White II, G.C., Lyman, S., Paddock, C., and Muller, W.A., 1990, PECAM-1 (CD31) cloning and relation to adhesion molecules of the immunoglobulin gene superfamily. *Science* 247:1219-1222.

Nose, A., Nagafuchi, A., and Takeichi, M., 1988, Expressed recombinant cadherins mediate cell sorting in model systems. *Cell* 54:993-1001.

Ozawa, M., Baribault, H., and Kemler, R., 1989, The cytoplasmatic domain of the cell adhesion molecule uvomorulin associates with three independent proteins structurally related in different species. *EMBO J.* 8:1711-1717.

Salomon, D., Ayalon, O., Patel-King, R., Hynes, R.O., and Geiger, B., 1992, Extrajunctional distribution of N-cadherin in cultured human endothelial cells. *J. Cell. Sci.* 102:1-11.

Schmelz, M. and Franke, W.W., 1993, Complexus adhaerentes, a new group of desmoplakin-containing junctions in endothelial cells: the syndesmos connecting retothelial cells of lymph nodes.*Eur. J. Cell. Biol.* 61:274-289.

Simionescu, N., 1991, Endothelial transport macromolecules: transcytosis and endocytosis. *Cell. Biol. Reviews* 25:5-80.

Simmons, D.L.,Walker, C., Power, C.,and Pigott, R., 1990, Molecular cloning of CD31, a putative intercellular adhesion molecule closely related to carcinoembryonic antigen. *J. Exp. Med.* 171:2147-2152.

Stockinger, H., Gadd, S.J., Eher, R., Majdic, O., Schreiber, W., Kasinrerk, W., Strass, B., Schnabl, E., and Knapp, W., 1990, Molecular characterization and functional analysis of the leukocyte surface protein CD31. *J. Immunol.* 145:3889-3897.

Takeda, T., Yamashita, Y., Shimazaki, S., and Mitsui, Y., 1992, Histamine decreases the permeability of an endothelial cell monolayer by stimulating cyclic AMP production through the H2-receptor. *J. Cell. Sci.* 101:745-750.

Takeichi, M., 1991, Cadherin cell adhesion receptors as a morphogenetic regulator. *Science* 251:1451-1455.

Tanaka, Y., Albelda, S.M., Horgan, K.J., van Seventer, G.A., Shimizu, Y., Newman, W., Hallan , J., Newman, P.J., Buck, C.A., and Shaw, S., 1992, CD31 expressed on distinctive T cell subsets is a preferential amplifier of beta1 integrin-mediated adhesion. *J. Exp. Med.* 176:245-253.

Tsukita, S., Tsukita, S., Nagafichi, A., and Yonemura, S., 1992, Molecular linkage between cadherins and actin filaments in cell-cell adherens junctions. *Curr. Opin. Cell. Biol.* 4:834-839.

Vaporcyan, A.A, DeLisser, H.M., Yan, H.C., Mendiguren, I.I., Thorm, S.R., Jones, M.L., Ward, P.A. and Albelda, S.M., 1993, Involvement of platelet-endothelial cell adhesion molecule-1 in neutrophil recruitment in vivo. *Science* 262:1580-1582.

Furuse, M., Hirase, T., Itoh, M., Nagafuchi, A., Yonemura, S., Tsukita, S., and Tsukita, S., 1993, Occludin: a novel integral membrane protein localizing at tight junctions. *J. Cell. Biol.* 123:1777-1788.

Goyert, S.M., Ferrero, E.M., Seremetis, S.V., Winchester, R.J., Silver, J., and Mattison, A.C., 1986, Biochemistry and expression of myelomonocytic antigens. *J. Immunol.* 137:3909-3914.

Haselton, F.R., Alexander, J.S., Mueller, S.N., and Fishman, A.P., 1992, Modulation of endothelial paracellular permeability: A mechanistic approach. In Simionescu N and Simionescu M (eds.) *Endothelial cell dysfunctions* 103-126.

Hynes, R.O., 1992, Integrins: Versatility, modulation and signaling in cell adhesion. *Cell* 69:11-25.

Joris, I., Cuenoud, H.F., Doern, G.V., Underwood, J.M., and Majno, G., 1992, Endothelial permeability in inflammation. The role of capillaries versus venules. In Simionescu N and Simionescu M (eds.) *Endothelial cell dysfunctions* 233-242.

Lampugnani, M.G., Resnati, M., Dejana, E., and Marchisio, P.C., 1991, The role of integrins in the maintenance of endothelial monolayer integrity. *J. Cell Biol.* 112:479-490.

Lampugnani, M.G., Resnati, M., Raiteri, M., Pigott, R., Pisacane, A., Houen, G., Ruco, L.P., and Dejana, E., 1992, A novel endothelial-specific membrane protein is a marker of cell-cell contacts. *J. Cell. Biol.* 118:239-246.

Leach, L., Clark, P., Lampugnani, M.G., Arroyo, A.G., Dejana, E., and Firth, J.A., 1993, Immunoelectron characterization of the inter-endothelial junctions of human term placenta. *J. Cell. Sci.*, in press.

Liaw, C.W., Cannon, C., Power, M.D., Kiboneka, P.K., and Rubin, L.L., 1990, Identification and cloning of two species of cadherins in bovine endothelial cells. *EMBO J.* 9:2701-2708.

Majno, G., Palade, G.E., and Schoefl, G.I., 1961, Studies on inflammation II. The site of action of histamine and serotonin along the vascular tree: A topographic study. *J. Biophys. Biochem. Cytol.* 11:607-626.

Miyatani, S., Shimamura, K., and Hatta, M. et al., 1989, Neural cadherin: role in selective cell-cell adhesion. *Science* 245:631-635.

Muller, W.A., Ratti, C.M., McDonnell, S.L., and Cohn, Z.A., 1989, A human endothelial cell-restricted, externally disposed plasmalemmal protein enriched in intercellular junctions. *J. Exp. Med.* 170:399-414.

Muller, W.A., Weigl, S.A., Deng, X., and Phillips, D.M., 1993, PECAM-1 is required for transendothelial migration of leukocytes. *J. Exp. Med.* 178:449-460.

Newman, P.J., Berndt, M.C., Gorski, J., White II, G.C., Lyman, S., Paddock, C., and Muller, W.A., 1990, PECAM-1 (CD31) cloning and relation to adhesion molecules of the immunoglobulin gene superfamily. *Science* 247:1219-1222.

Nose, A., Nagafuchi, A., and Takeichi, M., 1988, Expressed recombinant cadherins mediate cell sorting in model systems. *Cell* 54:993-1001.

Ozawa, M., Baribault, H., and Kemler, R., 1989, The cytoplasmic domain of the cell adhesion molecule uvomorulin associates with three independent proteins structurally related in different species. *EMBO J.* 8:1711-1717.

Salomon, D., Ayalon, O., Patel-King, R., Hynes, R.O., and Geiger, B., 1992, Extrajunctional distribution of N-cadherin in cultured human endothelial cells. *J. Cell. Sci.* 102:1-11.

Schmelz, M. and Franke, W.W., 1993, Complexus adhaerentes, a new group of desmoplakin-containing junctions in endothelial cells: the syndesmos connecting retothelial cells of lymph nodes. *Eur. J. Cell. Biol.* 61:274-289.

Simionescu, N., 1991, Endothelial transport macromolecules: transcytosis and endocytosis. *Cell Biol. Reviews* 25:5-80.

Simmons, D.L., Walker, C., Power, C., and Pigott, R., 1990, Molecular cloning of CD31, a putative intercellular adhesion molecule closely related to carcinoembryonic antigen. *J. Exp. Med.* 171:2147-2152.

Stockinger, H., Gadd, S.J., Eher, R., Majdic, O., Schreiber, W., Kasinrerk, W., Strass, P., Schnabl, E., and Knapp, W., 1990, Molecular characterization and functional analysis of the leukocyte surface protein CD31. *J. Immunol.* 145:3889-3897.

Takeda, T., Yamashita, Y., Shimazaki, S., and Mitsui, Y., 1992, Histamine decreases the permeability of an endothelial cell monolayer by stimulating cyclic AMP production through the H2-receptor. *J. Cell. Sci.* 101:745-750.

Takeichi, M., 1991, Cadherin cell adhesion receptors as a morphogenetic regulator. *Science* 251:1451-1455.

Tanaka, Y., Albelda, S.M., Horgan, K.J., van Seventer, G.A., Shimizu, Y., Newman, W., Hallam, J., Newman, P.J., Buck, C.A., and Shaw, S., 1992, CD31 expressed on distinctive T cell subsets is a preferential amplifier of beta1 integrin-mediated adhesion. *J. Exp. Med.* 176:245-253.

Tsukita, S., Tsukita, S., Nagafuchi, A., and Yonemura, S., 1992, Molecular linkage between cadherins and actin filaments in cell-cell adherens junctions. *Curr. Opin. Cell. Biol.* 4:834-839.

Vaporciyan, A.A., DeLisser, H.M., Yan, H.C., Mendiguren, I.I., Thom, S.R., Jones, M.L., Ward, P.A., and Albelda, S.M., 1993, Involvement of platelet-endothelial cell adhesion molecule-1 in neutrophil recruitment in vivo. *Science* 262:1580-1582.

V. PULMONARY HYPERTENSION

THE CELLULAR BASIS OF THE PATHOPHYSIOLOGY AND TREATMENT OF PULMONARY HYPERTENSION

Stuart Rich, M.D.

University of Illinois at Chicago
Section of Cardiology (M/C 787)
840 South Wood Street
Chicago, Illinois

Pulmonary hypertension has traditionally been categorized into primary and secondary types. Although primary pulmonary hypertension (PPH) is felt to be relatively unique as a cardiopulmonary disease, its distinction from secondary forms of pulmonary hypertension often solely depends on whether or not an identifiable etiology can be ascertained. Histologically it can be impossible to distinguish primary from secondary forms of pulmonary hypertension (*Pietra et al., 1989),* suggesting that the distinction may be in the mechanism(s) that produce the pulmonary vascular disease with pulmonary hypertension as a common end point.

Pulmonary hypertension secondary to congenital heart disease and left-to right shunts has been well characterized. Speculation regarding the mechanism has been that the increased pulmonary blood flow coming directly from the systemic circulation causes reactive vasoconstriction (*Hoffman et al., 1981*). Collagen vascular diseases have all been implicated in the development of pulmonary hypertension, presumably with an underlying vasculitis as the cause (*Unger et al., 1983*). The ingestion of agents such as the appetite suppressant aminorex (*Gurtner, 1985)* and the toxic rapeseed oil from Spain in the 1980's (*Gomez-Sanchez et al., 1989*) resulted in chronic pulmonary hypertension, presumably due to some direct toxic effect on the pulmonary vasculature. In these settings, however, the reactive pulmonary vasoconstriction occurs as an "inappropriate" response.

There are also several, even less understood, conditions that have been associated with pulmonary hypertension. For example, although pulmonary hypertension is not common in patients with cirrhosis and portal hypertension, the relationship has been well established but the mechanism remains uncertain (*McDonnel et al., 1983*). More recently, it has been reported that patients who are infected with the HIV virus can develop unexplained pulmonary hypertension (*Polos et al., 1992*). Histologic studies, however, do not implicate a direct viral infection of the pulmonary vascular bed as an underlying mechanism.

The fact that secondary forms of pulmonary hypertension can be present across a broad spectrum of severity that may appear proportional to the underlying disease suggests that the severity of the insult to the pulmonary vasculature bed determines the extent of the subsequent pulmonary hypertension. However, often the pulmonary hypertension bears no relationship to the duration or severity of the insult (aminorex exposure as a classic example). Moreover, since pulmonary hypertension occurred in a minority of patients exposed to aminorex, an underlying genetic predisposition may be also necessary for the development of the pulmonary vascular disease.

Vascular Endothelium: Responses to Injury
Edited by John D. Catravas *et al.*, Plenum Press, New York, 1996

Primary pulmonary hypertension has served as a template for histologic and physiologic studies of pulmonary hypertensive disease states. Histologically, pathologists have characterized the vascular changes seen in PPH under subclassifications, presumable to reflect the underlying disease process. Plexogenic arterioparthy is one pattern that is a consequence of chronic severe pulmonary hypertension from a variety of etiologies including congenital heart disease, aminorex ingestion and PPH (*Pietra and Ruttner, 1987*). The sequence of lesions appears to be medial hypertrophy, intimal proliferation and fibrosis, and the development of plexiform lesions, an out-pouching of the arteriolar wall which is a manifestation of disease. The appearance of these lesions has been presumed to reflect pulmonary arterial vasoconstriction as the mechanism for the elevation in the pulmonary artery pressure that is associated.

Recent data would support the conjecture that injury to the pulmonary vascular endothelium could directly result in the cascade of pathologic findings that produces plexogenic pulmonary arteriopathy. Histologic studies in patients with congenital heart disease and pulmonary hypertension have revealed ultrastructural abnormalities in the endothelial cells and heightened metabolic activities (*Rabinovitch et al., 1986*) as well as increased levels of von Willebrand factor antigenic activity (*Rabinovitch et al., 1987*). These abnormalities are seen, irrespective of the severity of the pulmonary vascular disease, but are not observed in patients with congenital heart disease without pulmonary hypertension. Shear stress has been described as a cause of endothelial cell dysfunction in systemic beds and is a plausible explanation for the mechanism of endothelial injury in patients with congenital heart disease and post tricuspid left-to-right shunts (*Miller et al., 1986*). However, pulmonary hypertension can occur in patients with congenital heart disease and increased pulmonary blood flow alone, as seen in the pre tricuspid shunts (atrial septal defects), although the incidence is quite low (*Yamaki et al., 1986).* This could suggest that, whatever perturbations high pulmonary blood flow can produce, it will lead to chronic pulmonary hypertension only in patients who are genetically predisposed.

One animal model of pulmonary hypertension that has been associated with toxin ingestion is the administration of the herbal plant Crotalaria to rats. Pulmonary vasoconstriction has been presumed to occur following the plant ingestionleading to vascular changes of intimal proliferation and medial hypertrophy (*Meyrick and Reid, 1979*). However, when studied metabolically, the earliest abnormality noted following the ingestion of Crotolaria is an increase in the metabolic activity of the pulmonary vascular endothelium which precedes an increase in pulmonary vascular smooth muscle proliferation by three to four weeks.

More recently, endothelin-1, an endogenous vasoactive peptide produced by vascular endothelium, has been shown to have several properties that could play a potential pathophysiologic role in pulmonary hypertension (*Luscher, 1992*). Its long half-life would allow subtle disturbances in production or release to predispose to sustained vasoconstriction. In addition, endothelin-1 has myogenic activity on vascular smooth muscle and fibroblasts consistent with a role in vascular remodeling, which is a common finding in pulmonary hypertension. Increased endothelin levels have been reported in patients with primary and secondary pulmonary hypertension, suggesting that endothelin may be produced as either a cause or effect of the disease (*Stewart et al., 1991*). Interestingly, increased tissue levels of endothelin-1 in the face of normal circulating levels had been noted in the Fawn-Hooded Rat the only animal model known to develop idiopathic pulmonary hypertension (*Stelzner et al., 1992*). Enhanced preproendothelin messenger RNA expression in neonatal Fawn-Hooded Rats prior to the development of right ventricular enlargement suggests that increased activity may be a cause rather than a response to the pulmonary vascular disease. Finally, prostacyclin, which is a potent vasodilator, has been found to be reduced in patients with pulmonary hypertension (*Christman et al., 1992*). Although the precise role that prostacyclin plays in the pulmonary vasculature is unclear, it may play a role in modulating pulmonary vascular tone.

In that regard, diminished prostacyclin production or release may allow pulmonary vasoconstriction to occur unopposed.

The other major histologic subset of pulmonary vascular changes in patients with PPH is one of a thrombotic pulmonary arteriopathy (*Pietra et al., 1989*). Histologically one finds predominantly excentric intimal fibrosis along with medial hypertrophy and fibro-elastic intimal pads in the arteries and arterioles. Scattered evidence of old recanalized thrombus is also found, usually appearing as fibrous webs in the small vascular channels. It is now established that endothelial injury results in a predisposition to local thrombosis in vascular beds (*Ryan, 1986*), which seems to be a ubiquitous feature in pulmonary hypertensive states. The normal vascular endothelium appears to play a critical role as a barrier to *in situ* thrombosis and hemostasis. The endothelium possesses an antithrombotic surface to the blood that resists the adherence of platelets and leukocytes and prevents the activation of coagulation and complement factors (*Tracy, 1991*). The endothelium also produces prostacyclin, which is a potent inhibitor of platelet aggregation, as well as a number of other proteins that are intimately involved in the coagulation and thrombolytic process. These active and passive antithrombotic properties of the vascular endothelium can be impaired through physical disruption by chemical, immunologic and mechanical processes. Thus, patients who develop pulmonary hypertension related to the ingestion of aminorex (a chemical injury), in association with a collagen vascular disease (an immunologic injury), or related to a congenital left-to-right shunt (a mechanical injury), have the substrate for the development of this form of pulmonary vascular disease. Von Willebrand factor, which is increased in patients with congenital heart disease and pulmonary hypertension, has procoagulant properties (*Rabinovitch et al., 1987*). Increased urinary metabolites of thromboxane in patients with primary and secondary pulmonary hypertension implies that activation of the coagulation cascade occurs in pulmonary hypertension. Serotonin, which is released from platelets, can also produce local vasoconstriction (*Houston and Vanhoutte, 1988*). Finally, elevations in Fibrinopeptide A levels in most patients with primary pulmonary hypertension points to abnormalities in endothelial cell function leading to the development of a procoagulant environment in the pulmonary vascular bed predisposing local thrombosis (*Eisenberg et al., 1990*). As in patients who develop plexogenic arteriopathy, the variation in the extent and duration of exposure to these noxious stimuli and the severity of the subsequent thrombotic pulmonary arteriopathy suggests a predisposing genetic factor is also playing a role in the development of pulmonary hypertension in affected patients.

Familial PPH has been well described as an autosomal dominant genetic trait with variable penetrance (*Loyd et al., 1984*). It occurs in only a minority of patients with PPH but clinically is indistinguishable from sporadic forms. Interestingly, the pathologic studies on patients with familial PPH have revealed both thrombotic and plexogenic arteriopathy of a varying degree in different families (*Loyd et al., 1988*). These observations, along with the other features of plexogenic and thrombotic pulmonary arteriopathies just described, would be consistent with the hypothesis that the development of clinical pulmonary hypertension is a result of an underlying genetic predisposition towards pulmonary vascular endothelial injury which, when combined with an exposure to some known pulmonary vascular noxious agent, results in the clinical expression of pulmonary hypertension and its associated manifestations. In this context, primary pulmonary hypertension would be a disease which occurs in the absence of an identifiable stimulus.

THE CURRENT TREATMENT OF PULMONARY HYPERTENSION

Vasodilators have been a major focus of the treatment of chronic pulmonary hypertension, both primary and secondary. Historically, acetylcholine was one of the first vasodilators tested and was shown to have acute vasodilatory properties in PPH (*Wood, 1958*). Although only a minority of patients with PPH respond to vasodilators, the histologic determinants of a pulmonary vasodilator response has been difficult to elucidate. It does appear, however, that the ability to respond to a vasodilator agent is an early feature of the disease (*Palevsky et al.,*

1989) which can become lost over time (*Samet and Bernstein, 1963*). Studies of patients with PPH and pulmonary hypertension associated with congenital heart disease point to a progressively impaired pulmonary vasodilator response manifested by a diminished response to endothelium dependent vasodilators initially, progressing to lack of response and a diminished response to endothelium independent vasodilators and progressing to lack of response to all types of vasodilators (*Celemajer et al., 1993*).

Several short acting vasodilators have been tested which may provide some insight into the nature of the pulmonary endothelial response in these patients. Adenosine is a potent vasodilator which works through specific vascular receptors. In normal patients the administration of adenosine in a continuous infusion results in increased pulmonary blood flow believed to be the result of endothelial mediated vascular smooth muscle relaxation (*Belardinelli et al., 1989*). In patients with PPH, adenosine has demonstrated marked potency with the ability to achieve a reduction in pulmonary vascular resistance that exceeds the calcium blockers (*Schrader et al., 1992*). Although the actions of adenosine in these patients is quite complex, it appears to work through an endothelium independent way.

Nitric oxide, believed to have an identical action to endothelial derived relaxing factor, has also been studied acutely in patients with pulmonary hypertension. Marked reductions in pulmonary vascular resistance have been noted to occur within minutes in patients with primary and secondary hypertension (*Pepke-Zaba et al., 1991; Roberts et al., 1993*). Like adenosine, the extremely short half-life allows a relatively preferential pulmonary vasodilator effect. The effectiveness of nitric oxide in patients with severe pulmonary vasodilator hypertension also suggests that it works independently of endothelial cell function. It is possible that deficiencies in agents such as adenosine and nitric oxide, which normally either stimulate the endothelial cell or are produced by it, result in unopposed vasoconstriction and smooth muscle migration which contributes to the progression of the pulmonary hypertensive state.

Vasodilators have also been employed as a chronic treatment of pulmonary hypertension. The most successful study to date has been the use of high doses of calcium channel blockers in patients with primary pulmonary hypertension (*Rich et al., 1992*). Although only about 25% of patients respond, the response is fairly dramatic and associated with a marked reduction in the level of pulmonary hypertension, regression of right ventricular hypertrophy and improved survival. It is interesting that it requires high doses of calcium blockers to produce these hemodynamic effects, and that patients often have better hemodynamics when studied serially after several years. It may be that the calcium blockers are also involved in vascular remodeling resulting in halting of progression and even regression of vascular disease. It is also interesting that the calcium blockers have not been similarly effective in patients with secondary forms of pulmonary hypertension (*Nootens et al., 1993*). Whether this relates to a specific mechanism inducing pulmonary hypertension in patients with PPH or the chronicity that is generally associated with secondary forms remains unknown.

One of the more fascinating chronic treatments of primary pulmonary hypertension has been the use of a continuous infusion of prostacyclin (*Rubin et al., 1990*). By protocol its use has been reserved to patients with advanced disease, refractory to oral vasodilators. Hemodynamically it produces modest reductions of pulmonary arterial pressure and pulmonary vascular resistance, but long term usage has generally shown improved exercise tolerance in most all patients who receive it. In addition, early studies now suggest that patients receiving prostacyclin have improved survival as well (*Cremona et al., 1991*).

The fact that the long term response to prostacyclin is not predicted by the acute hemodynamic effects suggests that prostacyclin may be working in ways other than vascular vasodilatation. It is possible that chronic prostacyclin promotes improvement in endothelial cell integrity, a reduction in smooth muscle cell hypertrophy and migration, and has effects on other vascular beds that have not yet been described. It also remains unknown how effective prostacyclin would be if it were given to patients with less severe pulmonary hypertension. The fact that the acute vasodilator response to calcium blockers seems to be predicted by the acute response to prostacyclin (*Reeves et al., 1986*) makes it possible that, when given to less severely affected patients, it could also result in regression of right ventricular hypertrophy and halting of the vascular disease process.

Given that histologic evidence of *in situ* thrombosis is very common in patients with PPH and secondary forms of pulmonary hypertension, it would seem intuitive that the chronic use of an anticoagulant would be of benefit to these patients. The lack of prospective, randomized trials, however, makes clinical documentation of this difficult. A retrospective review of the use of anticoagulants in patients with PPH at the Mayo Clinic over a 15 year period revealed that the use of warfarin was associated with an improved survival over three years (*Fuster et al., 1984*). A subsequent case report showed regression of pulmonary hypertension in one patient with PPH who was treated with chronic anticoagulant therapy, suggesting that in one patient an active thrombotic process was effectively arrested (*Cohen et al., 1986*). More recently the prospective assessment of the effects of anticoagulation in patients evaluated with high dose calcium blockers revealed that the chronic administration of warfarin to patients with PPH over five years was associated with a substantially improved survival (*Rich et al., 1992*). These observations strongly suggest that chronic anticoagulation can reduce arterial thrombosis and its influence on the development of further pulmonary vascular disease in PPH. Interestingly, prostacyclin also has antithrombotic properties which could contribute to the favorable, long term, clinical effectiveness in PPH as well.

Although there are no studies on the chronic effectiveness of the use of anticoagulation in patients with secondary pulmonary hypertension, it would appear logical that the use of anticoagulation in these patients should also provide some long term salutary clinical benefit. Whether the use of other anticoagulants, such as heparin, or specific antiplatelet agents would be efficacious over the long term is completely unknown.

CONCLUSIONS

Based on current data, it appears that PPH and secondary forms of pulmonary hypertension occurring at the arteriolar level result from an injury to the pulmonary vascular endothelium which then produces a spectrum of vasoconstriction and *in situ* thrombosis. The vasuclar response to this injury may relate to the nature or severity of the injury. In addition, it appears that the development of pulmonary hypertension requires or is governed by an underlying genetic predisposition. In patients who have primary pulmonary hypertension, the stimulus causing the expression of the pulmonary vascular disease remains unknown. As our understanding of the factors influencing the vascular response to injury improves, more focused treatments on patients with advanced pulmonary hypertension should be forthcoming.

REFERENCES

Belardinelli, I., Linder, J. and Berne, R.M., 1989, The cardiac effects of adenosine. *Prog. Cardiovas. Dis.* 32:73.

Celemajer, D.S., Cullen, S. and Deanfield, J.E., 1993, Impairment of endothelium-dependent pulmonary artery relaxation in children with congenital heart disease and abnormal pulmonary hemodynamics. *Circulation* 87:440.

Christman, B.W., McPherson, C.D., Newman, J.H., King, J.A., Bernard, G.R., Groves, B.M., and Loyd, J.E., 1992, An imbalance between the excretion of thromboxane and prostacyclin metabolites in pulmonary hypertension. *N. Engl. J. Med.* 327:70.

Cohen, M.D., Edwards, W.D., and Fuster, V., 1986, Regression in thromboembolic type of primary pulmonary hypertension during 2-1/2 years of antithrombotic therapy. *J. Am. Coll. Cardiol.* 7:172.

Cremona, G., Higenbottam, T.W., and Scott, J.P., 1991, Continuous intravenous infusion of prostacyclin (PGI_2) improves survival in primary pulmonary hypertension [abstract]. *Am. Rev. Respir. Dis.* 143:A180.

Eisenburg, P.R., Lucore, C., Kaufmann, L., Sobel, B.E., Jaffe, A.S. and Rich, S., 1990, Fibrinopeptide A levels indicative of pulmonary vascular thrombosis in patients with primary pulmonary hypertension. *Circulation* 82:841.

Fuster, V., Steele, P.M., Edwards, W.D., Gersh, B.J., McGooon, M.D. and Frye, R.L., 1984, Primary pulmonary hypertension: natural history and the importance of thrombosis. *Circulation* 70:580.

Gomez-Sanchez, M.A., Mestre de Juan, M., Gomez-Pajuelo, C., et al, 1989, Pulmonary hypertension due to toxic oil syndrome. *Chest.* 95:325.

Gurtner, H.P., 1985, Aminorex and pulmonary hypertension. *Cor. Vasa.* 27:160.
Hoffman, J.I.E., Rudolf, A.M. and Heymann, M.A., 1981, Pulmonar vascular disease with congenital heart lesions: Pathologic features and causes. *Circulation* 64:873.
Houston, D.S. and Vanhoutte, P.M., 1988, Comparison of serotonergic receptor subtypes on the smooth muscle and endothelin of the canine coronary artery. *J. Pharmacol. Exp. Ther.* 244:1.
Loyd, J.E., Atkinson, J.B., Pietra, G.G., Virmani, R. and Newman, J.H., 1988, Heterogeneity of pathologic lesions in familial primary pulmonary hypertension. *Am. Rev. Respir. Dis.* 138:952.
Loyd, J.E., Primm, R.K. and Newmann, J.H., 1984, Familial primary pulmonary hypertension: clinical patterns. *Am. Rev. Respir. Dis.* 129:194.
Luscher, T.F., 1992, Endothelin: Systemic arterial and pulmonary effects of a new peptide with potent biologic properties. *Am. Rev. Respir. Dis.* 146:556.
McDonnell, P.J., Toye, P.A. and Hutchins, G.M., 1983, Primary pulmonary hypertension and cirrhosis: Are they related? *Am. Rev. REspir. Dis.* 127:437.
Meyrick, B. and Reid, L., 1979, Development of pulmonary arterial changes in rats fed Crotalaria Spectabilis. *Am. J. Pathol.* 94:37.
Miller, V.M., Aahus, L.L. and Vanhoutte, P.M., 1986, Modulation of endothelium-dependent responses by chronic alterations of blood flow. *Am. J. Physiol.* 251:H520.
Nootens, M., Kaufmann, E. and Rich, S., 1993, Short-term effectiveness of nifedipine in secondary pulmonary hypertension. *Am. J. Cardiol.* 72:1475.
Palevsky, H.I., Schloo, B.L., Pietra, G.G., Weber, K.T., Janicki, J.S., Rubin, E., et al, 1989, Primary pulmonary hypertension: vascular structure, morphometry and responsiveness to vasodilator agents. *Circulation* 80:1207.
Pepke-Zaba, J., Higgenbottam, W., Tuan Ding-Xuan, A., Strong, D. and Wallwork, J., 1991, Inhaled nitric oxide as a cause of selective pulmonary vasodilation in pulmonary hypertension. *Lancet* 338:447.
Pietra, G.G., Edwards, W.D. and Kay, J.M., et al, 1989, Histopathology of pulmonary hypertension: a qualitative and quantitative study of pulmonary blood vessels from 58 patients in National Heart, Lung, and Blood Institute, Primary Pulmonary Hypertension Registry. *Circulation* 80:1198.
Pietra, G.G., and Ruttner, J.R., 1987, Specificity of pulmonary vascular lesions in primary pulmonary hypertension. *Respiration* 52:81.
Polos, P.G., Wolfe, D., Harley, R.A., Strange, C. and Sahn, S.A., 1992, Pulmonary hypertension and immunodeficiency virus infection. *Chest.* 101:474.
Rabinovitch, M., Andrew, M. and Thom, H., et al, 1987, Abnormal endothelial factor VIII associated pulmonary hypertension and congenital heart defects. *Circulation* 76:1043.
Rabinovitch, M., Bothwell, T. and Hayakawa, B.N., et al, 1986, Pulmonary artery endothelial abnormalities in patients with congenital heart defects and pulmonary hypertension. *Lab. Invest.* 55:632.
Reeves, J.T., Groves, B.M. and Turkevich, D., 1986, The case for treatment of selected patients with primary pulmonary hypertension. *Am. Rev. Respir. Dis.* 134:342.
Rich, S., Kaufmann, E. and Levy, P.S., 1992, The effect of high doses of calcium-channel blockers on survival in primary pulmonary hypertension. *N. Engl. J. Med.* 327:76.
Roberts, J.D., Lang, P., Bigatello, L.M., Vlahakes, G.J. and Zapol, W.M., 1993, Inhaled nitric oxide in congenital heart disease. *Circulation* 87:447.
Rubin, L.J., Mendoza, J., Hood, M., McGoon, M., Barst, R. and Williams, W.B., et al, 1990, Treatment of primary pulmonary hypertension with continuous intravenous prostacyclin (epoprostenol). *Ann. Intern. Med.* 112:485.
Ryan, U.S., 1986, The endothelial surface and responses to injury. *Fed. Proc.* 45:101.
Samet, P. and Bernstein, W.H., 1963, Loss of reactivity of the pulmonary vascular bed in primary pulmonary hypertension. *Am. Heart J.* 66:197.
Schrader, B.J., Inbar, S., Kaufmann, L., Vestal, R.E. and Rich, S., 1992, Comparison of the effects of adenosine and nifedipine in pulmonary hypertension. *J. Am. Coll. Cardiol.* 19:1060.
Stelzner, T.J., O'Brien, R.F., Yanagisawa, M., Sakurai, T., Sato, K., Webb, S., Zamora, M., McMurtry, I.F. and Fisher, J.H., 1992, Increased lung endothelin-1 production in rats with idiopathic pulmonary hypertension. *Am. J. Physiol.* L614.
Stewart, D.J., Levy, R.D., Cernacek, P. and Langleben, D., 1991, Increased plasma endothelin-1 in pulmonary hypertension: Marker or mediator of disease. *Ann. Intern. Med.* 114:464.
Tracy, P.B., 1991, The role of endothelium in the expression of procoagulant activity. *Coronary Artery Disease* 2:146.
Ungerer, R.G., Tashkin, D.P., Furst, D., et al, 1983, Prevalence and clinical correlates of pulmonary arterial hypertension in progressive systemic sclerosis. *Am. J. Med.* 75:65.
Wood, P., 1958 Pulmonary hypertension with special reference to the vasoconstrictive factor. *Br. Heart J.* 20:557.
Yamaki, S., Horiuchi, T., Miura, M., et al, 1986, Pulmonary vascular disease in secundum atrial septal defect with pulmonary hypertension. *Chest.* 89:694.

INHALED NITRIC OXIDE IN ACUTE RESPIRATORY FAILURE

Jean-Jacques Rouby

Surgical Critical Care Unit
Department of Anesthesiology
La Pitié-Salpétrière Hospital
University of Paris VI
83 boulevard de l'Hopital
75013 Paris, France

Gaseous nitric oxide (NO) and nitrogen dioxide (NO_2) are well known atmospheric outdoor pollutants. The toxicities of these pollutants were extensively studied in the seventies. In the eighties, inhaled NO was used in humans as an alternative to CO to measure the oxygen diffusing capacity of the lungs *(Meyer and Piiper, 1989; Moinard and Guenard, 1990)*. In 1987, one of the endothelium-derived relaxing factors was identified as NO *(Furchgott and Zawadski, 1980; Palmer et al., 1987)*. It was then demonstrated that NO was continuously released by the endothelial cells after activation of the constitutive NO synthase in response to shear stress. Later, NO was found to be involved in a large number of physiological systems *(Berdeaux, 1993; Moncada et al., 1991)* such as coagulation *(Macdonald et al., 1988; Alheid et al., 1987)*, neuronal transmission and inflammation.

The first description of inhaled NO as a therapeutic agent in humans was published in 1988 in an abstract form by Higenbottam's group *(Higenbottam et al., 1988)*. Inhaled NO was shown to reduce pulmonary pressure without any systemic effect in seven patients with severe primary pulmonary hypertension. This work was published as a full paper in 1991 *(Pepke-Zaba et al., 1991)*. Inhaled NO-induced selective pulmonary vasodilation was later confirmed in different experimental models of pulmonary hypertensions, such as, injection of an analogue of thromboxane A_2 or injection of heparin-protamine *(Fratacci et al., 1991)*, hypoxic pulmonary vasoconstriction *(Frostell et al., 1991; Pison et al., 1993; Roberts et al., 1992)*, LPS or streptococcal intravenous infusion *(Weitzberg, 1991; Dyar et al., 1993; Berger et al., 1993)*. Concomitant to the reduction in pulmonary artery pressure, a sustained inhaled NO-induced improvement in arterial oxygenation was described in the adult respiratory distress syndrome by Falke's group in 1993 *(Rossaint et al., 1993)* and then confirmed by other studies *(Bigatello et al., 1994; Puybasset et al., 1994; Puybasset, 1994)*. In addition, pulmonary hypertension secondary to cardiac surgery *(Girard et al., 1992; Rich et al., 1993)* or COPD *(Adnot et al., 1993; Moinard et al., 1994)* was shown to be also responsive to inhaled NO.

The main risk of an extensive use of NO in acute respiratory failure is its potential toxicity. Even if the beneficial effects of NO on arterial oxygenation and pulmonary pressure have been well demonstrated the potential pro- or antioxidant effects of NO are still a subject of

Vascular Endothelium: Responses to Injury
Edited by John D. Catravas *et al.*, Plenum Press, New York, 1996

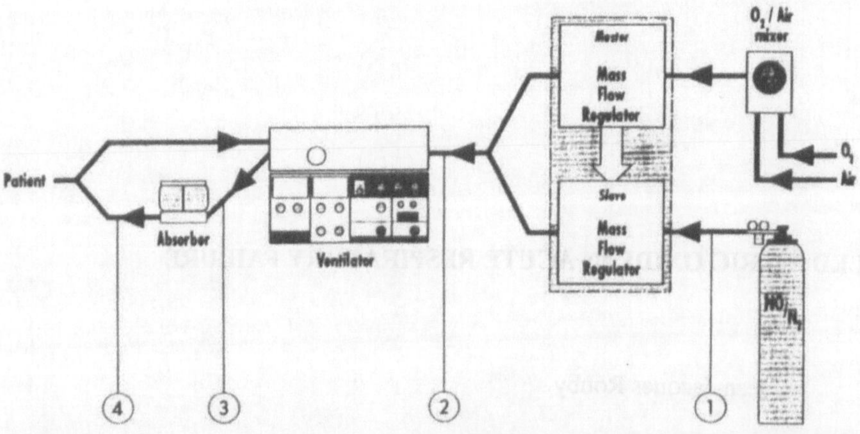

Figure 1. *NO delivery at the gas inlet.* The flow of NO delivered to the ventilator at the gas inlet is controlled by a mass flow regulator subordinated to the mass flow regulator delivering O_2 or air: any reduction in minute ventilation or oxygen supply will immediately lead to a reduction in NO flow and therefore avoid excessive NO intratracheal concentration. The long contact time between NO and O_2 can generate NO_2 and imposes the use of a soda lime absorber located in the inspiratory limb of the respiratory circuit (from reference *Stenqvist et al.,*1993).

controversy. The potential benefit of the long term administration of NO during acute respiratory failure is presently under investigation in controlled multicenter human studies.

ADMINISTRATION AND MONITORING

Methods of Administration

NO tanks used in the ICU are highly concentrated (ranging between 225 and 900 parts per million ppm) in order to limit the reduction in FIO_2 resulting from the administration of NO which is diluted in pure nitrogen. *In concentrations above 200 ppm, NO is cytotoxic and therefore should not be administered directly through an intratracheal catheter in contact with the tracheal mucosa.* NO should be diluted in the respiratory circuits before reaching the tracheobronchial tree, in order to obtain the desired concentration in patient's airways.

NO can be delivered in the respiratory circuit either before (Figure 1) or after (Figure 2) the ventilator. In the former type of administration mass flow regulators are necessary to precisely mix NO, air and oxygen. The main benefit of this technique, initially developed in Northern Europe *(Stenqvist et al., 1993)*, is in the precision and the stability of the NO concentration delivered to the patient *(Wessel et al., 1994)*. However, there are some drawbacks related firstly to the elevated price of mass flow regulators, secondly to the long contact time between NO and oxygen resulting in the generation of NO_2 and requiring the mandatory use of a soda lime absorber, and thirdly to the risk of oxidation of the internal circuits of the ventilator as they are permanently in contact with nitrogen oxides. NO can also be administered within the inspiratory circuit after the ventilator. It can be delivered either continuously during the entire respiratory cycle or discontinually only during inspiration. The inspiratory phase of the ventilator is detected either using the nebulizer signal as in the CFPO prototype *(Mourgeon et al., 1994)* or using an airway pressure measurement as in the Statice prototype *(Capellier et al., 1993)*.

Continuous and discontinuous administrations have been compared *(Mourgeon et al., 1994)* in patients with acute respiratory failure. In both cases, the mean NO concentration at the proximal tip of the endotracheal tube was set at 2.5 ppm. Mean NO concentrations were measured in the respiratory circuits, the endotracheal tube, and the proximal airways using a

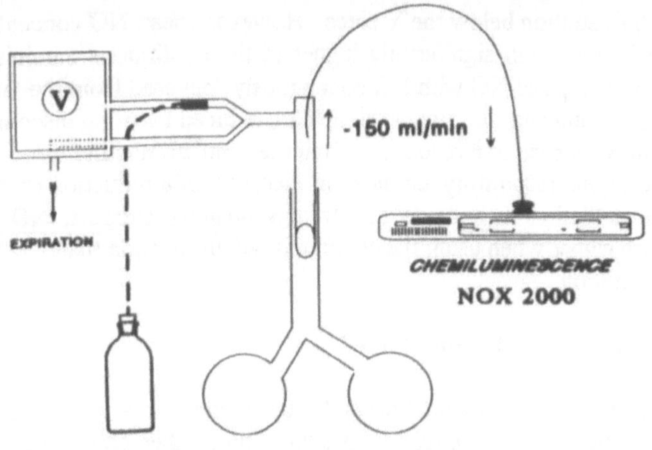

Figure 2. *NO delivery within the inspiratory limb.* NO is delivered within the inspiratory limb before the Y piece. Highly concentrated NO is diluted in the tidal volume before reaching the airways. NO can be delivered in the inspiratory circuit either continuously by a flow-meter or discontinuously only during the inspiratory phase of the respiratory cycle. NO and NO_2 concentrations are measured by a chemiluminescence apparatus (NOX 4000, Sérès, France) continously sampling tracheal gas at the proximal tip of the endotracheal tube and allowing the precise determination of inspired and expired NO concentrations.

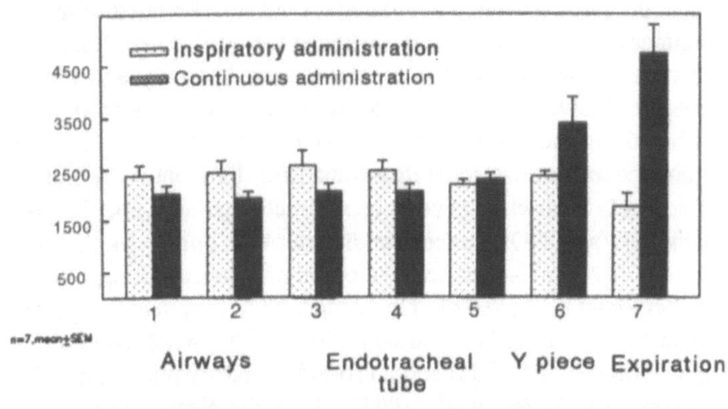

Figure 3. *Comparison of mean NO concentrations within the respiratory cirucits and the upper airways when using two different NO delivery systems.* Mean NO concentrations were measured by a chemiluminescence apparatus in 7 patients with acute respiratory failure. NO was delivered within the inspiratory limb of the ventilator just before the Y piece either continuously using a calibrated flow-meter or only during inspiration using a CFPO prototype which detects inspiration from nebulizer flow. In both modes, NO settings were adjusted in order to obtain a mean NO concentration at the proximal tip of endotracheal tube of 2500 ppb (=2.5 ppm). Sites 1 and 2 are located in the upper airways respectively 10 and 5 cms below the distal tip of the endotracheal tube. Sites 3 to 5 are located inside the endotracheal tube whereas site 6 is close to the Y piece and site 7 is within the expiratory limb, 40 cms distally from the Y piece. Expiratory concentrations are significantly lower when NO is administered only during inspiration as compared to a continuous administration (*Mourgeon et al., 1994*).

chemiluminescence apparatus (Figure 3). Mean NO concentrations were found in the same range in both modes of administration below the Y piece. However, mean NO concentrations in the expiratory circuit were found significantly higher in the continuous administration because of the summation of expired NO with NO continuously delivered from the tank. As a consequence, room air pollution and the cost of therapy are reduced by using discontinuous administration. Furthermore, in case of a reduction of minute ventilation either by accidental disconnection of a part of the respiratory circuits or because of a reduction of patient's contribution to minute ventilation in a mode of partial ventilatory support, NO airway concentrations will be much higher when using the continuous administration than when using the discontinuous administration.

Where Should NO Concentrations be Monitored?

The choice of where NO should be measured depends mainly on the site of administration. If NO is administered into the inspiratory circuit before the Y piece, then NO concentrations are identical inside the endotracheal tube, the lower trachea and the main bronchi (Figure 3). Therefore, the continuous measurement of NO and NO_2 can be easily performed at the proximal tip of the endotracheal tube in order to avoid obstruction by tracheal secretions (Figure 2).

Methods of Monitoring

Due to the potential toxicity of NO itself and of its oxidative derivative NO_2 it has been recently recommended to measure NO and NO_2 concentrations during long-term administration of inhaled NO in humans with acute respiratory failure *(Zapol et al., 1994)*. Because measurements are performed on gases saturated in humidity, only two methods of monitoring appear adequate, chemiluminescence and the electrochemical method. The infrared method implies analysis of dry gas, and, therefore, the interposition of a humidity absorber on the suctionning circuit is necessary. However, NO_2 is absorbed together with water and thus can no longer be measured when using the infrared method. The electrochemical method is less expensive than chemiluminescence but has some disadvantages: frequent recalibrations are necessary, the electrochemical cells themselves have to be replaced regularly, and the precision of the method is small for NO and NOx concentrations below 2 ppm. As far as NO is concerned, the concentration of 2 ppm is precisely the optimum concentration to be used in non-septic shock patients *(Puybasset et al., 1994)*. Concerning NO_2, 2 ppm is far greater than the acceptable threshold in humans. A bronchoconstrictor effect of NO_2 has been described for a concentration of 0.3 ppm *(Bauer et al., 1986)* and a pro-inflammatory pulmonary effect for a concentration of 2.25 ppm *(Sandström et al., 1991)*. Thus the electrochemical method appears less appropriate than chemiluminescence for clinical monitoring of NO concentrations during long-term administration of inhaled NO in humans.

The reference method chemiluminescence allows measurements of very low NO and NO_2 concentrations with a precision in the range of the part per billion (ppb). When designed for measuring atmospheric pollution the response time of chemiluminescence apparatus used in industry is low, around 40 sec (NOX 2000, Sérès, France). However the intrinsic time-response of chemiluminescence itself is high in the range of the millisecond. In fact, the observed time-response depends mainly on the aspirative flow and on the dead space of connecting and internal circuits. Chemiluminescence apparatus designed for medical use have a fast time-response, around 1 sec (NOX 4000, Sérès, France), and allow the accurate. measurement of inspired and expired concentrations of NO if a high aspirative flow is selected (*Law-Koune, et al., 1995*).

METABOLISM AND TOXICITY

Metabolism of Inhaled NO in Humans

Inhalation of NO at a concentration of 25 ppm in healthy volunteers increases methemoglobin levels and plasma nitrate concentrations *(Wennmaln et al., 1993)*. After crossing the alveolar-capillary barrier, NO can react with oxygen to generate nitrates or can bind to hemoglobin for which it has a very high affinity (1000 times more than CO). If NO binds to oxyhemoglobin, methemoglobin is generated together with nitrates (Figure 4); if NO binds to reduced hemoglobin, nitrosylhemoglobin is generated. Methemoglobin is detoxified back to hemoglobin by a slow endogenous process, and several hours are necessary to return to control values of methemoglobin after accidental inhalation of high NO concentrations. The concentrations of methemoglobin measured in 10 patients with acute respiratory failure, inhaling increasing concentrations of NO (from 0, 1 to 100 ppm) and pure oxygen, are represented in Figure 5 *(Law-Koune et al., 1994)*. The mean NO concentration of 10 ppm is the upper limit which should not be exceeded to maintain safe levels of methemoglobin.

Toxicity of Inhaled NO

Inhaled NO toxicity, apart from the generation of high levels of methemoglobin, can be due to NO itself and to its oxidative derivatives (NOx), particularly NO_2.

A. Toxicity of NO:

NO released by the macrophage after induction of the inducible NO synthase is a potent cytotoxic agent. NO also has a high mutagenic potential through DNA deamination *(Isomura et al., 1984; Arroyo et al., 1992; Nguyen et al., 1992; Wink et al., 1991)*. This last property could theoretically increase the incidence of cancers in patients receiving NO as a long-term therapy. Thereby, the registration of patients and neonates receiving long-term NO inhalation should be implemented. The NO toxicity can also be the result of the synthesis of peroxynitrite anion when NO reacts with superoxide anion, $NO + O_2^- \rightarrow ONOO^-$. This compound is a highly potent oxidant inducing lipidic peroxidation of cell membranes *(Beckman et al., 1990)* and structural changes of surfactant *(Haddad et al., 1994)*. NO, administered in dogs is lethal at a concentration of 20000 ppm. Death is caused by high levels of methemoglobin and non-cardiogenic pulmonary edema *(Greenbaum et al., 1967)*. The inhalation of low concentrations of NO during very long periods has been shown to be apparently free of toxicity. Ten ppm for six months *(Oda et al., 1976)* or 2 ppm for two years in mice *(Oda et al., 1980)* does not induce major pulmonary histologic changes. Emphysema-like lesions have been described in one study after inhalation of 20 ppm of NO for 6 weeks in rats *(Azoulay et al., 1977)*.

B. Toxicity induced by NO_2:

The amount of NO_2 $[NO_2]$ generated when NO is in presence of oxygen depends on the contact time, the FIO_2 and the NO concentration [NO] according to the following formula *(Foubert et al., 1992)*:

$$[NO_2] = k.t.\ [FIO_2].[NO]^2$$

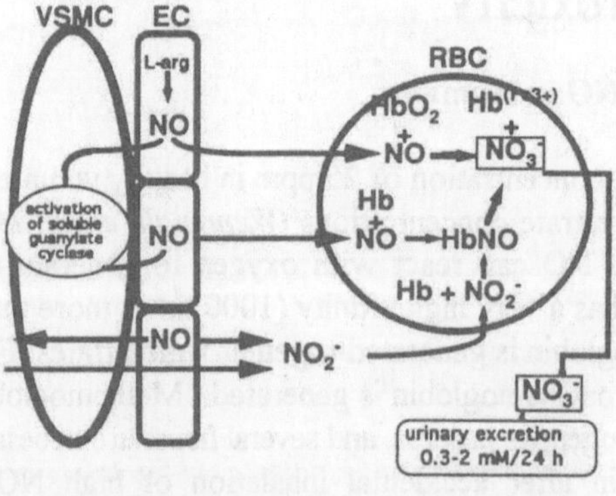

Figure 4. *Metabolism of inhaled NO.* A part of inhaled NO activates the soluble guanylate cyclase of the smooth muscle cell (VSMC) after crossing the alveolar epithelium. Another part reaches the capillary lumen and binds to hemoglobin in the red blood cell (RBC). When NO is fixed to oxyhemoglobin (HbO_2), methemoglobin ($HbFe^{3+}$), and nitrates (NO_2) are generated whereas when NO binds to reduced hemoglobin (Hb), nitrosyl hemoglobin (HbNO) and nitrates are generated. Nitrates have a renal clearance and methemoglobin is detoxified back to hemoglobin by slow endogenous mechanisms (*Wennmaln et al., 1993*).

Figure 5 represents mean intratracheal NO_2 concentrations measured by chemiluminescence at steady state in 10 patients with acute respiratory failure inhaling pure oxygen and increasing concentrations of NO in the range 0.1-1OO ppm. At an NO concentration of 10 ppm, the observed NO_2 concentration is above the toxic threshold of 0.3 ppm. Toxicity of NO_2 is different whether it is administered acutely or chronically.

Toxicity of Inhaled NO_2 Administered Acutely

Inhalation of NO_2 induces acute lung injury, increases bronchial reactivity and enhances the development of lung viral infections.

Increased alveolo-capillary membrane permeability. NO_2 inhaled at a concentration of 400 ppm for 1 hour in dogs induces interstitial and alveolar edema *(Man et al., 1990)*. Similar results have been described in other species such as guinea pigs *(Ranga et al., 1980)* and rats *(Guth and Mavis, 1985; Stavert and Lehnert, 1990)*. In the human, inflammatory cell response in bronchoalveolar lavage fluid have been observed after inhalation of NO_2 at concentrations as low as 2.25 ppm *(Sandström et al, 1991)*.

Increased bronchial reactivity. NO_2 increases the bronchomotor tone of isolated human bronchi *(Ben-Jebria et al., 1992)*. In asthmatics, inhaled NO_2 at a concentration of 0.3 ppm potentiates exercice induced bronchospasm *(Bauer et al., 1986)*. NO_2 can also increase the bronchial reactivity of healthy volunteers if inhaled at a concentration of 1.5 ppm for 3 hours *(Frampton et al., 1991)*.

Increased incidence of lung viral infections. Mice inhaling NO_2 have a concentration-dependent increase in the incidence of lung pneumonia after intratracheal instillation of a normally noninfectant viral inoculum. It requires 100 times less virus to induce a pneumonia in mice inhaling 5 ppm of NO_2 than in control animals *(Rose et al., 1988)*.

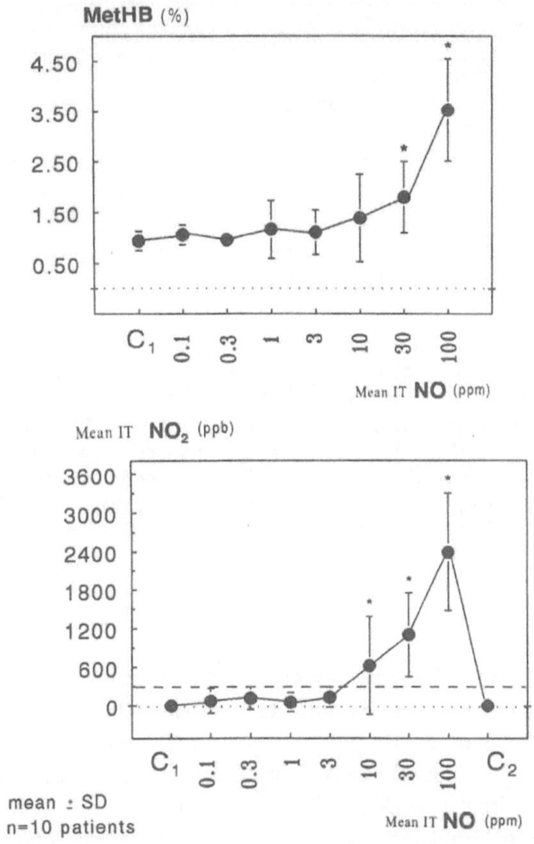

Figure 5. *Acute toxicity of inhaled NO in ARDS.* The figure represents the methemoglobin concentrations (upper panel) and the mean NO_2 intratracheal concentrations (lower panel) measured in 10 patients inhaling increasing concentrations of NO. The X axis is in a log-scale. The measured NO concentrations are mean of expired and inspired gas. The FIO_2 was maintained at 100% and the temperature at 37°C throughout the procedure. Measurements were performed after a steady state of 20 minutes. A significant increase in methemoglobin concentration was observed when mean NO concentrations delivered to the patient were above 10ppm. Mean NO_2 concentrations are above the toxicity threshold of 300 ppb(part per billion) as soon as mean NO concentrations delivered to the patient are higher or equal to 10 ppm (*Law-Koune et al., 1994*).

Toxicity of inhaled NO_2 administered chronically

It is very similar to that of oxygen or ozone with emphysema-like pulmonary lesions *(Mustafa and Tierney, 1978)*.

EFFECTS OF NO ON THE PULMONARY CIRCULATION

Mechanisms Involved

NO, either inhaled or produced by the vascular endothelium, relaxes the pulmonary vascular smooth muscle cell through an activation of the soluble guanylate cyclase. Contrary to the

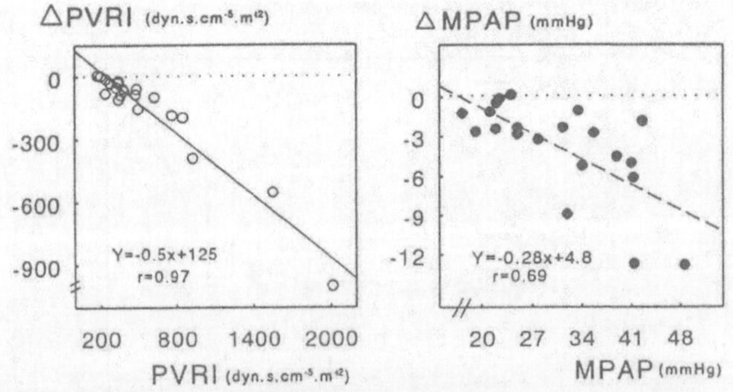

Figure 6. *Pulmonary vascular effects of inhaled NO: predictive factors.* Regression lines between basal pulmonary vascular resistance index (PVRI) and inhaled NO-induced decrease in PVRI (ΔPVRI) and between basal mean pulmonary arterial pressure (MPAP) and inhaled NO-induced decrease in MPAP (ΔMPAP) are represented. Each point is representative of a patient with severe ARDS and inhaled NO was administered to each patient at a concentration of 2 ppm (*Puybasset et al., 1995*).

other pulmonary vasodilators *(Radermacher et al., 1990; Radermacher et al., 1989)*, inhaled NO has no systemic effect because it is inactivated by fixation to hemoglobin when it reaches the vascular lumen. However, contrary to endogenous NO, which continuously vasorelaxes pulmonary vascular smooth muscle cells, inhaled NO has no effect on the normal, non-constricted, pulmonary circulation *(Frostell et al., 1993)*.

In acute respiratory failure, a large part of the lung parenchyma is no longer accessible to gas. Because NO is administered through the inhalation route, it can only reach pulmonary vessels located in ventilated parts of the lung. Therefore, inhaled NO-induced vasodilation is specific to the pulmonary circulation - (inhaled NO has no effect on systemic circulation) - and specific to the pulmonary vessels of the ventilated part of the lung. It does not reach the non-ventilated zone. This last property explains why inhaled NO increases PaO_2 and decreases pulmonary shunt through a redistribution of intrapulmonary blood flow.

The main determinant of the decrease in pulmonary vascular resistance index (PVRI) induced by inhaled NO is the basal level of PVRI *(Bigatello et al., 1994; Rich et al., 1993; Puybasset et al., 1995)*. The greater the baseline PVRI, the greater the inhaled NO-induced decrease in PVRI (Figure 6). In the absence of a previous constriction of the pulmonary vessels, no decrease in PVRI is observed. The NO-induced decrease in PVRI varies from 0%, in case of normal basal PVRI, to 50% if PVRI are around 2000 dynes.sec.$cm^{-5}.m^2$.

The increase in PVRI characterizing acute respiratory failure is partly related to the presence of circulatory inflammatory mediators such as thromboxane A_2, platelet activating factor or endothelin *(Langleben et al., 1993)* and partly related to fixed non-reversible mechanisms such as microthrombosis or increase in vascular thickness secondary to smooth muscle proliferation *(Zapol et al., 1985)*. The so-called "NO-responders" are patients with an increase in PVRI related to vasocontriction (> 200 dynes .sec.$cm^{-5}.m^2$). "NO non responders' are either patients with normal pulmonary vascular tone or patients with an increase in PVRI mainly related to thrombosis or to anatomical remodeling of the pulmonary vessels. The correlation between the basal level of PVRI and the NO-induced decrease in PVRI observed during ARDS *(Bigatello et al., 1994; Puybasset et al., 1995)* has been also described for

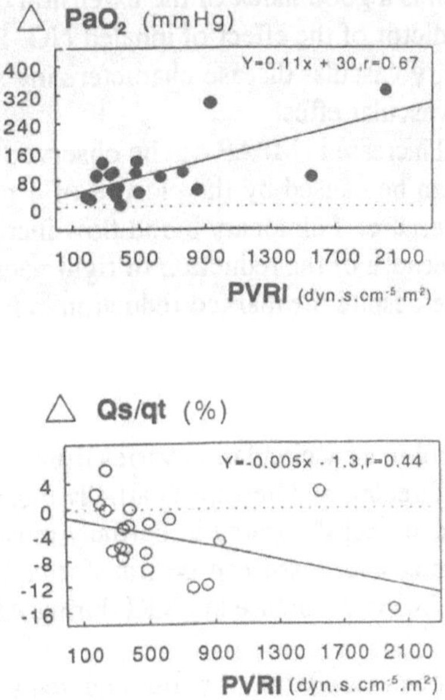

Figure 7. ***Effects of inhaled NO on arterial oxygenation:predictive factors.*** Regression lines between basal PVRI and inhaled NO-induces increase in PaO_2 and inhaled NO-induced decrease in Qs/Qt are represented. Each point is representative of a patient with severe ARDS and inhaled NO was administered to each patient at a concentration of 2 ppm (*Puybasset et al., 1995*).

pulmonary hypertensions complicating cardiac surgery *(Rich et al., 1993)*. The correlation between inhaled NO-induced decrease in mean pulmonary artery pressure (MPAP) and baseline MPAP, although stastistically significant, appears less tight than the correlation between NO-induced decrease in PVRI and baseline PVRI (Figure 6). At least two reasons can be given to explain this result. Firstly, pulmonary hypertension can be mainly flow-dependent without a marked increase in PVRI. As a consequence, NO will not reduce MPAP even if the baseline MPAP is elevated. Secondly, the inhalation of NO can indirectly increase cardiac output - for instance by unloading a failing right ventricle- and, thus, despite a decrease in PVRI, MPAP will not decrease or will decrease only slightly. Such an inhaled NO-induced increase in cardiac output is observed in presence of right heart failure either due to acute respiratory failure to another cause. If NO normalizes an initially low cardiac output, the observed decrease in MPAP will be proportionally less than the decrease in PVRI because the pressure-flow relationship of the pulmonary circulation is curvilinear for low cardiac output *(McGregor and Sniderman, 1985; Bshouty and Younes, 1990)*. NO is a pharmacologic tool to test the vasodilation reserve of the pulmonary circulation during acute respiratory failure. The correlation shown on Figure 6 was obtained in patients during the early phase of their ARDS. However, patients in whom a decrease in PVRI was observed during this early phase continued to show decreased PVRI after NO inhalation during the entire course of their disease as long as their basal PVRI remained elevated. This result supports the hypothesis that a vasodilation reserve persists during the course of ARDS and that the fixed part of elevated pulmonary pressure does not increase with time. In fact initially "NO-responder" patients become "NO-non responder" patients when their basal PVRI values return to normal in the recovery phase of acute lung injury. There is no real tachyphylaxia to inhaled NO *(Bigatello et al., 1994)*. Inhaled NO-induced decrease in PVRI does not seem to be correlated to the initial degree of hypoxemia and pulmonary shunt. This is due to the fact that in ARDS, arterial hypoxemia is depending on several factors, hypoxic pulmonary vasoconstriction in non-ventilated lung areas, extension of the alveolar disease and hemodynamic status. Therefore, it has no direct link with PVRI.

This also applies to static respiratory compliance which is a good index of the extension of the alveolar disease *(Gattinoni et al., 1987)* but not a predictor of the effect of inhaled NO. PVRI alone, which can be seen as an index of the pulmonary vascular disease characterizing acute lung injury, is predictive of NO-induced pulmonary vascular effect.

In some patients, no decrease or even a paradoxical increase in MPAP can be observed after NO inhalation despite a decrease in PVRI. This can be caused by the closure of a patent foramen ovale following the reduction of right atrial pressure. Pulmonary blood flow increases both because of the closure of intracardiac shunt and because of the reduction of right ventricle afterload. As a consequence, MPAP does not decrease despite the marked reduction in PVRI.

Effects on Pulmonary Arteries and Veins

The diameter of resistive arteries of the human pulmonary circulation varies from 200 to 500 μm. The muscular coat of the arteries is initially circular, and becomes partially muscular before it disappears at the capillary level. "Intermediate cells" located in partially muscular arteries can change their phenotype during acute lung injury and can be transformed into muscular cells. This mechanism partially explains the observed increase in PVRI during ARDS *(Jones et al., 1991)*.

Inhaled NO is potentially both an arterial and a venous vasodilator. Arteries and veins have different pathways inside the secondary pulmonary lobule: arteries follow the bronchovascular axis of the lobule whereas veins go back to left atrium with lymphatics within interlobular septa. As a consequence, arteries and veins are in contact with alveoli and therefore can be reached by NO administered through the inhalation route. Part of the increase in PVRI during late stages of ARDS is due to venous constriction and should be reversed after NO inhalation.

Experimental arguments in favor of inhaled NO-induced venodilation have been recently presented *(Lindeborg and Pearl, 1993)*.

EFFECTS ON GAS EXCHANGE

Effect on Arterial Oxygenation

In acute respiratory failure, inhaled NO improves arterial oxygenation by reducing lung areas characterized by low VA/Q ratio and by diverting pulmonary blood flow away from non-ventilated toward ventilated lung areas *(Rossaint et al., 1993)*. Inhaled NO-induced pulmonary vasodilation involves selectively ventilated lung territories and does not inhibit hypoxic pulmonary vasoconstriction characterizing non-ventilated lung areas. In contrast to intravenous vasodilators, inhaled NO has no effect on cardiac output and does not induce any vasodilation in pulmonary vessels perfusing non-ventilated lung areas *(Radermacher et al, 1990; Radermacher et al, 1989)*. As a consequence, pulmonary shunt does not increase and PaO_2 does not deteriorate following NO inhalation *(Rossaint et al., 1993)*. Inhaled NO-induced increase in PaO_2 and decrease in pulmonary shunt is depending on the initial level of PVRI *(Puybasset et al., 1995)*. The greater the baseline PVRI, the greater the inhaled NO-induced increase in PaO_2 and the greater the inhaled NO-induced decrease in pulmonary shunt (Figure 7).

The correlation between baseline PVRI and inhaled NO-induced increase in PaO_2, although statistically significant, is weak because PaO_2 is a complex parameter depending on many factors such as VA/Q at the alveolar level, cardiac output and oxygen consumption. Inhaled NO-induced increase in PaO_2 is depending on baseline PaO_2 only when the most hypoxemic patients are those who have the greatest pulmonary vascular resistance *(Bigatello et al., 1994; Puybasset et al., 1995)*.

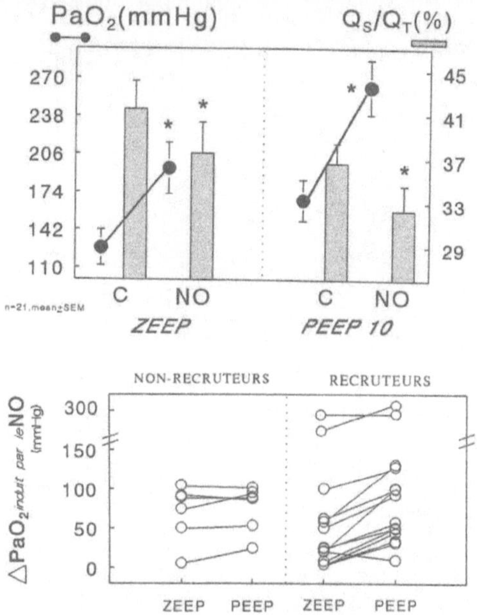

Figure 8. *Influence of PEEP and alveolar recruitment on inhaled NO-induced improvement in arterial oxygenation.* In 20 critically ill patients with severe ARDS, PEEP-induced alveolar recruitment was assessed using a high resolution thoracic CT scan. A significant alveolar recruitment was evidenced following PEEP 10 cm H_2O in 14 patients ("recruiters") whereas the absence of any alveolar recruitment was observed in 6 patients ("non-recruiters"). Inhaled NO-induced changes in PaO_2 (FIO_2=1) and pulmonary shunt (Qs/Qt) are more marked with PEEP 10 cm H_2O than without PEEP (ZEEP) $p<0.01$. Inhaled NO-induced individual changes in PaO_2 (ΔpaO_2) are more marked in PEEP than in ZEEP in recruiters only, suggesting that PEEP-induced alveolar recruitment rather than PEEP by itself enhances the effect of inhaled NO on arterial oxygention (*Puybasset et al., 1995*).

Alveolar recruitment resulting from the application of PEEP potentiates inhaled NO-induced improvement in arterial oxygenation *(Puybasset et al., 1995)*. This beneficial effect of PEEP is observed only when PEEP is associated with alveolar recruitment (Figure 8). When PEEP induces lung overdistention, the effect of inhaled NO on arterial oxygenation does not differ in PEEP and ZEEP conditions. As a consequence, the effect of inhaled NO on arterial oxygenation should be always tested after optimization of alveolar recruitment. Patients considered as non-responders in ZEEP conditions might become responders after PEEP administration.

PEEP-induced alveolar recruitment has an unforeseeable effect on inhaled NO-induced decrease in PVRI. Mean inhaled NO-induced decrease in pulmonary vascular resistance is similar in ZEEP and PEEP conditions and is not systematically potentiated by alveolar recruitment (Figure 9). As a consequence, inhaled NO-induced increase in PaO_2 is depending on alveolar recruitment whereas the vascular effect is not. A possible explanation for this dissociated effect could rely on an anatomical basis. Pulmonary vessels follow a long pathway within lung parenchyma: arteries are going along the bronchovascular axis of the secondary pulmonary lobules and veins are going back to the left atrium within interlobular septa. Because pulmonary lesions are non-homogeneously distributed in ARDS and because pulmonary vessels and alveolar spaces are in contact on a long distance, there are always

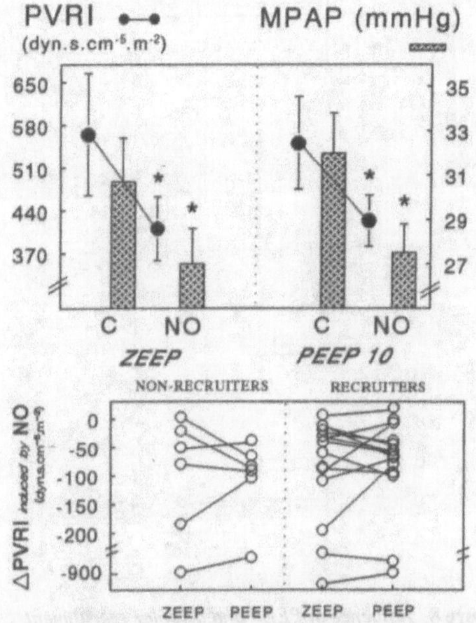

Figure 9. *Influence of PEEP and alveolar recruitment on inhaled NO-induced pulmonary vascular effects.* In the 20 patients of figure 8, inhaled NO-induced changes in pulmonary vascular resistance index (PVRI) and in mean pulmonary arterial pressure (MPAP) are similar in ZEEP (zero end expiratory pressure) and PEEP conditions suggesting that alveolar recruitment does not potentiate the pulmonary vascular effect of inhaled NO(*Puybasset et al., 1995).*

enough ventilated alveolar territories surrounding pulmonary vessels to allow inhaled NO to reach and relax constricted vascular smooth muscles. The primary effect of inhaled NO-induced arteriolar vasodilation is to increase pulmonary blood flow within the corresponding capillary network. The effect on gas exchange then depends on the VA/Q ratio at the alveolar level. For a given secondary pulmonary lobule, VA/Q and PaO_2 will remain unchanged if the increased capillary blood flow is equally distributed among ventilated and non-ventilated acini. VA/Q and PaO_2 will increase if the increased capillary blood flow is preferentially distributed to ventilated acini. The resulting effect on systemic arterial oxygenation will depend on the sum of each local effect. If this hypothesis is valid, inhaled NO should be associated, in a small number of patients, with a worsening of arterial oxygenation. It is also easy to explain why PEEP potentiates inhaled NO-induced improvement in arterial oxygenation: by inducing alveolar recruitment, PEEP reduces the proportion of non-ventilated alveolar territories and inhaled NO, by dilating pulmonary arterioles, increases capillary blood flow in these newly aerated alveoli.

Effect on Alveolar Dead Space

Most often NO inhalation induces an increase in end-tidal carbon dioxide tension ($PetCO_2$) simultaneously with the decrease in pulmonary arterial pressure (Figure 10) These changes are observed independently of any change in minute ventilation and cardiac output and are generally associated with a small but significant reduction in alveolar-arterial CO_2 gradient which corresponds to a marked decrease in alveolar dead space *(Puybasset et al., 1995;Stewart et al., 1994)*. The reduction of lung areas characterized by high VA/Q ratio, which is likely related to the re-opening of constricted pulmonary arteries in ventilated lung regions,

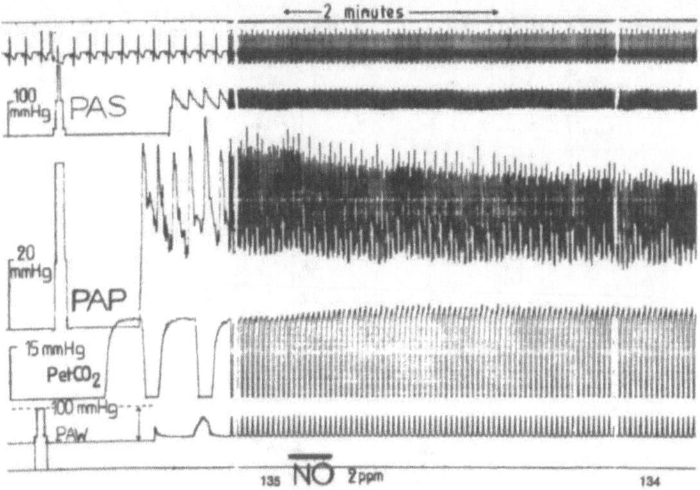

Figure 10. *Effect of inhaled NO on end-tidal CO_2.* Continuous recording of EKG, arterial pressure (PAS), pulmonary arterial pressure (PAP), end tidal CO_2 ($PetCO_2$), measured using a non-aspirative infrared capnometer and airway pressure (PAW) was obtained in a critically ill patient with severe ARDS. A few seconds following the inhalation of 2 ppm of NO (black line), a rise in $PetCO_2$ is observed simultaneously with the decrease in PAP. In this patient, inhaled NO increased PaO_2 from 184 to 324 mmHg (FIO_2=1) and decreased alveolar dead space from 42 to 37% *(Puybasset et al., 1994)*.

contributes to increase alveolar ventilation. The effect on alveolar dead space is independent on alveolar recruitment (Figure 11) and has a large interpatient variability. In a minority of patients, the decrease in $PaCO_2$ induced by the reduction in alveolar dead space can reach 10 mmHg and gives the possibility of reducing tidal volume and peak airway pressure *(Stewart et al., 1994)*.

OTHER EFFECTS OF POTENTIAL BENEFIT

Mechanisms other than the diversion of pulmonary blood flow induced by arterial and venous vasodilation can be invoked for explaining inhaled NO induced improvement in gas exchange. Because inhaled NO likely dilates both sides of the pulmonary circulation (arteries and veins), the reduction in pulmonary arterial pressure is associated with a reduction in hydrostatic capillary pulmonary pressure, which, in turn, could reduce the amount of pulmonary edema and improve arterial oxygenation. Because inhaled NO is a bronchodilator *(Dupuy et al., 1992)*, it could decrease venous admixture and reduce hypoxemia. However, their effect appears of limited importance in patients with ARDS (Figure 12).

Inhaled NO could also limit the remodeling of the pulmonary circulation which occurs following lung injury *(Jones et al., 1991)*. In response to various types of lung injury, vascular smooth muscle rapidly appears in distal units of the pulmonary bed increasing the proportion of vessels that are completely muscular. This anatomical remodeling of the pulmonary circulation forms the basis of the increase in pulmonary arterial pressure and pulmonary vascular resistance in ARDS. Endogenous NO released by pulmonary endothelial cells seems to exert a permanent anti-trophic effect on adjacent vascular smooth muscle fibers *(Garg and Hassid, 1989)*. Endothelial lesions observed in acute lung injury could annihilate the NO-induced anti-proliferative effect and enhance proliferation of vascular smooth muscle distally in the pulmonary vascular bed. Chronic inhalation of small concentrations of NO could re-establish the anti-proliferative effect and prevent structural remodeling of the pulmonary

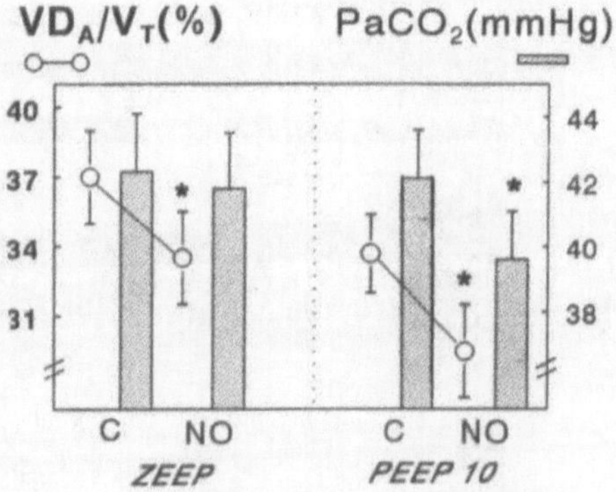

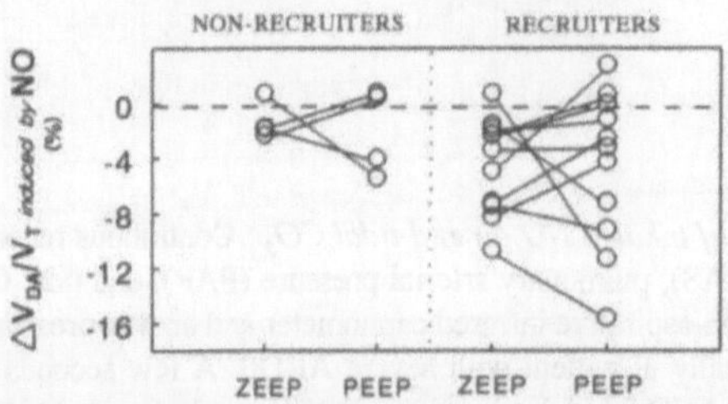

Figure 11. *Effects of inhaled NO on respiratory resistance in patients with ARDS.* Respiratory resistance (Rrs max) was measured using the end-inspiratory clamping technique in 11 patients with severe ARDS. Four different concentrations of inhaled NO were administered: 10, 20, 40, and 80 ppm. Control measurements were performed before NO administration (C1) and between each concentration (C2-C5). No bronchodilating effect is observed despite the increased basal respiratory resistance (normal values for respiratory resistance: 3-5 cm H_2O/l/s (*Puybasset et al., 1993*).

circulation during the course of ARDS thus reducing pulmonary arterial hypertension. Because it has been recently shown that inhaled NO markedly inhibits platelet aggregation in patients with ARDS (*Samama et al, 1995*) it could prevent the formation of pulmonary thromboemboli which is observed at the early stage of ARDS, and consequently, reduce alveolar dead space and pulmonary hypertension. Inhaled NO has been shown to prolong bleeding time in healthy volunteers *(Högman et al., 1993)* but not in patients with ARDS (*Samama et al, 1995*). It could also inhibit polymorphonuclear activation which seem to be a key point in the pathophysiology of ARDS *(Niu et al., 1994)*.

Experimentally, inhaled NO appears either to enhance or attenuate oxidant-induced acute lung injury. In isolated rabbit lungs submitted to an oxidative-stress, inhaled NO partially prevents the development of increased capillary permeability and pulmonary arterial hypertension *(Kavanach et al., 1994)* and dilates constricted pulmonary vessels in the presence of pulmonary endothelial injury *(Rimer and Gillis, 1992)*. On the other hand, it has been shown that endogenous NO likely mediates hyperoxia-induced damage of human lung cells *(White et al., 1994)* and is an essential intermediary in the production of acute lung injury due to paraquat in guinea pigs *(Berisha et al., 1994)*. N0 synthase inhibitors totally prevent lung injury due to paraquat and to xanthine oxidase. Therefore, the administration of inhaled NO for treating patients with paraquat-induced lung injury could be potentially harmful.

BRONCHIAL EFFECT

Inhaled NO inhibits metacholine-induced bronchoconstriction in guinea pigs *(Dupuy et al., 1992)* and rabbits *(Högman et al., 1993)*: the bronchodilating effect is dose-dependant in the range 5-150 ppm. Acute lung injury is most often associated with increased respiratory

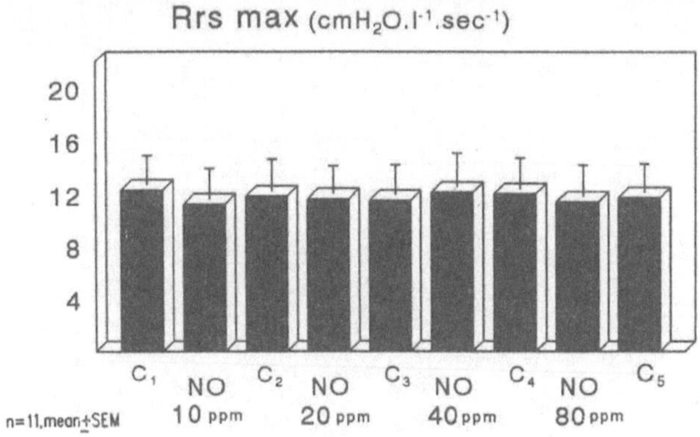

Figure 12. *Effects of inhaled NO on respiratory resistance in patients with ARDS.* Respiratory resistance (Rrs max) was measured using the end-inspiratory clamping technique in 11 patients with severe ARDS. Four different concentrations of inhaled NO were administered: 10, 20, 40, and 80 ppm. Control measurements were performed before NO administration (C1) and between each concentration (C2-C5). No bronchodilating effect is observed despite the increased basal respiratory resistance (normal values for respiratory resistance: 3-5 cm H_2O/l/s (*Puybasset et al., 1993*).

resistance which can be caused by two different mechanisms: a part of the increased respiratory resistance is fixed resulting from the external compression of bronchioles by the surrounding edematous lung parenchyma; another part is reversible, resulting from the bronchoconstriction induced by circulating inflammatory mediators, such as thromboxane A_2 and leucotriens.

However, in a recent study, inhaled NO-induced decrease in respiratory resistance could not be evidenced in patients with ARDS for NO concentrations ranging from 10 to 80 ppm (Figure 12) *(Puybasset et al., 1993)*. Reasons for this lack of bronchodilating effect remain speculative. In patients with ARDS, the increased respiratory resistance might be entirely due to the reduction in lung volume with the specific bronchial resistance remaining normal. Therefore, the absence of true bronchoconstriction could explain the lack of inhaled NO-induced bronchodilator effect. Another possibility could be that the inhaled NO-induced bronchodilator effect is offset by an NO_2-induced bronchoconstrictive effect *(Bauer et al., 1986; Frampton et al., 1991)*. When using NO concentrations greater than 10 ppm and FIO_2 ≥ 0.6, NO_2 concentrations might be above 300 ppb *(Law-Koune et al., 1994)* and could induce bronchoconstriction *(Bauer et al., 1986)*.

In spontaneously breathing asthmatic patients, inhaled NO at a concentration of 80 ppm 0exerts a bronchodilating effect much smaller than β_2 agonists *(Högman et al., 1993)*. Therefore, in humans with increased bronchial reactivity, inhaled NO appears as a poorly effective bronchodilating agent.

DOSE-RESPONSE CURVES

As previously demonstrated in animals *(Frostell et al., 1991)*, dose-response curves of inhaled NO in humans are depending on the pathophysiology of the pulmonary arterial hypertension. An illustrative example is provided by the different dose response curves obtained in patients with ARDS according to the presence or absence of associated septic shock During septic shock, the inducible NO synthase is activated leading, to the release by macrophages and vascular smooth muscles of endogenous NO which, in turn, contributes to the reduction of blood pressure *(Kilbourn et al., 1990; Kilbourn et al., 1990; Julou-Schaeffer et al., 1990)*. Despite the excess of endogenous NO production, critically ill patients with ARDS and septic shock still have pulmonary arterial hypertension. The presence of a septic shock associated with ARDS modifies the dose-response curve of inhaled NO *(Mourgeon et al., 1994)*.

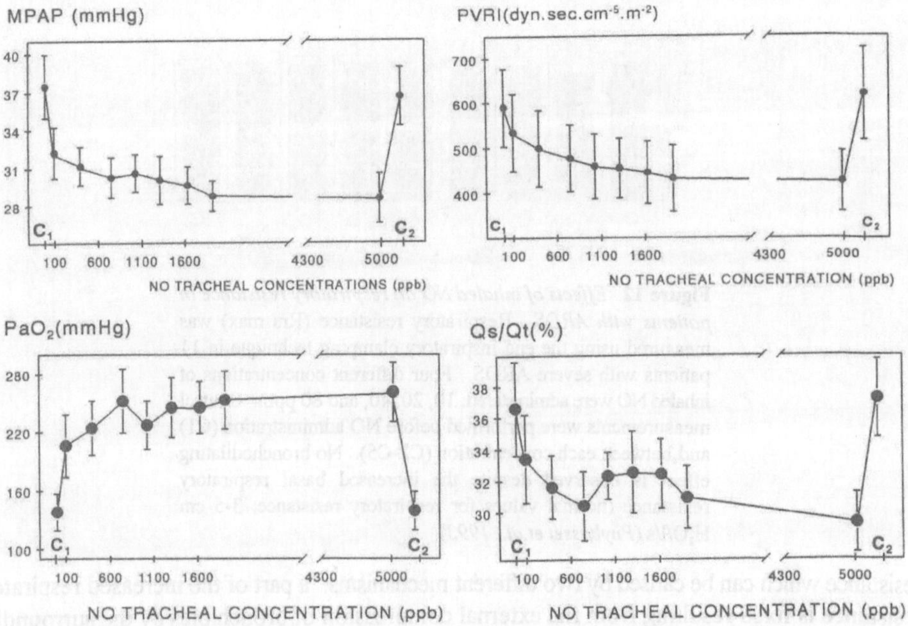

Figure 13. ***Dose-response curves of inhaled NO in patients with ARDS but without septic shock.*** **These curves were obtained in 6 patients with ARDS who received increasing concentrations of inhaled NO in a random order: 100, 400, 700, 1000, 1300, 1600, 1900, and 5000 ppb. The maximum effect on mean pulmonary arterial pressure (MPAP) and on pulmonary vascular resistance index (PVRI) is obtained for a mean intratracheal NO concentration of 1900 ppb. The maximum effect on PaO_2 (FIO_2=1) and pulmonary shunt (Qs/Qt) is observed for a mean intratracheal NO concentration of 900 ppb. Control measurements were obtained before (C1) and after (C2) NO administration(*Puybasset et al., 1994*).**

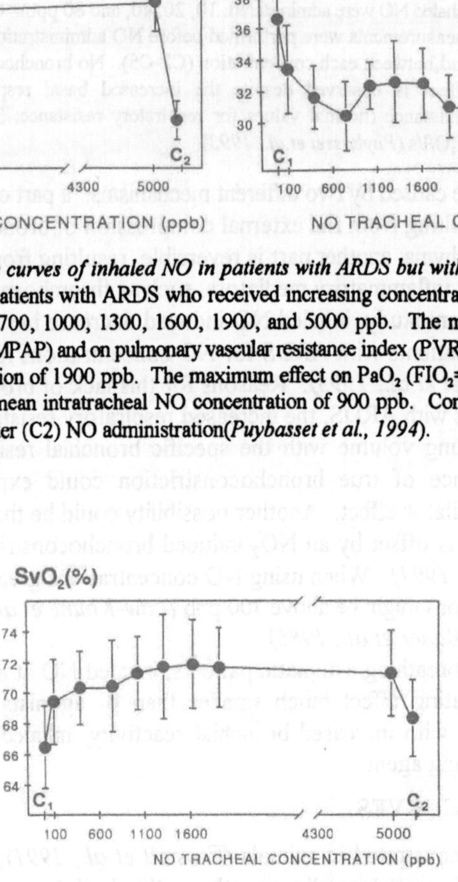

Figure 14. ***Effect of inhaled NO on $S\bar{v}O_2$.*** **$S\bar{v}O_2$ is measured at increasing concentrations of inhaled NO (100-5000 ppb) administered to 6 patients with severe ARDS. Maximum effect on $S\bar{v}O_2$ is obtained at a NO concentration of 900 ppb. In anesthetized patients, $S\bar{v}O_2$ closely reflects SaO_2 (*Puybasset et al., 1994*).**

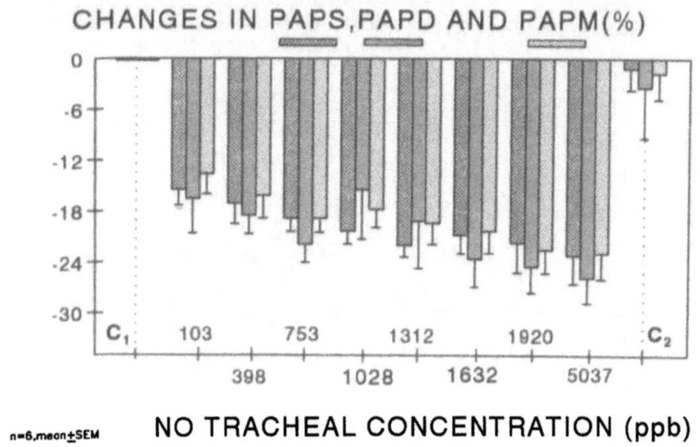

Figure 15. *Quantitative aspects of inhaled NO-induced pulmonary arterial vasodilation.* Decreases in systolic (PAPS), diastolic (PAPD) and mean pulmonary arterial pressure (PAPM) are expressed as percentage of variation of the control values at increasing concentrations of inhaled NO between 100 and 5000 ppb in 6 patients with ARDS with circulatory shock. As a mean, pulmonary arterial pressure decreases by 25% of control values (*Puybasset et al., 1994*).

ARDS Without Septic Shock

Minimal mean intratracheal concentrations of inhaled NO required for obtaining the maximum effect on the pulmonary circulation and on arterial oxygenation are respectively around 2 and 1 ppm *(Puybasset et al., 1994)*. As recently shown, inspiratory concentrations of inhaled NO are 1½ times greater than mean NO concentrations (*Law-Koune et al, 1995*). Dose-response curves for pulmonary arterial pressure, pulmonary vascular resistance, PaO_2, pulmonary shunt and $S\bar{v}O_2$ are shown in figures 13 and 14. In patients with ARDS treated by extracorporeal membrane oxygenation, dose-response curves for systemic oxygenation and pulmonary hypertension are in the range 0.1 - 100 ppm *(Gerlach et al., 1993)*. This different dose-response curves could be related to a permanent activation of the pulmonary circulation by ECMO. In some other patients, dose-response curves are in the range 20-100 ppb *(Puybasset et al., 1994; Gerlach et al., 1993)*. In a patient under general anesthesia, and if inhaled NO does not change cardiac output, changes in $S\bar{v}O_2$ are tightly related to changes in SaO_2. Consequently, continuous $S\bar{v}O_2$ monitoring is an easy mean for determining the minimal efficient concentration of inhaled NO at the bedside.

In critically ill patients with acute respiratory failure, systolic, diastolic and mean arterial pressure decrease by 25% following NO inhalation (Figure 15). In neonates with acute pulmonary hypertension, inhaled NO-induced decrease in arterial pressure is much more pronounced and reaches 34% of the control value *(Journois et al., 1994; Sellden et al., 1993)*. Dose-response curves for both oxygenation and pulmonary hypertension are between 0 and 5 ppm *(Tiner et al., 1994)*. Systolic, diastolic and mean pulmonary arterial pressure decrease in the same proportion excluding a predominant effect on pulmonary arterial compliance. Maximal increase in arterial oxygenation varies from 30 to 100% of the control value and is obtained for NO concentrations around 1 ppm in patients ventilated using conventional ventilatory support. For these very low NO concentrations, NO_2 concentrations remain low even if pure oxygen is administered and the risk of toxicity is limited (Figure 16).

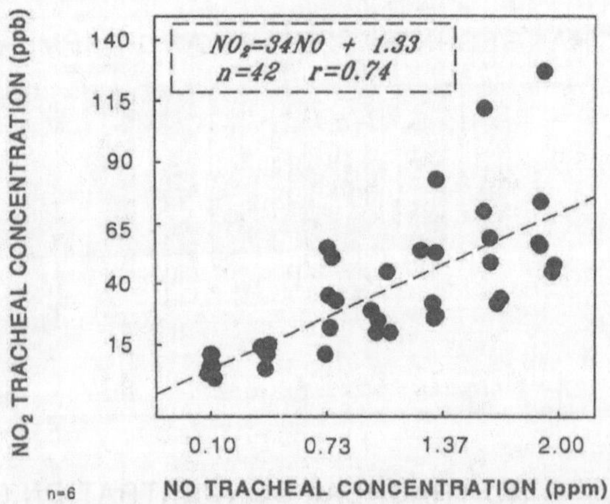

Figure 16. *Mean intratracheal concentrations of NO_2* following the administration of increasing concentrations of NO in the range of 0.1-2 ppm in 6 patients with ARDS ventilated using pure oxygen. There is a linear and significant relationship between NO_2 and NO concentrations. At 2 ppm of NO, the maximum NO_2 concentration was of 125 ppb (*Puybasset et al., 1994*).

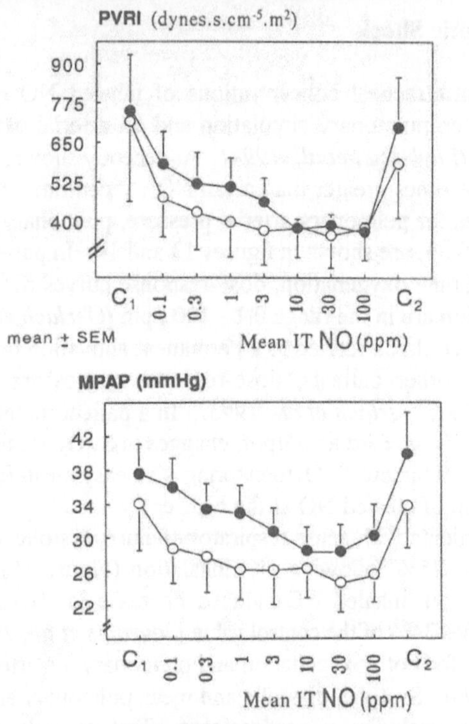

Figure 17. *Dose response curve of inhaled NO on pulmonary arterial pressure (MPAP) and pulmonary vascular resistance index (PVRI) in patients with ARDS according to the presence or absence of septic shock.* In patients with septic shock (•-•), the maximum effect is obtained for an inspiratory intratracheal concentration of NO around 15 ppm whereas in patients without septic shock (○-○), the maximum effect is obtained at an inspiratory concentration of 1.5 ppm. However, this difference does not reach statistical significance (*Mourgeon et al., 1994*).

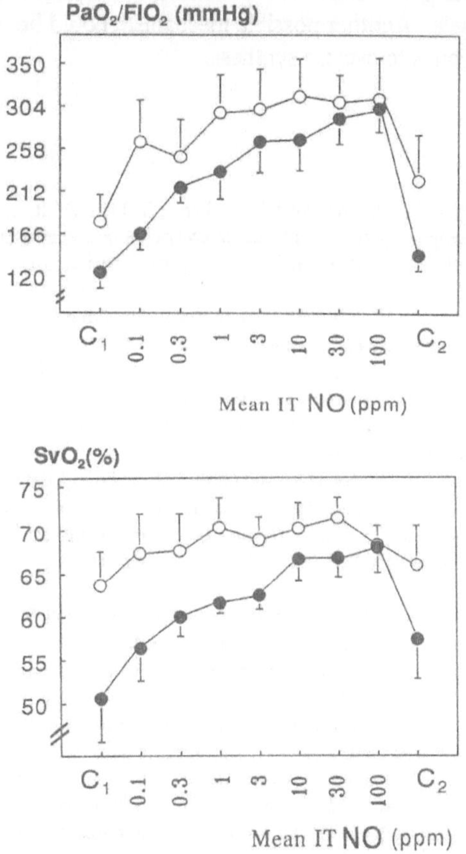

Figure 18. *Dose-response curve of inhaled NO on PaO_2 (FIO_2=0.85) and on mixed venous oxygen saturation ($S\bar{v}O_2$) in patients with ARDS according to the absence or presence of septic shock.* In patients with septic shock (•-•), the maximum effect is obtained on PaO_2 at an inspiratory NO concentration of 150 ppm whereas in patients without septic shock (○-○), the maximum effect is obtained at an inspiratory concentration around 1.5 ppm. This difference is statistically significant (*Mourgeon et al., 1994*).

ARDS Associated with Septic Shock

Dose-response curves of inhaled NO for arterial oxygenation and pulmonary arterial hypertension in patients with ARDS are modified by the presence of a septic shock (Figures 17 and 18). The maximum increase in PaO_2 is obtained at inspiratory NO concentrations around 1.5 ppm in patients without septic shock, and at inspiratory NO concentrations around 150 ppm in patients with septic shock. Dose-response curves in patients with septic shock or treated by extracorporeal membrane oxygenation are very similar (*Mourgeon et al., 1994; Gerlach et al., 1993*). The maximum reduction in pulmonary arterial pressure is obtained at inspiratory NO concentrations around 1.5 ppm in patients without septic shock and at inspiratory NO concentrations around 15 ppm in patients with septic shock. Reasons for these differences between patients with and without septic shock remain unknown. The coexistence of a septic shock with a pulmonary arterial constriction implies the existence of circulating vasoactive mediators which constrict pulmonary vessels despite the massive release of endogenous NO. Exogenous catecholamines administered to support blood pressure and circulating vasoconstrictors such as endothelin *(Langleben et al., 1993)* could play a role in this pulmonary arterial constriction. Increasing concentrations of inhaled NO would then be required to relax constricted pulmonary vessels. Another possible mechanism could be a septic shock-induced decrease in intracellular guanylate cyclase synthesis.

REFERENCES

Adnot, S., Kouyoumdjian, C., Defouilloy, C., Andrivet, P., Sediame, S., Herigauklt, R. and Fratacci, M.D., 1993, Hemodynamic and gas exchange responses to infusion of acetylcholine and inhalation of nitric oxide in patients with chronic obstructive lung disease and pulmonary hypertension. *Am. Rev. Resp. Dis.* 148:310-316.

Alheid, U., Frolich, J.C. and Fosterman, U., 1987, Endothelial derived relaxing factor from human endothelial cells inhibits aggregation of human platelets. *Thromb. Res.* 47:561-571.

Arroyo, P.L., Hatch-Pigott, V., Mower, H.F. and Cooney,1992, R.V. Mutagenecity of nitric oxide and its inhibition by antioxidants. *Mutation Research* 281:193-202.

Azoulay, E., Soler, P., Blayo, M.C. and Basset, F., 1977, Nitric oxide effects on lung structure and blood oxygen affinity in rats. *Bull. Europ. Physiopath. Resp.*13:629-644.

Bauer, M.A., Utell, M.J., Morrow, P.E., Speers, D.M. and Gibb, F.R., 1986, Inhalation of 0.30 ppm nitrogen dioxide potentiates exercice-induced bronchospasm in asthmatics. *Am. Rev. Resp. Dis.* 134:1203-1208.

Beckman, J.S., Beckman, T.W., Marshall, P.A. and Freeman, B.A., 1990, Apparent hydroxyl radical production by peroxynitrite: Implications for endothelial injury from nitric oxide and superoxide. *Proc. Natl. Acad. Sci.* 87:1620-1624.

Ben-Jebria, A., Marthan, R. and Savineau, M., 1992, Effect of *in vitro* nitrogen dioxide exposure on human bronchial smooth muscle response. *Am. Rev. Respir. Dis.* 146:378-382.

Berdeaux, A., 1993, Nitric oxide: an ubiquitous messenger. *Fundam. Clin. Pharmacol.* 1993:7:401-411.

Berger, J.I., Gibson, R.L., Redding, G.J., Standaert, T.A., Clarke, W.R. and Truog, W.E., 1993, Effect of inhaled nitric oxide during group B streptococcal sepsis in piglets. *Am. Rev. Resp. Dis.* 147: 1080-1086.

Berisha, H., Pakbaz, H., Absood, A. and Said, S.I., 1994, Excess nitric oxide (NO)production in acute lung injury: role of constitutive vs. inducible NO synthase. *Am. J. Resp. Crit. Care Med.* 149:A552 (Abstract).

Bigatello, L.M., Hurford, W.E., Kacmarek, R.M., Roberts, J.D. and Zapol, W.M., 1994, Prolonged inhalation of low concentrations of nitric oxide in patients with severe adult respiratory distress syndrome. *Anesthesiology* 80:761.770.

Bshouty, Z. and Younes, M., 1990, Distensibility and pressure flow relationships of the pulmonary circulation II multibranched model. *J. Appl. Physiol.* 68:1514-1527.

Capellier, G., Belle, F., Fouche, R., Maheu, M.R. and Barale, F., 1993, Interet d'une regulation de l-'administration de monoxyde d'azote. *Réan Soins Intens. Med. Urg.* 2:706 (Abstract).

Dupuy, P.M., Shore, S.A., Drazen, J.M., Frostell, C., Hill, W.A. and Zapol, W.M., 1992, Bronchodilator action of inhaled nitric oxide in guinea pigs. *J. Clin. Invest.* 90:421-428.

Dyar, 0., Young, J.D., Xiong, L. and John, S., 1993, Dose-response relationship for inhaled nitric oxide in experimental pulmonary hypertension in sheep. *Br. J. Anaesth.* 71:702-708.

Foubert, L., Fleming, B., Latimer, R., Jonas, M., Oduro, A., Borland, C., and Higenbottam, T., 1992, Safety guidelines for use of nitric oxide. *Lancet* 339:1615-1616.

Frampton, M.W., Morrow, P.E., Cox, C., Gibb, F.R., Speers, D.M. and Utell, M.J., 1991, Effects of nitrogen dioxide exposure on pulmonary function and airway reactivity in normal humans. *Am. Rev. Resp. Dis.*143:522-527.

Fratacci, M.D., Frostell, C.G., Chen, T.Y., Wain, J.C., Robinson, D.R. and Zapol, W.M., 1991, Inhaled nitric oxide: A selective pulmonary vasodilator of heparin-Protamine vasoconstriction in sheep. *Anesthesiology* 75:990-999.

Frostell, C.G., Blomqvist, H., Hedenstierna, G., Lundberg, J. and Zapol, W.M., 1993, Inhaled nitric oxide selectively reverses human hypoxic pulmonary vasoconstriction without causing systemic vasodilation. *Anesthesiology* 78:427-435.

Frostell, C., Fratacci, M.D., Wain, J.C., Jones, R. and Zapol, W.M., 1991, Inhaled nitric oxide: A selective pulmonary vasodilator reversing hypoxic pulmonary vasoconstriction. *Circulation* 83:203 8-2047.

Furchgott, R.F. and Zawadski, J.V., 1980, The obligatory role of endothelial cells in the relaxation of arterial smooth muscle by acetylcholine. *Nature* 288:373-376.

Garg, U.C. and Hassid, A., 1989, Nitric oxide-generating vasodilators and 8-bromo cyclic guanosine monophosphate inhibit mitogenesis and proliferation of cultured rat vascular smooth muscle cells. *J. Clin. Invest.* 83:1774-1777.

Gattinoni, L., Pesenti, A., Avalli, L., Rossi, F. and Bombino, M., 1987, Pressure-volume curve of total respiratory system in acute respiratory failure. Computed tomographic scan study. *Am. Rev. Resp. Dis.* 136:730-736.

Gerlach, H., Pappert. D., Lewandowski, K., Rossaint, R. and Falke, K.J., 1993, Long-term inhalation with evaluated low doses of nitric oxide for selective improvement in oxygenation in patients with adult respiratory distress syndrome. *Intensive Care Med* 19:443-449.

Gerlach, H., Rossaint, R., Pappert, D. and Falke, K.J., 1993, Time-course and dose-response of nitric oxide inhalation for systemic oxygenation and pulmonary hypertension in patients with adult respiratory distress syndrome. *Eur. J. Clin. Invest.* 23:499-502.

Girard, C., Lehot, J., Pannetier, J., Filley, S., French, P., Estanove, S., 1992, Inhaled nitric oxide after mitral valve replacement in patients with chronic pulmonary hypertension. *Anesthesiology* 77:880-883.

Greenbaum, R., Bay, J., Hargreaves M.D., Kain, M.L., Kelman, G.R. Nunn, J.F., Prys-Roberts, C., and Siebold, K., 1967, Effects of higher oxides of nitrogen in the anaesthetized dog. *Br. J. Anaesth.* 39:393-404.

Guth, D.J. and Mavis, R.D., 1985, Biochemical assessment of acute nitrogen dioxide toxicity in rat lung. *Toxicol. Appl. Pharmacol.* 81:128-138.

Haddad, L.Y., Crow, J.P., Yaozu, Y., Beckman, J. and Matalon, S., 1994, Concurrent generation of nitric oxide and superoxide damages surfactant protein A (SP-A). *Am. J. Resp. Crit. Care Med. Suppl* 149:A549 (Abstract).

Higenbottam, T., Pepke-Zaba, J., Scott, J., Woolman, P., Coutts, C. and Wallwork, J., 1988, Inhaled "endothelial derived-relaxing factor" (EDRF) in primary hypertension (PPH). *Am. Rev. Resp. Dis. Suppl.* 137:A107 (Abstract).

Hőgman, M., Frostell, C.G., Arnberg, H. and Hedenstierna, G., 1993, Bleeding time prolongation and NO inhalation. *Lancet* 341:1664-1665.

Hőgman, M., Frostell, C.G., Arnberg, H., Hedenstierna, G., 1993, Inhalation of nitric oxide modulates metacholine induced bronchoconstriction in the rabbit. *Eur. Respir. J.* 6:177-180.

Hőgman, M., Frostell, C.G., Hedenstrőm, H. and Hedenstierna, G., 1993, Inhalation of nitric oxide modulates adult bronchial tone. *Am. Rev. Resp. Dis.* 48:1474-1478.

Isomura, K., Chikahira, M., Terashini, K., Hamada, K., 1984, Induction of mutations and chromosome aberrations in lung cells following *in vivo* exposures to nitrogen oxides. *Mutation Research* 136:119-125.

Jones, R., Langleben, D. and Reid, L.M., 1991, Patterns and mechanisms of remodeling of the pulmonary circulation in acute and subacute lung injury. In: *The Pulmonary Circulation and Acute Lung Injury*, Said S.I. (ed), P-179-242, Futura Publishing Company, Inc., Mount Kisko, NY.

Journois, D., Pouard, P., Mauriat, P., Malhère, T., Vouhè, P. and Safran, D., 1994, Inhaled nitric oxide as a therapy for pulmonary hypertension after operations for congenital defects. *J. Thorac. Cardiovasc. Surg.* 107:1129-1135.

Julou-Schaeffer, G., Gray, G.A. Fleming, I., Schott, C., Parratt, J.R. and Stoclet, J.C., 1990, Loss of vascular responsiveness induced by endotoxin involves L-arginine pathway. *Am. J. Physiol.* 259 (Heart Circ. Physiol. 28):H1038-H1043.

Kavanach, B.P., Mouchawar, J., Goldsmith, J. and Pearl, R.G., 1994, Effects of inhaled NO and inhibition of endogenous NO synthesis in oxidant-induced acute lung injury. *J. Appl. Physiol.* 76:1324-1329.

Kilbourn, R.G., Jubran, S.S., Griffith, O.W., Levi, R., Adams, J. and Lodato, R.F., 1990, Reversal of endotoxin mediated shock by NG-Methyl-L-arginine, an inhibitor of nitric oxide synthesis. *Biochem. Biophys. Res. Commun.* 172:1132-1138.

Kilbourn, R.G., Gross, S.S., Jubran, S.S., Adams, J., Griffith, O.W., Levi, R. and Lodato, R.F., 1990, NG-m-ethyl-L-arginine inhibits tumor necrosis factor-induced hypotension: implications for the involvement of nitric oxide. *Proc. Natl. Acad. Sci.* 87:3629-3632.

Langleben, D., Demarchie, A., Laporta, D., Spanier, A.,H., Schlesinger, R.D. and Stewart, D.J., 1993, Endothelin-1 in acute lung injury and the adult respiratory distress syndrome. *Am. Rev. Resp. Dis.*148:1646-1650.

Law-Koune, J.D., Roche, S., Mourgeon, E., Lu, Q., Puybasset, L., Kalfon, P., Rouby, J.J. and Viars, P., 1994, Acute toxicity of inhaled NO in ARDS. *Anesthesiology* (Abstract).

Law-Koune, J.D., Rouby, J.J., Lu, Q., Mourgeon, E., Puybasset, L., Diaby, M., Viars, P., 1995, Measurement of inspiratory and expiratory concentrations of inhaled NO in patients with ARDS. *Br. J. Anesth.* 74:A404(Abstract).

Lindeborg, D.M. and Pearl, R.G., 1993, Inhaled nitric oxide and the longitudinal distribution of pulmonary vascular resistance. *Am. Rev. Respir. Dis. Suppl.*147:A225 (Abstract).

Macdonald, P.S., Read, M.A. and Dushing, G.J., 1988, Synergistic inhibition of platelet aggregation by endothelium-derived relaxing factor and prostacyclin *Thromb. Res.* 49:437-449.

Man, S.F.P., Williams, D.J., Amy, G.C., Man, G.C.W. and Lien, D.C., 1990, Sequential changes in canine pulmonary epithelial and endothelial cell functions after nitrogen dioxide. *Am. Rev. Resp. Dis.* 142:199-205.

McGregor, M. and Sniderman, A., 1985, On pulmonary vascular resistance: the need for more precise definition. *Am. J. Cardiol.* 55:217-227.

Meyer, M. and Piiper, J., 1989, Nitric oxide (NO), a new test gas for study of alveolar capillary diffusion. *Eur. Respir. J.* 2:494-496.

Moinard, J., Manier, G., Pillet, 0. and Castaing, Y., 1994, Effect of inhaled nitric oxide on hemodynamics and VA/Q inequalities in patients with chronic obstructive pulmonary disease. *Am. J. Resp. Crit. Care Med.* 149:1482-1487.

Moinard, J. and Guenard, H., 1990, Determination of lung capillary blood volume and membrane diffusing capacity in patients with COLD using the NO-CO method. *Eur. Respir. J.* 3:318-322.

Moncada, S., Palmer, R.M.J. and Higgs, E.A., 1991, Nitric oxide: physiology, pathophysiology and pharmacology. *Pharmacol. Rev.* 43:109-142.

Mourgeon, E., Roche, S., Lu, Q., Law-koune, J.D., Puybasset, L, Diaby, M., Bodin, L, Rouby, J.J. and Viars, P., 1994, NO inhalè dans le SDRA: courbes dose-rèsponse en prèsence ou en l'absence de choc septique. *Ann. Fr. Anesth. Rèanim.* 13(suppl):Abstract.

Mourgeon, E., Puybasset, L., Renaudin, M.H., Ravault, C., Bodin, L., Rouby, J.J. and Viars, P., 1994, Distribution of NO concentrations during 2 methods of NO inhalation. *Br. J. Anaest.* 72 (supplement 1):AI8 (Abstract).

Mustafa, M.G. and Tierney, D.F., 1978, Biochemical and metabolic changes in the lung with oxygen ozone and nitrogen dioxide toxicity. *Am. Rev. Resp. Dis.* 118:1061-1088.

Nguyen, T., Brunson, D., Crospi, C.L., Penman, B.W., Wishnok, J.S. and Tannenbaum, S.R., 1992, DNA damage and mutation in human cells exposed to nitric oxide in vitro. *Proc. Natl. Acad. Sci.* 89:3030-3034.

Niu, X-F, Smith, W. and Kubes, P., 1994, Intracellular oxidative stress induced by nitric oxide synthesis inhibition increases endothelial cell adhesion to neutrophils. *Circ. Res.* 74:1133-1140.

Oda, H., Nogami, H., Kusumotos, S., Nakajima, T., Kurata, A. and Imai, K., 1976, Long-term exposure to nitric oxide in mice. *J. Jpn. Soc. Air Pollut.* 11:150-160.

Oda, H., Nogami, H., Kusumoto, S., Nakajima, T. and Kurata, A., 1980, Lifetime exposure to 2.4 ppm nitric oxide in mice. *Env. Research* 22:254-263.

Palmer, R.M.J., Ferrige, A.G. and Moncada, S., 1987, Nitric oxide release accounts for the biological activity of endothelium-derived relaxing factor. *Nature* 327:524-526.

Pepke-Zaba, J., Higenbottam, T.W., Dinh-Xuan, A.T., Stone, D. and Wallwork, J., 1991, Inhaled nitric oxide as a cause of selective pulmonary vasodilation in pulmonary hypertension. *Lancet* 338:1173-1174.

Pison, U., Lopez, F.A., Heidelmeyer, C.F., Rossaint, R. and Falke, K., 1993, Inhaled nitric oxide reverses hypoxic pulmonary vasoconstriction without impairing gas exchange. *J. Appl. Physiol.* 74:1287-1292.

Puybasset, L., Rouby, J.J., Mourgeon, E., Cluzel, P., Souhil, Z., Law-Koune, J.D., Stewart, T., Devilliers, C., Lu, Q., Roche, S., Kalfon, P., Vicaut, E., Viars, P., 1995, Factors influencing cardiopulmonary effects of inhaled nitric oxide in acute respiratory failure. *Am. J. Respir. Crit. Care Med.* 152:318-328.

Puybasset, L., Mourgeon, E., Segal, E., Bodin, L., Rouby, J.J. and Viars, P., 1993, Effects of inhaled nitric oxide on respiratory resistance in patients with ARDS. *Anesthesiology* 79:supplement 3A:A299 (Abstract).

Puybasset, L., Rouby, J.J., Mourgeon, E., Stewart, T.E., Cluzel, P., Belin, M.F., Grenier, P., Arthaud, M., Poéte, P., Bodin, L., Korinek, A.M. and Viars, P., 1994, Inhaled nitric oxide in acute respiratory failure: Dose-response curves. *Intensive Care Med.* 20:319-327.

Puybasset, L., Stewart, T.E., Rouby, J.J., Cluzel, P., Mourgeon, E., Belin, M.F, Arthaud, M., Landault, C. and Viars, P., 1994, Inhaled nitric oxide reverses the increase in pulmonary vascular resistance induced by permissive hypercapnia in patients with ARDS. *Anesthesiology* 80:1254-1267.

Radermacher, P., Santak, B., Wüst, F.U., Tarnow, J. and Falke, K.J., 1990, Prostacyclin for the treatment of pulmonary hypertension in patients with acute respiratory distress syndrome: Effects on pulmonary capillary pressure and ventilation-perfusion distributions. *Anesthesiology* 72:238-244.

Radermacher, P., Santak, B., Becker H. and Falke, K.J., 1989, Prostaglandine El and nitroglycerin reduce pulmonary capillary pressure but worsen ventilation-perfusion distributions in patients with adult respiratory distress syndrome. *Anesthesiology* 70:601-606.

Ranga, V., Kleinerman, J., I.P., M.P.C. and Collins, A.M., 1980, The effect of nitrogen dioxide on tracheal uptake and transport of horseradish peroxidase in the guinea pig. *Am. Rev. Resp. Dis.* 122:483-490.

Rich, F.R., Murphy, G.D., Roos, C.M. and Johns, R.A., 1993, Inhaled nitric oxide: Selective pulmonary vasodilation in cardiac surgical patients. *Anesthesiology* 78:1028-1035.

Rimar, S. and Gillis, C.N., 1992, Pulmonary vasodilation by inhaled nitric oxide after endothelial injury. *J. Appl. Physiol.* 73:2179-2183.

Roberts, J.D., Polaner, D.M., Lang, P. and Zapol, W.M., 1992, Inhaled nitric oxide in persistent pulmonary hypertension of the newborn. *Lancet* 340:818-819.

Rose, R.M., Fuglestad, J.M., Skornik, W.A., Hammer, S.M., Wolfhal, S.F., Beck, B.D., Brain, J.D., 1988, The pathophysiology of enhanced susceptibility to murine cytomegalovirus respiratory infection during short-term exposure to 5 ppm nitrogen dioxide. *Am. Rev. Resp. Dis.* 137:912-917.

Rossaint, R., Falke, K.J., Lopez, F., Slama, K., Pison, U. and Zapol, W.M., 1993, Inhaled nitric oxide for the adult respiratory distress syndrome *N. Engl. J. Med.* 328:399-405.

Samama, Ch. M., Diaby, M., Fellahi, J.L., Mdhafar, A., Eyraud, D., Arock, M., Guillosson, J.J., Coriat, P., Rouby, J.J., 1995, Inhibition of platelet aggregation by inhaled NO in patients with adult respiratory distress syndrome. *Anesthesiology* 83:56-65.

Sandström, T., Stjernberg, N., Eklund, A., Ledin, M-C, Bjermer, L., Kolmodin-Hedman, B., Lindström, K., Rosenhall, L. and Angström, T., 1991, Inflammatory cell response in bronchoalveolar lavage fluid after nitrogen dioxide exposure of healthy subjects: a dose-response study. *Eur. Respir. J.* 3:332-339.

Sellden, H., Winberg, P., Gustafsson, L.E., Lundell, B.O., Book, K. and Frostell, C.G., 1993, Inhalation of nitric oxide reduced pulmonary hypertension after cardiac surgery in a 3.2 Kg infant. *Anesthesiology* 78:577-580.

Stavert, D.M. and Lehnert, B.E., 1990, Nitrogen oxide and nitrogen dioxide as inducers of acute pulmonary injury when inhaled at relatively high concentrations for brief periods *Inhal. Toxicol.* 2:53-57.

Stenqvist, 0., Kjelltoft, B. and Lundin, S., 1993, Evaluation of a new system for ventilatory administration of nitric oxide. *Acta Anaesthesiol. Scand.* 37:687-691.

Stewart, T.E., Rouby, J.J., Puybasset, L., Mourgeon, E., Descouls, A.M. and Viars, P., 1994, Inhaled NO reduces alveolar dead space in patients with ARDS and pulmonary hypertension. *Am. J. Resp. Crit. Care Med. Suppl.* 149:A424 (Abstract).

Tiner, N.N., Etches, R.C., Kamstra, B., Tierney, A.J., Peliowski, A. and Ryan, A., 1994, Inhaled nitric oxide in infants referred for extracorporeal membrane oxygenation : dose response. *J. Pediatr.* 124:302-308.

Weitzberg, E., Rudehill, A., Alving, K., and Lundberg, J.M., 1991, Nitric oxide inhalation selectively attenuates pulmonary hypertension and arterial hypoxia in porcine endotoxin shock. *Acta Physiol. Scand.* 143:451-452.

Wennmaln, A., Benthin, G., Edlund, A., Jungersten, L., Kieler-Jensen, N., Lundin, S, Nathorst Westfelt U., Petersson A. and Waagstein, F., 1993, Metabolism and excretion of nitric oxide in humans. An experimental and clinical study. *Cir. Res.*73:1121-1127.

Wessel, D.L., Adatia, I., Thompson, J.E. and Hickey, P.R., 1994, Delivery and monitoring of inhaled nitric oxide in patients with pulmonary hypertension. *Crit. Care Med.* 22:930-938.

White, C.W., Nguyen, D.H. and Tuder, R., 1994, Nitric oxide modulation of hyperoxic injury. *Am. J. Resp. Crit. Care Med. Suppl.* 49:A548 (Abstract).

Wink, D.A., Kasprzak, K.S., Maragos, C.M., Elespuru, R.K., Misra, M., Dunams, T.M., Cebula, T.A., Koch, W.H., Andrews, A.W., Allen, J.S. and Keefer, L.K., 1991, DNA deaminating ability and genotoxicity of nitric oxide and its progenitors. *Science* 254:1001-1003.

Zapol, W.M., Snider, M.T., Rie, M.A., Quinn, D.A. and Frikker, M., 1985, Pulmonary circulation during adult respiratory distress syndrome. In: *Acute Respiratory Failure*, Zapol WM and Falke KJ, (eds), P-209-226, New York: Dekker, Inc, (Lung Biology in Health and Disease; vol. 24)

Zapol, W.M., Rimar, S., Gillis, N. and Bosken, C.H., 1994, Nitric oxide and the lung. *Am. J. Respir. Crit. Care Med.* 149:1375-1380.

TISSUE ANGIOTENSIN CONVERTING ENZYME

Suzanne Oparil, Qingcheng Meng, Shuang-dan Sun, Yiu-Fai Chen, and Louis J. Dell'Italia

Vascular Biology and Hypertension Program
University of Alabama at Birmingham
Birmingham, Alabama, USA

INTRODUCTION

Biochemical, immunohistochemical, and molecular biologic evidence indicates that all components of the renin-angiotensin system (renin, angiotensinogen, angiotensin converting enzyme (ACE), angiotensin peptides and angiotensin receptors) are found in a variety of tissues, including heart, kidney, brain and vasculature. The renin-angiotensin system plays an important role in the maintenance of blood pressure and volume homeostasis and of cardiovascular structure in normotensive subjects without obvious cardiovascular disease and in various disease states, including systemic hypertension and congestive heart failure (*Oparil and Haber, 1974*; *Gibbons and Dzau, 1994*). Renin (EN 3.4.23.15), a highly selective serine protease, cleaves its substrate, angiotensinogen, to produce the biologically inactive decapeptide, angiotensin I (ANG I) (Figure 1). The peptidyl dipeptide hydrolase angiotensin-converting enzyme (ACE; EN 3.4.15.1) removes the carboxyl terminal His-Leu from ANG I to produce the octapeptide angiotensin II (ANG II). In addition, ACE inactivates bradykinin, a vasodilator peptide/mediator of inflammation, as well as substance P, enkephalins and endorphins. ANG I and ANG II are both rapidly hydrolyzed to inactive products by a number of proteolytic enzymes, the angiotensinases (*Regoli et al., 1975*).

ANG II is a potent pressor agent acting directly via specific G-protein-coupled receptors in resistance vessels and indirectly by stimulating the synthesis and release of the mineralocorticoid aldosterone from the adrenal glomerulosa and the release of noradrenaline from sympathetic nerve terminals. ANG II also has a mitogenic and/or hypertrophic effect on vascular smooth muscle cells and cardiac myocytes (*Naftilan et al., 1989*; *Owens, 1989*); these effects may contribute to the cardiovascular consequences of hypertension or may produce cardiac and/or vascular remodeling in the absence of blood pressure elevation.

The classical view that ANG II functions only as a circulating hormone has been challenged by the recent finding of a complete renin-angiotensin system in a variety of tissues (*Lee et al., 1993*; *Dzau and Hirsch, 1990*; *Lindpaintner et al., 1988*; *Dzau, 1988a*; and *Dzau, 1993*) and by the demonstration that ANG I and II are synthesized in cultured cells and isolated organ preparations. Further, it has been suggested that ACE inhibitors lower blood pressure by inhibiting the generation of ANG II in tissues rather than in circulating blood (*Lee et al., 1992*).

Vascular Endothelium: Responses to Injury
Edited by John D. Catravas *et al.*, Plenum Press, New York, 1996

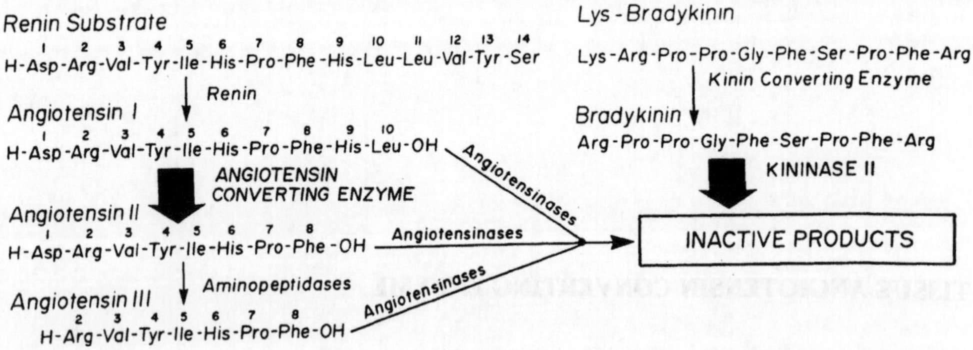

Figure 1. Biochemistry of the renin-angiotensin system and its interaction (via converting enzyme or kininase II) with the kallikrein-kinin system. Reproduced with permission from Oparil and Katholi, Chap 4:219, *Current Concepts in Cardiovascular Physiology, 1990*, Academic Press, Inc.

The observations that the ACE inhibitors are effective in lowering blood pressure in patients with low plasma renin activity and in nephrectomized patients and that the antihypertensive potency of the ACE inhibitors is better correlated with inhibition of tissue ACE than of plasma ACE are consistent with this interpretation. Application of the techniques of cell and molecular biology to cardiovascular research has demonstrated that the renin-angiotensin system operates as both a circulating (endocrine) and a tissue (autocrine/paracrine/intracrine) system. The endocrine component mediates acute cardiovascular homeostasis and fluid balance, while the tissue renin-angiotensin system appears to participate in the chronic regulation of cardiovascular function and structure (Figure 2). Further investigation is needed to define the mechanisms that regulate expression of components of the tissue renin-angiotensin system, particularly tissue ACE, at the gene level and to relate these to ANG II mediated control of cardiovascular function and structure in the normal subject and in cardiovascular disease.

ACE GENE REGULATION

Alterations in ACE activity *in vivo* and *in vitro* have been described in response to a number of hormones, pharmacologic agents and environmental factors including glucocorticoids (*Friedland et al., 1977*; *Mendelsohn et al., 1982*), thyroxine and triiodothyronine (*Yotsumoto et al., 1982*), ACE inhibitors (*Fyhrquist et al., 1980*; *Sakaguchi et al., 1988*) and hypoxia (*Oparil et al., 1988*; *King et al., 1989*). ACE can be induced by corticosteroids in cultured endothelial cells, alveolar macrophages and monocytes, and in lung. Hyperthyroidism elevates levels of circulating ACE. Both protein kinase C- and cAMP-dependent mechanisms appear to increase ACE activity in cultured endothelial cells, but only activators of protein kinase C increase the release of ACE into the medium (*Iwai et al., 1987*).

Although relatively little is known about the regulation of ACE mRNA expression (*Griendling et al., 1993*), the identification of sequences for a number of known promoters and regulatory elements (Figure 3) in the 5' flanking region of the endothelial form of the ACE gene allows us to make inferences about what factors can influence ACE gene transcription. This region contains consensus sequences for SP-1 binding sites (found in the promoter regions of many housekeeping genes) and sequences for AP-2 (protein kinase C Sensitive) and transcription factor IID binding sites (*Kumar et al., 1991*; *Griendling et al., 1993*). The more distal 5' regions of mouse and human ACE genes also contain sequence motifs resembling glucocorticoid- and cAMP-response elements (*Shai et al., 1990*).

Linkage analysis has shown that polymorphisms in the ACE gene are linked to the regulation of blood pressure in the rat (*Jacob et al., 1991*; *Hilbert et al., 1991*). However, sib-pair studies in human essential hypertension have failed to find a linkage between the ACE gene

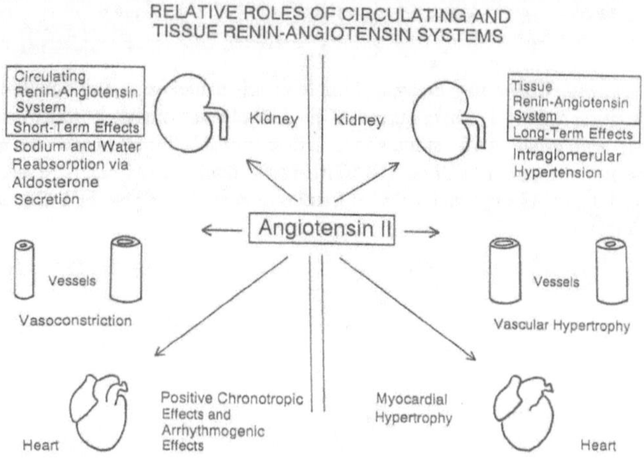

Figure 2. The traditional concept of the renin-angiotensin system is a circulating system in which the specific components of the system-renin, angiotensinogen, and ACE-are secreted from the kidney, liver, and lung, respectively. The product of the biochemical cascade, ANG II, circulates and acts on specific receptors in a variety of organs. It is now widely accepted that renin, angiotensinogen, and ACE are synthesized locally in many tissues. One important difference between tissue(autocrine/paracrine) systems and the circulating (endocrine) system is that the circulating system regulates short-term effects and the tissue systems regulate long-term function. Furthermore, the growth stimulating effects which lead to vascular and cardiac hypertrophy seem to be an effect of ANG II generated within the tissues. Reproduced with the permission from *A Slide Compendium*, 1993, from English Thornton Communications, Inc.

and increased blood pressure (*Jeunemaitre et al., 1992*; *Cambien et al., 1992*). A deletion (D)-insertion (I) polymorphism of the human ACE gene that appears to affect the level of serum ACE activity has been identified (*Rigat et al., 1990*; *Harrap et al., 1993*). The genotype (DD) with the highest ACE levels appears to be strongly associated with left ventricular hypertrophy (*Schunkert et al., 1994*) and with increased risk of myocardial infarction and of dilated and hypertrophic cardiomyopathy (*Cambien et al., 1992*; *Raynolds et al., 1993*; *Marian et al., 1993*). The mechanism by which the DD genotype of the ACE gene effects cardiac hypertrophy and accelerated coronary artery disease is unknown, but is unrelated to hypertension since the ACE polymorphism does not appear to be linked with blood pressure in humans (*Jeaunemaitre et al., 1992*; *Harrap et al., 1993*). These findings are consistent with the clinical observations that (1) ACE inhibitors appear to be superior to other antihypertensive agents in inducing regression of cardiac hypertrophy independent of their antihypertensive efficacy (*Dahlof et al., 1992*), and (2) ACE inhibitors promote better survival in heart failure than do other vasodilators (*Cohn et al., 1991*). Even in the absence of clear delineation of a phenotype relating the DD genotype of the ACE gene to increased cardiovascular risk, the presence of this linkage may provide a useful clinical test for cardiovascular risk assessment

HYPOXIA

Exposure to chronic hypoxia is associated with alterations in ACE activity in intact animals and cultured pulmonary artery endothelial cells (*Caldwell and Blatteis, 1983*; *Jackson et al., 1986*; *Keane et al., 1982*; *Krulewitz and Fanburg, 1984*; *Molteni et al., 1974*; *Oparil et al., 1988*). Both increased (*Krulewitz and Fanburg, 1984*; *Molteni et al., 1974*) and decreased (*Caldwell and Blatteis, 1983*; *Jackson et al., 1986*; *Keane et al., 1982*) ACE activity and

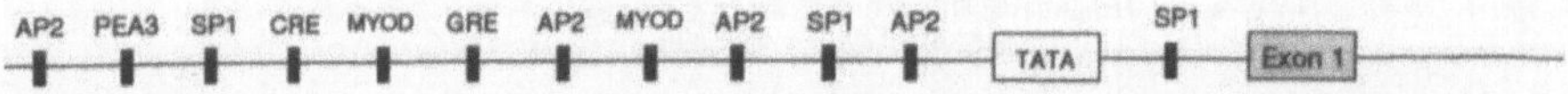

Figure 3. *Human Angiotensin Converting Enzyme*. Simplified schematic diagram of putative regulatory elements within the promoter region (-1.5kb) of human ACE. AP2-element mediates inducible expression during phorbol ester and cyclic AMP stimulation; CRE-cyclic AMP responsive element; GRE-glucocorticoid response element; MYOD-CANNTG(E-box) motif for MyoD, a muscle-specific transcription factor, binding; PEA3-element for PEA 3 transcription factor binding; SP1-GC rich element for SP1 protein binding.

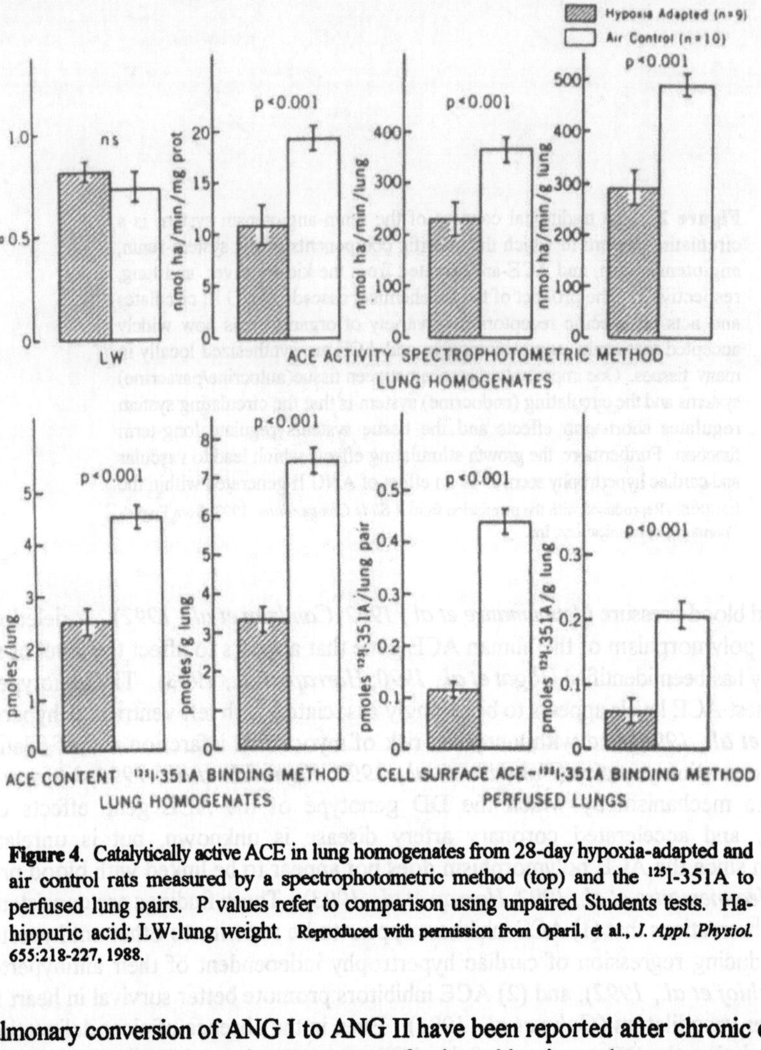

Figure 4. Catalytically active ACE in lung homogenates from 28-day hypoxia-adapted and air control rats measured by a spectrophotometric method (top) and the [125]I-351A to perfused lung pairs. P values refer to comparison using unpaired Students tests. Ha-hippuric acid; LW-lung weight. Reproduced with permission from Oparil, et al., *J. Appl. Physiol.* 655:218-227, 1988.

intrapulmonary conversion of ANG I to ANG II have been reported after chronic exposure to normobaric or hypobaric hypoxia. Exposure of cultured bovine pulmonary artery endothelial cells to hypoxia (3% O_2) for 24-48 hrs was associated with increased intracellular ACE activity without a concomitant increase in ACE activity in the culture medium (*Krulewitz and Fanburg, 1984*). The increase in endothelial cell ACE activity occurred without a significant alteration in cell number. In contrast, reduced pressor responses have been reported after bolus injections of ANG I into the right atria of rats exposed to high-altitude (5,500 m) hypoxia for 2 wk (*Caldwell and Blatteis, 1983*), and reduced lung ACE activity has been reported in rats

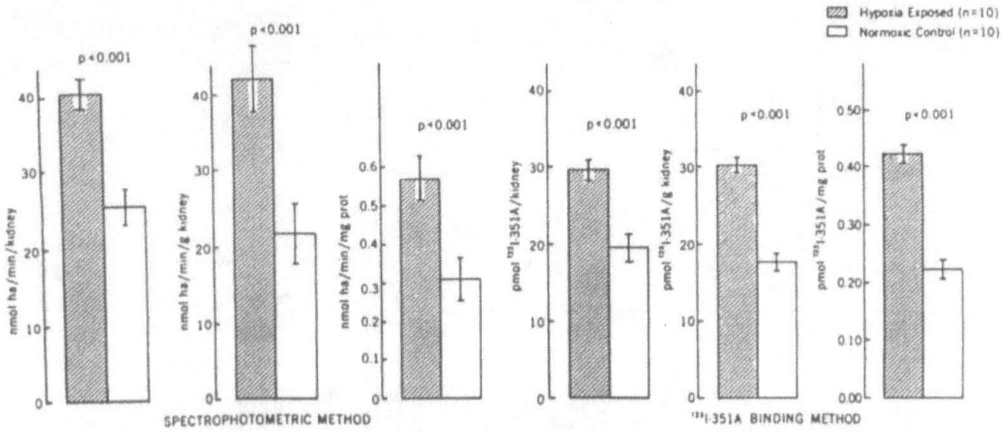

Figure 5. *Kidney ACE Content.* Catalytically active ACE in kidney homogenates from 14-day hypoxia-adapted and air control rats measured by spectrophotometric and ^{125}I-351A binding methods. P values refer to comparisons using unpaired Student's tests. Ha-Hippuric acid. Reproduced with permission from Oparil, et al., *J. Appl. Physiol.* 65:218-227, 1988.

exposed to hypobaric (380 Torr) hypoxia for 3 wk (*Keane et al., 1982*). Further, exposure of rats to 10% O_2 at atmospheric pressure for periods of 21-24 days has been shown to impair the conversion of ANG I to ANG II in a single pass through the pulmonary circulation (*Jackson et al., 1986*). Hypoxia-induced impairment of conversion was not related to altered kinetics of substrate exposure. The latter studies provide evidence for decreased intrapulmonary ACE activity in rats exposed to chronic hypoxia but do not provide a clear and consistent picture of the cellular and molecular mechanisms by which chronic hypoxia regulates ACE activity *in vivo* and *in vitro*. Further, these studies did not determine whether hypoxia-induced alterations in ACE activity reflect changes in number of catalytically active ACE molecules available for substrate hydrolysis and/or in total number of ACE molecules synthesized or whether adaptation to hypoxia modulates ACE activity directly by altering ACE synthesis or by some other effect on the cellular processing of the enzyme. Therefore, we tested the hypothesis that adaptation to chronic hypoxia inhibits ACE activity in lung and suppresses intrapulmonary conversion of ANG I to ANG II by reducing the number of ACE molecules available to bind to substrate (ANG I) (*Oparil et al., 1988*). Catalytically active ACE was quantitated by spectrophotometric assay, and active sites of ACE were estimated by measuring the binding of an active site-directed ACE inhibitor to lung homogenates and perfused lungs. In addition, since we had demonstrated that adaptation to hypoxia blunts the systemic pressor response to exogenous (intravenous) ANG I (*Jin et al., 1987*), suggesting a reduction in whole body conversion of ANG I to ANG II, extrapulmonary ACE activity and content were examined in hypoxia-adapted and control animals. ACE in kidney and brain was examined because ANG II generation in these organs is thought to play an important role in the regulation of local (*Navar et al., 1983*; *Pandey et al., 1984*) and systemic (*Ganten et al., 1984*) hemodynamics, and little is known about the sensitivity of ACE in these organs to regulation by chronic hypoxia. Testis was examined because it contains very large amounts of a form of ACE that shares catalytic and immunologic properties with ACE from other organs but is thought not to be involved in cardiovascular regulation. Finally, the reversibility of hypoxia-induced alterations in ACE was assessed by measuring ACE in hypoxia-recovered rats. ACE content of tissue homogenates and serum was assessed by quantitating the binding of ^{125}I-351A (Merck, Sharp & Dohme Research Laboratories, West Point, PA), a derivative of lisinopril [N-(1-carboxyl-3-phenylpropyl)-L-lysyl-L-proline], a highly specific active site-directed inhibitor of ACE. ACE levels at the endothelial cell surface were estimated by injecting ^{125}I-351A into an isolated perfused lung preparation and measuring binding of the ACE inhibitor to the lung tissue.

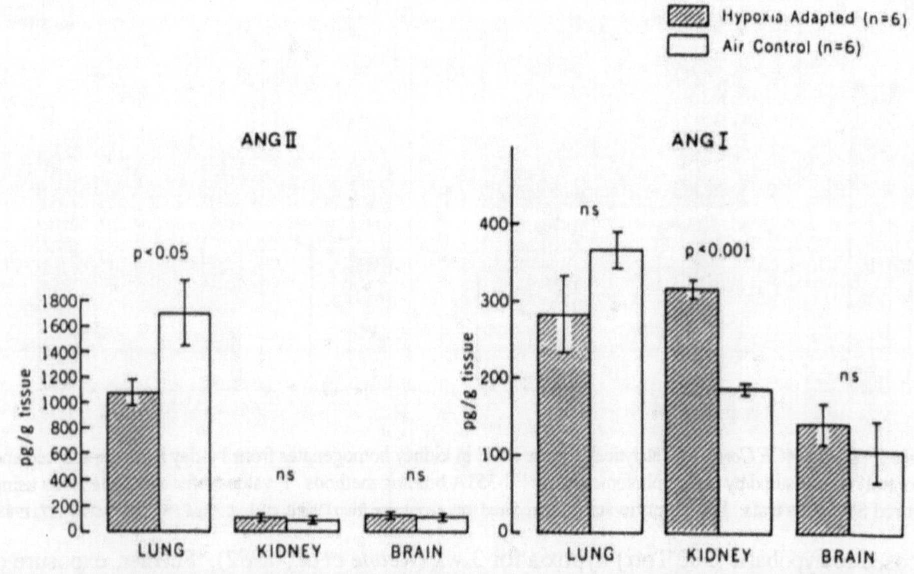

Figure 6. ANG I and ANG II levels in homogenates of lung, kidney, and brain from 14-day hypoxia-adapted and air control rats measured by radioimmunoassay. P values refer to comparisons using unpaired Student's test. Reproduced with permission from Oparil, et al., *J. Appl. Physiol.* 65:218-227, 1988.

Adaptation to hypoxia was associated with reductions in lung ACE activity and content after chronic (28 days) hypoxic exposure, ACE activity was reduced by 43% in lungs from hypoxia-adapted animals compared to air controls, while the catalytically active ACE content of lung homogenates from hypoxia adapted rats was reduced by 57% (Figure 4). Regression analysis revealed a linear relationship between ACE levels from the same pair of lungs measured by both the *in vitro* ^{125}I-351A binding method and the spectrophotometric method. ^{125}I-351A binding to isolated perfused lungs was reduced by 73% in hypoxia-adapted lungs compared to controls. Levels of ^{125}I-351A binding in the perfusion study were ~1/100 of those observed in the *in vitro* membrane-binding study, suggesting that under the conditions of these experiments, only 1/100 of total intrapulmonary stores of ACE were easily accessible to circulating substrate. Assessment of ACE binding sites by use of brief (15 s) periods of perfusion of ^{125}I-351A into isolated perfused lungs was likely carried out under nonequilibrium conditions and therefore probably underestimated the number of ACE binding sites at the endothelial cell surface. These data suggest that both intracellular and cell surface-associated ACE were suppressed in lungs from hypoxia-adapted rats.

In contrast to lung, kidney ACE activity and content were significantly increased in hypoxia-exposed animals compared to air controls (Figure 5). This was true whether results were expressed per g of kidney weight, per mg of protein or per kidney. Brain and testicular ACE contents were unaffected by hypoxic exposure. Lung and kidney ACE levels returned to control values in hypoxia-adapted rats after restoration of a normoxic environment. When hypoxia-adapted animals were maintained in room air for a period equivalent in duration to the hypoxia exposure, ACE levels in both lung and kidney returned to air control values. These relationships held whether data were expressed per whole organ or per g of tissue. Thus the hypoxia-induced alterations in pulmonary and renal ACE were reversible over time with restoration of a normoxic environment.

After chronic hypoxic exposure, ANG II levels in lung were significantly decreased compared to air controls (Figure 6), suggesting that the reduction in intrapulmonary ACE activity was associated with reduced local generation of ANG II. Lung ANG I levels were not significantly different in hypoxia-adapted vs. control lungs. Neither kidney nor brain ANG II

Table I. PRA, ANG I and II levels, and serum ACE content in hypoxia-adapted and control rats.

	Hypoxia Adapted, 14 days	Air control, 14 days
PRA,ng/ml/h	9.3 ± 1.5(10)	6.0 ± 1.0(10)
Plasma ANG I, pg/ml	415 ± 131*(8)	170 ± 40(9)
Plasma ANG II, pg/ml	20 ± 4(9)	20 ± 9(9)
Serum ACE, nmol ^{125}I-351A/ml	134 ± 4(10)	143 ± 3(10)

Values are means ± SE with nos. of rats in parentheses. PRA, plasma renin activity; ANG I and II, angiotensin I and II; ACE; angiotensin converting enzyme. *$P<0.05$ compared with air control by students unpaired t test.

nor brain ANG I values differed between groups, whereas kidney ANG I stores were significantly greater in the hypoxia-adapted group. Plasma ANG II levels, renin activity, and serum ACE content were not significantly different in hypoxia-adapted animals than in air controls, whereas plasma ANG I levels were significantly elevated in the hypoxia-adapted group compared to air controls (Table 1). The finding that both intracellular and cell surface-associated ACE levels were reduced in hypoxia-adapted lungs coupled with the finding that circulating (serum) ACE levels were not significantly different in hypoxia-adapted vs. air control animals suggests that the production of active ACE was reduced in the hypoxia-adapted lung and that the reduced availability of ACE at the cell surface was likely a reflection of decreased production of active ACE by the hypoxia-adapted lung rather than of a defect in transport of ACE from its intracellular sites of synthesis to its site of action at the cell surface. Whether this reflects hypoxia-induced synthesis of an inactive precursor of ACE with impairment of an activating enzyme, synthesis of an ACE inhibitor, or some other mechanism remains to be determined.

This study revealed remarkable organ specificity of the effects of hypoxia on ACE levels, i.e., reductions in lung, increases in kidney and no change in brain and testis. This organ specificity may relate to differences in cell types of origin of ACE in the various organs: predominantly endothelial cells in lung; endothelial cells plus proximal tubular and juxtaglomerular cells in kidney; endothelial cells plus neurons and glial cells in brain; and Leydig cells in testis. ACE isolated from lung, kidney and brain, as well as from a number of other organs, has common catalytic and antigenic properties and a molecular weight of ~150 kDa. Testicular ACE shares catalytic and immunologic properties with ACE from other organs but has a lower molecular weight (~90 kDa). Despite these similarities in structure and catalytic function, the regulation of ACE synthesis, intracellular processing, and release may differ among cell types, perhaps due to the presence of cell specific transcription factors that can influence ACE gene transcription. For example, synthesis of testicular ACE, but not of ACE in other organs, is androgen dependent and stores of ACE in testes increase greatly at the time of puberty (*El-Dorry et al., 1982*). The regulatory effects of hypoxia on ACE gene regulation in cultured renal tubular cells, neurons, and Leydig cells remain to be explored.

A subsequent study assessed directly the effects of hypoxia on ACE synthesis in cultured porcine pulmonary artery endothelial cells(*King et al., 1989*). Confluent cultures were exposed to hypoxia (2.5% O_2 at 1 atmosphere); controls were cultured in normoxic conditions. After exposure to hypoxic or normoxic conditions, followed by incubation with [^{35}S] methionine for an additional 24 hrs under the same conditions, newly synthesized radiolabeled

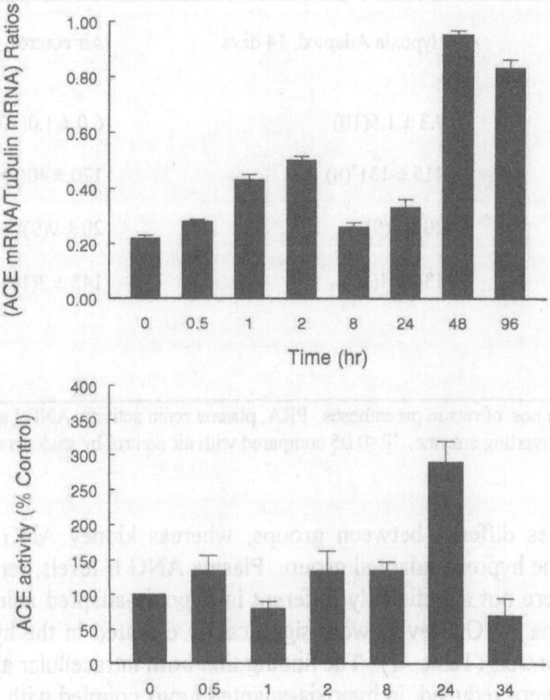

Figure 7. TOP: Slot-blot analysis of ACE mRNA from first passage porcine pulmonary artery endothelial cells exposed to 100μM lisinopril for times indicated. Values represent ACE MRNA levels standardized to α-tubulin mRNA. Bars represent means ± SE (n=4). BOTTOM: ACE activity for first passage porcine pulmonary artery endothelial cells exposed to 100μM captopril for times indicated. Bars represent means ± SE(n=4). ACE activity is expressed as percent of time O control. Reproduced with permission from King and Oparil, *Am. J. Physiol.* 263:C743-749, 1992.

ACE was isolated by an immunobead procedure using either anti-ACE (porcine lung) immunoglobin G (IgG) or nonimmune IgG and quantitated. A single radiolabeled peak (150 kDa) with the same electrophoretic mobility as purified porcine lung ACE was observed. There was a significant time-dependent increase (2.5 fold) in endothelial cell ACE antigen synthesis without a concomitant change in either cell number or total trichloracetic acid-precipitable protein in hypoxic cells compared to normoxic controls. In contrast, ACE activity was unchanged in cultures exposed to hypoxia and there was no detectable change in ACE antigen in the conditioned growth medium. After hypoxic exposure, ACE activity at the cell surface, within the cell and in the conditioned growth medium was not significantly different from normoxic controls. This suggests that an inactive form of ACE is synthesized by cultured pulmonary artery endothelial cells under hypoxic conditions.

This dissociation between the increase in ACE antigen and the unchanged ACE activity in hypoxia adapted endothelial cells can be explained by several possible mechanisms: 1) ACE may be synthesized as an inactive precursor that requires further processing for activation. Hypoxia may inhibit or downregulate the processing of ACE from the inactive precursor to the catalytically active form of the enzyme. 2) Hypoxia may stimulate the synthesis of an endogenous inhibitor of ACE that blocks the activity of the newly synthesized enzyme. 3) Hypoxia may increase the turnover of ACE in endothelial cells, stimulating both synthesis of

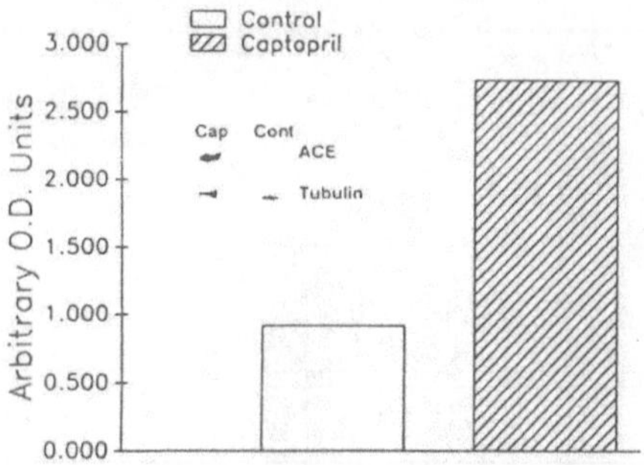

Figure 8. Nuclear runoff assay of ACE gene transcription in nuclei isolated from first-passage porcine pulmonary artery endothelial cells exposed to 100μM ACE inhibitor (lisinopril or captopril) or control medium for 48 h. Assay was performed once with each inhibitor; results were similar for both. Values represent ACE mRNA levels standardized to α-tubulin mRNA from captopril assay. Inset: representative autoradiograms. Reproduced with permission form King and Oparil, *Am. J. Phsiol.* 263:C743-749, 1992.

new ACE and destruction of the active form of the enzyme. Further study of regulation of ACE gene expression on the molecular level is needed to distinguish among these alternatives.

ACE INHIBITORS

ACE is found in highest concentrations on the luminal surface of endothelial cells (*Ryan et al., 1976*) but is also found in several other cell types, including epithelial cells of the kidney proximal tubules (*Hall et al., 1976*), gut (*Ward et al., 1980*), germinal epithelial cells of the testis (*Velletri et al., 1985*), neurons and glia (*Pickel et al., 1986*). Active site-directed inhibitors of ACE have proved clinically useful in the chronic treatment of patients with hypertension and congestive heart failure, as well as in studies of the pathogenesis of both disorders. The antihypertensive and afterload reducing effects of the ACE inhibitors are thought to relate primarily to inhibition of ANG II production by the vasculature, although other mechanisms, such as potentiation of vasodilator kinins and prostaglandins, are also thought to play a role. Remarkably, the ACE inhibitors appear to be effective in nearly all hypertensive patients, even those with low circulating renin and angiotensin levels, and they do not appear to lose their antihypertensive vasodilator potency during prolonged use.

Chronic exposure to the ACE inhibitors induces increased ACE activity *in vivo* and *in vitro.* Administration of the ACE inhibitors to patients and experimental animals is associated with increased ACE activity in the circulation and in a variety of tissues (*Fyhrquist et al., 1983; Kokubu et al., 1980; Larochelle et al., 1979*). Forslund et al., *(1984)* demonstrated an increase in ACE activity in lung homogenates and in purified membranes from lung after chronic ACE inhibitor treatment in the rat with no increase in ACE activity in either tissue homogenates or plasma membrane fractions from kidney, intestine, testis, or epididymis. They concluded that the most important source of increased ACE biosynthesis during chronic ACE inhibitor treatment is the lung. Addition of the ACE inhibitor captopril to the culture medium of human umbilical artery endothelial cells has been associated with significant increases in both ACE activity in the medium and immunoreactive ACE in the cell sheet (*Fyhrquist et al., 1982*). Further, Schunkert *et al., (1993)* recently reported that 3 days of ACE inhibition resulted in significant increases in lung and testis ACE mRNA levels, accompanied by suppression of tissue

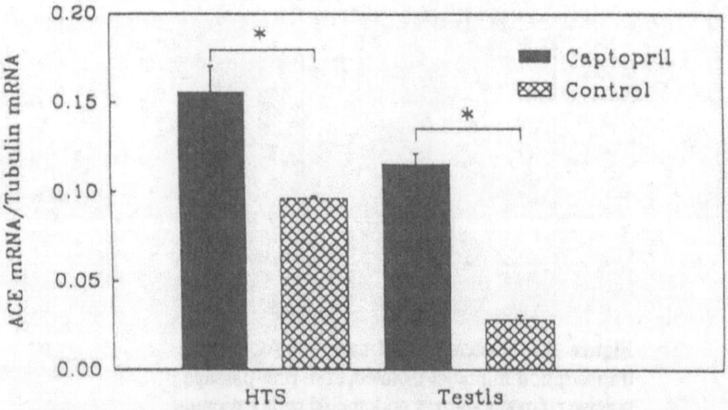

Figure 9. ACE mRNA in HTS brain region and testis of lifetime captopril-treated SHR (n=5) vs vehicle-treated SHR (n=6). *P<0.01 for SHR captopril vs SHR control. Values represent ACE mRNA levels standardized to α-tubulin mRNA. HTS-hypothalamus-thalamus-septal block. Reproduced with permission form King, et al., *Mol. Cell. Neurosci.* 2:13-20, 1991.

and serum ACE activity in male Sprague-Dawley rats. Discontinuation of the ACE inhibitor resulted in a marked increase in serum ACE activity, indicating that ACE synthesis was enhanced after exposure to the ACE inhibitor. Infusion of ANG II for 3 days decreased ACE mRNA and ACE activity in lung and testis compared to controls; serum ACE activity was unchanged. These data suggest that active site-directed ACE inhibitors may stimulate the synthesis and release of ACE *in vivo* and *in vitro* and that the *in vivo* effect may be related to removal of end-product inhibition by ANG II.

Consistent with this conclusion is the preliminary observation in our laboratory that ANG II produced a dose-dependent reduction in ACE mRNA levels in glial cultures obtained from SHR and WKY rats (*King et al., 1992*). SHR-derived glia appeared to be less sensitive to this end-product inhibition than did cells derived from WKY. This effect was blocked by the type 2 ANG II receptor (AT_1) antagonist losartan, but not by the type 2 ANG II receptor (AT_2) antagonist EXP655, suggesting that the effect of ANG II on ACE gene expression is mediated by the AT_1 receptor (*Berecek et al., 1992*). Mechanisms by which ANG II could affect the expression of the gene responsible for its synthesis include a direct action via the ANG II receptor(s) on cis-acting regulatory sequences in the ACE gene and ANG II-induced synthesis of transacting factors that act on regulatory sequences in the ACE gene. Further study is needed to identify and characterize such factors and responsive sequences in the ACE gene.

Further studies in our laboratory have examined the effects of ACE inhibition on ACE gene expression in cultured porcine pulmonary artery endothelial cells and neuronal enriched cultures from rat brain *in vitro* (*King et al., 1991*; *King and Oparil, 1992*). We observed that exposure of cultured porcine pulmonary artery endothelial cells to the active site-directed ACE inhibitors lisinopril and captopril was associated with increased ACE mRNA levels and concomitant increases in ACE activity (Figure 7). These effects were both concentration and time dependent. ACE mRNA levels began to increase within 30 min of ACE inhibitor exposure and showed an early peak at 2 hrs and a higher, delayed peak at 48 hrs. ACE activity peaked at 24 hrs. Both ACE mRNA levels and activity were highest during incubation with 100 μM ACE inhibitor. Nuclear runoff assays carried out in nuclei isolated from first-passage porcine pulmonary artery endothelial cells indicated that 48 hrs of exposure to 100 μM of either captopril or lisinopril increased ACE gene transcription approximately threefold relative to a tubulin control (Figure 8), a level comparable to the increases in ACE mRNA levels and activity observed during ACE inhibitor exposure. These findings support the hypothesis that ACE gene transcription in endothelial cells is stimulated by active site-directed ACE inhibitors *in vitro*.

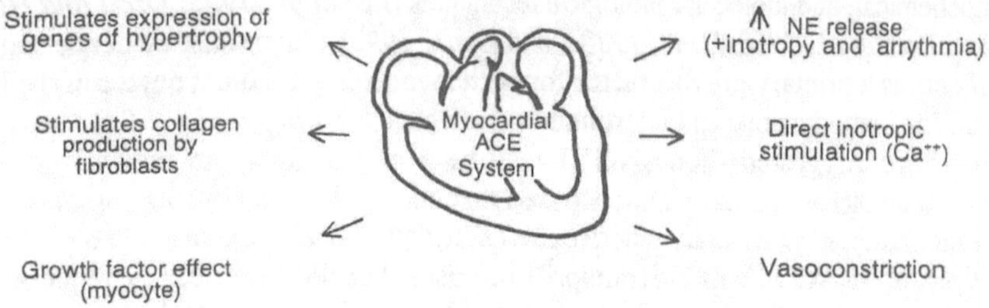

Figure 10. *Regional Wall Stress Activates Local Renin-Angiotensin System.* Schematic representation of the pathophysiologic consequences of activation of the intracardiac renin-angiotensin system by increases in refined wall stress.

This provides a molecular mechanism for the observations that plasma and tissue ACE activity and ANG II levels *in vivo* are increased toward or above pretreatment levels during chronic exposure to ACE inhibitors. Gavras and associates have reported that plasma ANG II concentration falls to very low levels after acute administration of enalapril but returns to normal during chronic ACE inhibitor treatment despite maintenance of the antihypertensive effect (*Biollaz et al., 1982*). Similarly, normal or increased stores of ANG II have been demonstrated in the brain of spontaneously hypertensive rats (SHR) treated with ACE inhibitors (*Phillips and Kimura, 1986*). This apparent overriding of ACE inhibition occurs in the presence of a persistent, stable antihypertensive effect of the ACE inhibitor. The persistence of the antihypertensive effects of the ACE inhibitors in the absence of sustained blockade of the enzyme may relate to the non-angiotensin-dependent actions of these drugs, such as potentiation of vasodilator kinins and prostaglandins (*Zusman, 1984*) or to an effect on vascular growth and remodeling that persists beyond the period of inhibition of ANG II production. Taken together with *in vitro* data from our laboratory, these findings suggest that chronic administration of ACE inhibitors *in vivo* induces enhanced ACE gene expression and production of catalytically active ACE, some of which escapes blockade by the inhibitor and gives rise to ANG II formation.

In separate studies, we have demonstrated significant increases in ACE mRNA levels and activity in hypothalamus-thalamus-septal (HTS)blocks of brain and in ACE mRNA levels in testis of SHR in response to chronic ACE inhibition with captopril (*King et al., 1991*) (Figure 9). These whole animal studies did not, however, identify the cell(s) of origin of the increased ACE during chronic ACE inhibitor treatment. To test the hypothesis that the increased ACE found in whole brain following ACE inhibitor treatment was of neuronal origin, we measured ACE mRNA levels and activity in neuronal-enriched cultures from 1-day-old SHR pups exposed to captopril throughout gestation. ACE mRNA levels were increased in cultures obtained from offspring of dams treated with captopril compared to cultures obtained from vehicle-treated controls. These increases in ACE mRNA persisted in neuronal cells after 12-14 days in culture without addition of ACE inhibitor to the media, suggesting that increases in brain ACE mRNA levels seen following chronic captopril treatment are at least partially of neuronal origin and can be sustained in the absence of continued ACE inhibitor exposure. Further study is needed to determine whether ACE gene expression in other cell types, including germinal cells of the testis, epithelial cells of the kidney proximal tubules, and gut epithelium, manifests similar sensitivity to regulation by ACE inhibitors.

HEART

All of the components of the renin-angiotensin system, including renin, angiotensinogen, ACE, ANG II, and ANG II receptors, have been demonstrated in heart by biochemical, immunohistochemical, and molecular biological techniques (*Lee et al., 1993*; *Dzau and Hirsch, 1990*; *Lindpaintner et al., 1988*; *Dzau, 1988a*; and *Dzau, 1993*). Increasing evidence suggests that ANG II acts as a primary growth factor for both myocytes and nonmyocytes in the heart, independent of its hemodynamic and neurohumoral effects (*Schelling et al., 1991*). Several *in vivo* studies have suggested that ANG II may be a critical factor in mediating cardiac hypertrophy. Khairallah *et al.*, have demonstrated that chronic infusion of subpressor doses of ANG II in rats causes ventricular hypertrophy (*Khairallah and Kanabus, 1983*). Further, administration of an ACE inhibitor (ramipril) in doses that do not affect systemic arterial pressure can reverse cardiac hypertrophy induced by banding the aorta in a rat model of pressure overload (*Linz et al., 1989*). These investigators also demonstrated the superiority of an ACE inhibitor over a calcium channel blocker and hydralazine in the regression of cardiac hypertrophy after normalization of blood pressure with each agent. These indirect data suggest that ANG II synthesized in the myocardium may be important in mediating cardiac hypertrophy.

Direct evidence in support of this concept has come from the work of Aceto *et al.*, who have shown that ANG II increases protein synthesis by chick myocytes *in vitro* (*Aceto and Baker, 1990*) and of Dostal *et al.*, who have demonstrated that both myocytes and fibroblasts from ventricles of neonatal rat hearts are capable of synthesizing ANG I and II and ACE (*Dostal et al., 1990 and 1992*). In a rat model of cardiac hypertrophy induced by abdominal aortic constriction, Baker *et al.*, demonstrated an increase in plasma renin and ANG I within the first three days following aortic constriction, but these levels returned to control values as cardiac hypertrophy developed over the ensuing 7-15 days (*Baker et al., 1990*). In contrast, aortic constriction resulted in a significant increase in steady state angiotensinogen mRNA levels in the left ventricle that was sustained for up to 15 days after surgery. Subpressor doses of enalapril completely prevented the increase in left ventricular mass in abdominal aortic-constricted rats even though carotid artery pressures were not different in conscious, awake aortic-constricted animals receiving and not receiving enalapril. These data suggest a direct growth effect of ANG II on the left ventricle as a result of increased local production because circulating plasma renin activity and thus, by inference, circulating ANG II levels, were not different from control values at three days after surgery. These experiments did not identify the specific cell type(s) that were responsible for production of angiotensinogen mRNA, i.e. myocytes, endothelial cells, or fibroblasts, however.

In contrast, subpressor doses of ACE inhibitors did not promote regression of left ventricular hypertrophy induced by placement of an aortic constrictor close to the aortic valve of the rat heart (*Zierhut et al., 1991*). Proximal aortic constriction produces a greater hemodynamic load than descending aortic constriction providing such a potent mechanical stimulus for hypertrophy that regression could not be achieved by ACE inhibition alone. Left ventricular hypertrophy produced by chronic experimental aortic stenosis in the rat has been associated with increased intracardiac ACE activity and mRNA expression, however (*Schunkert et al., 1990*). Intracardiac conversion of ANG I to ANG II was increased two-fold in hypertrophied hearts compared to controls, and ACE activity and mRNA levels were increased four-fold in left ventricular tissue from hypertrophied hearts compared to controls. Evaluation of myocardial function in these isolated hearts demonstrated that the higher ANG II levels in the hypertrophied hearts retarded diastolic relaxation compared to control hearts. In a rat model of myocardial infarction, ACE activity and steady state ACE mRNA levels were increased in noninfarcted myocardium and levels correlated positively with infarct size (*Hirsch et al., 1991*). The increased ACE levels were presumably induced by higher wall stress on noninfarcted myocardium imposed by larger areas of infarction.

Although few studies relate intracardiac ANG II to functional parameters, there is a large body of clinical data demonstrating the beneficial effects of ACE inhibitors in reducing mortality, rate of recurrent myocardial infarction and subsequent development of heart failure in patients after myocardial infarction (*Acute Infarction Ramipril Efficacy (AIRE) Study Investigators, 1993*; *Pfeffer et al., 1992*) and in patients with chronic congestive heart failure

(*CONSENSUS Trial Study Group, 1987*; *SOLVD Investigators, 1991*; *SOLVD Investigators, 1992*). These beneficial effects of the ACE inhibitors did not appear to be accounted for by their antihypertensive action (*Yusuf et al., 1992*). This clinical information, coupled with the evidence for the existence of a functional intracardiac renin-angiotensin system in the rat, has led to the hypothesis that the intracardiac renin-angiotensin system is activated locally in response to increases in regional wall stress (Figure 10). In contrast, the circulating renin-angiotensin system remains unaltered in patients with asymptomatic left ventricular dysfunction. The continued presence of intracardiac ANG II in these patients could have important pathophysiologic consequences, i.e., fibrosis, vasoconstriction of the small coronary arteries, arrhythmogenesis, and diastolic dysfunction mediated by effects on Ca^{++} homeostasis (Figure 10). However, the failing heart may also be dependent on local ANG II production to provide inotropic support (*Freer et al., 1976*; *Meulemans et al., 1990*; *Moravec et al., 1990*) and to promote myocyte hypertrophy in order to minimize wall stress.

Early treatment of myocardial infarction in rats resulted in greater left ventricular dilatation and lower mass (increased wall thinning) compared to rats treated greater than one month after infarction (*Schoemaker et al., 1991*). In humans, early therapy of myocardial infarction using intravenous enalaprilat within 24 hours resulted in a slightly greater mortality than in untreated patients (*CONSENSUS II Investigators, 1992*). Although this was attributed by the investigators to a higher incidence of hypotension with acute therapy, an alternative explanation is that important ANG II mediated effects on myocyte hypertrophy may result in larger left ventricular volumes and higher wall stress. The question of optimal timing of ACE inhibitor treatment following myocardial infarction is currently being addressed in the HEART trial, in which patients are being randomized to early vs. late therapy with an ACE inhibitor.

Further study is needed to assess the role of the intracardiac renin-angiotensin system in the pathogenesis of heart failure since there are major differences between species in the level of expression of the renin-angiotensin system in the heart, e.g., ACE activity and steady state ACE mRNA levels are much lower in human heart and in the hearts of many larger mammals than in the rat. Few systematic studies of expression of the components of the renin-angiotensin system in hearts of larger mammals, including humans, have been carried out, and the level of responsiveness of this system to various forms of hemodynamic stress in larger mammals has not been well studied.

Evidence obtained using the quantitative polymerase chain reaction (PCR) technique has demonstrated a two-fold higher steady state level of ACE mRNA in patients with end-stage heart failure than in potential heart donors whose (presumably normal) hearts were not used for transplantation (*Studer et al., 1994*). It was postulated that enhanced diastolic and/or systolic wall stress is the mechanism triggering ACE expression. However, the variability of expression of ACE mRNA in donor and failing hearts was very large, perhaps related to problems in sampling, the intrinsic limitations of the quantitative PCR technique, and the low levels of expression of ACE mRNA in human heart (*Studer et al., 1994*; *Urata et al., 1990a*). Further, ACE mRNA and activity levels have not been related to levels of ANG I and ANG II in human hearts. In addition, *in vitro* studies in human heart tissue have demonstrated that captopril inhibited ANG II formation by only 11±4% in normal hearts, by 11±3% in idiopathic cardiomyopathic hearts, and by 4±3% in ischemic cardiomyopathic hearts, suggesting that ACE is not the major ANG II-forming enzyme in the human heart (*Urata et al., 1990a*). Preliminary data from our own laboratory have demonstrated that ANG I and ANG II levels were elevated two- to four-fold in hearts from dogs with experimentally induced mitral regurgitation and that these levels were directly related to diastolic wall stress (*Dell'Italia et al., 1994*). The findings of low levels of ACE mRNA expression and catalytic activity in the presence of elevated ANG II stores in the normal dog heart and of major increases in ANG II stores in the absence of concomitant activation of ACE in hearts of dogs with experimentally induced mitral regurgitation suggests the presence of an alternative ANG II-generating pathway that is

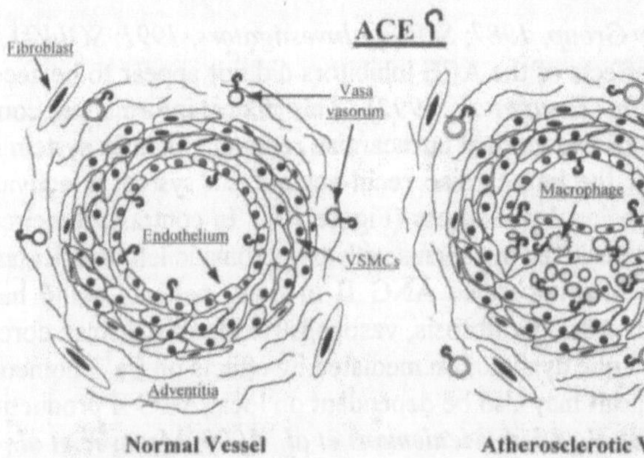

Figure 11. LEFT: Schematic representation of the cross section of a normal artery depicting localization of ACE in endothelium and vasa vasorum. VSMC- vascular smooth muscle cells. RIGHT: Schematic representation of the cross section of an artery with a developing atheroma depicting localization of ACE in various cell types, including macrophages and smooth muscle cells, as well as endothelium and vasa vasorum.

activated by the stimulus of volume overload. Current efforts in the laboratory are focused on characterizing this alternate pathway.

One candidate for this alternative intracardiac ANG II-generating pathway is the "chymostatin-sensitive angiotensin generating enzyme" (heart chymase) that has been isolated from human heart, purified and characterized as a neutral serine protease (*Husain, 1993*; *Urata et al., 1990b*). This human cardiac enzyme has high specificity for conversion of ANG I to ANG II (*Kinoshita et al., 1991*). The heart chymase gene has been cloned and sequenced (*Urata et al., 1991*). Its genomic structure indicates homology to the chymase enzyme family and it is considered to be a mast cell-specific enzyme. This "human heart chymase" (hHC) represents approximately 90% of the ANG II-forming capacity in tissue extracts from human myocardium (*Urata et al., 1990b*). Chymase is insensitive to ACE inhibition and does not degrade bradykinin. In situ hybridization in the human heart demonstrates that chymase mRNA is expressed in endothelial cells and in interstitial cells, including mast cells (*Urata et al., 1993*). Electron microscope immunocytochemistry using an anti-human chymase antibody demonstrates chymase-like immunoreactivity in the cardiac interstitium and in cytosolic granules of mast cells, endothelial cells, and some mesenchymal cells. Tissue chymase levels are approximately two-fold higher in ventricles than in atria, whereas ACE levels are higher in the atria than in the ventricles.

Most of the studies of the functional role of the intracardiac renin-angiotensin system and its sensitivity to ACE inhibition have been performed in the rat, a species in which an ANG II-forming enzyme analogous to hHC appears not to exist. Thus, caution must be used in extrapolating from experimental data in the rat to the clinical situation. Our preliminary findings in the normal dog and in the dog with chronic mitral regurgitation suggest that intracardiac ANG II forming pathways are very similar to those recently described in the human heart. The availability of a large animal model of volume overload hypertrophy and heart failure will make possible studies of the role of intracardiac ANG II in heart disease that are impossible both in human subjects because of the difficulty in quantitating hemodynamic forces and in obtaining tissue for assay and in rodents because of the small size of the heart and the questionable relevance of findings in the rodent to the human. Planned studies in the dog mitral regurgitation model will define the molecular mechanisms of ANG II production in the normal and diseased dog heart, will relate intracardiac ANG II production and expression of the intracardiac renin-angiotensin system and heart chymase to clinical markers of heart failure, regional wall stress (and thereby state of compensation) and ultrastructural markers of hypertrophy and decompensation. Further studies will evaluate the hemodynamic and molecular responses of the volume overloaded heart to therapy with ACE inhibitors and ANG II antagonists.

THE VASCULATURE

Gene transcripts of the components of the renin-angiotensin system and their protein products are expressed differentially in the various layers of the arterial wall. ACE is expressed primarily in endothelium at the luminal surface and in the vasa vasorum of large vessels (Figure 11) (*Ryan et al., 1976*; *Soubrier et al., 1988*; *Shiota et al., 1992*); angiotensinogen, in the medial and adventitial layers and the surrounding fat (*Campbell, 1987*; *Cassis et al., 1988*; *Naftilan et al., 1991*); renin, in all three layers, and the ANG II receptors, primarily in the medial layer. The autocrine/paracrine function of the vascular renin-angiotensin system has been demonstrated by the observation of local ANG I and II synthesis and release *in vivo* and *in vitro* (*Oliver and Sciacca, 1984*; *Kifor and Dzau, 1987*; *Mizuno et al., 1988*). Locally synthesized ANG II may be an important determinant of vascular tone and blood pressure by a direct vasoconstrictor effect (*Oliver and Sciacca, 1984*) and by enhancing sympathetic noradrenergic nerve transmission (*Malik and Nasjletti, 1976*; *Zimmerman, 1981*). ANG II is also an important determinant of vascular structure via mitogenic and trophic actions on vascular smooth muscle cells that are mediated by stimulation of platelet-derived growth factor (PDGF) A-chain and the proto-oncogenes, c-myc and c-fos (*Geisterfer et al., 1988*; *Naftilan et al., 1989a*; *Taubman et al., 1989*). In addition, renin-like activity and immunoreactive ANG I and II have been detected in the plasma of anephric subjects, suggesting that renin or renin-like enzymes synthesized in extra-renal tissues may contribute to the generation of circulating ANG II (*Catt et al., 1967*; *Weinberger et al., 1977*; *Sealey et al., 1977*).

Whether the ANG II generated in the vasculature is the product of locally synthesized renin and angiotensinogen or of renin taken up from the systemic circulation acting on locally synthesized angiotensinogen is a topic of continued debate (*Cassis et al., 1988*; *Gohlke et al., 1992*). While renin mRNA has been detected in blood vessel walls and in cultured endothelial cells (*Lilly et al., 1985*; *Samani et al., 1988*), immunoreactive renin in the aorta has been shown to fall to unmeasurable levels following bilateral nephrectomy (*Thurston et al., 1979*) and to increase following renin infusion (*Loudon et al., 1983*). In both of these situations, blood pressure responses to inhibitors of the renin-angiotensin system correlated with vascular renin rather than plasma renin. Infused renin was taken up predominantly into the media, an optimal site for locally generated ANG II to produce vasoconstriction (*Swales et al., 1983*). These observations, plus the finding that levels of mRNA in aorta are much lower than those in kidney and extrarenal tissues such as brain, liver and adrenal, are compatible with the interpretation that most of the vascular renin is taken up from the circulation rather than synthesized locally.

Whether ANG I generated in the vasculature is converted to ANG II by endothelial or extra endothelial ACE or via an alternate pathway such as chymostatin-sensitive ANG II generating enzyme or sequential action of carboxypeptidases is a second point of controversy (*Okunishi et al., 1984*; *Gohlke et al., 1992*). A chymostatin-sensitive, captopril-resistant ANG II generating enzyme has been described in the adventitia of dog, monkey and human aorta (*Okunishi et al., 1987*), but the relative importance of this enzyme vs. ACE in the generation of vascular ANG II *in vivo* is uncertain and may be species dependent. Recent studies in an isolated rabbit aorta preparation have shown that conversion of exogenous radiolabeled ANG I to ANG II is dependent on an intact endothelium and is completely blocked by ACE inhibitors, giving strong evidence that endothelial ACE is the most important site of vascular ANG II generation and an important target site for ACE inhibitors (*Luckhoff et al., 1985*; *Gohlke et al., 1992*). In contrast, de-endothelialized aortic rings derived from rabbit and rat contract in response to addition of ANG I to the organ bath, indicating that extra-endothelial generation of ANG II does take place (*Saye et al., 1984*; *Gohlke et al., 1992*). Further, as much as 70% of total ACE activity has been shown to remain in rat aorta following removal of the endothelium (*Pipili et al., 1989*). Further investigation using in situ hybridization and immunocytochemistry to monitor cellular levels of mRNA for renin and angiotensinogen and techniques for measuring ANG II concentrations at the local receptor level are needed to resolve this controversy and to establish the functional and pathophysiological significance of the vascular renin-angiotensin system.

The pathophysiological significance of the vascular renin-angiotensin system in the development and maintenance of hypertension is a topic of intense interest and controversy. The observations that ACE inhibitors and ANG II receptor antagonists can lower blood pressure in hypertensive patients or animal models with normal or low plasma renin activity and even in anephric subjects have been adduced as evidence for a functional role of the vascular renin-angiotensin system in hypertension (*Riegger et al., 1977*; *Dzau, 1988b*). The duration and magnitude of blood pressure reduction correlate better with inhibition of ACE activity in aorta and kidney than with inhibition of serum ACE activity (*Cohen and Kurz, 1982*; *Unger et al., 1984&1985*). For example, 24 hours after withdrawal of long-term ACE inhibitor therapy, when serum ACE activity and the pressor response to exogenous ANG I have returned to normal, blood pressure and tissue ACE activity remain depressed, suggesting the persistence of ACE inhibition in tissues. Further, studies in humans have shown that captopril, which has a short serum half-life, is an effective antihypertensive agent when given once daily (*Garanin, 1984*). Thus, the principal mechanism of the chronic antihypertensive action of this class of drugs appears to be inhibition of ACE in the blood vessels and perhaps in other tissues (e.g., heart, brain and kidney). The major contribution of the circulating renin-angiotensin system to the depressor effect of ACE inhibition is the short-term, rather than the long-term, response (Figure 2). During long-term therapy, the magnitude and duration of action of an ACE inhibitor appears to be determined by its concentration in tissue, and the serum half-life of the ACE inhibitor likely has little bearing on the actual duration of the pharmacologic effect of the drug (*Dzau, 1988b*).

Direct measurements of components of the renin-angiotensin system in the vasculature have given additional evidence for activation of the system in hypertensive subjects. Studies in hypertensive models, including two-kidney, one clip (2K1C) rats (*Okamura et al., 1986*) and dogs (*Miyazaki et al., 1987*), one-kidney, one clip (1K1C) rats (*Miyazaki et al., 1988*), and SHR (*Okunishi et al., 1991*), have demonstrated consistently elevated vascular ACE activity in the presence of normal or suppressed plasma renin and ACE activities. Elevated vascular tissue ACE activity is a common feature of all of these models despite major differences in the pathogenesis of the various forms of hypertension, suggesting that increased ACE in the blood vessels sustains the hypertension via enhanced local generation of ANG II. Further, depressor responses to ACE inhibitors and ANG II antagonists in these models are linearly related to the increases in vascular ACE activity (*Okunishi et al., 1991*), and the development of hypertension in the SHR coincides with an increase in aortic ACE activity confirming the presence of a relationship between vascular ACE activity and the pathogenesis of hypertension.

More recently, elevations in steady state aortic ACE mRNA levels have been described in 2K1C hypertensive rats (*Shiota et al., 1992*) and SHR (*Sun et al., 1993*). In the 2K1C model, aortic ACE activity increased in parallel with aortic ACE mRNA levels and blood pressure, but no significant changes occurred in lung ACE mRNA or in plasma ACE activity. Further, mechanical decompression of the aorta in 2K1C rats and SHR in the chronic stages of hypertension caused aortic ACE activity to return to prehypertensive levels (*Shiota et al., 1992*). The stimulus for the selective increase in ACE gene expression in vascular tissues and heart may relate to increased wall tension, as previously described in the hypertrophied ventricles of rats with experimentally induced heart failure (*Hirsch et al., 1991*) and aortic stenosis (*Schunkert et al., 1990*). Hypertension (and therefore increased wall tension) can increase mRNA levels of the platelet-derived growth factor β receptor (*Sarzani et al., 1991*) and fibronectin (*Takasaki et al., 1990*) in aorta, and fibroblast growth factor can stimulate growth and induce ACE activity in endothelial cells (*Okabe et al., 1988*). Thus, growth factors or vasoactive agents may mediate the hypertension-induced ACE gene expression in the vascular wall (*Shiota et al., 1992*). Whether the hypertension-induced increase in ACE mRNA levels is related to increased gene transcription or increased stability of the mRNA remains to be determined. Further, aortic angiotensinogen and renal renin mRNA levels and aortic and plasma renin activity are increased early in the development of 2K1C hypertension but return to normal in the established phase of hypertension, suggesting that enhanced ANG II generation in the vascular wall plays a role in the development but not the maintenance of this form of hypertension (*Shiota et al., 1992*).

The hypothesis that ACE gene expression may play a role in the pathogenesis of hypertension was supported by linkage studies performed in crosses between stroke-prone SHR (SHRSP) and Wistar-Kyoto rats from Heidelberg. Three groups of investigators identified a gene locus which cosegregated with hypertension and localized the ACE gene to the region containing this locus (*Hilbert et al., 1991*; *Jacob et al., 1991*; *Nara et al., 1991*). Although other genes mapping to this locus could also explain these findings, ACE is an attractive candidate. A recent study tested the hypothesis that vascular ACE is increased in blood vessels of SHRSP by measuring the conversion of ANG I to ANG II in perfused isolated hind limbs from SHRSP and WKY from the same colony that was used for the linkage studies (*Hilgers et al., 1993*). Conversion rates of ANG I to ANG II did not differ between SHRSP and WKY, and captopril completely prevented the formation of ANG II in both strains. Extraction of ANG I from the perfusate was significantly decreased in SHRSP compared to WKY. Thus, the metabolism of ANG is decreased in SHRSP but vascular ACE does not appear to be responsible for high blood pressure in this rodent model of hypertension. These findings suggest that other genes close to the ACE locus or the increased expression of ACE in extravascular loci may contribute to hypertension in these rats. The dissociation between vascular ACE gene expression in the SHRSP and hypertension is consistent with the observed absence of linkage between the ACE gene and increased blood pressure in humans (*Jeunemaitre et al., 1992*; *Cambien et al., 1992*; *Harrap et al., 1993*).

ACE and its products, ANG II and bradykinin, have been implicated in the response to vascular injury, inflammation and atheroma formation (Figure 11). Evidence that ANG II and/or bradykinin participates in the proliferative response of the vascular wall to arterial injury first came from the observation that inhibition of ACE reduces the intimal thickening that occurs in carotid arteries of rats subjected to endothelial denudation and injury by balloon catheterization (*Powell et al., 1989*). ACE inhibitor administration was subsequently shown to reduce intimal hyperplasia in a rabbit vein graft model (*O'Donohoe et al., 1991*). Increased ACE activity and mRNA levels were later described in areas of injury-induced intimal hyperplasia, apparently in proliferating smooth muscle cells in the carotid injury model (*Dzau and Pratt, 1992*; *Johnston, 1994*). Similar increases in ACE activity were observed in vein grafts (*O'Donohoe et al., 1994*). Injury of the rabbit carotid artery by placement of a flexible, non-occlusive collar causes ACE expression to shift from the intima and adventitia to the extra-adventitial inflammatory tissue (*Zambetis-Bellesis, 1991*). ANG II receptor binding is present throughout the entire media and in extra-adventitial inflammatory tissue in injured vessels but is confined to the outer media in controls. The observations that both AT_1 and AT_2 receptor antagonists reduce intimal proliferation after balloon injury suggest that ANG II may be one of the growth factors responsible for the development of intimal hyperplasia following vascular injury, at least in rodent models (*Kauffman et al., 1991; Janiak et al., 1992*). The importance of ANG II in mediating vascular responses to injury appears to be species dependent, since ACE inhibitors are ineffective in preventing this response in monkeys (*Hanson et al., 1991*) and in preventing restenosis after balloon angioplasty in humans (*MERCATOR Study Group, 1992*).

This may relate to the higher concentrations of ACE in blood vessels of the rat compared to humans and other primates (*Johnston et al., 1994*).

The only evidence to date supporting a relationship between ACE and vascular pathology in humans is the observation from the EVA Study on vascular aging that plasma ACE activity is positively correlated with carotid artery thickness in elderly subjects (*Bonithon-Kopp, 1994*). This relationship was significant only when subjects taking lipid lowering drugs and with antecedent atheromatous plaques were excluded. Further study is needed to rigorously test the hypothesis that chronic exposure to high levels of plasma ACE results in vascular wall thickening and to assess the role of ACE in vascular aging.

The appearance of ACE and ANG II binding sites in extra-adventitial tissue of injured vessels has been attributed to infiltration by monocytes and macrophages, which express ACE activity (*Rohrbach, 1984*) and ANG II binding sites (*Shimada and Yazaki, 1978*; *Weinstock and Kassab, 1984*). Further, loss of ACE from intimal and adventitial sites in carotid arteries with developing atheroma-like lesions has been interpreted as a consequence of changes in endothelial cells and fibroblasts due to cell proliferation and tissue remodeling, while the appearance of ACE in abnormal sites (extra-adventitial inflammatory tissue) could contribute to ANG II production in the vessel wall. These findings and the observation that ANG II can stimulate both growth and matrix elaboration of vascular smooth muscle cells (*Naftilan et al., 1989a&b*; *Scott-Burden et al., 1990*; *Powell et al., 1990)* suggest that the vascular renin-angiotensin system plays a role in the development of atheroma-like lesions.

Atheroma formation is, at least in part, an inflammatory process that includes neointimal hyperplasia in its pathogenesis. ACE inhibitors can prevent atheroma formation in a variety of animal models, including Watanabe hyperlipidemic rabbits (*Chobanian et al., 1990*) and cholesterol fed monkeys and rabbits (*Aberg and Ferrer, 1990*; *Campbell et al., 1992*). This effect may be mediated by inhibition of macrophage ACE and thus inhibition of inflammation (*Johnston et al., 1994*). Precedents exist for such actions of ACE inhibitors in other systems, as captopril has been shown to reduce the inflammatory response in experimental granuloma formation (response to BCG vaccination) (*Schrier et al., 1982*). Further, ACE has been implicated in the processing of antigens by major histocompatibility complex class I-restricted T lymphocytes (*Eisenlohr et al., 1992*). Vasoprotective/antiatherosclerotic properties have not yet been demonstrated for ACE inhibitors in humans, but appropriate clinical trials have not been carried out.

The endothelial dysfunction that is associated with hypertension and atheroma formation has been related to altered function of vascular ACE. In hypertension and atheroma, the response to endothelium-dependent vasodilators is blunted (*Lockette et al., 1986*; *Lundmer et al., 1986*), suggesting endothelial dysfunction; treatment with ACE inhibitors restores these responses (*Mombouli et al., 1991*; *Clozel et al., 1990*). The mechanism of this effect may involve bradykinin, which can stimulate the release of both vasodilator prostaglandins and nitric oxide from endothelial cells (*Boulanger et al., 1990*). ACE inhibitors, by inhibiting bradykinin breakdown, restore the endothelial vasodilator response.

Evidence implicating bradykinin in the vascular effects of the ACE inhibitors includes the observations that: (1) concomitant administration of the B_2 kinin receptor antagonist HOE 140 (icatibant) attenuates the antiproliferative effects of the ACE inhibitor ramipril in the rat balloon injury model (*Farhy et al., 1992*); (2) HOE 140 attenuates the hypotensive affects of ACE inhibitors in rats with experimental hypertension (*Cachefeiro et al., 1992*); and (3) HOE 140 suppresses cGMP formation and abolishes the increase in PGI_2 release observed following treatment of cultured endothelial cells with ACE inhibitors (*Wiemer et al., 1991*). Further study is needed to evaluate the importance of bradykinin potentiation in the long term antihypertensive effects of the ACE inhibitors in humans.

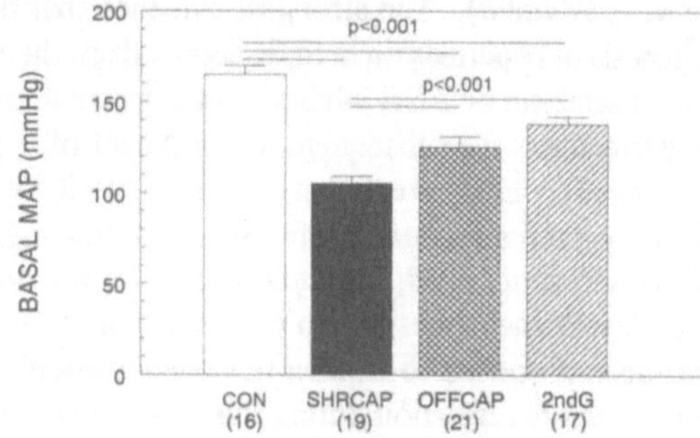

Figure 12. Bar graph shows basal mean arterial pressure (MAP) of four rat groups at 4 to 6 months of age. Basal MAP was monitored for 4 consecutive days in conscious, freely moving rats. Daily MAP values were averaged for each rat. Data are mean ± SEM analyzed by analysis of variance with Bonferroni's posttest procedure. Number of animals is in parentheses. CON, control rats; SHRCAP, spontaneously hypertensive rats that receive captopril; OFFCAP, SHRCAP taken off captopril at 2 months; 2nd G, first offspring of OFFCAP rats. Reproduced with permission from WU and Berecek, *Hypertension* 22:139-146, 1993.

BRAIN

There is strong evidence for an intrinsic renin-angiotensin system and local ANG II generation in brain (*Harding et al., 1988*). Components of the renin-angiotensin system including mRNA for angiotensinogen, ACE, renin, and ANG II receptors have been localized in brain (*Raizada et al., 1993*). ANG II-immunoreactive cell bodies and fibers, as well as specific ANG II receptors, have been identified in areas of the brain that participate in cardiovascular regulation including the circumventricular organs, paraventricular nucleus of the hypothalamus, locus coeruleus, nuclear tractus solitarius (NTS), area postrema and medulla (*Lind et al., 1985*). The presence of components of the renin-angiotensin system in neuroblastoma cells (*Fishman et al., 1981*) and primary neuronal enriched cell cultures (*Raizada et al., 1984a&b*) and the finding that these cells synthesize immunoprecipitable ANG II add further support to the concept of local synthesis of ANG II in brain.

Brain ANG II participates in blood pressure and fluid and electrolyte regulation through mechanisms distinct from those in the periphery. These mechanisms involve enhancement of sympathetic outflow, blunting of baroreceptor reflex gain, stimulation of AVP and ACTH release and increased catecholamine biosynthesis and turnover (*Ganten et al., 1978*). In addition, brain ANG II elicits two behaviors, thirst and salt appetite, which are critical in body fluid and electrolyte regulation (*Epstein et al., 1970*). ANG II acts as a neurotransmitter in the CNS, and specific, high affinity ANG II receptors are found on nerve terminals (synaptosomes) in brain *(Bradford, 1986)*. Iontophoresis of ANG II enhances the rate of discharge of 75% of the neurons in the subfornical organ: this effect is dose dependent and antagonized by saralasin. ANG II also activates neurons of the supraoptic and paraventricular nuclei, areas that synthesize and release AVP (*Sladek et al., 1987*).

Brain ANG II plays a major role in the pathogenesis of hypertension in the SHR. SHR show increased sympathetic tone and increased AVP and ACTH secretion compared to normotensive WKY controls (*Dietz et al., 1978*). SHR also show increased levels of ANG II-like material in brain and cerebrospinal fluid, increased turnover of ANG II in brain *(Hermann et al., 1984)* and increased pressor and neuronal responsiveness to central administration of ANG II (*Casto and Phillips, 1985*). Comparison of primary neuronal enriched cultures from newborn SHR and WKY has revealed differences in ANG II stores, metabolism and binding between strains (*Raizada et al., 1984a&b*). The strongest evidence that brain ANG II plays a functional role in the pathogenesis of hypertension in SHR comes from the findings that acute or chronic ICV administration of saralasin or ACE inhibitors in doses that are ineffective when given systemically significantly attenuates the subsequent development of hypertension (*Unger et al., 1981*; *Berecek et al., 1983*). Early treatment of young SHR with ACE inhibitors prevents the full expression of hypertension later in life even after treatment is stopped, thus "curing" the hypertension (*Giudicelli et al., 1980*; *Harrap et al., 1990*; *Wu and Berecek, 1993*).

A possible critical phase of development sensitive to pharmacological interference with the renin-angiotensin system has been proposed to explain this phenomenon (*Unger and Rettig, 1990*). Interference with blood pressure regulation during this phase would alter the long-term course of the disease. Berecek *et al.* have treated mating pairs of SHR with captopril and maintained their pups on captopril until experimentation. Captopril treatment "cured" the hypertension in these rats: animals treated with captopril throughout fetal and neonatal life had normal blood pressures, while animals that had been removed from captopril and the offspring of treated rats had significantly lower blood pressures than sham treated controls (Figure 11). The mechanisms underlying the chronic antihypertensive effect of early treatment of SHR with ACE inhibitors are not known with certainty. The observation that the antihypertensive effect was associated with a decrease in CNS responses to ANG I and II suggests that captopril may permanently alter the development of hypertension in SHR through an alteration in the brain renin-angiotensin system. Other possible mechanisms include the remodeling of cardiac and blood vessel structure, an alteration in factors in the pregnant female SHR which may contribute to the full expression of hypertension in their offspring, or a down-regulation of ANG II receptors in the brain.

Regulation of ANG II receptors in SHR is abnormal in a number of tissues including brain (*Raizada et al., 1984a&b*; *Wilson et al., 1988*). ACE inhibitor treatment causes down regulation of ANG II receptors in brain and primary neuronal cell cultures from SHR but not from normotensive rat strains (*Wilson et al., 1988*; *Berecek et al., 1992*). In contrast, there was an increase in ANG II binding to vascular smooth muscle cells associated with increased vascular reactivity to ANG II in ACE inhibitor treated SHR. These data are consistent with previous reports that brain ANG II receptors are regulated differently from those of vascular smooth muscle cells and that ANG II receptors in SHR are regulated differently from those in WKY rats (*Catt et al., 1979*; *Mendelsohn et al., 1984*; *Raizada et al., 1984a&b*). Functional evidence in support of this hypothesis was provided by Schelling and Felix, who reported an increase in neuronal firing rate in response to microiontophoresis of ANG II in SHR compared to WKY rats (*Schelling and Felix, 1983*). The increase in septal ANG II receptor sensitivity in SHRs was abolished when the rats were placed on oral captopril treatment in the early weaning period. Therefore, captopril appears to cross the blood brain barrier and decrease responsiveness to ANG II by altering ANG II receptor number or affinity. Recent studies from our own laboratory indicate that endogenous ANG II in the anterior hypothalamic area of salt sensitive SHR (SHR-S) is tonically active in the maintenance of hypertension and suggest that upregulation of AT_1 receptors is involved in this process (*Yang et al., 1992a&b*). Blockade of AT_1 receptors in anterior hypothalamic area by local microinjection of losartan (2-*n*-butyl-4-chloro-5(hydroxymethyl)-1-[2'-(1H-tetrazol-5-yl)biphenyl-4-yl)methyl]imidazole, potassium salt), a highly selective nonpeptide antagonist, caused significant dose-related decreases in mean arterial pressure with unchanged heart rate in SHR-S but not in WKY rats (Figure 12). The depressor response was enhanced by dietary NaCl supplementation. Further, microinjection of ANG II into the anterior hypothalamic area of SHR-S caused a pressor response that was enhanced by dietary NaCl supplementation and blocked by pretreatment with the losartan metabolite EXP 3174. Microinjection of the selective type 2 angiotensin II (AT_2) receptor antagonist PD 123319 into the anterior hypothalamic area did not affect blood pressure or heart rate in SHR-S. These data demonstrate that endogenous ANG II in the anterior hypothalamic area participates in the tonic control of blood pressure in SHR and that this effect is mediated by AT_1 receptors. These findings further suggest that endogenous ANG II and AT_1 receptors in the anterior hypothalamic area are involved in the pathogenesis of NaCl-sensitive hypertension in this model.

The renin-angiotensin system in brain has also been implicated in the regulation of cognitive function (*Barnes et al., 1990*). ANG II directly inhibits release of acetylcholine from cerebral cortical neurons thus leading to impairment of learning and memory. In entorhinal cortex slices from both human and rat subjects, *in vitro* application of ANG II inhibits the potassium-stimulated release of acetylcholine from nerve terminals (*Barnes et al., 1992, Wyss and van Groen, 1994*). This inhibitory effect of ANG II is blocked by pretreatment of the slice with an ANG II receptor antagonist. Thus, it is hypothesized that ANG II is a tonic inhibitor of acetylcholine release in the cerebral cortex. Further, expression of ACE in the brain has been linked to disorders of learning and memory. ACE activity in the human cerebral cortex has been shown to increase with age (*McGeer and Singh, 1992*) and ACE has been shown to be overexpressed in the brains of Alzheimer's patients (*Barnes et al., 1991a&b*). Binding studies using the ACE inhibitor [^{3}H] ceranapril suggest that ACE is elevated in the temporal lobe of the cerebral cortex, but not in the frontal cortex or the cerebellum of Alzheimer's patients compared to age-matched control subjects. Further, increased ACE activity has been demonstrated in the temporal pole of the cortex in patients with Alzheimer's disease compared to unaffected, age-matched control subjects (*Wyss et al., 1993*). It has been postulated that increased ACE activity and the resultant overproduction of ANG II may contribute to the pathogenesis of Alzheimer's disease, at least in part, by reducing the activity of cholinergic neurons in the temporal cortex. Accordingly, the hypothesis that ACE inhibition can improve cognitive function in Alzheimer's patients has been tested: the ACE inhibitor ceranapril (0, 5, or 40 mg twice daily) was administered in a preliminary study for 4 weeks to normotensive patients with Alzheimer's disease (*Sudilovsky et al., 1993*). ACE inhibition did not significantly alter the overall performance of these patients on cognitive test batteries, but a post hoc analysis demonstrated that the largest number of patients received their highest scores after treatment with the highest dose of ceranapril. Further study is needed to assess the usefulness of ACE inhibitors in the treatment of Alzheimer's disease.

ACE inhibitor treatment of hypertensive patients has also been shown in some studies to improve cognitive performance, particularly when compared to other antihypertensive agents. In a study of 30 middle-aged hypertensive patients, Blumenthal *et al.* showed that neither atenolol nor the ACE inhibitor enalapril had adverse effects on quality of life, but that enalapril improved performance on memory tests while atenolol impaired memory (*Blumenthal et al., 1990*). Further, both captopril and enalapril appear to facilitate the digit span tests compared to atenolol and propranolol (*Steiner et al., 1990*). Thus, of the various classes of antihypertensive drugs, ACE inhibitors are least likely to affect cognitive function adversely (*Bulpitt and Fletcher, 1992*). The effect of different ACE inhibitors on cognitive function in hypertension has been considered in a double blind study in which 479 male hypertensive patients (55 to 79 years old) were treated with either captopril or enalapril for 24 weeks (*Testa et al., 1993*). Patients on captopril reported significant improvement in quality of life compared to both their initial baseline and to the final reports of the patients who had received enalapril for 24 weeks. The authors concluded from these data that although these two ACE inhibitors are virtually indistinguishable in their cardiovascular effects, there are meaningful differences in their effects on cognitive function. Several investigators have questioned the interpretation of the results underscoring the need for better standardization of protocols that test cognitive ability and quality of life in individuals undergoing antihypertensive therapy.

Chronic hypertension can contribute to cerebrovascular disease, stroke and myocardial infarction and thus has significant deleterious effects on cognitive function, especially learning and memory (*Wyss and van Groen, 1994*). Hypertension per se may directly impair cognitive performance in human subjects and animal models (*Elias and Elias, 1993*). An age-related disorganization of the midline retrosplenial cortex that has the potential to cause disconnection

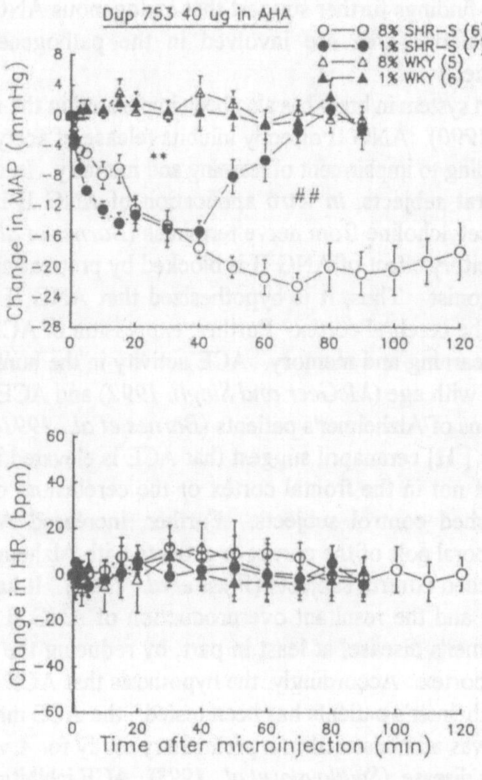

Figure 13. Effects of microinjection of losartan into the anterior hypothalamic area (AHA) on mean arterial pressure (MAP) and heart rate (HR) in NaCl-sensitive SHR (SHR-S) and WKY rats on 1% NaCl and 8% NaCl diets. **p less than 0.01, comparison between WKY and SHR-S on either diet 3 to 50 minutes after injection. ##p less than 0.01, comparison between 8% NaCl fed SHR-S and any of other three groups 50 to 90 minutes after injection. Reproduced with permission from Yang et al., *Hypertension* 20:755-762, 1992.

of the hippocampal formation (an area important for memory encoding) from other regions of the brain has been demonstrated in SHR (*Wyss and van Groen, 1992*). Normotensive rats manifest similar lesions, but much later in life. Coincident with the development of this anatomic lesion, mature SHR lose their ability to learn a radial arm maze test and retain the memory (Figure 13, Top) (*Wyss et al., 1992*). Lifetime captopril treatment prevents both the age-related learning and memory loss and the breakdown of the neurons in the retrosplenial cortex in mature SHRs (Figure 13, Bottom) (*Wyss and van Groen, 1992*; *Wyss et al., 1992*).

Age-related memory and learning dysfunction also occurs in human subjects. This deficit typically appears first in complex tasks that require the subject to use information from the environment, such as mapping the location of familiar objects in a spatial environment (*Wyss et al., 1992*). The age-dependent decline in cognitive function may be accelerated by the coexistence of hypertension in human subjects, as in the SHR. Deficits in cognitive function, particularly in the performance of tasks that require rapid cognitive processing, use of short-term memory and rapid reaction times, have been reported in hypertensive subjects compared to age-matched normotensive controls (*Blumenthal et al., 1993*). The relationship between hypertension and impaired cognitive function appears to be most pronounced in poorly educated persons, perhaps because the availability of alternate learning and memory strategies may mask losses of cognitive function in better educated individuals (*Perez-Stable et al., 1992*; *Wyss and van Groen, 1994*). To fully assess the effect of hypertension on cognitive function, both controlled and uncontrolled hypertensive patients must be studied in a longitudinal design with attention to the confounding variables of age, gender, level of education and genetic background.

KIDNEY

It is now generally accepted that there is an endogenous renin-angiotensin system in kidney and that ANG II generated within the kidney may play an important role in the regulation of renal hemodynamics and sodium excretion (*Abe et al., 1979*; *Mendelsohn et al., 1982*; *Campbell, 1987*; *Phillips et al., 1993*; *Rosivall et al., 1987*; *Cogan, 1990*). Expression of mRNA for renin, angiotensinogen and ACE has been demonstrated in kidney (*Campbell et al., 1986*; *Dzau et al., 1987*), and all components of the renin-angiotensin system have been identified in kidney using immunocytochemical techniques (*Taugner et al., 1984*; *Inagami et al., 1986*).

In the adult kidney, the components of the renin-angiotensin system are anatomically compartmentalized. Under normal conditions, renin is localized in the juxtaglomerular cells, modified smooth-muscle cells in the afferent arteriole. Renin-containing cells are also found in the glomeruli, particularly in the superficial glomeruli, and in the efferent arterioles. Angiotensinogen mRNA is localized predominantly in the proximal tubules, with lesser quantities in glomeruli, distal tubules, and intrarenal vessels (*Ingelfinger et al., 1990*). ACE is found in greatest concentrations in the proximal tubule, with smaller amounts in arterioles and glomeruli (*Marchetti et al., 1987*).

Both brush border and basolateral membranes of the proximal tubule contain ACE, but the highest concentrations are found on brush border (*Burns et al., 1993*). ACE activity is 10-fold higher in brush border membranes from inner cortex, compared to outer cortex, and immunofluorescence for ACE is greatest in the proximal tubule of the inner cortex.

The regulation of ACE activity and gene expression in the proximal tubule has not been studied, but whole rat kidney ACE activity has been shown to be significantly increased by dietary salt deprivation but not suppressed by a high salt diet (*Fox et al., 1992*). Renal ACE activity was reduced by only 58% by doses of enalaprilat that completely abolished plasma ACE activity, suggesting that renal tissue ACE is relatively resistant to the effects of systemically administered ACE inhibitors. Preliminary data from our own laboratory suggest that steady state mRNA levels for ACE are increased in both renal cortex and renal medulla in WKY rats on a high salt diet: no effect was seen in salt supplemented SHR (*Sun et al., 1993*). In contrast, dietary salt deprivation did not alter renal cortical or medullary ACE gene transcript levels in either strain. WKY had higher renal cortical ACE mRNA levels than salt sensitive SHR on high, basal and low salt intakes; there was no difference between strains in renal medullary ACE mRNA levels during basal salt intake. These data suggest that expression of the renal ACE gene is sensitive to dietary salt intake and that this gene is regulated differentially in hypertensive vs. normotensive subjects. Further study is needed to elucidate the molecular mechanisms by which dietary salt and other environmental factors modulate intrarenal ACE gene expression and the basis for the difference in responsiveness in hypertensive vs. normotensive subjects.

ANG II receptors, predominantly of the AT_1 subtype, are found in highest concentrations on the arterioles, glomeruli and vasa recta (*Yamada et al., 1990*). ANG II also binds to numerous nephron segments: the greatest concentration of tubule receptors is found in proximal tubule (*Mujais et al., 1986*). A renal paracrine system has been proposed in which renin is filtered in the glomerulus, where it cleaves angiotensinogen to form ANG I, which is converted to ANG II by ACE on the proximal tubule brush border membrane (*Paul et al., 1992*). The intrarenal ANG II thus generated then binds to AT_1 receptors in the locations described above to regulate renal hemodynamics and fluid and electrolyte handling.

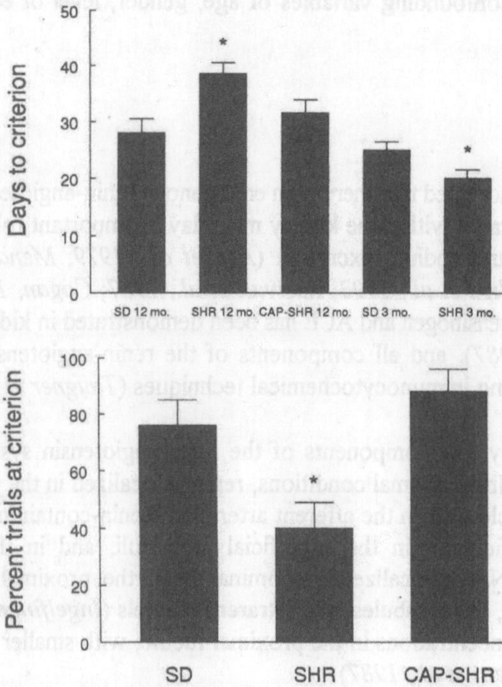

Figure 14. TOP: Number of days to reach the criterion of at least 5 correct responses out of the first 6 choices on 4 consecutive days in an 8 arm radial maze. BOTTOM: Percentage of trials on which the 12-month-old rats within each group met criterion in the re-test. The re-test was begun after a 14-day hiatus that started on the day each rat reached criterion in the original task. Bars represent means ± SEM for each group. *$p < 0.05$ compared to all other groups SD: Sprague-Dawley rats; CAP-SHR: lifetime captopril treated SHR. Reproduced with permission from Wyss et al., *Brain Research*, 592:135-140, 1992.

The relative importance of the intrarenal vs. systemic renin-angiotensin system in the regulation of renal function is a topic of intense investigation and controversy (*Navar et al., 1983*; *Burns et al., 1993*). Levels of ANG II in renal lymph (*Bailie et al., 1971*), proximal tubule fluid (*Seikaly et al., 1990*) and renal tissue (*Mendelsohn, 1982*) are higher than circulating levels indicating net intrarenal production, and kidney ACE plays an important role in the maintenance of intrarenal ANG II levels (*Campbell et al., 1991*). Numerous studies in experimental animals and in humans have documented alterations in glomerular hemodynamics and function after administration of ACE inhibitors, providing indirect evidence for the functional significance of the intrarenal renin-angiotensin system. These studies have suggested that intrarenal ANG II has effects on the glomerulus distinct from its effects on systemic blood pressure. In some cases, as in diabetes mellitus, ACE inhibitors lower intraglomerular pressure at times when neither plasma renin activity nor systemic blood pressure is elevated. However, because ACE inhibitors may have other effects, such as increasing bradykinin levels, determination of the role of intrarenal ANG II in regulating glomerular hemodynamics must await results of studies with specific ANG II receptor antagonists (*Burns et al., 1993*).

Several lines of evidence have suggested that enhanced activity of the renal renin-angiotensin system plays a role in the pathogenesis and maintenance of hypertension (*Rosenfeld and Zabludowski 1989*; *Pratt et al., 1989*; *Kost and Jackson, 1993*). The development of hypertension in the SHR is critically dependent on both a renal defect and a functional renin-angiotensin system, since (1) transplantation of kidneys from SHR donors to normotensive recipients causes hypertension in recipients (*Rettig et al., 1990 & 1991*) and (2) hypertension in the SHR can be prevented or reversed by interrupting the renin-angiotensin system with ACE inhibitors, AT_1 receptor antagonists, renin inhibitors or active immunization against renin (*Kost and Jackson, 1993*). This is true despite normal or suppressed circulating renin levels in SHR. The renal defect may involve enhanced responsiveness to ANG II, as evidenced by increased vasoconstriction and sodium retention in response to in situ autoperfusion with ANG II and an enhanced renal vasodilator response to an AT_1 antagonist in SHR (vs. WKY) (*Li and Jackson, 1989; Fenoy et al., 1991*). Further, intrarenal infusion of an AT_1 receptor antagonist in doses too small to affect blood pressure when administered intravenously lowers blood pressure in SHR (*Wood et al., 1992*), suggesting that endogenous ANG II in kidney contributes to the maintenance of elevated renal vascular resistance and blood pressure in SHR via tonic activation of the AT_1 receptor. The mechanism of enhanced renal responsiveness to ANG II in SHR does not appear to involve altered AT_1 receptor density or affinity, but may instead relate to a defect in vasodilator prostaglandins or a postreceptor signaling defect in SHR kidney (*Chatziantoniou and Arendshorst, 1991*; *Osanai and Dunn, 1992*).

Enhanced renal responsiveness to ANG II may contribute to the development of hypertension in human subjects as well. Low dose intravenous infusions of ANG II have been shown to reduce renal plasma flow and increase renal vascular resistance in normotensive young men with a positive family history of hypertension (presumably prehypertensive subjects) but not in normotensive controls with a negative family history of hypertension (*Widgren et al., 1992*). Further, acute administration of the ACE inhibitor captopril has been shown to produce a significantly greater reduction in renal vascular resistance in diabetics with a positive family history of hypertension than in those with a negative family history (*Hannedouche et al., 1992*). This differential responsiveness was not seen in response to a calcium channel blocker suggesting that the renal defect is selective for the renin-angiotensin system.

There is evidence for aberrant regulation of the intrarenal renin-angiotensin system in other forms of cardiovascular disease, including congestive heart failure (*Paul et al., 1992*). Intrarenal infusion of ACE inhibitors has been shown to normalize renal perfusion pressures and single nephron glomerular filtration rates in rats with chronic heart failure (*Ichikawa et al., 1984*). The observation that steady state intrarenal angiotensinogen levels are increased in rats with chronic heart failure (*Schunkert et al., 1989*) suggests that increased tissue-specific expression of the renin-angiotensin system in kidney may contribute to the pathogenesis of heart failure. Further study of the intrarenal renin-angiotensin system in various forms of cardiovascular disease is needed to enhance our understanding of the pathogenesis of these disorders and to develop new therapeutic strategies for dealing with them.

REFERENCES

Abe, Y., Okahara, T., Kishimoto, T. and Yamamoto, K., 1979, Intrarenal role of renin-angiotensin system in the regulation of renal hemodynamics. *Jpn. J. Pharmacol*, 29:325-333.

Aberg, G. and Ferrer, P., 1990, Effects of captopril on atherosclerosis in cynomolgus monkeys. *J.Cardiovasc. Pharmacol.* 15:S65-S72.

Aceto, J.F. and Baker, K.M., 1990, [Sar[1]]angiotensin II receptor-mediated stimulation of protein synthesis in chick heart cells. *Am. J. Physiol.* 258 (*Heart Circ. Physiol.* 27):H806-H813.

Acute Infarction Ramipril Efficacy (AIRE) Study Investigators, 1993, Effect of ramipril on mortality and morbidity of survivors of acute myocardial infarction with clinical evidence of heart failure. *Lancet* 342:821-828.

Bailie, M.D., Rector, F.C. and Seldin, D.W., 1971, Angiotensin II in arterial and renal venous plasma and renal lymph in the dog. *J. Clin. Invest.* 50:1119-1126.

Baker, K.M., Chernin, M.I., Wixon, S.K. and Aceto, J.F., 1990, Renin-angiotensin system involvement in pressure-overload cardiac hypertrophy in rats. *Am. J. Physiol.* 259 (*Heart Circ. Physiol.* 28):H324-H332.

Barnes, J.M., Barnes N.M., Costall, B., Coughlan, J., Kelly, M.E., Naylor, R.J., Tomkins, D.M. and Williams, T.J., 1992, Angiotensin-converting enzyme inhibition, angiotensin, and cognition. *J. Cardiovasc. Pharmacol.* 19:S63-S71.

Barnes, J.M., Barnes, N.M., Costall, B., Horovitz, Z.P., Ironside, J.W., Naylor, R.J. and Williams, T.J., 1990, Angiotensin II inhibits cortical cholinergic function: implications for cognition. *J. Cardiovasc. Pharmacol.* 16:234-238.

Barnes, N.M., Cheung, C.H.K., Costall, B., Naylor, R.J., Williams, T.J. and Wischik, C.M., 1991a, Angiotensin converting enzyme density is increased in temporal cortex from patients with Alzheimer's disease. *Eur. J. Pharmacol.* 200:289-292.

Barnes, N.M., Costall, B., Egli, P., Horovitz, Z.P., Ironside, J.W., Naylor, R.J. and Williams, T.J., 1991b, Characterization of [^{3}H] ceranapril recognition sites in rat and human brain tissue. *Neuropharmacology* 30:907-914.

Berecek, K.H., King, S.J., Wu, J-N., Oparil, S., Rydzewski, B. and Raizada, M., 1992, Effects of angiotensin II (AII) in the presence and absence of subtype-specific AII blockers on angiotensin converting enzyme (ACE) gene expression in central nervous system tissue. (Abstract). *FASEB J.* 6:A1578.

Berecek, K.H., Okuno, T., Nagahama, S. and Oparil, S., 1983, Altered vascular reactivity and baroreflex sensitivity induced by chronic central administration of captopril in the spontaneously hypertensive rat. *Hypertension* 5:689-700.

Biollaz, J., Brunner, H.R., Gavras, I., Waeber, B. and Gavras, H., 1982, Antihypertensive therapy with MK421: angiotensin II-renin relationships to evaluate efficacy of converting enzyme blockade. *J. Cardiovasc. Pharmacol.* 4:966-972.

Blumenthal, J.A., Ekelund, L.G. and Emery, C.F., 1990, Quality of life among hypertensive patients with a diuretic background who are taking atenolol and enalapril. *Clin. Pharmacol. Ther.* 48:447-454.

Blumenthal, J.A., Madden D.J., Pierce, T.W., Siegel, W.C. and Appelbaum, M., 1993, Hypertension affects neurobehavioral functioning. *Psychosom. Med.* 55:44-50.

Bonithon-Kopp, C., Ducimetiere, P., Touboul, P-J., Feve, J-M., Billaud, E., Courbon, D. and Heraud, V., 1994, Plasma angiotensin-converting enzyme activity and carotid wall thickening. *Circulation* 89:952-954.

Boulanger, C., Schini, V.B., Moncada, S. and Vanhoutte, P.M., 1990, Release of EDRF and exogenous nitric oxide stimulate the production of cyclic AMP in cultured endothelial cells. *Br. J. Pharmacol.* 101:152-156.

Bradford, H.F., 1986, Neuropeptides: The rising generation of neuroactive substances. In: Chemical Neurobiology, WH Freeman and Co., NY, 265-310.

Bulpitt, C.J. and Fletcher, A.E., 1992, Cognitive function and angiotensin-converting enzyme inhibitors in comparison with other antihypertensive drugs. *J. Cardiovasc. Pharmacol.*, 19(suppl 6):S100-S104.

Burns, K.D., Homma T. and Harris R.C., 1993, The intrarenal renin-angiotensin system. *Seminars in Nephrol.*, Vol. 13. No 1:13-30.

Cachefeiro, V., Sakakibara, T. and Nasjletti, A., 1992, Kinins, nitric oxide, and the hypotensive effect of captopril and ramiprilat in hypertension. *Hypertension* 19:138-145.

Caldwell, R.W. and Blatteis, C.M., 1983, Effect of chronic hypoxia on angiotensin-induced pulmonary vasoconstriction and converting enzyme activity in the rat. *Proc. Soc. Exp. Biol. Med.*, 172:346-350.

Calhoun, D.A., Zhu, S., Wyss J.M. and Oparil, S., 1994, Diurnal blood pressure variation and dietary salt in spontaneously hypertensive rats. *Hypertension* 24:1-7.

Cambien, R., Poirier, O., Lecerf, L., Evans, A., Cambou, J-P., Arveiler, D., Luc, G., Bard, J-M.,Bara, L., Ricard, S., Tiret, L., Amouyel, P., Alhenc-Gelas, F. and Soubrier, F., 1992, Deletion polymorphism in the gene for angiotensin-converting enzyme is a potent risk factor for myocardial infarction. *Nature* 359:641-644.

Campbell, D.J., 1987, Tissue renin-angiotensin system: sites of angiotensin formation. *J. Cardiovasc. Pharmacol.* 10(suppl 7):S1-S8.

Campbell, D.J. and Habener, J.F., 1986, Angiotensinogen gene is expressed and differentially regulated in multiple tissue of the rat. *J. Clin. Invest.* 78:31-39.

Campbell, D.J. and Habener, J.F., 1987, Cellular localization of angiotensinogen gene expression in brown adipose tissue and mesentery: quantification of messenger ribonucleic acid abundance using hybridization in situ. *Endocrinology* 121:1616-1626.

Campbell, J., Fenessey, P. and Campbell, G.R., 1992, Effect of perindopril on the development of atherosclerosis in the cholesterol fed rabbit. *Clin. Exp. Pharmacol. Physiol.* 19(suppl 19):13-18.

Campbell, D.J., Lawrence, A.C., Towrie A., et al., 1991, Differential regulation of angiotensin peptide levels in plasma and kidney of the rat. *Hypertension* 18:763-773.

Cassis, L.A., Lynch, K.R. and Peach, M.J., 1988, Localization of angiotensinogen messenger RNA in rat aorta. *Circ. Res.* 62:1259-1262.

Casto, R. and Phillips, M.I., 1985, Neuropeptide action in nucleus tractus solitarius: angiotensin specificity and hypertensive rats. *Am. J. Physiol.* 249:R341-347.

Catt, K.J., Cain, M.D. and Coghlan, J.P., 1967, Measurement of angiotensin II in blood. *Lancet* 2:1005-1007.

Catt, K.J., Harwood J.P., Aguilera, G. and Dufau, M.L., 1979, Hormonal regulation of peptide receptors and target cell responses. *Nature* 280:109-116.

Chatziantoniou, C. and Arendshorst, W.J., 1991, Angiotensin and thromboxane in genetically hypertensive rats: renal blood flow and receptor studies. *Am. J. Physiol.* 261:F238-F247.

Chobanian, A.V., Haudenschild, C.C., Nickerson, C. and Drago, R., 1990, Antiatherogenic effect of captopril in the Watanabe heritable hyperlipidemic rabbit. *Hypertension* 15:327-331.

Clozel, M., Kuhn, H. and Hefti, F., 1990, Effects of angiotensin converting enzyme inhibitors and of hydralazine on endothelial function in hypertensive rats. *Hypertension* 16:532-540.

Cogan, M.G., 1990, Angiotensin II: a powerful controller of sodium transport in the early proximal tubule. *Hypertension* 15:451-458.

Cohen, L. and Kurz, K.D., 1982, Angiotensin converting enzyme inhibition in tissues from spontaneously hypertensive rats after treatment with captopril or MK-421. *J. Pharmacol. Exp. Ther.* 220:63-69.

Cohn, J.N., Johnson, G., Ziesche, S., Cobb, F., et al., 1991, A comparison of enalapril with hydralazine-isosorbide dinitrate in the treatment of chronic congestive heart failure. *N. Engl. J. Med.* 325:303-310.

CONSENSUS Trial Study Group, 1987, Effects of enalapril on mortality in severe congestive heart failure: results of the Cooperative North Scandinavian Enalapril Survival Study (CONSENSUS). *N. Engl. J. Med.* 316:1429-1435.

CONSENSUS II Study Group, 1992, Effects of the early administration of enalapril on mortality in patients with acute myocardial infarction: results of the Cooperative New Scandinavian Enalapril Survival Study II (CONSENSUS II). *N. Engl. J. Med.*, 327:678-684.

Dahlof. B., Pennert, K. and Hansson. L., 1992, Regression of left ventricular hypertrophy - a meta-analysis. *Clin. Exp. Hypertens.* (A), 14:173-180.

Dell'Italia, L.J., Meng, Q.C., Straeter-Knowlen, I., Hankes, G.H., Oparil, S. and Elton, T.S., 1994, Increased angiotensin peptide levels in cardiac tissue of dogs with chronic mitral regurgitation. *Circulation,* in press.

Dietz, R., Schomig, A., Haebara, H., Mann, J.F.E., Rascher, W., Luth, JB., Grunherz, N. and Gross, F., 1978, Studies on the pathogenesis of spontaneous hypertension of rats. *Circ. Res.* 43:98-106.

Dostal, D.E., Rothblum, K.N., Chernin, M.I., Cooper, G.R. and Baker, K.M., 1990, Intracardiac detection of angiotensin and renin: a localized renin-angiotensin system in neonatal rat heart. *Am. J. Physiol.* 263(*Cell Physiol.* 32):C838-C850.

Dostal, D.E., Rothblum, K.N., Conrad, K.M., Cooper, G.R. and Baker, K.M., 1992, Detection of angiotensin I and II in cultured rat cardiac myocytes and fibroblasts. *Am. J. Physiol.* 263 (*Cell Physiol.* 32):C851-C863.

Dzau, V.J., 1988a, Tissue renin-angiotensin system: Physiologic and pharmacologic implications. *Circulation* 77(suppl 1):11-13I.

Dzau, V.J., 1988b, Vascular renin-angiotensin system in hypertension: new insights into the mechanism of action of angiotensin converting enzyme inhibitors. *Am. J. Med.* 84(suppl4A):4-8.

Dzau, V.J., 1993, Local expression and pathophysiological role of renin-angiotensin in the blood vessels and heart. *Basic Res. in Cardiol.* 88(suppl 1):2-14.

Dzau, V.J., Ellison, K.E., Brody, T., Ingelfinger, J. and Pratt, R.E., 1987, A comparative study of the distribu tions of renin and angiotensinogen messenger ribonucleic acids in rat and mouse tissues. *Endocrinology* 120:2334-2338.

Dzau, V.J. and Hirsch, A.T., 1990, Emerging role of the tissue renin-angiotensin systems in congestive heart failure. *Europ. Heart J.* 11(suppl B):65-71.

Dzau, V.J. and Pratt, R., 1992, Tissue renin angiotensin system in experimental restenosis after vascular injury: evidence for local activation. *J. Cardiovasc Pharmacol.* 20(suppl B):S28-S32.

Eggena, P., Clegg, K., Barrett, J.D., Fredal, A.M. and Sambhi, M.P., 1986, A comparison of the renin-angiotensin system in aortic tissue of spontaneously hypertensive and Wistar-Kyoto rats. *J. Hypertension* 4:5365-5367.

Eggena, P., Krall, F., Eggena, M.P., Clegg, K., Fittingoff, M. and Barrett, J.D., 1990, Production of angiotensinogen by cultured rat aortic smooth muscle cells. *Clin. Exp. Hypertens. (A),* 12:1175-1189.

Eisenlohr, L.C., Bacik, I., Bennink, J.R., Bernstein, K. and Yewdell, J.W., 1992, Expressing of a membrane protease enhances presentation of endogenous antigen to MHC class I-restricted T lymphocytes. *Cell* 71:963-972.

El-Dorry, H.A., Pickett, C.B., MacGregor, J.S. and Soffer, R.L., 1982, Tissue-specific expression of mRNAs for dipeptidyl carboxypeptidase isoenzymes. *Proc. Natl. Acad. Sci. USA* 79:4295-4297.

Elias, M.F. and Elias, P.K., 1993, Hypertension affects neurobehaviorial functioning: so what's new? *Psychosom. Med.* 55:51-54.

Epstein, A.N., Fitzsimons, J.T. and Rolls, B.J., 1970, Drinking induced by injection of angiotensin into the brain of the rat. *J. Physiol.* 210:457-474.

Farhy, R.D., Ho, K.L., Carretero, O.A. and Scicli, A.G., 1992, Kinins mediate the antiproliferative effect of ramipril in rat carotid artery. *Biochem. Biophys. Res. Commun.* 182:283-288.

Fenoy, F.J., Milicic, I., Smith, R.D., Wong, P.C., Timmermans P.B.M.W.M. and Roman, R. , 1991. Effects of DuP 753 on renal function of normotensive and spontaneously hypertensive rats (SHR). (abst) *Am. J. Hypertension* 4:321S-326S.

Fishman, M.C., Zimmerman, E.A. and Slater, E.E., 1981, Renin and angiotensin: the complete system within the neuroblastoma x glioma cell. *Science* 214:921-923.

Fordis, C.M., Megorden, J.S., Rupchak, T.G. and Keiser, H.R., 1983, Absence of renin-like activity in rat aorta and microvessels. *Hypertension* 5:635-641.

Forslund, T., Konvonen, I. and Fyhrquist, F., 1984, Tissue distribution of angiotensin converting enzyme in the rat: effect of captopril treatment. *Acta Pharmacol. Toxicol.* 54:124-128.

Fox, J. Guan, S., Hymel, A.A., et al., 1992, Dietary Na and ACE inhibition effects on renal tissue angiotensin I and II and ACE activity in rats. *Am. J. Physiol.* 262:F902-F909.

Freer, R.J., Pappano, A.J., Peach, M.J., Bing, K.T., McLean, M.J., Vogel, S. and Spereleakis, N., 1976, Mechanism for the positive inotropic effect of angiotensin II on isolated cardiac muscle. *Circ. Res.* 39:178-183.

Friedland, J., Setton, C. and Silverstein, E., 1977, Angiotensin converting enzyme: induction by steroids in rabbit alveolar macrophages in culture. *Science* 197:64-65.

Fyhrquist, F., Forslund, T., Tikkanen, I. and Gronhagen-Riska, C., 1980, Induction of angiotensin-converting enzyme in rat lung with captopril (SQ 14225). *Eur. J. Pharmacol.* 67:473-475.

Fyhrquist, F., Gronhagen-Riska, C., Hortling, L., Forslund, T., Tikkanen, I. and Klockers, M., 1983, The induction of angiotensin-converting enzyme by its inhibitors. *Clin. Exp. Hypertens. Part A Theory Pract.* 5:1319-1330.

Fyhrquist, F., Hortling, L. and Gronhagen-Riska, C., 1982, Introduction of angiotensin I-converting enzyme by captopril in cultured human endothelial cells. *J. Clin. Endocrinol. Metab.* 55:783-786.

Ganten, D., Fuxe, K., Phillips, M.I., Mann, J.F.E. and Ganten, U., 1978, The brain isorenin-angiotensin system: biochemistry, localization, and possible role in drinking and blood pressure regulation in: Ganong WF and Martini L (eds.); Frontiers in Neuroendocrinology, Raven Press, New York, NY, 61-99.

Ganten, D., Lang, R.E., Lehmann, E. and Unger, T., 1984, Brain angiotensin: on the way to becoming a well-studied neuropeptide system. *Biochem. Pharmacol.* 33:3523-3528.

Garanin, G., 1986, A comparison of once daily antihypertensive therapy with captopril and enalapril. *Kidney Int.* 21:969-983.

Geisterfer, A.A., Peach, M.J. and Owens, G,K., 1988, Angiotensin II induces hypertrophy not hyperplasia of cultured rat aortic smooth muscle cells. *Circ. Res,* 62:749-756.

Gibbons, G.H. and Dzau, V.J., 1994, The emerging concept of vascular remodeling. *N. Engl. J. Med.* 330:1431-1438.

Giudicelli, J.F., Freslon, J.L., Glasson, S. and Richer, C., 1980, Captopril and hypertension development in the SHR. *Clin. Exper. Hypertension* 2:1083-1096.

Gohlke, P., Bunning, P. and Unger, T., 1992, Distribution and metabolism of angiotensin I and II in the blood vessel wall. *Hypertension* 20:151-157.

Griendling, K.K., Murphy, T.J. and Alexander, P.W., 1993, Molecular biology of the renin-angiotensin system. *Circulation* 87:1816-1828.

Haber, E., Zusman, R., Burton, J., Dzau, V.J., Barger, A.C., 1983, Is renin factor in the etiology of essential hypertension? *Hypertension* 5(suppl V):V8-V15.

Hall, E.R., Kato, J., Erdos, E.G., Robinson, C.J.G. and Oshima, G., 1976, Angiotensin I-converting enzyme in the nephron. *Life Sci.* 18:1299-1303.

Hannedouche, T.P., Marques, L., Natov, S., Delgado, A.G., Boitard, C., Lacour, B. and Grunfeld, J., 1992, Renal abnormalities in normotensive insulin-dependent diabetic offspring of hypertensive parents. *Hypertension* 19:378-384.

Hanson, S.R., Powell, J.S., Dodson, T., Lumsden, A., Kelly, A.B., Anderson, J.S., Clowes, A.W. and Harker, L.A., 1991, Effects of angiotensin converting enzyme inhibition with cilazapril on intimal hyperplasia in injured arteris and vascular grafts in the baboon. *Hypertension* 19(suppl II):II-70-II-76.

Harding, J.W., Wright, J.W., Speth, R.C. and Barnes, C.D., 1988, Angiotensin and Blood Pressure Regulation, Harding, Wright, Speth and Barnes (eds), Academic Press, Inc., San Diego, CA.

Harrap, S.B., Davidson, H.R., Connor, J.M. et al., 1993, The angiotensin I converting enzyme gene and predisposition to high blood pressure. *Hypertension* 21:455-460.

Harrap, S.B., Van der Merwe, W.M., Griffin, S.A., Macpherson, F. and Lever, A.F., 1990, Brief angiotensin converting enzyme inhibitor treatment in young spontaneously hypertensive rats reduces blood pressure long-term. *Hypertension* 16:603-614.

Hermann, K., McDonald, W., Unger, T., Lang, R.E. and Ganten, D., 1984, Angiotensin biosynthesis and concentrations in brain of normotensive and hypertensive rats. *J. Physiologie* 79:471-480.

Hilbert, P., Lindpaintner, K., Beckmann, J.S., Serikawa, T., Soubrier, F., Dubay, C., Cartwright, P., DeGouyon, B., Julier, C., Takahasi, S., Vincent, M., Ganten, D., Georges, M. and Lanthrop, G.M., 1991, Chromosomal mapping of two genetic loci associated with blood pressure regulation in hereditary hypertensive rats. *Nature* 353:521-529.

Hilgers, K.F., Veelken, R., Mai, M., Ganten, U., Ganten, D., Luft, F.C. and Mann, J.F.E., 1993, Vascular Conversion of angiotensin I in stroke-prone spontaneously hypertensive and Wistar-Kyoto rats. *J. Hypertens.* 11:1053-1059.

Hirsch, A.T., Talsness, C.E., Schunkert, H., Paul, M. and Dzau, V.J., 1991, Tissue-specific activation of cardiac angiotensin converting enzyme in experimental heart failure. *Circ. Res.* 69:475-482.

Husain, A., 1993, The chymase-angiotensin system in humans. *J. Hypertension* 11:1155-1159.

Ichikawa, I., Pfeffer, J.M., Pfeffer, M.A., Hostetter, T.H. and Brenner, B.M., 1984, Role of angiotensin II in the altered renal function of congestive heart failure. *Circ Res*, 55:669-675.

Inagami, T., Kawamura, M., Naruse, K. and Okamura, T., 1986, Localization of components of the renin-angiotensin system within the kidney. *Federation Proc.* 45:1414-1419.

Ingelfinger, J.R., Min Zuo, W., Fon, E.A., et al., 1990, In situ hybridization evidence for angiotensinogen messenger RNA in the rat proximal tubule. *J. Clin. Invest.* 85:417-423.

Iwai, N., Matsunaga, M., Kita, T., Tei, M. and Kawai, C., 1987, Regulation of angiotensin converting enzyme activity in cultured human vascular endothelial cells. *Biochem. Biophys. Res. Commun.* 149:1179-1185.

Jackson, R.M., Narkates, A.J. and Oparil, S., 1986, Impaired pulmonary conversion of angiotensin I to angiotensin II in rats exposed to chronic hypoxia. *J. Appl. Physiol.* 60:1121-1127.

Jacob, H.J., Lindpainter, K., Lincoln, S.E., Kusumi, K., Bunker, R.K., Mao, Y.P., Ganten, D., Dzau, V.J. and Lander, E.S., 1991, Genetic mapping of a gene causing hypertension in the stroke-prone spontaneously hypertensive rat. *Cell* 677:213-224.

Janiak, P., Pillon, A., Prost, J.F. and Vilaine, J. P., 1992, Role of angiotensin subtype 2 receptor in neointima formation after vascular injury. *Hypertension* 20:737-745.

Jeunemaitre, X., Lifton, R.P., Hunt, S.C., Williams, R.R. and Lalouel, J.M., 1992, Absence of linkage between the angiotensin converting enzyme locus and human essential hypertension. *Nat. Genet.* 1:72-75.

Jin, H., Oparil, S., Ann, H.S., Yang, A.R. and Jackson, R.M., 1987, Hypoxia-induced inhibition of converting enzyme activity: Role in vascular regulation. *J. Appl. Physiol.* 63:1012-1018.

Johnston, C.I., 1994, Tissue angiotensin converting enzyme in cardiac and vascular hypertrophy, repair, and remodeling. 23:258-268.

Kauffman, R.F., Bean, J.S., Zimmerman, K.M., Brown, R.F. and Steinberg, M.I., 1991, Losartan, a nonpeptide angiotensin II (Ang II) receptor antagonist inhibits neointima formation following balloon injury to rat carotid arteries. *Life Sci.* 49:223-228.

Keane, P.M., Kay, J.M., Suyama, K.L., Gauthier, D. and Andrew, K., 1982, Lung angiotensin converting enzyme activity in rats with pulmonary hypertension. *Thorax* 37:198-204.

Khairallah, P.A. and Kanabus, J., 1983, Angiotensin and myocardial protein synthesis. *Perspect.Cardiovasc. Res.* 8:337-347.

Kifor, I. and Dzau, V.J., 1987, Endothelial renin-angiotensin pathway: evidence for intracellular synthesis and secretion of angiotensins. *Circ. Res.* 60:422-428.

King, S.J., Berecek, K.H., Oparil, S., Rydzewski, B. and Raizada, M., 1992, Angiotensin-converting enzyme (ACE) gene expression is decreased after exposure to angiotensin II (ANG II) *in vitro* (Abstract). *FASEB J.* 6:A946.

King, S.J., Booyse, F.M., Lin, P-H., Traylor, M., Narkates, A.J. and Oparil, S., 1989, Hypoxia stimulates endothelial cell angiotensin-converting enzyme antigen synthesis. *Am. J. Physiol.* 256(*Cell Physiol.* 25):C1231-C1238.

King, S.J. and Oparil, S., 1992, Converting-enzyme (CE) inhibitors increase converting-enzyme mRNA and activity in endothelial cells. *Am. J. Physiol.: Cell Physiol.* 263:C743-C749.

King, S.J., Oparil, S. and Berecek, K.H., 1991, Neuronal angiotensin-converting enzyme (ACE) gene expression is increased by converting enzyme inhibitors (CEI). *J. Mol. Cell Neurosci.* 2:13-20.

Kinoshita, A., Urata, H., Bumpus, F.M. and Husain, A., 1991, Multiple determinants for the high substrate specificity of an angiotensin-forming chymase from the human heart. *J. Bio. Chem.* 266:19192-19197.

Kokubu, T., Ueda, E., Ono, M., Kawabe, T., Hayashi, Y. and Kan, T., 1980, Effects of captopril (SQ14,255) on the renin-angiotensin-aldosterone system in normal rats. *Eur. J. Pharmacol.* 62:269-275.

Kost, C.K., Jr. and Jackson, E.K., 1993, Enhanced renal angiotensin II subtype 1 receptor responses in the spontaneously hypertensive rat. *Hypertension* 21:420-430.

Krulewitz, A.H. and Fanburg, B.L., 1984, The effect of oxygen tension on the *in vitro* production and release of angiotensin-converting enzyme by bovine pulmonary artery endothelial cells. *Am. Rev. Respir. Dis.* 130:866-869.

Kumar, R.S., Thekkumkara, T.J. and Sen, G.C., 1991, The mRNAs encoding the two angiotensin-converting isozymes are transcribed from the same gene by a tissue-specific choice of alternative transcription initiation sites. *J. Bio. Chem.* 266:3854-3862.

Larochelle, P., Genest, J., Juchel, O., Bocher, R., Gutkowska, Y. and McKinstry, D., 1979, Effect of captopril (SQ 14225) on blood pressure, plasma renin activity and angiotensin I converting enzyme activity. *Can. Med. Assoc. J.* 121:309-316.

Lee, M.A., Bohm, M., Paul, M. and Ganten, D., 1993, Tissue renin-angiotensin systems. Their role in cardiovascular disease. *Circulation* 87[suppl V]:IV-7-IV-13.

Lee, M.A.E., Paul, M., Bohm, M. and Ganten, D., 1992, Effects of angiotensin-converting enzyme inhibitors on tissue renin-angiotensin systems. *Am. J. Cardiol.* 70:12C-19C.

Li, P. and Jackson, E.K., 1989, Enhanced slow-pressor response to angiotensin II in spontaneously hypertensive rats. *J. Pharmacol. Exp. Ther.* 251:909-921.

Lilly, L.S., Pratt, R.E., Alexander, W., Larson, D.M., Ellison K.E., Gimbrone, J.A., Jr. and Dzau, V.J., 1985, Renin expression by vascular endothelial cells in culture. *Circ. Res.*57:312-318.

Lind, R.W., Swanson, L.W. and Ganten, D., 1985, Organization of angiotensin II immunoreactive cells and fibers in the rat central nervous system. An immunohistochemical study. *Neuroendocrinol.* 40(1):2-24.

Lindpaintner, K., Jin, M., Wilhelm, M.J., Suzuki, F., Linz, W., Schoelkens, B.A. and Ganten, D., 1988, Intracardiac generation of angiotensin and its physiologic role. *Circulation* 77(suppl I):1-18.

Linz, W., Scholkens, B.A. and Ganten, D., 1989, Converting enzyme inhibition specifically prevents the development and induces regression of cardiac hypertrophy in rats. *Clin. Exper. Hyper. Theory and Practice* A11:1325-1350.

Lockette, A., Otsuka, Y. and Carretero, O., 1986, The loss of endothelium dependent vascular relaxation in hypertension. *Hypertension* 8(suppl II):II-61-II-66.

Loudon, M., Bing R.F., Thurston, H. and Swales, J.D., 1983, Arterial wall uptake of renal renin and blood pressure control. *Hypertension* 5:629-634.

Luckhoff, A., Budek, W., Holtz, J. and Busse, R., 1985, Role of the endothelium in the vasomotor effects of angiotensin I in isolated arteries with or without inhibition by MK 422. *J. Hypertens.* 3(suppl 3):S279-S281.

Lundmer, P.L., Selwyn, A.P., Shoot, T.L., Wayne, R.R., Gilbert, B.S., Mudge, G.H., Alexander, R.W. and Ganz, P., 1986, Paradoxical vasoconstriction induced by acetylcholine in atherosclerotic coronary arteries. *N. Engl. J. Med.* 315:1046-1051.

Malik, K.U. and Nasjletti, A., 1976, Facilitation of adrenergic transmission by locally generated angiotensin II in rat mesenteric arteries. *Circ. Res.* 38:26-30.

Marchetti, J., Roseau, S. and Alhenc-Gelas, F., 1987, Angiotensin I converting enzyme and kinin-hydrolyzing enzyme along the rabbit nephron. *Kidney Int.* 31:744-751.

Marian, A.J., Yu, Q-T., Workman, R., Greve, G. and Roberts, R., 1993, Angiotensin-converting enzyme polymorphism in hypertrophic cardiomyopathy and sudden cardiac death. *Lancet* 342:1085-1086.

McGeer, E.G. and Singh, E.A., 1992, Angiotensin-converting enzyme in cortical tissue in Alzheimer's and some other neurological diseases. *Dementia* 3:299-303.

Mendelsohn, F.A.O., 1982, Angiotensin II: evidence for its role as an intrarenal hormone. *Kidney Int.* 22(suppl 12):S78-S-81.

Mendelsohn, F.A.O., Lloyd, C.J., Katchel, C. and Funder, J.W., 1982, Induction by glucocorticoids of angiotensin-converting enzyme production from bovine endothelial cells in culture and rat lung in two. *J. Clin. Invest.* 70:684-692.

Mendelsohn, F.A.O., Quirion, R., Saavedra, J.M., Aguilera, G. and Catt, K.J., 1984, Autoradiographic localization of angiotensin II receptors in rat brain. *Proc. Natl. Acad. Sci. U.S.A.* 81:1575-1579.

Meulemans, A.L., Andries, L.J. and Brutsaert, D.L., 1990, Does endocardial endothelium mediate positive inotropic response to angiotensin I and angiotensin II? *Circ. Res.* 66:1591-1601.

Miyazaki, M., Okunishi, H., Okamura, T., Toda, N., 1987, Elevated vascular angiotensin converting enzyme in chronic two-kidney, one clip hypertension in the dog. *J. Hypertens.* 5:155-160.

Mizuno, K., Nakamura, M., Higashimori, K. and Inagami, T., 1988, Local generation and release of angiotensin II in peripheral vascular tissue. *Hypertension* 11:223-229.

Mizuno, K., Tani, M., Niimura, S. and Fukuchi, S., 1991, Effect of delapril on the vascular angiotensin II release in isolated hind legs of the spontaneously hypertensive rat: Evidence for potential relevance of vascular angiotensin II to the maintenance of hypertension. *Clin. Exp. Pharmacol. Physiol.* 18:619-625.

Molteni, A., Zakheim, R.M., Mullis, K.B. and Mattioli, L., 1974, The effect of chronic alveolar hypoxia on lung and serum angiotensin I converting enzyme activity. *Proc. Soc. Exp. Biol. Med.* 147:263-265.

Mombouli, J.V., Nephtali, M. and Vanhoutte, P.M., 1991, Effect of the converting enzyme cilazaprilat in endothelium-dependent responses. *Hypertension* 18(suppl II):II-22-II-29.

Moravec, C.S., Schluchter, M.D., Parnandi, L., Czerska, B., Stewart, R.W., Rosenkranz, E. and Bond, M., 1990, Inotropic effects of angiotensin II on human cardiac muscle *in vitro*. *Circulation* 82:1973-1984.

Mujais, S.K., Kauffman, S. and Katz, A.I., 1986, Angiotensin II binding sites in individual segments of the rat nephron. *J. Clin. Invest.* 77:315-318.

The Multicenter European Research Trial With Cilazapril After Angioplasty to Prevent Transluminal Coronary Obstruction and Restenosis (MERCATOR) Study Group, 1992,. Does the new angiotensin converting enzyme inhibitor cilazapril prevent restenosis after percutaneous transluminal coronary angioplasty? Results of the MERCATOR study: A Multicenter, Randomized, Double-Blind Placebo-Controlled Trial. *Circulation* 86:100-110.

Naftilan, A.J., Pratt, R.E. and Dzau, V.J., 1989a, Induction of PDGF A-chain and c-myc gene expression by angiotensin II in cultured rat vascular smooth muscle cells. *J. Clin. Invest.* 83:1419-1424.

Naftilan, A.J., Pratt, R.E., Eldridge, C.S., Lin, H.L. and Dzau, V.J., 1989b, Angiotensin II induces c-fos expression in smooth muscle via transcriptional control. *Hypertension* 13:706-711.

Naftilan, A.J., Zuo, W.M., Inglefinger, J., Ryan, T.J., Jr., Pratt, R.E. and Dzau, V.J., 1991, Localization and differential regulation of angiotensinogen mRNA expression in the vessel wall. *J. Clin. Invest,* 87:1300-1311.

Nakamura, Y., Nakamura, K., Matsukura, T. and Nakamura, K., 1988, Vascular angiotensin converting enzyme activity in spontaneously hypertensive rats and its inhibition by ciliazopril. *J. Hypertens.* 6:105-110.

Nara, Y., Nabika, T., Ikeda, K., Sawamura, M., Endo, J. and Yamori, Y., 1991, Blood pressure cosegregates with a microsatellite of angiotensin I converting enzyme in F_2 generation from a cross between original normotensive Wistar-Kyoto rat and stroke-prone spontaneously hypertensive rat. *Biochem. Biophys. Res. Commun.* 181:941-946.

Navar, L.G., Evan, A.P. and Rosivall, L., 1983, Microcirculation of the kidney. In: Physiology and Pharmacol ogy of Microcirculation. N.A. Mortillaro (ed), Academic Press, NY, Vol 1, 397-488.

O'Donohoe, M.K., Schwartz, L.B., Radic, Z.S., Mikat, E.M., McCann, R.L. and Hagen, P-O., 1991, Chronic ACE inhibition reduces intimal hyperplasia in experimental vein grafts. *Ann. Surg.* 214:727-732.

O'Donohoe, M.K., Davies, M.G., Radic, Z.S., Mikat, E.M. and Hagen, P-O., 1994, Increased concentration of angiotensin-converting enzyme in the intimal hyperplasia of experimental vein grafts. *J. Cardiovasc. Pharmacol.* 23:594-601.

Okabe, T., Yamagata, K., Fujisawa, M., Takaku, F., Hidaka, H. and Umezawa, Y., 1987, Induction by fibroblast growth factor of angiotensin converting enzyme in vascular endothelial cells *in vitro*. *Biochem. Biophys. Res. Commun.* 145:1211-1216.

Okamura, T., Miyazaki, M., Inagami, T. and Toda, N., 1986, Vascular renin-angiotensin system in two-kidney, one clip hypertensive rats. *Hypertension* 8:560-565.

Okunishi, H., Kawamoto, T., Kurobe, Y., Oka, Y., Ishii, K., Tanaka, T. and Miyazaki, M., 1991, Pathogenic role of vascular angiotensin-converting enzyme in the spontaneously hypertensive rat. *Clin. Exp. Pharmacol. Physiol.* 18:649-659.

Okunishi, H., Miyazaki, M., Okamura, T. and Toda, N., 1984, Evidence for a putatively new angiotensin II-generating enzyme in the vascular wall. *J. Hypertens.* 2:277-284.

Okunishi, H., Myiyazki, M., Okamura, T. and Toda, N., 1987, Different distribution of two types of angiotensin II-generating enzymes in the aortic wall. *Biochem. Biophys. Res. Commun.* 149:1186-1192.

Oliver, J.A. and Sciacca, R.R., 1984, Local generation of angiotensin II as a mechanism of regulation of peripheral vascular tone in the rat. *J Clin Invest,* 74:1247-1251.

Oparil, S. and Haber, E., 1974, The renin-angiotensin system. *N. Engl. J. Med.* 291:389-401 and 446-457.

Oparil, S., Narkates, A.J., Jackson, R. and Ann, H.S., 1988, Altered angiotensin-converting enzyme in lung and extrapulmonary tissues of hypoxia-adapted rats. *J. Appl. Physiol.* 65:218-227.

Osanai, T. and Dunn, M.J., 1992, Phospholipase C responses in cells from spontaneously hypertensive rats. *Hypertension* 19:446-455.

Owens, G.K., 1989, Control of hypertrophic versus hyperplastic growth of vascular smooth muscle cells. *Am. J. Physiol.* 257:H1755-H1765.

Pandey, K.N., Misono, K.S. and Inagami, T., 1984, Evidence for intracellular formation of angiotensins: coexistence of renin and angiotensin-converting enzyme in Leydig cells of rat testis. *Biochem. Biophys. Res. Commun.* 122:1337-1343.

Paul, M., Bachmann, J. and Ganten, D., 1992, The tissue renin-angiotensin systems in cardiovascular disease. *Trends Cardiovasc. Med.* 2:94-99.

Perez-Stable, E.J., Coates, T.J., Halliday, R., Gardiner, P.S. and Hauck, W.W., 1992, The effects of mild diastolic hypertension on the results of tests of cognitive function in adults 22 to 59 years of age. *J. Gen. Intern. Med.* 7:19-25.

Pfeffer, M.A., Braunwald, E., Moye, L.A., et al., 1992, Effect of captopril on mortality and morbidity in patients with left ventricular dysfunction after myocardial infarction. *N. Engl. J. Med.* 327:669-677.

Phillips, M.I. and Kimura, B., 1986, Converting enzyme inhibitors and brain angiotensin. *J. Cardiovasc. Pharmacol.* 8(Suppl 10):S82-S90.

Phillips, M.I., Speakman, E.A. and Kimura, B., 1993, Levels of angiotensin and molecular biology of the tissue renin angiotensin systems. *Regulatory Peptide* 43:1-20.

Pickel, V.M., Chan, J. and Ganten, D., 1986, Dual peroxidase and colloidal gold-labeling study of angiotensin converting enzyme and angiotensin-like immunoreactivity in rat subfornical organ. *J. Neurosci.* 6:2457-2469.

Pipili, E., Manolopoulos, V.G., Catravas, J.D. and Maragoudakis, M.E., 1989, Angiotensin converting enzyme activity is present in the endothelium-denuded aorta. *Br. J. Pharmacol.* 98:333-335.

Powell, J.S., Clozel, J-P., Muller, R.K.M. et al., 1989, Inhibitors of angiotensin-converting enzyme prevent myomtimal proliferation after vascular injury. *Science* 245:186-188.

Powell, J.S., Muller, R.K.M., Rouge, M., Kuhn, H., Hefti, H. and Baumgartner, H.R., 1990, The proliferative response to vascular injury is suppressed by angiotensin-converting enzyme inhibition. *J. Cardiovasc. Pharmacol.* 16(suppl):S42-S49.

Pratt, R.E., Zou W.M., Naftilan A.J., Ingelfinger, J.R. and Dzau, V.J., 1989, Altered sodium regulation of renal angiotensinogen mRNA in the spontaneously hypertensive rat. *Am. J. Physiol.* 256:H469-H474.

Raizada, M.K., Muther, T.F. and Sumners, C., 1984a, Increased angiotensin II receptors in neuronal cultures from hypertensive rat brain. *Am. J. Physiol.* 247:C364-372.

Raizada, M.K., Phillips, M.I. and Sumners, C., 1993, Cellular and Molecular Biology of the Renin-Angiotensin System. Raizada, Phillips, Summers (eds), CRC Press, Inc., Boca Raton, FL.

Raizada, M.K., Stenstrom, B., Phillips, M.I. and Sumners, C., 1984b, Angiotensin II in neuronal cultures from brains of normotensive and hypertensive rats. *Am. J. Physiol.* 247:C115-C119.

Raynolds, M.V., Bristow, M.R., Bush, E.W. et al., 1993, Angiotensin-converting enzyme DD genotype in patients with ischemic or idiopathic dilated cardiomyopathy. *Lancet* 342:1073-1075.

Regoli, D., Park, W.K., Rioux, F. and Magnan, J., 1975, Metabolism of angiotensin in vascular smooth muscle, In: Peptides: Chemistry, Structure and Biology, R. Walter and J. Meienhofer (eds); Ann Arbor Science Publishers, Ann Arbor, MI, 617-624.

Rettig, R., Folberth, C., Graf, C., Kopf, D., Stauss, H. and Unger, T., 1991, Are renal mechanisms involved in primary hypertension? Evidence from kidney transplantation studies in rats. *Klin Wochenschr* 69:597-602.

Rettig, R., Folberth, C., Stauss, H., Kopf, D., Waldherr R. and Unger, T., 1990, Role of the kidney in primary hypertension: A renal transplantation study in rats. *Am. J. Physiol.* 258:F606-F611.

Riegger, A.J.G, Lever, A.F., Millar, J.J. and Slack, B., 1977, Correction of renal hypertension in the rat by prolonged infusion of saralasin inhibitors. *Lancet* II:1317-1319.

Rigat, B., Hubert, C., Alhenc-Gelas, F., Cambien, F., Corvol, P., and Soubrier, F., 1990, An insertion/deletion polymorphism in the angiotensin I-converting enzyme gene accounting for half the variance of serum enzyme levels. *J. Clin. Invest.* 86:1343-1346.

Rohrbach, M.S., 1984, Metabolism and subcellular localization of angiotensin converting enzyme in cultured human monocytes. *Biochem. Biophys. Res. Comm.* 124:843-849.

Rosenfeld, J.B. and Zabludowski, J., 1989, The renin-angiotensin system, the kidney, and hypertension.(Review) *J. Cardiovasc. Pharmacol.* 13(Suppl 3)S22-S26.

Rosivall, L., Narkates, A.J., Oparil, S. and Navar, L.G., 1987, De novo intrarenal formation of angiotensin II during control and enhanced renin secretion. *Am. J. Physiol.* 252:F-1118-F1123.

Ryan, U.S., Ryan, J.W., Whitaker, C. and Chiu, A., 1976, Localization of angiotensin converting enzyme (kininase II). II. Immunocytochemistry immunofluorescence. *Tissue & Cell* 8:125-145.

Sakaguchi, K., Siew-Yeen, C., Jackson, B., Johnson, C.I. and Mendelsohn, F.O.A., 1988, Differential angiotensin-converting enzyme inhibition in brain after oral administration of perindopril demonstrated by quantitative *in vitro* autoradiography. *Neuroendocrinology* 48:223-228.

Samani, N.J. and Swales, J.D., 1991, Molecular biology of the vascular renin-angiotensin system. *BloodVessels* 28:210-216.

Samani, N.J., Swales, J.D. and Brammar, W.J, 1988, Expression of the renin gene in extra-renal tissues of the rat. *Biochem. J.* 253:907-910.

Samani, N.J., Swales, J.D. and Brammar, W.J., 1989, A wide-spread abnormality of renin gene expression in the spontaneously hypertensive rat: modulation in some tissues with the development of hypertension. *Clin. Sci.* 77:629-636.

Sarzani, R., Arnaldi, G. and Chobanian, AV., 1991, Hypertension-induced changes of platelet-derived growth factor receptor expression in rat aorta and heart. *Hypertension* 17:888-895.

Saye, J.A., Singer, H.A. and Peach, M.J., 1984, Role of endothelium in conversion of angiotensin I to angiotensin II in rabbit aorta. *Hypertension* 6:216-221.

Schelling, P. and Felix, D., 1983, Influence of captopril treatment on angiotensin II receptors and angiotensinogen in the brain of spontaneously hypertensive rats. *Hypertension* 5:935-942.

Schelling, P., Fischer, H. and Ganten, D., 1991, Angiotensin and cell growth: a link to cardiovascular hypertrophy? *J. Hypertens.* 9:3-15.

Schoemaker, R.G., Debets, J.J.M., Struyker-Boudier, H.A.J. and Smits, J.F.M., 1991, Delayed but not immediate captopril therapy improves cardiac function in conscious rats, following myocardial infarction. *Mol. Cell. Cardiol.* 23:187-197.

Schrier, D.J., Ripani, L.M., Katzenstein, A. and Moore, V.L., 1982, Role of angiotensin converting enzyme in bacille calmette-guerin-induced granulomatous inflammation. *J. Clin. Invest.* 69:651-657.

Schunkert, H., Hirsch, A.T., Mankodi, S., Talsness, C., Dzau, V.J. and Ingelfinger, J.R., 1989, Renal angiotensinogen gene expression in experimental heart failure: effect of angiotensin converting enzyme inhibition (abst). *Clin. Res.* 37:584A.

Schunkert, H., Dzau, V.J,, Tang, S.S., Hirsch, A.T., Apstein, C.S. and Lorell, B.H., 1990, Increased rat cardiac angiotensin converting enzyme activity and mRNA expression in pressure overload left ventricular hypertrophy. *J. Clin. Invest.* 86:1913-1920.

Schunkert, H., Ingelfinger, J.R., Hirsch, A.T., Pinto, Y., Remme, W.J., Jacob, H. and Dzau, V.J., 1993, Feedback regulation of angiotensin converting enzyme activity and mRNA levels by angiotensin II. *Circ. Res.* 72:312-318.

Schunkert, H., Hense, H-W., Holmer, S.R., Stender, M., Perz, S., Keil, U., Lorell, B.H. and Riegger, G.A.J., 1994, Association between a deletion polymorphism of the angiotensin-converting-enzyme gene and left ventricular hypertrophy. *N. Engl. J. Med.* 330:1634-1638.

Scott-Burden, T., Hahn, A.W.A., Resink, T.J. and Buhler, F.R., 1990, Modulation of extracellular matrix by angiotensin II: stimulated glycoconjugate synthesis and growth in vascular smooth muscle cells. *J. Cardiovasc. Pharmacol.* 16(suppl):S36-S41.

Sealey, J.E., White, R.P., Laragh, J.H. and Rubin, A.L., 1977, Plasma Pro-renin and renin in anephric patients. *Circ. Res.* 41(suppl II):II-17-II-2.

Seikaly, M.G., Arant, B.S., Jr. and Seney, F.D., Jr., 1990, Endogenous angiotensin concentrations in specific intrarenal fluid compartments in the rat. *J. Clin. Invest.* 86:1352-1347.

Shai, S.Y., Langford, K.G., Martin, B.M. and Bernstein, K.E., 1990, Genomic DNA 5' to the mouse and human angiotensin-converting enzyme genes contains two distinct regions of conserved sequence. *Biochem. Biophys. Res. Commun.* 167:1128-1133.

Shimada, K. and Yazaki, Y., 1978, Binding sites of angiotensin II in human mononuclear leukocytes. *J. Biochem.* 84:1013-1015.

Shiota, N., Miyazaki, M. and Okunishi, H., 1992, Increase of angiotensin converting enzyme gene expression in the hypertensive aorta. *Hypertension* 20:168-174.

Sladek, C.D. and Armstrong, W.E., 1987, Effect of neurotransmitters and neuropeptides on vasopressin release in: Sladek CD, Armstrong WE, Gash DM and Boer GJ (eds.). Vasopressin: principles and properties. Plenum Press, NY, 275-333.

SOLVD Investigators, 1991, Effect of enalapril on survival in patients with reduced left ventriuclar ejection fractions and congestive heart failure. *N. Engl. J. Med.* 325:293-302.

SOLVD Investigators, 1992, Effect of enalapril on mortality and the development of heart failure in asymptom atic patients with reduced left ventricular ejection fractions. *N. Engl. J. Med.* 327:685-691.

Soubrier, F., Alhenc-Gelas, F., Hubert, C., Allelgrini, J., John, M. et al., 1988, Two putative active centers in human angiotensin I converting enzyme revealed by molecular cloning. *Proceedings of the National Academy of Sciences USA* 84:9386-9390.

Steiner, S., Friedhoff, A.J., Wilson, B.L., Wecker, J.R. and Santo, J.P., 1990, Antihypertensive therapy and quality of life: A comparison of atenolol, captopril, enalapril and propanolol. *J. Hum. Hypertens.* 4:217-225.

Studer, R., Reinecke, H., Muller, B., Holtz, J., Just, H., and Drexler, H., 1994, Increased angiotensin-1 converting enzyme gene expression in the failing human heart. *J. Clin. Invest.* 94:301-310.

Sudilovsky, A., Cutler, N.R., Sramek, J.J., Wardle, T., Veroff, A.E., Mickelson, W., Markowitz, J. and Repetti, S., 1993, A pilot clinical trial of the angiotensin-converting enzyme inhibitor ceranapril in Alzheimer disease. *Alzheimer Dis. Assoc. Disord.* 7:105-111.

Sun, S.D., Oparil, S., Elton, T.S. and Chen, Y.F., 1993, Differential expression of vascular and renal angiotensin converting enzyme mRNA in normotensive and hypertensive rats. *Clin. Research* 41:793A.

Swales, J.D., Abramovici, A., Beck, F., Bing, R.F., Loudon, M. and Thurston, H., 1983, Arterial wall renin. *J. Hypertens.* 1(suppl1):17-23.

Takasaki, I., Saouaf, R., Sarzani, R., Chobanian, A.V. and Brecher, P., 1990, Hypertension-induced changes of fibronectin expression in the rat aorta. (abstract) *Clin. Res.* 38:242A.

Taubman, M.B., Berk, B.C., Izumo, S., Tsuda, T., Alexander, R.W. and Nadal-Ginard, B., 1989, Angiotensin II induces c-*fos* mRNA in aortic smooth muscle. *J. Biol. Chem.* 264:526-530.

Taugner, R., Mannek, E., Nobiling, R., Buhrle, C.P., Hackenthal, E., Ganten, D., Inagami, T. and Schroder, H., 1984, Coexistence of renin and angiotensin II in epitheloid cell secretory granules of rat kidney. *Histo chemistry* 81:39-45.

Testa, M.A., Anderson, R.B., Nackley, J.F. and Hollenberg, N.K., 1993, Quality of life and antihypertensive therapy in men: a comparison of captopril with enalapril. *N. Engl. J.Med.* 328:907-913.

Thurston, H., Swales, J.D., Bing R.F., Hurst, B.C. and Marks, E.S., 1979, Vascular renin-like activity and blood pressure maintenance in the rat: studies on the effect of changes in sodium balance hypertension and nephrectomy. *Hypertension* 1:643-649.

Unger, T., Ganten, D., Lang, R.E. and Scholkens, B.A., 1984, Is tissue converting inhibition a determinant of the antihypertensive efficacy of converting enzyme inhibitors? Studies with the two different compounds, Hoe 398 and MK 421, in spontaneously typertensive rats. *J. Cardiovasc. Pharmacol.* 6:872-880.

Unger, T., Ganten, D., Lang, R.E. and Scholkens, B.A., 1985, Persistent tissue converting enzyme inhibition following treatment with Hoe49 and MK421 in spontaneously hypertensive rats. *J. Cardiovasc. Pharmacol.* 7:36-41.

Unger, T., Kaufmann-Buhler, I., Scholkens, B. and Ganten, D., 1981, Brain converting enzyme inhibition: a possible mechanism for the antihypertensive action of captopril in spontaneously hypertensive rats. *Eur. J. Pharmacol.*, 70:467-478.

Unger, T. and Rettig, R., 1990, Development of genetic hypertension. Is there a "critical phase"? *Hypertension* 16:615-616.

Urata, H., Healy, B., Stewart, R.W., Bumpus, F.M. and Husain, A., 1990a, Angiotensin II-forming pathways in normal and failing human hearts. *Circ. Res.* 66:883-890.

Urata, H., Kinoshita, A., Misono, K.S., Bumpus, F.M., and Husain, A., 1990b, Identification of a highly specific chymase as the major angiotensin II-forming enzyme in the human heart. *J. Bio. Chem.* 265(36):22348-22357.

Urata, H., Kinoshita, A., Perez, D.M., Misono, K.S., Bumpus, F.M., Graham, R.M. and Husain, A., 1991, Cloning of the gene and cDNA for human heart chymase. *J. Biol. Chem.* 266:17173-17179.

Urata, H., Boehm, K.D., Philip, A., Kinoshita, A., Gabrovsek, J., Bumpus, F.M. and Husain, A., 1993, Cellular localization and regional distribution of an angiotensin II-forming chymase in the heart. *J. Clin. Invest.* 91:1269-1281.

Velletri, P.A., Aquilano, D.R., Bruckwick, E., Tsai-Morris, C.H., Dufau, M.L. and Lovenberg, W., 1985, Endocrinological control and cellular localization of rat testicular angiotensin-converting enzyme (EC 3.4.15.1). *Endocrinology* 116:2516-2522.

Ward, P.E., Sheridan, M.A., Hammon, K.J. and Erdos, E.G., 1980, Angiotensin I converting enzyme(kininase II) of the brush border of human and swine intestine. *Biochem. Pharmacol.* 29:1525-1529.

Weinberger, M.H., Wade, M.B., Aoi, W., et al., 1977, An extrarenal source of "renin-like" activity in anephric man. *Circ. Res.* 40(suppl I):I1-I14.

Weinstock, J.V. and Kassab, J.T., 1984, Functional angiotensin II receptors on macrophages fromisolated liver granulomas of murine. *Schistosoma mansoni. J. Immunol.* 132:2598-2602.

Widgren, B.R., Herlitz, H., Aurell, M., Berglund, G., Wikstrand, J. and Andersson, O.K., 1992, Increased systemic and renal vascular sensitivity to angiotensin II in normotensive men with positive family histories of hypertension. *Am. J. Hypertens.* 5:167-174.

Wiemer, G., Scholkens, B.A., Becker, R.H.A. and Busse, R:, 1991, Ramiprilat enhances endothelial autacoid formation by inhibiting breakdown of endothelium-derived bradykinin. *Hypertension* 18:558-563.

Wilson, K.M., Magargal, W. and Berecek, K.H. Long-term captopril treatment. Angiotensin II receptors and responses. *Hypertension* 11:I148-152, 1988.

Wilson, S.K. , Lynch, D.R. and Snyder, S.H., 1987, Angiotensin-converting enzyme labeled with ^{3}H-captopril: tissue localization and changes in different models of hypertension in the rat. *J. Clin. Invest.* 80:841-851.

Wood, J.M., Schnell, C.R. and Levens, N.R., 1992, The kidney is an important target for the antihypertensive action of an angiotensin II (ANG II) receptor antagonist in spontaneously hypertensive rats (SHR), (abstract) *Hypertension* 20:436.

Wu, J.N., and Berecek, K.H., 1993, Prevention of genetic hypertension by early treatment of spontaneously hypertensive rats with the angiotensin converting enzyme inhibitor captopril. *Hypertension* 22:139-146.

Wyss, J.M. and van Groen, T., 1992, Early breakdown of dendritic bundles in the retrosplenial granular cortex of hypertensive rats: prevention by antihypertensive therapy. *Cereb. Cortex* 2:468-476.

Wyss, J.M., Fisk G. and van Groen, T., 1992, Impaired learning and memory in mature spontaneously hypertensive rats. *Brain Res.* 592:135-140.

Wyss, J.M. and van Groen, T., 1994, Neurologic consequences of hypertension and antihypertensive drug therapy. *Curr. Opin. Nephrol. Hypertens.* 3:228-235.

Wyss, J.M., van Groen, T., Meng, Q. and Oparil, S., 1993, Angiotensin converting enzyme (ACE) activity is selectively elevated in the temporal cortex of Alzheimer's patients. *Soc. Neurosci. Abstr.* 19:191.

Yamada, H., Sexton, P.M., Chai, S.Y, et al., 1990, Angiotensin II receptors in the kidney. Localization and physiological significance. *Am. J. Hypertension* 3:250-255.

Yang, R.H., Jin, H., Wyss, M. and Oparil, S., 1992a, Depressor effect of blocking angiotensin subtype 1 receptors in anterior hypothalamus. *Hypertension* 19:475-481.

Yang, R.H., Jin, H., Chen, S.J., Wyss, J.M. and Oparil, S., 1992b, Blocking hypothalamic AT1 receptors lowers blood pressure in salt-sensitive rats. *Hypertension* 20:755-62.

Yotsumoto, H., Imai, I., Kuzuya, N., Uchimura, H. and Matsuzaki, F, 1982, Increased levels of serum angiotensin-converting enzyme activity in hyperthyroidism. *Ann. Intern. Med.* 96:326-328.

Yusuf, S., Pepine, C.J., Carces, C. et al., 1992, Effect of enalapril on myocardial infarction and unstable angina in patients with low ejection fractions. *Lancet* 340:1173-1178.

Zambetis-Bellesis, M., Dusting, G.J., Mendelsohn, F.A.O. and Richardson, K., 1991, Autoradiographic localization of angiotensin-converting enzyme and angiotensin II binding sites in early atheroma-like lesions in rabbit arteries. *Clin. Exper. Pharmacol. Physiol.* 18:337-340 .

Zierhut, W., Zimmer,G. and Gerdes, A.M., 1991, Effect of angiotensin converting enzyme inhibition on pressure-induced left ventricular hypertrophy in rats. *Circ. Res.* 69:609-617.

Zimmerman, B.G., 1981, Adrenergic facilitation by angiotensin. Does it serve a physiological function? *Clin. Sci.* 60:343-348.
Zusman, R.M., 1984, Renin and non-renin-mediated antihypertensive actions of converting enzyme inhibitors. *Kidney Int.* 25:969-983.

VI. ABSTRACTS OF ORAL AND POSTER PRESENTATIONS

ISOLATION OF AN 86 KDa HEMORRHAGIC TOXIN FROM THE VENOM OF THE EASTERN DIAMONDBACK RATTLESNAKE *(CROTALUS ADAMANTEUS)* AND ITS EFFECT OF RAT AORTA ENDOTHELIAL CELLS IN CULTURE

Steffan G. Anderson and Charlott L. Ownby

Department of Physiological Sciences
College of Veterinary Medicine
Oklahoma State University
Stillwater, Oklahoma 74078-0350 USA

The venoms of many poisonous snakes, particularly those of the families *Viperidae* and *Crotalidae*, exhibit a hemorrhagic activity that is attributable to a family of zinc-dependent, proteolytic, hemorrhagic glycoproteins. *In vivo* studies on the pathogenesis of snake venom hemorrhagic toxins (HTs) induced hemorrhage have revealed that hemorrhage occurs *per rhexis*. In this process the capillary endothelial basal lumina is disorganized and there is endothelial cytoplasmic swelling and vacuolation, plasmalemmal blebbing and rupture and nuclear pyknosis.

Previous *in vitro* have investigated the action of HTs on isolated basal lamina proteins, but until recently no work on the direct effect of HTs on endothelial cells had been published. In this study an 86 KDa HT from the venom of the Eastern diamondback rattlesnake *(Crotalus adamanteus)* were isolated using four liquid chromatographic steps: Open column Sephasose S-200 HR SEC, Superdex 75 FPLC SEC, DEAE anion exchange HPLC and Sulfoproplycation exchange HPLC. The cytotoxic effect of this toxin was then studied using cultured monolayers of Rat Aorta Endothelial Cells (RAECs) by measuring trypan blue dye uptake and lactate dehydrogenase release and by observing morphologic changes following treatment with the toxin.

Doses of HT of 1, 0.1, and 0.01 μg/ml were cytotoxic, yielding LDH release and trypan blue uptake of about 90% of the 0.1% Triton X-100 control. These groups also exhibited severe "blebbing", cytoplasmic vacuolation and nuclear pyknosis when viewed with inverted phase contrast microscopy.

THE ROLE OF ENDOTHELIUM AND OF EXTRACELLULAR AND INTRACELLULAR ACIDOSIS FOR HYPERCAPNIA INDUCED INHIBITION OF TENSION OF ISOLATED RAT CEREBRAL ARTERIES

Rong Tian, Pia Vogel, *Niels A. Lessen, Michael J. Mulvany, Frederick Andreasen and Christian Aalkjær

Department of Pharmacology
University of Aarhus
DK-8000 Aarhus C Denmark

*Department of Clinical Physiology
Bispebjerg
Copenhagen, Denmark

The importance of changes in extracellular and intracellular pH (pH_0 and pH_i) for hypercapnic vasodilation was evaluated in vitro. Segments of the middle cerebral artery of the rat were mounted in a myograph for isometric tension recording and in all studies basal tension and tension in the presence of arginine vasopressin (AVP) was measured. pH_i was determined by loading the smooth muscle cells with the fluorescent dye bis-carboxy-ethyl-carboxy fluorescein. Control solution was a physiological saline solution with 25 mM HCO_3^- and 5% CO_2 in the bubbling air (pH 7.45-7.50). Induction of hypercapnic acidosis (10% CO_2) or normocapnic acidosis (15 mM HCO_3^-) reduced pH_i as well as pH_0 and both caused significant reduction in AVP induced tension as well as in basal tension. Selective reduction in pH_0 obtained by reducing [HCO_c^-] to 9 mM and CO_2 to 2.5% caused significant reduction in tensions induced by 1 as well as 2 U/l of AVP and in basal tension. Selective reduction in pH_i obtained by increasing [HCO_3^-] to 65 mM and CO_2 to 15% decreased AVP tension at 1 U/l but not at 2 U/l and was accompanied by an increase in basal tension. The same pattern in the tension development following selective pH_0 and pH_i reduction was found after removal of the endothelium whether 1 μM sodium nitroprusside was present or not. L-NNA significantly inhibited the relaxation of the AVP response induced by hypercapnic acidosis. Responses to selective reductions in pH_0 or pH_i in the presence of L-NNA were, however, qualitatively similar to responses seen in solutions without L-NNA.

Conclusions: (1) For the vasodilatory response to hypercapnic acidosis pH_0 is important whereas pH_i has little effect. (2) Even pronounced hypercapnia may be without vasodilating properties, in particular if endothelial function is damaged.

EFFECTS OF TITANOCENE DICHLORIDE ON ENDOTHELIAL CELLS *IN VITRO*

M. Bastaki, E. Missirlis, J.C. Murray and M.E. Maragoudakis

Department of Pharmacology
University of Patras Medical School
Patras, GREECE

Endothelium responds to tumor factors by migration and proliferation, which can lead to angiogenesis *(Hobson and Denekamp, 1984)*. Angiogenesis is an essential process for solid tumor growth and metastases and has therefore been considered as a possible target for tumor therapy *(Blood and Zetter, 1990)*. Indeed, it has been shown that inhibition of angiogenesis can inhibit solid tumor growth *(Folkman, 1972; Maragoudakis et al., 1990)*. Extracellular matrix (ECM) is a regulator of the angiogenic process *(Adams and Watt, 1993)* and, as such, it can be targeted in order to manipulate angiogenesis. We have investigated titanocene dichloride (TD), a novel antitumor agent with remarkably low toxicity *in vivo (Kopf-Majer and Kopf, 1986)* and strong antiangiogenic properties *(Bastaki et al., 1994)* for its effects on ECM secretion and deposition by endothelial cells (EC), as a possible mechanism for its antiangiogenic action. An *in vitro* system was employed, based on microvascular endothelial cells cultured on collagen I in medium with or without tumor-conditioned medium. the production of fibronectin (FN) and laminin (LM), two of the major components of the ECM was then examined in the medium, intracellular phase and subcellular matrix by ELISA.

It was demonstrated that TD reduces the deposition of FN in the subcellular matrix while increases it in the culture supernatant. This effect was more significant in the presence of TCM. TD also increased LM in the medium, particularly when TCM was present, without any significant change in the intracellular LM.

These results indicate that at least part of TD action may be confined in the alteration of ECM deposition. Reduced FN deposition may disrupt the migration of endothelial cells during angiogenesis *(McAuslan et al., 1980)*, thus inhibiting the process. Moreover, FN acts as a stabilizing component of the ECM *(Austria and Couchman, 1991)* and decreased levels may lead to fragile vessels. In conclusion, the already fragile vasculature of the tumors may be further disturbed after TD treatment, causing extensive tumor damage.

REFERENCES

Adams, J.C. and Watt, F.M., 1993, Regulation of development and differentiation by the extracellular matrix. *Development* 117:1183-1198.

Austria, M.R. and Couchman, J.R., 1991, Enhanced assembly of basement membrane matrix by endodermal cells in response to fibronectin substrate. *J. Cell Science* 99:443-451.

Bastaki, M., Missirlis, E., Klouras, N., Karakloulakis, G. and Maragoudakis, M.E., 1994, Suppression of angiogenesis by the antitumor agent titanocene dichloride. *Europ. J. Pharmac.* 251:263-269.

Blood, C.H. and Zetter, B.R., 1990, Tumor interactions with the vasculature: angiogenesis and tumor metastasis. *Biochimica et Biophysica Acia* 1032:89-118.

Folkman, J., 1972, Anti-angiogenesis: new concept for therapy of solid tumors. *Ann. Surg.* 175:409-416.

Hobson, B. and Denekamp, J., 1984, Endothelial proliferation in tumors and normal tissues: continuous labeling studies. *Br. J. Cancer* 49:405-413.

Kopf-Majer, P. and Kopf, H., 1986, Metallocene complexes: organometallic antitumor agents. *Drugs of the Future* 11, No.4:297-319.

Maragoudakis, M.E., Missirlis, E., Sarmonika, M., Panoutsacopoulou, M. and Karakiulakis, G., 1990, Basement membrane biosynthesis as a target to tumor therapy. *J. Pharm. and Exp. Ther.* 252(2):753-757.

McAuslan, B.R., Hannan, G.N., Reilly, W. and Stewart, F.H.C., 1980, Variant endothelial cells. Fibronectin as a transducer of signals for migration and neovascularization. *J. Cellular Physiology* 104:177-186.

AGE RELATED CHANGES IN THE NON-ENZYMATIC GLACIATION OF HUMAN AORTIC ELASTIN

Stephan Baydanoff, E. Konova, I. Benchev, M. Atanasova, I. Tzakov

University School of Medicine
Pleven, Bulgaria

According to one of the recent theories, age related changes are induced by the action of free radicals and Maillard reactions. The aim of this study is to investigate non-enzymatic glaciation of human aortic elastins, obtained from different age groups, by measurement of AGEs, using fluorescence method, and also to determine the capacity of different age-elastins to form fructosamine and AGEs under glaciation *in vitro*.

Human elastins were purified from thoracic aortas, obtained from 15 age-groups of healthy accident victims. By fluorescence we established that the level of AGEs increases with aging. Furthermore "young elastin" has a higher capacity to form fructosamine and AGEs *in vitro*, compared to "old elastin".

These results suggest, that human aortic elastin, similarly to other connective tissue proteins, is subjected to glaciation with aging.

GLUCOCORTICOID MODULATION OF AHESION MOLECULE EXPRESSION ON HUMAN ENDOTHELIAL CELLS

Anne C. Burke-Gaffney and Paul G. Hellewell

Department of Applied Pharmacology
National Heart and Lung Institute
Dovehouse Street
London SW3 6LY
UNITED KINGDOM

Glucocorticoids are amongst the most potent and widely used antiinflammatory agents and it is well known that administration of these drugs to humans results in inhibition of leukocyte migration to sites of inflammation *(Zweiman et al., 1976)*. However, there is conflicting evidence as to whether this effect may in part be attributable to down-regulation of endothelial cell (EC) adhesion molecules such as intercellular adhesion molecule-1 (ICAM-1) *(Bochner et al., 1987; Cronstein et al., 1992)*. In this study, we compared the effects of a combination of tumor necrosis factor-α (TNF-α) and interferon-γ (IFN-γ) with those of lipopolysaccharide (LPS) on the induction of ICAM-1 on the human EC line EAhy926 *(Edgell et al., 1983)*. In addition, we sought to investigate whether dexamethasone altered expression of this adhesion molecule induced either by TNFα/IFNγ or by LPS alone or in combination with IFN-γ. Finally, we questioned whether the effects of dexamethasone on cytokine- or LPS-induced ICAM-1 are paralleled by similar effects on adhesion of human neutrophils (PMN) to EAhy926.

ICAM-1 was detected by ELISA using a mouse anti-human ICAM-1 primary antibody (RR1/1) and a peroxidase-linked goat anti-mouse secondary antibody. Binding of antibody was quantified using a peroxidase sensitive substrate, 2, 2'-azino-bis (3-ethylbenz-thiazoline-6-sulfonic acid) and chromophore development determined by measuring optical density (O.D.) at 405nm. Results were expressed as fold increase (± S.E.M.) in O.D. (ie. O.Ds. resulting from stimuli-induced ICAM-1 divided by that reflecting basal ICAM-1). Adhesion of human peripheral blood PMN (separated by plasma/Percoll gradient centrifugation) was determined by incubating [^{111}In]-labelled PMN with non-activated or cytokine/LPS-activated EC monolayers. The number of adherent cells was determined and results were expressed as a percentage (± S.E.M.) of cytokine- or LPS- stimulated adhesion. Statistical analysis was performed using one-way ANOVA followed by the Tukey-Kramer or Dunnetts Multiple Comparison Test where appropriate. Results were deemed significant if $P < 0.05$.

Stimulation of EAhy926 with a combination of TNFα and IFNγ resulted in a dose- and time- dependent increase in ICAM-1 expression. Similarly, dose- and time- dependent induction was also observed on incubation with LPS, although the kinetics of expression varied from those following TNFα/IFNγ stimulation. In addition, it was shown that IFNγ in the presence of a maximally effective dose of LPS (0.5 μg/ml), produced a concentration-dependent increase in ICAM-1. TNFα (10 ng/ml)/IFNγ (40 ng/ml)-induced ICAM-1 expression at 24h (15.9 ± 0.8 fold; n=6) was not significantly reduced by dexamethasone (1000 nM, 2h pretreatment; 15.7 ± 0.6 fold; n=6; $P > 0.05$), however LPS (0.5 μg/ml)-induced ICAM-1 expression (3.6 ± 0.3 fold; n=5) was significantly reduced under similar conditions (2.0 ± 0.3 fold; n=5; $P < 0.01$). The effect of dexamethasone on LPS-induced ICAM-1 was dose-dependent with significant reductions using concentrations as low as 0.1 nM dexamethasone. LPS/IFNγ-induced ICAM-1 expression was also reduced with high concentrations of dexamethasone (100 and 1000 nM) but lower concentrations of this agent appeared to have no effect. In contrast to the effects of dexamethasone on ICAM-1 expression, low concentrations of this glucocorticoid (0.1-10 nM) increased adhesion of PMN to monolayers of EAhy926 activated with TNFα/IFNγ or LPS. However, 1000 nM dexamethasone significantly reduced ($P < 0.05$) adhesion of PMN to LPS-stimulated EC lthough it had no effect on adhesion to cytokine-activated EC.

In summary, TNFα/IFNγ and LPS both induced dose- and time- dependent increases in ICAM-1. However, only LPS-or LPS/IFN-induced expression was inhibited by dexamethasone. The reduction in LPS-induced ICAM-1 expression was paralleled by a reduction in PMN adheson to LPS-stimulated EC only at high concentrations of dexamethasone. The stimulatory effect of low concentrations of dexamethasone on PMN adhesion is currently under investigation. In conclusion, our results suggest that modulation by dexamethasone of ICAM-1 expression on, or PMN adhesion to EC, depends on the stimulus triggering induction. *Support of the Clinical Research Committee of the Royal Brompton National Heart and Lung Hospital, the National Asthma Campaign and the Wellcome Trust are gratefully acknowledged.*

REFERENCES

Bochner, B.S., Rutlege, B.K. and Schleimer, R.P., 1987, *J. Immunol.* 139:2303-2307.

Cronstein, B.N., Kimmel, S.C., Levin, R.I. et al., 1992, *Proc. Natl. Acad. Sci. USA* 89:9991-9995.

Edgell, C-J.S., McDonald, C.C. and Graham, J.B., 1983, *Proc. Natl. Acad Sci USA* 80:3734-3737.

Zweiman, B., Slott, R.I. and Atkins, P.C., 1976, *Allergy Clin. Immunol.* 58:657-995.

ASTROCYTE REGULATION OF MICROVASCULAR ENDOTHELIAL CELL GROWTH, DIFFERENTIATION, AND PROGRAMMED CELL DEATH

M.A. Behzadian, J. Li, Z. Zhang and R.B. Caldwell

Cellular Biology and Anatomy
The Medical College of Georgia
Augusta, Georgia USA

Growth or regression of microvessels in response to long term changes in tissue needs for blood supply is critical for normal development and recovery from injury or disease. Our analyses are focused on determining the cellular mechanisms that control this process. Since astroglial cells are known to be in physiological communication with both neurons and microvessels, we hypothesized that they have a role in microvascular growth control. Our previous analyses using retinal models have supported this hypothesis and suggest that astrocytes regulate microvascular development by generating soluble factors that inhibit endothelial cell growth and induce their morphogenic differentiation *(Jiang et al., 1993; Jiang et al., in press; Behzadian et al., 1994)*. Here we report on experients designed to identify these factors and determine the mechanisms of their actions. Two different endothelial cells, bovine retinal microvascular (BRE) and aortic (BAE) cells were used for these experiments. Mink lung epithelial CCL-64 cell line was used as a control for assaying the activity of transforming growth factor β (TGF-β). Primary cultures of rat brain astrocytes prepared as confluent monolayers were maintained in serum-free medium for 72 hours. This astrocyte-conditioned medium (ACM) was then added to exponentially growing target cells for 3 days and effects on cell growth were tested. ACM was found to inhibit the growth of all three cell types in a dose dependent manner. Furthermore, the inhibitory activity of ACM was significantly increased by transient acidification (pH=2, 10-20 min). This treatment is known to activate the latent form of TGF-β. Adsorption of the ACM using an anti-TGF-β polyclonal antibody column reduced its growth inhibiting activity for CCL cells by 96%. By contrast, the growth inhibitory effect was reduced by only 50% for both endothelial cell types, indicating that astrocytes express an endothelial cell-specific growth inhibitory activity in addition to a TGF-β like activity. Fractionation of ACM showed that it contains multiple factors with varying effects on endothelial cell growth, differentiation or programmed cell death. Cultures of endothelial cells incubated with fractionated ACM showed DNA fragmentation and programmed cell death in some cells while other cells in the culture differentiated morphologically. The identity of these factors and the mechanisms targeting endothelial cells for differentiation or death following growth inhibition are under investigation.

REFERENCES

Behzadian, M.A., Zhang, Z.Q., Jiang, B., Caldwell, R.B., 1994, Astrocyte retulation of angiogenesis: an in vitro model. *Invest. Ophthalmol. Vis. Sci.* 35 (Suppl):1719.

Jiang, B., Liou, G.I., Caldwell, R.B., 1993, Astrocytes modulated endothelial cell growth, differentiation, and fibronectin expression. *Invest. Ophthalmol. Vis. Sci.* 34(Suppl):899.

Jiang, B., Liou, G.I. Behzadian, M.A., Caldwell, R.B., in press, Astrocytes induce retinal vasculogenesis: effects on fibronectin expression. *J. Cell Sci.*

A NOVEL PEPTIDE INHIBITS INDUCTION OF NITRIC OXIDE SYNTHASE

G. Abou-Mohamed, A. Papapetropoulos and R.W. Caldwell

Department of Pharmacology and Toxicology
Medical College of Georgia
Augusta, Georgia 30912 USA

Alpha-melanocyte stimulating hormone has been shown to prevent endotoxin shock. A heptapeptide analog (HPA) has recently been synthesized and shown to be an even more potent protective agent. Since the hypotensive and toxic actions of endotoxin lipopolysaccharide (LPS) appear to involve the induction of nitric oxide synthase, we have examined the actions of HPA on induction of iNOS using an endotoxemia model in conscious rats. In vascular smooth cells, we also determined whether the HPA could directly alter the actions of LPS, TNFα or Il-1β on NOS induction. Rats were instrumented for measurement of blood pressure and i.v. administration of agents prior to the day of experiment. In the conscious state, rats were given E. coli LPS (5 mg/kg, i.v.) and monitored for 6 hrs. One group of rats received HPA (30 μg/kg) 0.5 hr before LPS. Using L-NAME-sensitive cGMP production as an estimate of NOS activity, we determined that LPS increases iNOS and that HPA pretreatment markedly reduces this response. In isolated aortic cells, LPS and Il-1β, but not that of LPS. This former action was reversed by L-arginine. In conclusion, HPA can prevent the *in vivo* induction of NOS by LPS. It does not appear to directly prevent this action of LPS. Possibly, a metabolite of HPA or a secondary mediator is involved.

AN EXAMINATION OF ENDOTHELIAL CELL FUNCTION ON SUBSTRATES MODIFIED WITH IMMOBILIZED BIOACTIVE PEPTIDES

Sean M. Carroll, Kay C. Dee, Thomas T. Anderson* and R. Bizios

Department of Biomedical Engineering
Rensselaer Polytechnic Institute
Troy, New York 12180, USA

*Department of Biochemistry
Albany Medical College
Albany, New York 12208, USA

Recent advances in molecular/cellular biology and biochemistry have elucidated the role that functional domains of adhesive proteins (such as Arginine-Glycine-Aspartic Acid-Serine (RGDS) of fibronectin, and Tyrosine-Isoleucine-Glycine-Serine-Arginine-Glycine (YIGSRG) of laminin) play in mediating cell adhesion to substrates. Interaction of anchorage-dependent cells (via integrin receptors) with adhesive peptides of chemical components of the extracellular matrix trigger intracellular signal pathways which, subsequently, promote specific cellular functions such as motility and proliferation, which are crucial to endothelialization of a vascular implant material. Biomaterials for implantation should promote, maintain and, generally simulate the critical, beneficial conditions of the physiological milieu present during the wound healing process. A vascular implant, for example, should be designed to promote cellular functions which synergistically lead to the desirable end result of complete endothelialization within a clinically acceptable time period while remaining free of thrombotic complications.

The objective of the present study was to investigate the effects of soluble basic Fibroblast Growth Factor (bFGF) on the function of bovine pulmonary artery endothelial cells cultured on model substrates modified with covalently bound bioactive peptides (specifically YIGSRG and RDGS). For that purpose adhesion, cell population motility, and proliferation were studied under standard cell culture conditions in the presence or absence of bFGF (1ng/mL). Compared to substrates with covalently bound EGDS or to reference substrates (aminophase glass and plain borosilicate glass) endothelial cell population motility and proliferation on substrates with covalently bound YIGSRG were significantly ($p<0.05$) increased in the presence of bFGF. Cell adhesion, however, was not significantly affected by either substrate modification, presence of bFGF, or both.

This research constitutes an *in vitro* investigation of cellular events which are crucial to the patency of vascular grafts. Because endothelialization of a substrate is not dependent on adhesion alone, improved biomaterials surfaces should enhance ensuing cellular functions as well. The results of this study provide evidence that advanced biomaterials for vascular implantation must combine the influences of adhesive peptides and mitogenic chemical growth factors in order to promote and sustain endothelialization in a timely manner.

CHARACTERIZATION OF NOVEL HUMAN ENDOTHELIAL CELL LINES

J.E. Cartwright, G. St J. Whitley, A.P. Johnstone

Department of Cellular and Molecular Sciences
St. George's Hospital Medical School
London, SW17 ORE U.K.

Leukocyte-endothelial cell interactions are centra to the normal re-circulation of lymphocytes and to pathological situations such as inflammation. In vitro studies aimed at dissecting molecular processes underlying these interactions would benefit greatly from a stable, reproducible, homogeneous source of human endothelial cells. Here we report the characterization of a novel series of human endothelial cell lines (designated "SGHEC") regarding the expression and release of adhesion molecules and their binding of lymphocytes. SGHEC expressed significant levels of intercellular adhesion molecule-1 (ICAM-1) which increased after cytokine stimulation. Vascular cell adhesion molecule-1 (VCAM-1) and E-selectin were not detectable on unstimulated SGHEC but substantial levels were expressed after stimulation with either TNFα or IL-1β but not with IFN-γ. The increased expression of ICAM-1 and VCAM-1 was evident after 4h stimulation and was even higher after 24h; E-selectin was maximal after 4h and returned almost to basal levels by 24h. Substantial quantities of immunoreactive ICAM-1 and VCAM-1 also accumulated as soluble material in the supernatants of TNFα-stimulated SGHEC (VCAM-1) was substantially higher than ICAM-1), but E-selectin remained below the limits of detection. The quantities suggested that this release is stimulated by the cytokine and is not merely a loss of a fixed proportion of the cell-surface content, thus providing support for the idea of a physiological function for these soluble molecules.

Primary human lymphocytes bound to SGHEC; this binding increased substantially after activation of either cell type. The binding was inhibited by monoclonal antibodies against LFA-1 and, to a lesser extent, ICAM-1, thus demonstrating the importance of these molecules in the observe binding; neither anti-VCAM-1 nor anti-E-selectin antibodies affected the binding. From these various data, we conclude that LFA-1/ICAM-1 interactions are required but are probably not sufficient for lymphocytes to bind to endothelium. Our data also provide further evidence of the stability and normality of the differentiated endothelial functions of SGHEC, and support their potential use in investigating leukocyte-endothelial interactions and the mechanism of release of soluble adhesion molecules.

A NEW FUNCTION FOR THE LDL RECEPTOR: TRANSCYTOSIS OF LDL THROUGH THE BLOOD-BRAIN BARRIER

Bénédicte Dehouck, Marie-Pierre Dehouck, Jean-Charles Fruchart and Roméo Cecchelli

Serlia, Inserm U.325
Institut Pasteur
Lille, France

Lipoprotein transport across the blood brain barrier is of critical importance for delivery of essential lipids to the brain cells. The occurrence of a low density lipoprotein (LDL) receptor on the BBB has recently been demonstrated. To examine further the function of this receptor we have developed a model system in which brain capillary endothelial cells and astrocytes are co-cultured on opposing sides of a filter that prevents physical interaction between the two cell types.

Using this model, we have shown that in contrast to acetylated LDL which do not cross the blood-brain barrier, LDL is specifically transcytosed across the monolayer of brain capillary endothelial cells. This phenomenon is temperature dependent. The C-7 monoclonal antibody known to interact with the LDL receptor binding domain, totally blocked the transcytosis of LDL, while control irrevelant IgG had no effect, suggesting that the transcytosis is mediated by the receptor. Furthermore, radiolabeled LDL was not degraded during the transcytosis in contrast to what was observed with acetylated LDL, indicating that the transcytotic pathway in brain capillary endothelial cells is different from the LDL receptor classical pathway. We have recently shown that the lipid requirement of astrocytes increases the number of LDL receptors at the luminal side of brain capillary endothelial cells. That is why we next investigated the capacity of astrocytes to modulate LDL transcytosis. When astrocytes were preincubated in lipoprotein deficient media prior to coculture, the rate of LDL transcytosis was increased 10 fold, compared to that observed when endothelial cells cocultured with astrocytes pre-incubated in complete media. This result confirms that the transcytosis of LDL through the blood-brain barrier is receptor mediated.

RECEPTOR-MEDIATED PATHWAYS OF ENDOTHELIUM ACTIVITY IN EXPERIMENTAL ATHEROSCLEROSIS

Alessandro Chinellato, Eugenio Ragazzi, Luisa Pandolfo, Mariella De Biasi, Guglielmina Froldi, Laura Caparrotta and Giuliana Fassina

Department of Pharmacology
Largo E. Meneghetti
2 I-35131 Padova, Italy

Endothelium has been indicated as an independent tissue, the function of which has a considerable influence on the arterial wall. Endothelium modulates vascular tone by releasing many mediators, and among these nitric oxide (NO) is a potent vasodilator and inhibitor of platelet aggregation. Nitric oxide has been identified with the endothelium-derived relaxing factor (EDRF) which plays an important protective role in preventing vasoconstriction and thrombogenesis. Because of its strategical anatomical position, endothelium is a primary target for injurious stimuli and cardiovascular risk factors, such as hyperlipidemia: atherosclerosis has therefore been termed "endotheliopathy".

The pharmacological characterization of endothelial receptor activity during the progression of atherosclerosis in Watanabe heritable hyperlipidemic (WHHL) rabbits was examined.

Muscarinic receptor activity is progressively and gradually reduced by chronic exposure to hypercholesterolemia, being almost completely abolished in animals older than 12 months of age. Adenosine 5'-triphosphate (ATP) produces unimpaired relaxation in 1-year-old WHHL rabbit aorta. The potent P_{2y} -dependent relaxing drug 2-methyl-thio=ATP did not induce any relaxation, indicating that ATP-mediated relaxation is independent on P_{2y} -purinoceptor relaxant activity and that such activity is completely impaired in 12-month-old WHHL rabbit aorta as a consequence of the atherosclerotic process. In 3- and 12-month-old WHHL rabbit, the aortic relaxation due to UTP was comparable to that obtained in normolipemic rabbits, suggesting that *nucleotide receptor* activity is well maintained even after 1 year of chronic exposure to high lipid levels. Nucleotide-mediated relaxation was decreased only in a later stage, in 30-month-old WHHL rabbits. The data presented show that progression of atherosclerosis affects endothelial receptor functionality in different ways. *Nucleotide receptor* activity is better preserved than that of the other endothelial receptors, i.e. muscarinic and P_{2y} -purinoceptors. It is therefore hypothesized that atherosclerosis in genetically hyperlipidemic animals selectively influences the receptor arrangement of endothelium. Endothelial cell receptors may become a new interesting target for the development of drugs to treat atherosclerosis.

INTERACTIONS OF SUPEROXIDE DISMUTASES WITH NITRIC OXIDE

James D. Crapo, M.D.

Department of Medicine
Duke University Medical Center
349 Bell Building
Durham, NC 27707-3162, USA

The superoxide radical (O_2^-) is produced as a consequence of both spontaneous and enzymatic oxidations that occurs both within cells and on their outer surfaces. O_2^- is chemically reactive and it may contribute to the production of even more reactive entities, such as the hydroxyl radical (HO·). Indeed, were it not controlled, O_2^-, and its progeny, would rend the macromolecular fabric of cells and would make aerobic life all but impossible. There is evidence showing that O_2^- is a key player in a variety of vascular inflammatory diseases, in reperfusion injuries, in hemorrhagic shock, in oxygen toxicities, in infections, in neoplasia, and in apoptosis.

We manage to live with air, and to utilize oxygen, by virtue of a coordinated system of defenses. First among these defenses are the superoxide dismutases (SODs), a family to metalloenzymes which catalyze the conversion of O_2^- into H_2O_2 plus O_2. Mammals produce three distinct SODs. One is a dimeric copper- and zinc-containing enzyme (CuZn SOD) found in the cytosol of all cells. A second is a tetrameric manganese-containing SOD (MnSOD) found within mitochondria. the third is a tetrameric, glycosylated enzyme found in the extracellular fluids - extracellular superoxide dismutase (EC-SOD). EC-SOD activity is particularly high in the extracellular matrix surrounding the vasculature. The enzyme has a hydrophobic properties and contains a cluster of positively charged amino acids near the carboxy terminal of the protein which give this protein a heparin or glycosaminoglycan (GAG) binding property. EC-SOD has been speculated to partition between cell surfaces and extracellular fluids where it is thought to be a primary defense against O_2^- produced by inflammatory cells.

The discoveries that the endothelium derived relaxing factor is NO· and that NO· reacts with O_2^- at a diffusion-limited rate, have clarified one function of the EC-SOD. NO· exerts a hypotensive effect by causing relaxing of the smooth muscle in the walls of blood vessels and is a critical element of the normal regulation of blood pressure on a minute by minute basis. Thus, endothelial cells can sense the velocity of blood flow and regulate the production of NO· in responses to changes in that velocity. Endothelium also produces O_2^-, as do circulating neutrophils and eosinophils. This O_2^- decreases the half life of NO· and therefore exerts a hypertensive effect. Finally, EC-SOD, by removing the O_2^-, increases the half life of NO· and hence exerts a hypotensive effect. This hypotensive effect of EC-SOD has been demonstrated on genetically hypertensive rodents.

The Role of NO· and O_2^- in Inflammation: NO· has been shown to inhibit neutrophil adhesion and to inhibit the NADPH oxidase on neutrophil membranes. Thus NO· has the potential to act as a powerful anti-inflammatory agent. In the extracellular spaces the primary species that will react with NO· is O_2^-. The radical-radical reaction between O_2^- and NO· is one of the fastest reactions known ($\geq 10^9 M^{-1} \cdot s^{-1}$). This reaction converts NO· to peroxynitrite

($ONOO^-$). $ONOO^-$ has little or no smooth muscle relaxing properties and is a strong pro-inflammatory mediator. Thus it is critical that, where inflammation is to be controlled, the reaction of O_2^- with NO· must be prevented. High EC-SOD levels act to preserve NO· and thus create strong anti-inflammatory microenvironments.

REFERENCES

Clancy, R.M., Leszczynska-Piziak, J. and Abramson, S.B., 1992, Nitric oxide, an endothelial cell relaxation factor, inhibits neutrophil superoxide anion production via a direct action on the NADPH oxidase. *J. Clin. Invest.* 90:1116-1121.

Freeman, B.A. and Crapo, J.D., 1982, Biology of disease: free radicals and tissue injury. *Lab. Invest.* 47:412-426.

Gryglewski, R.J., Palmer, R.M.J. and Moncada, S., 1986, Superoxide anion is involved in the breakdown of endothelium-derived relaxing factors. *Nature* 320:454-456.

Kubes, P. and Granger, D.N., 1992, Nitric oxide modulates microvascular permeability. *Am. J. Physiol.* 262:H611-H615.

Marklund, S.L.,1984, Extracellular superoxide dismutase in human tissues and human cell lines. *J. Clin. Invest.*74:1398-1403.

Oury, T.D., Chang, L., Marklund, S.L., Day, B.J. and Crapo, J.D.,submitted 1993, Immunocytochemical localization of extracellular superoxide dismutase in human lung. *Lab. Invest.*

Oury, T.D., Ho, Y.S., Piantadosi, C.A. and Crapo, J.D., 1991, Extracellular superoxide disumutase, nitric oxide and central nervous system O_2 toxicity. *Proc. Natl. Acad. Sci USA* 89:9715-9719.

Radi, R., Beckman, J.S., Buck, K.M. and Moncada, S., 1991, Peroxynitrite-induced membrane lipid peroxidation: the cytotoxic potential of superoxide and nitric oxide. *Arch. Biochem. Biphys.* 288:481-487.

Saran, M., Michel, C. and Bors, W., 1990 Reaction of NO with $O_2^{-\cdot}$. Implication for the action of endothelium-derived relaxing factor (EDRF). *Free Radical Res. Commun.* 10:221-226.

MODULATION OF CULTURED HUMAN ENDOTHELIAL CELLS BY HYPOXIA AND REOXYGENATION

L.M. Cunha-Ribeiro[1] and K.S. Sakariassen[2]

[1]Departments of Physiology and Immunohematology
Porto Medical School
Porto, Portugal

[2]Nycomed Drug Research
Gaustadallen 21
Oslo, Norway

INTRODUCTION

Previous studies indicated that hypoxia is associated with a prothrombotic tendency. Exposure of bovine aortic endothelial cells (BAECs) to hypoxia elicits a specific biosynthetic response, including a factor X activator (Xact). Another striking effect of hypoxia on BAECs is the decrease on thrombomodulin on the EC surface. So far, the behavior of human ECs under such conditions has not been reported.

We have exposed cultured human umbilical vein-endothelial cells (HUVEC) to low concentration of oxygen (14 mmHg), followed by reoxygenation, and the expression of a potential procoagulant, as well as von Willebrand factor (vWF), tissue plasminogen activator (tPA), Plasminogen activator inhibitor type-1 (PAI-1) and endothelin was measured either to the luminal side, either the abluminal side of the ECs.

Material and Methods

HUVEC were isolated from human umbilical veins, and cultured according to Jaffe et al (JCK 52:2745, 1973) with some minor modifications (Exp. Cell Res. 139:191, 1982). Cells were made hypoxic by transferring them to a CO^2 incubator with a PO_2 of 14 mmHg, for variable periods of time (12-72h). Some of those cells were placed in an ambient air atmosphere (reoxygenation) for 24h.

The procoagulant properties were studied using a coagulant assay. Thrombomodulin expression was measured by the formation of activated protein C (aPC) using a chromogenic assay. PAI-1, tPA, vWF and endothelin were measured using commercial Elisa Kits. These studies employed cultures grown on polycarbonate membranes mounted on polystyrene inserts (Transwell plates, Costar™).

Results

In opposite to BAECs submitted to hypoxemic conditions no Xact was found on HUVEC. On these conditions, tissue factor (TF) is also not expressed. However hypoxia led to a fall in cell surface thrombomodulin (11.5 ± 2.3 μg/ml/10^5 cells vs 32.7 ± 5.2 μg/ml/10^5 cells, which partially recovered after 24h reoxygenation (18.7 ± 3.2 μg/ml/10^5 cells).

vWF release increases on the abluminal side under hypoxia (176.2 ± 52.0 ng/ml vs 309.2 ± 20.2 ng/ml), but increases on the luminal side after reoxygenation (296.2 ± 13.7 vs 241.1 ± 15.1 ng/ml).

Hypoxia induce an increase of tPA release on the abluminal side (1.35 ± 0.12 vs 0.98 ± 0.18 ng/ml/72h), but reoxygenation does not induce any significant change. By the contrary, PAI-1 release is not changed by hypoxia, but reoxygenation induces a significant reduction either in the luminal side either in the abluminal side. Endothelin release levels are also not affected by hypoxia, however reoxygenation induces an increased release in both sides.

Conclusion

These studies demonstrate that hypoxia and reoxygenation can perturb endothelial functions. Although a direct procoagulant factor has been demonstrate on BAECs, such as evidence was not found on human ECs. Decreased expression of thrombomodulin during hypoxia can contribute to the prothrombotic state observed under these experimental conditions. The release of tPA and vWF to the abluminal side is not easily understood and deserves further investigation. During reoxygenation the most striking finding are the reduction of PAI-1 release to the abluminal side and the increase release of endothelin. Although the in vivo physiologic relevance of these findings in cell culture has yet to be determined, they can contribute to a better understanding of the pathophysiology of the reperfusion injury.

MODULATION OF VASCULAR PROPERTIES OF CULTURED HUMAN ENDOTHELIUM BY HYPOXIA AND REOXYGENATION

L.M. Cunha-Ribeiro[1] and K.S. Sakariassen[2]

[1]Departments of Physiology and Immunohematology
Porto Medical School
Porto, Portugal

[2]Nycomed Bioreg AS
Guastadallen 21
Oslo, NORWAY

INTRODUCTION

Previous studies have indicated that hypoxia is associated with a prothrombotic tendency. Exposure of bovine aortic endothelial cells (BAECs) to hypoxia elicits a specific biosynthetic response, including a factor X activator (Xact) *(Ogawa et al., 1990; Ogawa et al., 1990; Ogawa et al., 1993)*. Another striking effect of hypoxia on BAECs is the decrease in thrombomodulin on the cell surface. So far, the behavior of human endothelium under such conditions has not been reported.

We have exposed cultured human umbilical vein-endothelial cells (HUVEC) to low concentration of oxygen (14 mmHg), followed by reoxygenation, and measured the expression of thrombomodulin and of tissue factor. Release of von Willebrand factor (vWF), tissue plasminogen activator (tPA), plasminogen activator inhibitor type-1 (PAI-1) and endothelin were measured on the luminal side, and on the abluminal side.

Materials and Methods

HUVEC were isolated from human umbilical veins, and cultured according to Jaffe *et al* with some minor modifications *(Willems et al., 1982)*. Cells were made hypoxic by transferring them to a CO_2 incubator with a PO_2 of 14 mmHg, for variable periods of time (12-72h). Some of those cells were exposed to atmosphere air (reoxygenation) for 24h.

The procoagulant properties were studied by coagulant assay. thrombomodulin expression was measured by the formation of activated protein C (aPC) using a chromogenic assay. PAI-1, tPA, vWF and endothelin were measured by commercial Elisa Kits. These studies employed cultures grown on polycarbonate membranes mounted on polystyrene inserts (Transwell plates, Costar™).

Results

In contrast to BAECs exposed to hypoxemic condition no Xact was found on HUVEC. At this condition, tissue factor (TF) was not expressed. However hypoxia led to a fall in cell surface thrombomodulin (11.5 ± 2.3 μg/ml/10^5 cells vs 32.7 ± 5.2 μg/ml/10^5 cells, which partially recovered after 24h of reoxygenation (18.7 ± 3.2 μg/ml/10^5 cells).

vWF release increase on the abluminal side under hypoxia (176.2 ± 52.0 ng/ml vs 309.2

± 20.2 ng/ml), but increase on the luminal side after reoxygenation (296.2 ± 13.7 vs 241.1 ± 15.1 ng/ml).

Hypoxia induced an increase of tPA release on the abluminal side (1.35 ± 0.12 vs 0.98 ± 0.18 ng/ml/72h), but reoxygenation did not induce any significant change. By the contrary, PAI-1 release was not changed by hypoxia, but reoxygenation induced a significant reduction either in the luminal side either in the abluminal side. Endothelin release levels were not affected by hypoxia. However, reoxygenation induced an increased release on both sides.

Conclusion

These studies demonstrate that hypoxia and reoxygenation can perturb endothelial functions. Although a direct procoagulant effect has been demonstrated on BAECs, such evidence was not found on human ECs. However, decreased expression of thrombomodulin during hypoxia can contribute to the prothrombotic state observed under these experimental conditions. the release of tPA and vWF to the abluminal side is not easily understood and deserves further investigation. During reoxygenation the most striking finding are the reduction of PAI-1 release to the luminal side and the increase in release of endothelin. Although the in vivo pathophysiological relevance of these findings in cell culture has yet to be determined, they may contribute to a better understanding of the pathophysiology of the reperfusion injury.

REFERENCES

Jaffe, E.A., Nachmann, R.L., Becker, C.G., and Minick, C.R., 1973, Culture of human endothelial cells derived from umbilical veins. Identification by morphological and immunologic criteria. *J. Clin. Invest.* 52:2745-2753.

Ogawa, S., Shreeniwas, R., Butura, C., Brett, J., and Stern, D.M., 1990, Modulation of endothelial function by hypoxia: perturbation of barrier and anticoagulant function, and induction of a novel factor X activator. *Adv. Exp. Med. Biol.* 281:303-312.

Ogawa, S., Gerlach, H., Esposito, C., Pasagian-Macaulay, A., Brett, J., and Stern, D.M., 1990, Hypoxia modulates the barrier and coagulant function of cultured bovine endothelium. Increased monolayer permeability and induction of procoagulant properties. *J. Clin. Invest.* 85:1090-1098.

Ogawa, S. Shreeniwas, R., Butura, C., Brett, J., and Stern, D.M., 1993, Modulation of endothelial function by hipoxia: perturbation of barrier and anticoagulant function, and induction of a novel factor X activator, *in:* Fibrinogen, Thrombosis, Coagulation and Fibrinolysis. C.Y. Liu and S. Chien (eds), Plenum Press, New York (1993).

Willems, C.H., Astaldi, C.R., de Groot, P.G., Jansen, M.C., Gonsalves, M.D., Zeylemaker, W.P, van Mourik, J.A., and van Aken, W.G., 1982, Media conditioned by cultured human vascular endothelial cells inhibit the growth of vascular smooth muscle cells. *Exp. Cell Res.* 139:191-197.

QUANTIFICATION OF PULMONARY CAPILLARY SURFACE AREA IN PATIENTS UNDERGOING CORONARY ARTERIAL BYPASS GRAFTING

Attila Cziraki, Joseph W. Rubin, Norman R. Finney and John D. Catravas

Medical College of Georgia
1120 15th Street
Augusta, Georgia 30912, USA

Inhibition of pulmonary capillary endothelium-bound (PCEB) ACE has been suggested as the locus of the antihypertensive action of ACE inhibitors. The purpose of this study was to contrast the enalaprilat-induced inhibition of PCEB and serum (SE) ACE activities. For PCEB ACE measurements, the specific ACE substrate, ^{3}H-benzoyl-Phe-Ala-Pro (40µCi or 2 nmol), was injected as a bolus into a central vein and immediately blood was withdrawn from a radial artery catheter. Plasma concentrations of surviving substrate (S) and product (P: ^{3}H-benzoyl-Phe) were estimated and enzyme activity, $v=1n[(S+P)/S]$, was calculated over the entire arterial outflow concentration curve. In chronically treated patients (n=6; 10 mg/day enalapril, p.o.) both PCEB and SE ACE were significantly reduced compared to a group (n=7) of untreated patients (0.54±0.1 vs 1.35±0.17 and 1.3±0.1 U/ml vs. 2.75±0.2 U/ml). In a second, acute study, similar parameters were estimated at baseline (T1) and at 15 min (T2) and 2 h (T3) after iv. administration of 1.5 µg/kg enalaprilat (n=6; patients undergoing diagnostic video assisted thoracic surgery). This treatment had no significant effect on mean arterial pressure (91±3 vs. 86±3 mmHg for T1, T2 and T3, respectively), but significantly decreased PCEB and SE ACE activities. When normalized to pre-drug (T1) activity levels, enalaprilat inhibited PCEB and ACE activity at T2 by 74±3% and 67±2%, respectively. Two hours after enalaprilat, however (T3), PCEB ACE inhibition was 66±2%, whereas SE ACE inhibition was reduced to 32±3%, suggesting tissue specificity for the inhibitory actions of enalaprilat. This procedure can be used to a) estimate changes in PCEB ACE activity in the lung, b) distinguish between SE and tissue effects of ACE inhibitors, and c) aid in the development of tissue-specific ACE inhibitors.

ESTIMATION OF PERFUSED CORONARY CAPILLARY SURFACE AREA IN PATIENTS UNDERGOING CORONARY ARTERIAL BYPASS GRAFTING

Attila Cziraki, Joseph W. Rubin, Norman R. Finney and John D. Catravas

Medical College of Georgia
Augusta, GA 30912, USA

We utilized indicator-dilution techniques to estimate the activity of capillary endothelium-bound angiotensin converting enzyme (ACE) in twelve patients (age: 47-72 yrs) undergoing coronary arterial bypass graft (CABG) surgery. Measurements were performed before and after graft connection in the coronary (C) vascular bed and, at the same time, similar measurements were performed in the pulmonary (P) vascular bed. The specific ACE substrate, ^{3}H-benzoyl-Phe-Ala-Pro (40μCi or 2 nmol), was injected as a bolus into the root of aorta or central vein and immediately blood was withdrawn from the coronary sinus or radial artery catheter, for C and P measurements, respectively. The procedure did not produce noticeable changes in routinely monitored cardiopulmonary parameters. Plasma concentrations of surviving substrate (S) and product (P: ^{3}H-benzoyl-Phe) were estimated and substrate metabolism (M=P/(S+P)), enzyme activity (v=1n(S+P)/S)) and the parameter A_{max}/K_m (proportional to perfused capillary surface area) were calculated over the entire arterial outflow concentration curve. After CABG, coronary blood flow increased by 40.6% (354±32 to 498±42 ml/min), whereas pulmonary blood flow remained unchanged (4.9±0.2 to 5.2±0.3 l/min). Perfused pulmonary capillary surface area also remained unaltered. However, perfused coronary capillary surface area increased significantly 1.7-, 2.3-, 4.1- and 3.9-fold in patients receiving 2(n=3), 3(n=4), 4(n=3) and 5(n=2) grafts, respectively. Overall, transpulmonary M and v values remained unaltered, whereas transcoronary M and v values increased (49.3±3% to 77.2±1.4% and 0.67±0.12 to 1.43±0.14, respectively). This procedure can provide quantitative diagnosis of coronary occlusion and an index of reperfusion after CABG.

ULTRAVIOLET B (UVB) RADIATION INDUCES PEROXINITRITE ($ONOO^-$) FORMATION BY HUMAN ENDOTHELIAL CELLS

G. Deliconstantinos, V. Villiotou and J.C. Stavrides

Department of Experimental Physiology
University of Athens Medical School
Athens, Greece, GR 11527

Ultraviolet B radiation (UVB 290-320 nm) up to 100 mJ/cm^2 of human endothelial cells increased nitric oxide (NO), cGMP, superoxide O_2- and hydroperoxide (H_2O_2) production. UVB irradiated endothelial cells elicited luminol -dependent chemiluminescence (CL) that increased linearly by the addition of L-Arg. N-mono-methyl-L-arginine (L-NMMA) (1 mM) inhibited CL by up to 70%. SOD inhibited CL response indicating that O_2- was a key intermediate for chemiexcitation. Neither NO nor O_2- alone were capable in directly inducing luminol CL suggesting that $ONOO^-$ was responsible for the CL of luminol. Cysteine (3 mM), and desferioxamine (200 μM) reduced the peak CL by 90% while mannitol (100 mM) and DMSO (100 mM) was ineffective. A similar CL response was also obtained using authentic $ONOO^-$. UV spectra were recorded 1 hour after UVB radiation (100 mJ/cm^2) of endothelial cells in a Hank's Balanced Salt Solution (HBSS) made basic (pH: 8.0) by Na_2CO_3. The concentration of $ONOO^-$ formed by irradiated endothelial cells based on an ε_{382} nm value of 1670 $<^{-1}$ cm^{-1} was estimated to be 50 nmol ONOO per 10^6 cells. Spectroscopy of human erythrocyte lysates showed absorbance at 540 and 576 nm. When intact erythrocytes (10^6 cells) were cultured with UVB (100 mJ/cm^2) irradiated endothelial cells (10^8 cells) the absorption spectrum of this lysate demonstrated a decrease (80%) of these peaks, findings consistent with those obtained using authentic $ONOO^-$. ONOO (up to 100 μM) increased four folds the soluble guanylate cyclase activity of endothelial cells. This effect was reduced by approximately 90% in the presence of desferioxamine and by approximately 50% by bicarbonate. UVB radiation causes lipid peroxidation as evidenced by the formation of thiobarbituric acid reactive substances (TBARS) and the stimulation of endothelial cells oxygen consumption. $ONOO^-$ initiated lipid peroxidation of endothelial cells in an extent dependent on $ONOO^-$ concentration, with the greatest yield occurring at 200 μM $ONOO^-$. Oxygen consumption was conducted using a Clark electrode, in a 1.8 ml mater-jacked chamber maintained at 37°C. In UVB irradiated endothelial cells a slow phase reaction of oxygen consumption lasting for 15 min was observed. $ONOO^-$ decomposition in HBSS resulted in oxygen evolution in the medium. Oxygen evolution from the decomposition of $ONOO^-$ was complete within 5-10 sec. In $ONOO^-$ endothelial cells reaction an initial rapid phase of oxygen consumption lasting for 10 min was observed. Our results showed that ultraviolet radiation of human endothelial cells results in the elaboration of NO and O_2- leading to the production of $ONOO^-$. Radiation of the human skin may lead to the production of $ONOO^-$ by endothelial cells resulting in inflammation that accompanies human sunburn reactions.

CELLULAR DIVERSITY OF HUMAN PLACENTAL VESSEL WALL

Ramazan Demir[1], A. Yasemin Demir[2], Gabriele Kohnen[3] and Peter Kaufmann[3]

[1]Departments of Histology and Embryology
Faculty of Medicine
Akdeniz University
Antalya, TURKEY

[2]Biochemistry
Faculty of Medicine
Akdeniz University
Antalya, TURKEY

[3]Department of Anatomy
RWTH
Aachen, GERMANY

The basic cell type of the placental villus core is undifferentiated stromal cells. Beginning in the early periods of vascularization, these show a differentiation towards the whole variety of villus stromal cells *(Benirschke and Kaufmann, 1990; Demir and Erbengi, 1984)*, and form the vessels *(Demir et al., 1989)*. From these new cell varieties; fibroblasts, reticulum cells, pericytes, endothelial cells, myofibroblasts, Hofbauer cells and embyronic plasma cells and the entire stromal architecture is constructed. From the early periods of pregnancy, many developmental changes and differentiations, which continue in various directions, take place in the villus core *(Benirschke and Kaufmann, 1990; Demir et al., 1989)*. The cellular variability of placental vessels wall is very rarely examined depending on the solution embryonal cell potence and differentiation processes.

The purpose of this study is; to describe the cellular diversity of human placental vessel wall in different periods of pregnancy, and to examine the three-dimensional configuration of the villous vessels and their relation with stromal architecture which are broken under different conditions. Placental samples were obtained from the curretage and hysterectomy cases for social and medical indications and spontaenous deliveries. Tissue samples were prepared separately for transmission and scanning electron microscopic observations.

We have identified various stromal cell varieties in placental villi core throughout the gestation. These are fibroblast, reticulum cells, Hofbauer cells, smooth muscle cells, angiogenetic cells, hemopoietic cells, pericytes, endothelial cells, myofibroblasts, glycogen-rich cells. The intrastromal impulse and regulation centers of these differentiations are multipotent mesenchymal cells. These cells are the predominant cell variety until the end of 6-8th week of gestation *(Benirschke and Kaufmann, 1990; Kaufmann, 1982)*. the transformation process has been showed in our previous study *(Demir et al., 1984)* depending on the mesenchymal cell differentiation to angiogenetic cells, presumptive pericytes, primitive endothelial cells and hemocytoblast cell, and these forms mature endothelial layer by ultrastructurally as has been showed by Kohnen, 1994, using immunocytochemical methods in placestal villi core. The cell composition of vessel wal is endothelium, fibroblast and smooth muscle cells are situated in subendothelium and in medial layer, myofibroblast-like cells distributed around the medial and in adventitial layers, glycogen-rich cell and mesenchymal cell dispersed extravascular areas and subtrophoblastic regions of placental villi. Smooth muscle cells are dominantly distributed in medial layer of large vessels and around capillaries in stromal architecture of placental villi. Endotheliogenesis and pericyte formation processes have been detailed in vasculogenesis and angiogenesis *(Demir et al., 1989)*.

The cellular diversity of vessel's medial and adventitial layers are different from each other. the cellular differentiation begins from stem cell and continues respectly to glicogen-rich cell, fibroblast, myofibroblast and finally smooth muscle cell which form medial layer. This differentiation process shows an immunological step due to the structural variabilities of the cells. The steps of reactions are with ; vimentin (V), V+desmin (D), V+D+å-smooth muscle actin (A), V+A+D+specific antibody (S), placental vessel wall shows a serial of transformations

depending on ultrastructural and immunocytochemical reactions. So, extravascular cell population has been made up as a source for vessel wall cell varieties.

Fibroblasts which are fixed connective tissue cells of the compact fibrous or older type stroma, charateristically observed in placental stem villi. They are also observed in terminal villi and in immature intermediate villi. Reticulum cells which are the predominant cell type in immature intermediate villi and in reticular types of stroma *(Benirschke and Kaufmann, 1990; Demir and Erbengi, 1984)*, observed in different villi segments. In more later periods, differentiations and developmental variations come about by means of specialized cells in the stromal architecture which are related with vessels formation, in a reciprocal manner *(Benirschke and Kaufmann, 1990; Demir et al., 1989)*. For example, there is an effective reciprocal relation between the number of Hofbauer cells and the vascularization of the villus *(Demir and Erbengi, 1984)*. At an advanced level of the vascularization of the chorionic villus in which collagen production is increased, a reduction is seen in the number of the Hofbauer cells *Demir and Erbengi, 1984; Demir et al., 1989)*.

According to the criteria, myofibroblast-like cells were identified only in the stroma of placental stem villi. Myofibroblast-like cells increased in triangular areas interposed between two large vessels *(Demir et al., 1992)*. In the vessel wall, this cell variety show a differentiation to smooth muscle cell in the medial layer. Development and differentiation of placental vessel wall has changed due to the structural and immunorective cell diversity and three-dimensional configurations. All of these cells were discussed according to their structural and immunocytochemical pecularities.

In the full term placentas, in order to provide for the increasing physiological requirements to the fetus, some histofunctional modifications were seen in the placental barrier. The basement membranes which belonged to the syncytiotrophoblast layer and to the fetal capillary endothelium welded with one another. In these areas of the syncytial layer are deprived of both the nucleus and the microvilli, and the maternal and fetal blood have very nearly approached to one another. The transportation of the substances and gases, that is a necessity for fetal life, take place in these areas as quick as possible. It has been published that these histofunctional modifications serve as a "nephropneumonic" function *(Demir and Erbengi, 1984)*.

In conclusion, placental vessel wall has several cell populations. This cellular diversity change due to the ultrastructural and immunocytochemical variabilities of the cells, and villi development and differentiation periods. Cellular diversity is decreased by means of the placental villi maturation and vascular development.

REFERENCES

Benirschke, K. and Kaufmann, P., 1990, Pathology of human placenta, P.P.-15-65, Springer-Verlag, New York, Heidelberg, Berlin, London.

Demir, R. and Erbengi, T., 1984, Some new findings about Hofbauer cellsin the chorionic villi of human placenta. *Acta. Anat.* 119(1):18-24.

Demir, R., et al., 1989, Fetalvasculogenesis and angiogenesis in human placental villi. *Acta. Anat.* 136:190-203.

Demir, R., and Erbengi, T., 1992, Structural features of nephropneumonic-like units in the full term human placental barrier. *J. Scanning Microscopy* 14:71.

Demir, et al., 1992, Distribution of myofibroblast-like cells in human placental stem villi. *Electron Microscopy* Vol. 3:409-510.

Kaufmann, P., 1982, Development and differentiation of the human placental villous tree. *Bibl. Anat.* 22:29-39.

Kohnen, G., 1994, Immunohistochemische charakterisierung der myofibroblasten diferenzierung in der menschlichen placenta. *Doctora pHD thesis, Aachen.*

ULTRASTRUCTURAL LOCALIZATION OF NITRIC OXIDE SYNTHASE IN VASCULAR ENDOTHELIAL CELLS

K. Dikranian

Department of Anatomy and Histology
Medical University Varna
9002 Varna, BULGARIA

Nitric oxide synthase (NOS), the enzyme generating NO from L-arginine *(Ignaro, 1990; Moncada et al., 1991; Vallance et al., 1989)* was studied in different vessels by pre- and post-embedding immunocytochemistry. Rat femoral arteries and veins, mesenteric, coronary and middle cerebral arteries, as well as human umbilical cord arteries and veins were investigated.

We found that endothelial cells of these vessels showed differential immunoreactivity to NOS. The majority of endothelial cells by systemic arteries as well as the umbilical vein were immunopositive to NOS, while only a small proportion of endothelial cells of the femoral veins and umbilical arteries were positive.

Peroxidase labelling was distributed in the perinuclear and peripheral regions of the cells and was localized on membranes of subcellular organelles as well as in the cytoplasmic matrix. Immnogold labelling was localized mainly in the cellular matrix with no apparent association with organelles.

The differential distribution of NOS-immunoreactivity in arterial and venous vessels is discussed *(Moncada et al., 1991; Sessa, 1994; Myatt et al.)*.

REFERENCES

Ignarro, L.J., 1990, Biosynthesis and metabolism of endothelium-derived nitric oxide. *Annu. Rev. Pharmacol. Toxicol.* 30:535-560.

Moncada, S., Palmer, R., Higgs, E., 1991, Nitric oxide: physiology, pathophysiology, and pharmacology. *Pharmacol. Rev.* 43:109-142.

Myatt, L., Brockman, D., Langdon, G., Pollock, J., Constitutive calcium-dependent isoform of nitric oxide synthase in the human placental villous vascular tree.

Sessa, W., 1994, The nitric oxide synthase family of proteins. *J. Vasc. Res.* 31:131-143.

Vallance, P., Collier, J., Moncada, S., 1989, Nitric oxide synthesized from l-arginine mediates endothelium-dependent dilatation of human veins. *Cardiovasc. Res.* 23:1053-1057.

CARDIAC DERIVED NEUTROPHIL CHEMOTACTIC FACTOR (NOURIN-1) STIMULATES THE SECRETION OF ADHESION MOLECULES BY NEUTROPHILS (LECAM-1) AND VASCULAR ENDOTHELIAL CELLS (ELAM-1 AND ICAM-2)

Salwa A. Elgebaly, Ph.D., Edward Tyles, Ph.D., Stuart L. Houser, M.D., FACS, Naveed Shams, M.D., Ph.D. and Catherine A. Consoli, B.A.

Surgical Research Center
Hartford Hospital
Hartford, Connecticut USA

University of Connecticut
School of Medicine
Hartford, Connecticut USA

High levels of a neutrophil chemotactic factor (NCF) were detected in coronary sinus effluents (CSE) collected during cold cardioplegic arrest of patients undergoing cardiopulmonary bypass for myocardial revascularization. The factor, which is rapidly released (i.e., 5-10 minutes), was found to be associated with neutrophil accumulation in ischemic myocardium. It was recently purified from CSE and found to be a 3 KDa peptide with an apparent charge of neutral to negative and an isoelectric point of 7-8. This previously unidentified factor, unrelated to interleukin-8, leukotriene B_4 and complement component C5a, is tentatively referred to as "Nourin-1". The present study tests the hypothesis that cardiac-derived NCF will stimulate the secretion of the adhesion molecule LECAM-1 by neutrophils and ELAM-1 and ICAM-1 by vascular endothelial cells. Under lipopolysaccharide (LPS) free conditions, neutrophils ($6.56 \pm 0.72 \times 10^6$ cells/ml)) isolated from peripheral blood of healthy volunteers (n = 5) were incubated with culture medium (RPMI) or with RPMI plus cardioplegic solution (CS). Cultured human aortic endothelial cells (HAEC) were also incubated with 300 ul concentrated CSE/ml endothelial growth media (EGM™), 100 ng LPS/ml media, or media for 6 hours, 24 h, 48 h, and 72 h. The levels of LECAM-1, ELAM-1 and ICAM-1 were measured in supernatant solutions using standard ELISA methods. As shown in Figure 1, high levels of LECAM-1 (mean ± S.E.M.) were secreted by neutrophils incubated with crude Nourin-1 for 4 hours, with partial reduction by 24 hours. Secretion of ELAM-1 and ICAM-1 by vascular endothelial cells treated with crude Nourin-1, purified Nourin-1, or LPS was also observed (Figure 2 and 3). However, the stimulated secretion kinetics of ELAM-1 and ICAM-1 by crude or purified Nourin-1 differed from LPS stimulated secretion in that Nourin-1 had an earlier response while the LPS response typically took longer. The results of this study support our hypothesis and demonstrate the capability of cardiac Nourin-1 to stimulate the secretion of adhesion molecules by neutrophils and vascular endothelial cells; an important facet in the cascade of an inflammatory response.

ANTISENSE OLIGONUCLEOTIDES TO TENASCIN OR FGFR1 INHIBIT RESTENOSIS IN THE RAT CAROTID ANGIOPLASTY MODEL

L. Denner, D.L. Stacy, A.A. Rege, L. Sturgis, C. Munsch, C. Kagan, C. Steffan, N. Kint, S.G. Eskin, and R.A.F. Dixon

Texas Biotechnology Corporation
Houston, Texas 77030 USA

Antisense oligonucleotides (ODN) are potentially an efficacious treatment for the restenosis which commonly follows interventional cardiovascular procedures *(Simons et al., 1992; Morishita et al., 1993; Abe et al., 1994)*. We synthesized phosphorothioate ODN *(Milligan et al., 1993)* and screened more than 500 ODN for their inhibition of proliferation of smooth muscle cells (SMC) derived from rat carotid artery, normal human aorta, or diseased human arteries. For *in vivo* experiments, ODN or vehicle was perfused into balloon-injured carotid arteries of Sprague-Dawley rats (1 mM, 60 ul in DMEM) and incubated for 15 min. Two weeks later, the vasculature was perfusion-fixed with formalin. the carotid arteries were processed for light microscopy, and intimal thickening expressed as the ratio of intimal area to medial area. Antisense ODN to tenascin or FGF receptor 1 (FGFR1) inhibited intimal thickening. In representative experiments, intimal/medial ratio for the control rats was 0.668 +/- 0.141 (mean +/- S.D., n = 15), whereas antisense ODN to FGFR1 inhibited intimal thickening approximately 50% (0.346 +/- 0.043, n = 5). Tenascin antisense ODN treated rats had an intimal/medial ratio of 0.342 +/-0.159 (n = 3) for 58% inhibition, and venicle 0.820 +/- 0.339 (n = 5). Antisense ODN to many other genes were ineffective. In addition we have shown induction of expression of the mRNAs for these genes in response to balloon angioplasty using RT-PCR. Treatment of cultured SMC with FGFR1 ODN resulted in down regulation of FGFR1 mRNA and decreased binding of bFGF to these cells.

REFERENCES

Abe, J. et al., 1994, Suppression of neotintimal smooth muscle cell accumulation in vivo antisense cdc2 and cdk2 oligonucleotides in rat carotid artery. *B.B.R.C.* 198:16-24.

Milligan, J.F. et al., 1993, Current concepts in antisense drug design. *J. Med. Chem.* 36:1923-1237.

Morishita, R. et al., 1993, Single intraluminal delivery of antisense cdc2 kinase and proliferating-cell nuclear antigen oligonucleotides results in chronic inhibition of proliferation. *P.N.A.S.* 90:8474-8478.

Simons, M. et al., 1992, Antisense c-myb oligonucleotides inhibit intimal arterial smooth muscle cell accumulation in vivo. *Nature* 359:67-60.

THROMBIN MEDIATION OF ENDOTHELIUM IN INJURY TO HEALING

John W. Fenton II, Ph.D.

FACB
NYSD Health - WCLR
POB 509
Albany, New York 12201-0509 USA

Thrombin is generally recognized for its ability to convert fibrinogen into clottable fibrin in the formation of thrombi. However, this is the only one of many important functions of thrombin from injury to wound healing. Thrombin is atypical when compared to other serine proteinases involved in thrombosis and thrombolysis: I) Its circulating zymogen (prothrombin) has a more highly evolved gene, and thrombin has unique topigraphical features which enable it to participate in various biological functions (e.g., fibrinogen recognition exosite, heparin exosite, carbohydrate attachment, chemotactic domain, RGD domain). ii) Thrombin controls its own generation throughout the amplification steps by activating factors V and VIII (or activating cellular surfaces). iii) thrombin is produced in larger amounts than the preceding activated coagulation factors but availability of thrombin in blood is kept low because of blood flow dilution, inactivation by blood-borne proteinase inhibitors (e.g., antithrombin III, α_2-macroglobulin, heparin cofactor II), and partitioning within thrombi or behind cellular barriers. Extravascular thrombin generation may involve pathways distinct from those of blood coagulation and extravascular thrombin functions appear to be directed primarily at the cellular proliferative level (e.g., scar information, atherosclerosis, certain cancers). In this sense, thrombin may have evolved from a growth regulatory proteinase with developmental functions (e.g., nerve cell differentiation) and may have been more recently captured into processes of hemostasis.

Upon initiation of prothrombin activation, very low thrombin concentrations should stimulate nitric oxide production by endothelial cells causing vessel relaxation. Increasing concentrations should stimulate platelets, as well as cause shape change of endothelial cells and increased transport across the endothelial barrier. Thrombin converts of fibrinogen into clottable fibrin, binds to intermediates during fibrin polymerization, and becomes incorporated into fibrin structures. Thrombi have architecture, consisting of a course fibrile core and entrapped blood cells, which are capped by a fine fibrile canopy, and its contents are protected from blood and circulatory conditions. Notably, thrombin within thrombi is shielded from proteinase inhibitors (e.g., antithrombin III, heparin coefactor II) and can remain for days. Neutrophil activation marks a turning point in thrombi maturation. Activated neutrophils secrete elastase and cathepsin g, and these proteinases partially degrade fibrin increasing fluidity within thrombi and releasing thrombin. Such thrombin becomes available to simulate thrombus-contacted cells (e.g., endothelial, smooth muscle cells). Both proteinases also convert α- into ϵ- and ζ- thrombins, respectively, which are thermodynamically less stable than α- thrombin and undergo protein denaturation. Such denaturation exposes the hidden RGD sequence in α- thrombin and the RGD ligand binds the vitronectin receptor on cells and may promote cell spreading. Thrombin also has another nonenzymic functions function of a different protein domain which simulates leucocyte chemotaxis. Nevertheless, the majority of thrombin biological activities require enzymatically active thrombin forms. the recently cloned thrombin receptor is proteolytically cleaved to form the tethered ligand which subsequently activates the receptor. Here, receptor expression cells becomes an important regulatory means and thrombin may have additional stimulatory pathways.

Because prothrombin may cross blood barriers and perhaps selectively, extravascular thrombin generation may take place and may proceed by novel mechanisms. Thrombin is a potent mitogenic substance for most cell types (also an angiogenic substance) implicating its participation in scar formation and certain disease conditions. It (or an ancestrally related enzyme) may participate in differiation processes. It is a T-cell stimulus, may substitute, in the alternative complement activation pathway, and stimulates interleukin 6 from smooth muscle and epithelial cells, suggesting roles in immunity and inflammation. Thrombin is clearly one of the more important regulant in vascular medicine.

ROLE OF G-PROTEINS IN MECHANICAL SIGNAL TRANSDUCTION

Siva R.P. Gudi, Nan Q. Li and John Frangos

Department of Chemical Engineering
The Pennsylvania State University
University Park, Pennsylvania 16802, USA

Cells have highly specialized forced sensing mechanisms such as those involved in hearing and touch which can convert a mechanical stimulus into electrical and biochemical responses. Fluid flow itself is another such stimulus that exists in the cellular environment. Indirect evidence suggests that flow regulation of vascular tone via the endothelium and smooth muscle cells involves mechanical signal transduction via G proteins. Human Umbilical Vein Endothelial cells (hUVECs) grown on glass slides were subjected to fluid flow in parallel plate flow chambers. Fluid shear-induced GTP binding in hUVECs was assayed by preincubating the cells with a GTP photoreactive analogue, 8-azido GTP (α-32_p) and subjecting to UV crosslinking after fluid flow (10 dynes/cm^2) for 10 sec. Autoradiographic analysis of these cells lysate show a 42 kDa protein being labeled by GTP upon flow stimulation. Further this protein was identified with anti Gα subunits Gαq and Gαi3 may be involved in the flow induced signal transduction pathway in endothelial cells. In other supporting experiments, Gproteins purified from bovine brain were reconstituted into PE:PS liposomes and subjected to fluid shear. There was an increase in Gprotein mediated GTP hydolysis upon the onset of shear. These experiments suggest the involvement of Gproteins as mechanochemical signal transducers. *This work is supported by HL40696-07.*

HUMAN ATHEROSCLEROSIS: COMPARATIVE STUDY OF MONONUCLEAR INFILTRATION IN AORTA, CORONARY AND CEREBRAL ARTERIES

P. Ghenev, I. Krasnaliev and G. Chaldakov*

Department of General and Clinical Pathology
Medical University
Varna, BULGARIA

*Laboratory of Electron Microscopy
Medical University
Varna, BULGARIA

Atherosclerotic involvement of aorta and large arteries depends on a variety of factors. In the last decade, literary data have accumulated about the role of mononuclear cellular infiltration (MCI) in atherogenesis *(Kohchi et al., 1985; Masuda and Ross, 1990; Ross, 1993; Stary, 1990; Stratford et al., 1986)*. The aim of the present study is to parallel intimal and adventitial MCI in human atherosclerotic aorta, coronary and cerebral arteries. In 59 autopsy cases (21 women and 38 men, between 30 and 85 years of age), segments from aorta, left anterior descending branch, basilar and each middle cerebral arteries were processed and stained routinely for light microscopy and analyzed for MCI location and density. In the 472 arteries harvested from all 59 autopsy cases a total of 243, both early and late atherosclerotic lesions were observed, among them: fatty streaks - 46(18, 63%), fibrofatty change - 33 (13, 58%) and fibroatheroma - 164 (67, 79%). MCI in the intima are presented either as single cells, or as small groups of cells beneath the endothelium. Adventitial MCI are found usually as small groups; in aorta they often encircle vasa vasorum; in the coronary arteries MCI also involve the perivascular nerve fibers, or the newly formed blood vessels. In the cerebral arteries MCI are not widespread. The presence of MCI does not correlate with the lesion type. Most of the lesions are not accompanied by any MCI and, on the contrary, massive MCI are found in early lesions and occasionally even in normal vessels. The semiquantitive analysis shows most frequent involvement of aorta, almost equally for intima and adventitia; while in coronary and cerebral arteries the adventitia is predominantly involved. As the present material demonstrates the most severe and fatal forms of human atherosclerosis it is tempting to suggest that MCI are responsible at least for the activation of the atherosclerotic process. In this regard, the adventitial MCI as more widespread, appear to be at least as equally important as the intimal. As for the role of vasa vasorum - related MCI, findings presented here further support this view.

REFERENCES

Kohchi, K., Takebayashi, S., Hiroki, T. and Nobuyoshi, M., 1985, Significance of adventitial inflammation of the coronary artery in patient with unstable angina: results at autopsy. *Circulation* 71:709-716.

Masuda, J. and Ross, R., 1990, Atherogenesis during low level hypercholesterolemia i the nonhuman primate I. Fatty streak formation; II. Fatty streak conversion to fibrous plaque. *Arteriosclerosis* 10:164-187.

Stary, H., 1990, The sequence of cell and matrix changes in atherosclerotic lesions of coronary arteries in the first forty years of life. *Eur. Heart J.* 11(Suppl E):3-19.

Stratford, N., Britten, K. and Gallagher, P., 1986, Inflammatory infiltrates in human coronary atherosclerosis. *Atherosclerosis* 59:271-276.

PROTEIN SYNTHESIS INHIBITION INDUCES NO SYNTHASE ACTIVITY IN VASCULAR SMOOTH MUSCLE CELLS

Carolyn Go, Nandor Marczin, Ilona Dienes-Kadi, Connie Snead
and John D. Catravas

Department of Pharmacology and Toxicology
Medical College of Georgia
Augusta, Georgia 30912, USA

The role of newly synthesized proteins in the activation of the NO pathway was investigated by evaluating the effects of the protein synthesis inhibitor cycloheximide (CHX) on basal and lipopolysaccharide (LPS-) induced cGMP accumulation in rat aortic smooth muscle cells (RASM). Pretreatment of RASM with 10 μM CHX for 24 h resulted in a 10-fold increase in cGMP. Inclusion of an inhibitor of NO synthase N^G-monomethyl-L-arginine (LNMMA, 1 mM), inhibited CHX-induced cGMPG formation (85±8 vs 266±55) pmol/mg protein). L-arginine (10 mM), the substrate of NO synthase, potentiated cGMP accumulation and reversed the inhibitory effect of LNMMA (981±87 and 79±49 pmol/mg protein, respectively), but had no effect on baseline cGMP. LPS produced a time-dependent increase in cGMP levels with maximal stimulation at 6 h and a decline to baseline at 24 h. CHX attenuated the LPS-induced cGMP accumulation at 3 and 6 hours, however, cGMP levels were superinduced by 10-fold at later time points. CHX also shifted the LPS concentration curve 20 times to the left. The concentration-dependence of cGMP stimulation by CHX was biphasic both in the absence and presence of LPS, with maximal stimulation at 10 μM and inhibition at higher concentrations. These results show the dual role of protein synthesis in NO synthase activation and suggest the involvement of labile repressors in the regulation of NO pathway in aortic smooth muscle cells. (Supported by HL 31422).

NITRIC OXIDE INHIBITS ENDOTHELIAL CELL IN CYTOMEGALOVIRUS REPLICATION

Keith J. Gooch

The Pennsylvania State University
158 Fenske Laboratory
University Park, Pennsylvania 16802-4400, USA

Nitric oxide (NO) has diverse biological functions in many different physiological systems. the endothelium releases NO as a result of fluid flow or agonist stimulation. It is established that this NO plays an important role in the regulation of vascular tone and platelet function. This poster presents several additional roles NO may play within the endothelium.

Studies of nonendothelial cell types reveal that NO quenches the tyrosyl radical in ribonucleotide reductase (RR) *(Lepoivre et al., 1992)*, an enzyme which catalyzes the conversion of ribonucleotides to their corresponding deoxyribonucleotides. The tyrosyl radical is necessary for activity of this enzyme *(Graslund et al., 1982)*, which catalyzes a rate-limiting step in DNA synthesis. Based on NO inhibition of DNA synthesis, four hypotheses with relevance to the vascular endothelium are as follows:

-NO inhibits the proliferation of endothelial cells (EC).
-NO inhibits the replication of DNA viruses within EC.
-Fluid flow inhibits the proliferation of EC.
-Fluid flow inhibits the replication of DNA viruses within EC.

Experimental evidence from the literature and our work are discussed in light of these hypotheses.

REFERENCES

Graslund, A., Ehrenberg, A. and Thelander, L. *J. Bio. Chem.* 257:5711-5715.

Lepoivre, M., Flaman, J. and Henry, Y., 1992, Early loss of the tyrosyl radical in ribonucleotide reductase of adenocarcinoma cells producing nitric oxide, *The Journal of Biological Chemistry* 267.32, 22994-23000.

FUNDAMENTAL DIFFERENCES IN ACTIONS OF ARACHIDONATE MEDIATORS AND NO IN THE PULMONARY CIRCULATION

G.H. Gurtner, K. Gupta and S. DeLeon

Division of Pulmonary/Critical Care Medicine
and Department of Physiology
New York Medical College
Vlhalla, New York 10595, USA

We are interested in the actions of arachidonic acid and NO mediators on the pulmonary circulation. AA mediators can directly cause vasoconstriction or dilatation and peptide leukotrienes can directly cause increased permeability. There appear to be tonic production of prostacyclin and NO in the pulmonary circulation (Am Rev Resp Dis 137:1343-1349, 1988) although it is not known if either is necessary for low perfusion pressure. To study the actions of EDRF-NO in the pulmonary circulation, we administered the NO-synthase inhibitor, N-Nitro-L-Arginine (NLA) to a well characterized isolated lung preparation (ibid) perfused with K-H buffer. At a perfusate concentration of 150uM, NLA caused pulmonary hypertension and edema formation. An index of capillary filtration coefficient was significantly increased. These changes could be prevented with pretreatment with 300uM L-Arginine or a combination of pre-treatment and treatment with 8-bromo-cAMP. Although we did not study L-Arginine in arachidonate lung injury, 8-bromo-cAMP did not ameliorate lung injury as measured by capillary filtration coefficient. In keeping with a fundamental different mechanism of injury, cAMP did not ameliorate NLA lung injury but is very effective in inhibiting AA mediator induced oxidant lung injury.

THE ROLE OF SUPEROXIDE NITRIC OXIDE AND PEROXYNITRITE IN NEUTROPHIL-MEDIATED HUMAN MICROVASCULAR CELL INJURY

M.M. Hardy*, A.G. Flickinger*, C.S. Schasteen*, and U.S. Ryan

*Department of Cardiovascular Disease Research
G.D. Searle, Inc.
St Louis, Missouri 63167, USA

T-Cell Sciences
38 Sidney Street
Cambridge, Massachusetts 02139, USA

Neutrophil-derived oxygen radicals have been implicated in many pathophysiological states, such as the tissue injury associated with inflammation, organ ischemia associated with thrombosis, adult respiratory distress syndrome, rheumatoid arthritis, atherosclerosis, and asthma. The identity of the oxygen species responsible for oxidative cell injury is still largely in dispute. Evidence suggests that it may differ for each disease state, depend on the intrinsic antioxidant defenses of the target cell and the mode of neutrophil activation. In this study, we show that peroxynitrite, formed upon the interaction of superoxide and nitric oxide, may be the primary toxic oxygen radical species generated *in vitro* by neutrophils primed with TNF-*alpha*, then stimulated with C5a or N-formyl-Met-Leu-Phe (fMLP). Phorbol myristate acetate (PMA) induced cell injury also appears to cause the generation of peroxynitrite, however, superoxide and hydrogen peroxide may also play a role. Immune complex induced injury appears not to be mediated by peroxynitrite in this system. Neutrophils stimulated by TNF-*alpha*/C5a, TNF-*alpha*/fMLP, or PMA generated superoxide and nitrate (a product of peroxynitrite decomposition), both of which correlated with human dermal microvascular endothelial (HDME) cell injury. The addition of authentic nitric oxide caused a dose-dependent amplification of injury only when the neutrophils were stimulated to generate superoxide and nitrate. Furthermore, superoxide dismutase (SOD) mimics or oxygemoglobin (oxyHb), a nitric oxide scavenger, attenuated TNF-*alpha*/C5a or TNF-*alpha*/fMLP stimulated neutrophil-mediated injury. PMA-induced cell injury was inhibited by SOD mimic or a combination of catalase and SOD mimic, but not oxyHb. Conversely, immune complex stimulated neutrophils caused significant HDME cell injury that directly correlated with superoxide generation, but not nitrate generation. Furthermore, oxyhemoglobin was ineffective at inhibiting immune complex-induced cell injury. These data suggest that TNF-*alpha*/C5a or TNF-*alpha*/fMLP stimulated neutrophils induce HDME cell injury *in vitro* by generating both superoxide and nitric oxide, which react to form peroxynitrite. PMA induced cell injury may be mediated by a combination of superoxide, hydrogen peroxide and peroxynitrite. Immune complex stimulated neutrophils appear to induce injury through a superoxide-mediated pathway other than peroxynitrite.

REFERENCES

Chapman-Kirkland, E.S., Wasvary, J.S. and Seligmann, B.E., 1991, *J. Immunol. Meth.* 142:95-104.

Damiani, P. and Burini, G., 1986, *Talanta* 33:649-652.

Huie, R.E. and Padmaja, S. *Free Rad. Res. Comm.* 18:195-199, 1993.

Ischiropoulos, H., Zhu, L., Chen, J., Tsai, M., Martin, J.C., Smith, C.D. and Beckman, J.S. , 1992, *Arch. Biochem. Biophys.* 298:431-437.

Look, D.C., Rapp, S.R., Keller, B.T. and Holtzman, M.J., 1992, Selective induction of intercellular adhesion molecule-1 by interferon-gamma in human airway epithelial cells. *Am. J. Physiol.* 263;L79-L87.

Marks, R.M., Czerniecki, M. and Penny, R., 1985, Human dermal microvascular cells: an improved method for tissue culture and a description of some singular properties in culture. *In*: Vit. Cell. Devel. Biol. 21:627-635.

Misko, T.P., Moore, W.M., Kasten, T.P., Nickols, G.A., Corbett, J.A., Tilton, R.G., McDaniel, M.L., Williamson, J.R. and Currie, M.G., 1993, *European J. Pharmacol.* 233:119-125.

Radi, R., Beckman, J.S., Bush, K.M. and Freeman, B.A.,1991, *J. Biol. Chem.* 266:4244-4250.

Radi, R., Beckman, J.S., Bush, K.M. and Freeman, B.A., 1991, *Arch. Biochem. Biophys.* 288:481-487.

IMMUNOGOLD LOCALIZATION OF ENDOTHELIN-1 IN ENDOTHELIAL CELLS

V.J. Harrison, *K. Barnes, *A.J. Turner, R. Corder and J.R. Vane

The William Harvey Research Institute
St Bartholomew's Hospital Medical College
London, EC1M 6BQ, UK

*Membrane Peptidase Research Group
Department of Biochemistry and Molecular Biology
University of Leeds
Leeds LS2 9JT, UK

Cultured endothelial cells secrete endothelin-1 (ET-1), a 21 amino-acid peptide, via the constitutive pathway of protein secretion. Constitutive secretory vesicle facilitate the transport of newly synthesized peptides from the *trans* Golgi network to the cell surface. In addition to their role in intracellular peptide transport, secretory vesicle serve as an important site for peptide processing. ET-1 is processed from its larger precursor molecule, Big ET-1, by proteolytic cleavage of the Trp^{21}-Val^{22} bond. This unusal cleavage is thought to require a novel protease which is referred to as endothelin-converting-enzyme (ECE) (*Yanagisawa et al.*, 1988). However this endopeptidase has proved technically difficult to identify (*Opgenorth et al.*, 1992). As intracellular processing of other peptide prohormones occurs in secretory vesicles, we hypothesized that Big ET-1 is converted to ET-1 in secretory vesicles in endothelial cells. In this study ET-1 constitutive secretory vesicles were isolated from bovine aortic endothelial cells (BAEC) using discontinuous sucrose density gradient ultracentrifugation (*Harrison et al.*, 1993). The vesicle-containing fraction was identified using a specific ET-1 [16-21] radioimmunoassay (*Corder et al.*, 1993).

Morphological examination of the vesicle fraction embedded in Durcapan, using transmission electron microscopy, identified clusters of vesicles of approximately 100nm in diameter. Ultrathin cryosections of the vesicle preparation were prepared using a modification of the technique described by Tokuyasu (1989) (*Barnes et al.*, 1993). Immunostaining with affinity purified anti (ET-1 [16-21]) sheep IgG identified ET-1 in small vesicles (≈100nm) detected by 15nm immunogold labelling. Single gold particles were observed within or attached to the membrane bound vesicles. Likewise immunogold staining of Big ET-1 using the polyclonal rabbit antiserum Big ET-1 (22-39] identified clusters of 15 nm gold particles in similar vesicles.

In parallel, immunostaining of whole endothelial cells in culture with anti-(ET-1 [16-21]) sheep IgG showed a significant amount of intracellular staining detected by fluorescein isothiocynate (FITC)-labelled secondary antibody. Confocal microscopy confirmed that the immunostaining was localized within intracellular vesicular compartments. Immunofluorescent staining using the polyclonal antiserum to Big ET-1 revealed defined perinuclear localization of precursor molecular. The specificity of the immunolabelling was tested at the light microscopic level by omission of the primary antibody, by replacement of primary antibody with normal serum, and by preabsorption of the primary antibody (24 h at 4°C) with 10^{-5}M ET-1.

In summary, we have evidence from immunoelectron microscopy and immunofluorescent light microscopy that Big ET-1 and ET-1 are localized within intracellular vesicles suggesting that these subcellular compartments are an important site of Big ET-1 processing by ECE.

REFERENCES

Barnes K., Turner A.J. & Kenny A.J., 1993, An immunoelectron microscopic study of pig substantia nigra shows co-localization of endopeptidase-24l1 l with substance P. *Neuroscience* 53:1073-1082.

Corder R., Harrison V.J., Khan N., Anggard E.E., and Vane, J.R., 1993, Effects of phosphoramidon in endothelial cell cultures on the endogenous synthesis of endothelin-1 and on conversion of exogenous big endothelin-1 to endothelin-1. *Journal of Cardiovascular Pharmacology* 22(suppl 8):S73-S76.

Harrison V.J., Corder R., Anggard E.E. and Vane, J.R., 1993, Evidence for vesicles that transport endothelin-1 in bovine aortic endothelial cells. *Journal of Cardiovascular Pharmacology* 22(Suppl 8):S57-S60.

Opgenorth T.J., Wu-Wong, J.R. and Shiosaki, K., 1992, Endothelin-converting enzymes. *The FASEB Journal* 6:2653-2659.

Tokuyasu K.T. Use of poly(vinylpyrrolidone) and poly(vinylalcohol) for cryoultramicrotomy. *Histochemistry Journal* 21:163-171.

Yanagisawa, M., Kurihara, H., Kimara, S., Tomote, Y., Koboyashi, M., Mitsui, Y., Yazaki, Y., Goto, K. & Masaki, T., 1988, A novel potent vasoconstrictor peptide produced by vascular endothelial cells. *Nature* 332:411-415.

ENDOTHELIAL INTEGRITY AS A MAJOR FACTOR IN COLD PRESERVATION OF LUNG FOR TRANSPLANTATION

M.A. Hidalgo, S. Manek, B.J. Fuller and C.J. Green

Section of Surgical Research
Northwick Park Hospital
Watford Road
Harrow, Middlesex HA1 3UJ, UNITED KINGDOM

The short period of safe storage of pulmonary tissue is limiting the number of lungs transplants in current clinical practice. Standard procedure for solid organ storage involves flushing the lungs with an especially designed preservation solution followed by hypothermic storage in that same solution. Despite the hypertonicity of the preservation solution, edema develops during ischemic storage of rat lungs and especially soon after revascularization, reducing lung viability to a few hours. the regulation of traffic of solutes, macromolecules and fluid across the blood vessel wall relies primarily on integrity of the vascular endothelium. Damage to the endothelial cell layer leads to several processes including edema of the underlying tissue. We studied the effects of cold-ischemic storage in whole lung tissue, using an *in vivo* rat model of organ transplantation, and in human cultures of EC *in vitro*. In the whole organ model.intraalveolar edema developed in lungs stored for 24 or 48 hours, reaching a peak of acute inflammation 48 hours after revascularization. Four weeks after transplantation mild and medium damaged pulmonary tissue recovered nearly normal histological appearance while severely damaged tissue developed fibrotic scarring. Recovery presented a patchy pattern, with chronic inflammation found only in fibrotic areas. In the cell culture model, confluent monolayers of human vein endothelial cells developed gaps when subjected to hypothermic-ischemic storage, destroying monolayer continuity. Morphological alteration of endothelial cells resulted in spherical shape and formation of surface blebs. We conclude that pulmonary edema found after hypothermic storage of lungs of transplantation could be due, at least in part, to several damage to the endothelial monolayer during the storage period. Designing a preservation solution to give better EC protection might improve viability of pulmonary tissue for transplantation.

LIPOPOLYSACCHARIDE (LPS)-INDUCED E-SELECTIN EXPRESSION AND NF-κB ACTIVATION IN ENDOTHELIAL CELLS REQUIRE CONTINUOUS PRESENCE OF LPS

Kun Huang, Dianne M. Fishwild, Hsiu-Mei Wu and Russell L. Dedrick*

Departments of Immunology and *Biological Chemistry
XOMA Corporation
Berkeley, California 94710, USA

Exposure of cultured human umbilical vein endothelial cells to lipopolysaccharide (LPS) causes increased expression of E-selectin (endothelial-leukocyte adhesion molecule-1, ELAM-1- by endothelial cells and consequently increased adherence of peripheral blood neutrophils. After administration of a bolus of LPS, the induction of E-selectin mRNA and cell surface protein in endothelial cells and their adhesiveness for neutrophils could be detected within 1 to 2 h. These induced adhesive activities peak at 4 to 6 h and gradually return to basal level by 24 h *(Pober et al., 1986; Pohlman et al., 1986; Bevilacqua et al., 1987)*, a kinetic profile typical of endothelial cell activation *in vitro*. While the precise molecular events that transmit the LPS signal are not defined, it is known that the activation of endothelial cells by LPS requires soluble CD14 *(Frey et al., 1992; Pugin et al., 1993)*. Furthermore, LPS-induced transcriptional activation of the E-selectin gene has been shown to require a nuclear factor κB (NF-κB) regulatory element in the E-selectin promoter region *(Montgomery et al., 1991; Pugin et al., 1993)*. LPS-mediated NF-κB activation in endothelial cells occurs within 1 h and lasts for at least 4 h.

Bactericidal/permeability increasing protein (BPI) is a naturally occurring bactericidal protein found in neutrophil granules *(Weiss et al., 1978)*, and it binds to and neutralizes LPS *(Gazzano-Santoro et al., 1992; Marra et al., 1990)*. It has been shown that recombinant proteins corresponding to the N-terminal BPI, designated as $rBPI_{23}$ and $rBPI_{21}$, inhibit biological responses to LPS both *in vitro* and *in vivo (Weiss et al., 1992; Mészáros et al., 1993; Kohn et al., 1993; Ammons et al., 1993; Lin, et al., 1994; Huang et al., 1994)*. When added to endothelial cells at the same time a LPS addition, $rBPI_{23}$ inhibited LPS-induced E-selectin expression and neutrophil adhesion. Moreover, LPS-induced neutrophil adhesion was inhibited when the BPI proteins were added to endothelial cells within 2 h of LPS exposure. In the present study, we found that $rBPI_{21}$ added up to 2 or 4 h after LPS exposure still greatly reduced the cell surface expression of E-selectin at 6,8, or 10 j. As expected, E-selectin mRNA induction was prevented by $rBPI_{21}$ added at the same time as LPS. However, $rBPI_{21}$ added 2, 4, or even 6 h after LPS exposure markedly inhibited LPS-induced E-selectin mRNA levels detected at later time points. Furthermore, delayed administration of $rBPI_{21}$ also reduced LPS-mediated activation of NF-κB. Two to 4 h following incubation of endothelial cells with LPS, at the time after NF-κ was activated, addition of $rBPI_{21}$ resulted in marked reduction of NF-κB activation at 4 or 6 h.

Our results suggest that, by neutralizing LPS, $rBPI_{21}$ prevents/reduces the induction of E-selectin expression and NF-κB activation when added during the initiation phase of the

induction. When $rBPI_{21}$ is added while the endothelial cells are fully activated, it acts to reverse LPS-mediated endothelial cell activation requires continuous presence of LPS. Furthermore, since NF-κB is a pleiotropic transcription factor involved in activation of many LPS inducible genes, and its activation by LPS is inhibited by $rBPI_{21}$, our results also suggest that $rBPI_{21}$ will be effective as an inhibitor of LPS activation of genes in endothelial and other cells.

REFERENCES

Ammons, W.S., et al., 1993, *Circ. Shock* 41:176.
Bevilacqua, M.P., et al., 1987, *Proc. Natl. Acad. Sci. USA* 84:9238.
Frey, E.A., et al., 1992, *J. Exp. Med.* 176:1665, 1992.
Gazzano-Santoro, H., et al., 1992. *Infect. Immun.* 60:4754.
Huang, K., et al., 1994. *Shock* 1:81, 1994.
Kohn, F.R., et al., 1993. *J. Infect. Dis.* 168:1307.
Lin, Y. et al., 1994. *Biochem. Pharmacol.* 43:1553.
Marra, M.N., et al., 1990, *J. Immunol.* 144:662.
Meszaros, K., et al., 1993, *J. Leuk. Biol.* 54:558.
Montgomery, K.F., et al., 1991, *Proc. Natl. Acad. Sci. USA* 88:6523.
Pober, J.S., et al., 1986, *J. Immunol.* 136:1680.
Pohlman, T.H., et al., 1986. *J. Immunol.* 136:4548.
Pugin, J., et al., 1993, *Proc. Natl. Acad. Sci. USA*.90:2744.
Weiss, J., et al., 1978, *J. Biol. Chem.* 253:2664.
Weiss, J., et al., 1992, *J. Clin. Invest.* 90:1122.
Whelan, J., et al.1991, *Nuc. Acids Res.* 19:2645.

EFFECTS OF OXIDANT INJURY ON ENDOTHELIAL CELL BARRIER FUNCTION

Alan K. Keenan

Department of Pharmacology
University College Dublin
Belfield, Dublin 4, IRELAND

Many vascular disease states and pathologies are associated with the overproduction of reactive oxygen species and resultant damage to the vascular endothelium including compromised barrier function. This study investigates the effects of oxidant injury (using H_2O_2) on endothelial monolayer permeability (to trypan blue labelled albumin, TB-BSA) and examines the effects of vasoactive mediators on the modulation of permeability by H_2O_2.

Endothelial cells isolated from bovine pulmonary artery were maintained and subcultured in minimum essential medium using standard procedures. Cells of passage numbers 6-15 were used in experiments. Endothelial barrier function was assessed by measuring the transfer of 4% trypan blue labelled bovine serum albumin (TB-BSA) across monolayers grown on transwell membrane assemblies. Results were calculated as % clearance of TB-BSA, or as the fold increase in % clearance over basal values.

A 1 hr exposure to H_2O_2 (0.2 mM) increased the basal clearance of TB-BSA (8.07 ± 0.9 %) to 17.97 ± 1.72 %). In separate experiments the effect of 0.2 mM H_2O_2 over time peaked at 20 min. When cells were examined for basal permeability 48 hr post H_2O_2 treatment, recovery of 'basal' barrier function was apparent in those cells treated for 20 min (but not for 1 hr).

The possible modulation of H_2O_2-induced damage by ANF was next investigated. Preincubation of BPAEC with 0.1 μM rANF(99-126) before 20 min treatment with H_2O_2 did not alter the response to the latter, nor the profile of recovery. The role of NO in modulation of H_2O_2 effects was examined indirectly, using (i) NOS inhibitors and (ii) an NO donor. NOS activity (measured as the production of [^{3}H]L-citrulline from [^{3}H]L-arginine) was 0.98 ± 1.5 and 0.19 ± 0.08 pmol citrulline/min/mg in particulate and soluble cellular fractions respectively. Cotreatment of cells with 0.2 mM H_2O_2 and L-NAME or L-Nω nitro-L-arginine (both 100 μM) did not affect the response to H_2O_2 nor the 48 hr recovery. In contrast however, while the NO donor sodium nitroprusside (SNP, 100 μM) did not itself increase permeability, it significantly increased the response to H_2O_2 (from 2.37 to 4.72 fold). Finally, the possible effect on permeability of endothelin-1 (ET-1), which is known to be associated with certain aspects of oxidant injury, was examined. Treatment of monolayers with ET-1 (10 nM) for up to 18 hr did not significantly increased TB-BSA clearance, however 100 nM ET-1 significantly enhanced the permeability response to H_2O_2 and this effect was abolished by 100 μM L-NAME.

In conclusion, exposure of endothelial monolayers to H_2O_2 disrupts endothelial barrier function reversibly by 20 min, irreversibly by 1 hr. Cytoprotection is not afforded by ANF-stimulated cellular cGMP production and SNP or ET-1 in combination with H_2O_2 alone. It can be concluded that both SNP and ET-1 generate NO which in the presence of H_2O_2 generates a potentially cytotoxic species capable of compromising endothelial barrier function. *Supported by grants from the Wellcome Trust and the Irish Health Research Board.*

REGULATION OF THE INFLAMMATORY RESPONSE BY THE α3 CHAIN OF TYPE IV COLLAGEN

Jean-Claude Monboisse+, Roselyne Garnotel+, Georges Bellon+, Nobuko Ohno*, Corinne Perreau+, Jacques P. Borel+ and Nicholas A. Kefalides*

+Laboratory of Biochemistry
University of Reims
Reims, France

*Department of Medicine
University of Pennsylvania
and University City Science Center
Philadelphia, Pennsylvania USA

Our initial observation *(Monboisse et al., 1987; Monboisse et al., 1990)* that type I collagen activates polymorphonuclear leukocytes (PMN) prompted the testing of the activating potential of type IV collagen. It was noted, however, that type IV collagen isolated from bovine lens capsule did not activate PMN but rather prevented their stimulation by N-formyl-methionyl-leucyl-phenylalanine (f-Met-Leu-Phe), phorbol myristate acetate (PMA) or type I collagen (3). This observation led to the present study (4) which demonstrates that the inhibitory effect of lens capsule type IV collagen resides in the non-collagenous (NC1) domain of the α3 chain and specifically in the region comprising residues 185-203 of the NC1 domain of both the human and bovine molecules. Synthetic peptides from the same region of the NC1 domains of the α1, α2, α4 and α5 chains did not possess the inhibitory effect seen with the α3 chain. The sequence S-N-S (residues 189-191) is unique to the peptide of the α3 chain and substitution of either serine with alanine abolishes the inhibition. Type IV collagen isolated from the mouse EHS tumor, a molecule that lacks the α3 chain, did not prevent PMN activation but instead stimulated the secretion of elastase and type IV collagenase. Incubation of PMN with intact lens capsule type IV collagen or a peptide comprising residues 185-203 of the α3(IV) chain resulted in a 3-fold increase of intracellular cAMP, whereas, Ca++ levels remained unchanged. Incubating PMN with forskolin or with dibutyryl-cAMP resulted in the inhibition of both O_{2-} production and degranulation by PMN, thus mimicking the effects of type IV collagen and the α3(IV) 185-203 peptide. Pertussis toxin abolished the inhibitory effect of type IV collagen and its peptide. The data suggest that type IV collagen, through its α3 chain down regulates PMN activation and thus decreases the potential for damage as these cells traverse the capillary wall. It is further suggested that this inhibition is mediated through the action of cAMP. Our in vitro experiments suggest that the higher the content of the α3(IV) chain is in a basement membrane, the wider would be its capacity for self-protection. *(Supported by NIH Grants AR-20553, HL-29492 and by grants from CNRS-ERX 084 and the University of Reims Contract-DRED).*

REFERENCES

Monboisse, J.C., Bellon, G., Dufer, J., Randoux, A. and Borel, J.P., 1987, *Biochem. J.* 246:599-603.

Monboisse, J.C., Bellon, G., Randoux, A., Dufer, J. and Borel, J.P., 1990, *Biochem. J.* 270:459-462.

Monboisse, J.C., Bellon, G., Perreau, C., Garnotel, R. and Borel, J.P., 1991, *FEBS Lett.* 294:129-1341.

Monboisse, J.C., Garnotel, R., Bellon, G., Ohno, N., Corinne, P., Borel, J.P. and Kefalides, N.A., 1994, *J. Biol. Chem.* In press.

MULTIPLE SIGNALLING SYSTEMS FOR ENDOTHELINS IN PULMONARY VASCULAR SMOOTH MUSCLE CELLS

A. Kent, G.P. Love, S. Kelleher and A.K. Keenan

Department of Pharmacology
University College Dublin
Belfield, Dublin 4, IRELAND

Endothelin (ET) has been shown to be linked via G-protein coupled receptors (designated ET-A and ET-B) to many intracellular signalling systems in vascular smooth muscle *(Huggins et al., 1993)*. Furthermore it has been proposed that long term exposure to ET may contribute to vascular remodelling in certain pathophysiological states *(Battistini et al., 1993)*.

In this study smooth muscle cells were cultured from bovine pulmonary artery by the explant method and regular passaging was performed using a cell scraper (cells of passage numbers 3-10 were used in experiments). Displacement analysis of ET receptor binding was performed by incubating cells with ^{125}I-ET-1 (10 pM) in the presence of increasing concentrations of the unlabelled ligands ET-1, ET-3, or sarafotoxin 6C, for 2 hr at 37°C. Total inositol phosphate formation was measured by exposing [^{3}H]-inositol-loaded cells (4μCi/ml for 48 hr) to ET-1 at 37°C in the presence of 20 mM Li+. ^{45} Ca^{2+} -efflux was measured in cells equilibrated with $^{45}Ca^{2+}$ (10μCi/ml for 16-18 hr). Plasma membranes prepared from cells were used to measure adenylyl cyclase activity according to the method of Johnson et al *(Johnson and Solomon, 1991)*, except that the assay concentrations of Mg^{2+} and Na^+ were 2 and 100 mM respectively. Possible mitogenic effects were investigated by monitoring the incorporation of [^{3}H]-thymidine into the DNA of quiescent cells exposed to ET-1.

Displacement of specifically bound ^{125}I-ET-1 by ET-B receptor selective ligands ET-3 and sarafotoxin 6C yielded biphasic curves (Hill slopes<1) indicating the presence of two receptor populations, while ET-1 produced a monophasic displacement (Hill slope not significantly different from 1). ET-1 dose dependently increased total inositol phosphate formation (1.45 ± 0.04 fold over basal with 10 nM ET-1 at 10 min). This increase was significantly inhibited by the ET-A antagonist FR 139317 (1μM) and was unaffected by the ET-B antagonist IRL1038 (1μM). ET-1 stimulated $^{45}Ca^{2+}$-efflux from cells in a concentration dependent manner with 67.7 ± 8.3% remaining associated with the cells at 10 min compared with basal (96.0 ± 7.8%). Basal or forskolin (10μM) stimulated adenylyl cyclase activities were inhibited by ET-1; IRL1038 but not FR139317 (both 1μM) reversed this inhibition. The effects of ET-1 (10nM) on [^{3}H]-thymidine incorporation were monitored over an 18 hr period (single or repeated administration), but no significant mitogenic effect of the peptide was observed.

These results indicate that ET is linked to multiple signalling systems through at least two receptor subtypes in this cell model. However no apparent mitogenic role could be ascribed to ET under the conditions examined. *Supported by EOLAS, (The Irish Science and Technology Agency), the Irish Heart Foundation and the Irish Health Research Board).*

REFERENCES

Battistini, B., Challier, P., D'Orleans-Just, P., Briere, N., Sirois, P. , 1993, *Peptides* 14:385-399.
Huggins, J.P., Pelton, J.T., Miller, R.C., 1993, *Pharmac. Ther.* 59:55-123.
Johnson, R.A., Solomon, Y., 1991, *Methods Enzymol.* 195:3-21.

ANTIOXIDANTS IN RESTENOSIS FOLLOWING VASCULAR INJURY

M.K. Konneh, *G.A. Ferns, E.E. Änggård

William Harvey Research Institute
St. Bartholomew's Hospital Medical College
London EC1 M, 6BQ

*Department of Chemical Pathology
University of Leicester
Glenfield General Hospital
Leicester LE3, 9PQ

Percutaneous transluminal coronary angioplasty (PTCA), has associated with it high a incidence of restenosis *(Leimgruber et. al., 1986)*.

Following injury oxygen free radicals are known to be released *(Coghlan et. al., 1991)* and may play a role in initiating smooth muscle cell proliferation *(Rao et. al., 1992)*. Although SMC's form the major constituent of the occluding restenotic lesions, macrophage derived foam cells are usually also present in high abundance. We have studied the effects of lipid soluble antioxidants (1). Probucol in the balloon-catheterized carotid artery of the cholesterol fed rabbit and (2) Vitamin E in the cholesterol fed rat, to evaluate the importance of oxidative processes in restenosis.

Method

Rabbits were fed; control chow; n=10, 2% cholesterol; n=11, 2% cholesterol + 1% Probucol; n=9, after 1 week balloon catheter de-endothelialization was carried. The animals were allowed to recover and kept on the diet for a further 4 weeks. Rats were fed; control; n=7, 1% cholesterol; n=8 and 1% Cholesterol + 0.5% Vitamin E; n=8. Animals were maintained on diet for 2 months until steady state levels of Vitamin E had been reached, followed by carotid denudation as above. Rats were sacrificed ten days after.

Results

Probucol inhibited neointimal macrophage accumulation by 68% (P,0.001), reduced absolute intimal size by 51% (P,0.05) and reduced intima/media thickness ratio by 51%. In the rat, absolute intimal thickness and the intimal/medial thickness ratio were reduced by 30% ($P<0.025$), in the vitamin E treated rats compared to cholesterol fed controls.

Conclusion

The above data suggest that reactive oxygen species may be involved in the intimal response to injury and that antioxidants such as probucol and Vitamin E may be therapeutically useful as inhibitors of restenosis.

ACKNOWLEDGEMENTS

We are grateful to the ONO pharmaceutical company, Japan and the British Heart Foundation for financial support of this project.

REFERENCES

Coghlan, J.G., Flitter, W.D., Holley, A.E., Mitchell, A.G., Illsley, C.D. and Slater, T.F., 1991, Detection of free radicals and lipid hydroperoxides in blood taken from the coronary sinus of man during percutaneous transluminal angioplasty, *Free. Rad. Res. Comm.* 14:409-417.

Leimgruber, P.P., Roubin G.S., Hollman, Cotsonis, G.A., Meier, B., Douglas, J.S., King, S.B. and Gruentzig, A.R., 1986, Restenosis after successful coronary angioplasty in patients with single vessel disease, *Circ.* 73:710-715.

Rao, G.N. and Berk, B.C., 1992, Active oxygen species stimulate vascular smooth muscle cell growth and proto-oncogene expression. *Circn. Res.* 70:593-599.

PATTERNS OF ^{3}H-BPAP METABOLISM IN HUMAN PRIMARY PULMONARY HYPERTENSION

David Langleben, Stylianos Orfanos, Joseph Khoury and John D. Catravas

Lady Davis Institute
Sir Mortimer B. Davis Jewish General Hospital
McGill University
Montreal, Canada

Evangelismos Hospital
University of Athens
Greece

Department of Pharmacology and Toxicology
Medical College of Georgia
Augusta, Georgia USA

Primary pulmonary hypertension (PPH) is a disorder characterized by idiopathic progressive narrowing and loss of functional pulmonary microvasculature. The processes contributing to the narrowing may include increased vascular tone, proliferation of cells in the vascular wall, fibrosis and thrombosis. Endothelial dysfunction may be the initiating cause, and it contributes to progression of the vascular narrowing.

The pulmonary endothelium is a major metabolic organ, converting angiotensin I to angiotensin II and deactivating kinins via the angiotensin converting enzyme (ACE), amongst other functions. It is not known how this important ACE function is affected by PPH. The effects on ACE activity, of obliteration and loss of functional pulmonary vasculature, are also known.

By measuring first-pass pulmonary metabolism of the ACE substrate ^{3}H-BPAP, in PPH patients, we have begun to identify patterns of endothelial dysfunction and lost functional surface area.

Examples of two patterns, from 2 patients, are presented:

Patient 1 had a pulmonary arterial pressure of 89/43 mm Hg, with a cardiac output of 4.7 L/min, and a pulmonary vascular resistance of 10.2 Wood units.

Patient 2 had a pulmonary arterial pressure of 109/52 mm Hg, with a cardiac output of 1.4 L/min, and a pulmonary vascular resistance of 44 Wood units, representing more advanced disease.

Patient 1 had reduced % BPAP metabolism (44.3%, with the normal level being approximately 60%), and the Amax/Km, a measure of functional endothelial surface area or enzyme mass was 1974 ml/min, with the normal being 3800-4000 ml/min.

Patient 2, despite very abnormal pulmonary hemodynamics, had a normal % metabolism, at 62.6%, but an Amax/Km of 1004 ml/min.

These preliminary results suggest that patients with PPH and moderately reduced cardiac outputs have reduced pulmonary ACE metabolism, and indices suggesting reduced functional endothelial surface area. In patients with advanced disease and extremely low pulmonary blood flow, the slow transit time may permit more complete ACE metabolism, but Amax/Km indices suggest a severe loss of functional endothelial surface area. *Supported by the Fonds de la Recherche en Sante du Quebec and the Heart and Stroke Foundation of Quebec.*

ROLE OF ENDOTHELIUM-DERIVED PLATELET ACTIVATING FACTOR IN BIOLOGICAL RESPONSES TO THROMBOLYTIC AGENTS

E. Lupia, G. Montrucchio, G. Alloatti, S. Bergerone, F. Bussolino, F. Mariano, L. Silvetro, A. De Martino, G. Emanuelli, G. Camussi

Department of Clinic Patho-Physiology
Chair of Nephrology
University of Turin, Second University of Naples
Naples, Italy

One limitation of administration of thrombolytic agents in patients with acute myocardial infarction (AMI) is that they can cause hypotension as well as alterations of platelet (PLT) functions *(Ganz et al., 1984)*, which may limit the response to this therapy *(Fitzgerald et al., 1988)*. The aim of the present study is to investigate whether platelet activating factor (PAF), an ether lipid mediator with multiple potent biological activities, including direct cardiovascular effects, is synthesized during the thrombolytic therapy. After streptokinase (SK) infusion (666 units/Kg/min for 60 minutes) in rabbits, we evaluated *in vivo:*

- the generation of PAF in blood samples and cardiac and aortic perfusates;
- the effect of two structurally unrelated PAF receptors antagonists, SDZ 63-675 and WEB 2170, on hypotension, thrombocytopenia and PLT hyperaggregability.

  In addition, we studied the intravascular release of PAF in patients with AMI treated or untreated with intravenous SK or plasmin.

Results

In rabbits, the hypotension and the decrease in PLT counts, induced by SK infusion, were prevented by the pretreatment with SDZ 6E3-675 and WEB 2170, which did not affect the SK induced changes in plasma concentrations of fibrinogen and α2-antiplasmin.

After 30 minutes, the infusion of SK caused a significant increase in PLT aggregation to ADP and PAF, PAF antagonists completely abrogated the increased sensitivity of PLTs to ADP caused by SK.
Significant amounts of PAF were detected in the cardiac and aortic perfusates obtained from rabbits treated with SK, but not from control rabbits. PAF was never present in blood samples.

PAF was detected in blood samples from 10 to 14 patients with AMI treated, and from none of 11 patients with AMI untreated with intravenous SK. PAF began to be detectable 60 to 90 min. after SK infusion and disappeared from the circulation within 120 to 180 min. The percent variation of PLT count over basal values correlated negatively with the amount of PAF present in the circulation at 90 and 120 min.

HUVEC synthesized PAF in a dose-dependent manner in response to SK and plasmin, with a synthesis that peaked at 15 min. and persisted up to 30 min. for SK and 2 hours for plasmin.

In conclusion, the observation that PAF was detectable in the blood of patients with AMI only after treatment with SK suggested that SK stimulates the synthesis of this mediator either directly or via plasmin generation. Indeed, HUVEC synthesized PAF after stimulation with both SK and plasmin. The effect of PAF receptor antagonists observed in the rabbit experimental model suggested that the hypotension and PLT activation induced by SK infusion are mediated by PAF. These events may limit the beneficial effects of thrombolytic therapy[2].

REFERENCES

Fitzgerald, D.J. et al., 1988, *Circulation* 77:142-150.
Ganz, W. et al., 1984, *Am. J. Cardiol.* 53:1209-1216.

PROTECTION OF CHINESE HAMSTER OVARY CELLS FROM COMPLEMENT-MEDIATED LYSIS BY HUMAN CD35

S.C. Makrides, S.M. Scesney, P.J. Ford, and H.C. Marsh, Jr.

T Cell Sciences, Inc.
38 Sidney Street
Cambridge, Massachusetts 02139, USA

INTRODUCTION

Full-length recombinant human complement receptor type 1 (CR1) was expressed in Chinese hamster ovary (CHO) cells, and shown to inhibit complement-mediated lysis of the host cells. The efficiency of inhibition was directly proportional to the number of receptors per cell. Purified polyclonal anti-sCR1 antibody was capable of abrogating the complement inhibitory activity of CR1 in a dose dependent manner. The complement lysis of untransfected CHO cells loaded with entrapped fluorescent dye was not inhibited by the addition of transfected CHO cells expressing CR1 but lacking dye. CR1 has long been recognized to regulate extrinsic complement activation, that is, to bind to and promote the degradation of fluid phase C3b and of C3b attached to immune complex. These results demonstrate that CR1 inhibits complement-mediated lysis of the cell on which it is expressed, but does not protect neighboring cells.

The protection of autologous cells from complement-induced destruction is mediated by cell surface-expressed proteins including Decay Accelerating Factor (DAF), Membrane Cofactor Protein (MCP), and CD59. CR1 possesses both convertase decay-accelerating factor activity and factor I cofactor activity in both the classical and alternative complement pathways *(Fearon, 1979)*, combining the functions of DAF and MCP. The soluble recombinant protein, sCR1, has been shown to significantly prolong graft survival in a sensitized cardiac allograft model in the rat *(Pruitt and Bollinger, 1991)* and in guinea pig-to-rat *(Pruitt et al., 1991)* and pig-to-primate *(Pruitt et al., 1994)* discordant cardiac xenograft models. Our data suggest that surface-expressed CR1 might provide another approach for inhibiting hyperacute rejection occurring in transplantation. Transgenic animals expressing CR1, especially on endothelial cells, might serve as a suitable source of vascularized organ grafts for xenotransplantation.

ACKNOWLEDGEMENTS

We would like to thank The American Society for Biochemistry and Molecular Biology for permission to reproduce selected data from *Makrides et al., J. Biol. Chem.* 267:24754, 1992.

REFERENCES

Fearon, D.T., 1979, *Proc. Natl. Acad. Sci. USA* 76:5867-5871.
Pruitt, S.K., and Bollinger, R.R., 1991, *J. Surg. Res.* 50:350-355.
Pruitt, S.K., et al., 1991, *Transplantation* 52:868-873.
Pruitt, S.K. et al., 1994, *Transplantation* 57:363-370.

CONCURRENT GENERATION OF NITRIC OXIDE AND SUPEROXIDE DAMAGES THE PULMONARY SURFACTANT SYSTEM

Imad Y. Haddad and Sadis Matalon

Departments of Pediatrics, Anesthesiology, Physiology and Biophysics
University of Alabama at Birmingham
Birmingham, Alabama 35233-6810, USA

Activated alveolar macrophages and epithelial type II cells release both nitric oxide and superoxide which react at near diffusion-limited rate ($6.7 \times 10^9\ M^{-1}s^{-1}$) to form peroxynitrite, a potent oxidant capable of damaging the alveolar epithelium and pulmonary surfactant. Peroxynitrite, but not nitric oxide or superoxide, readily nitrates phenolic rings including tyrosine. We quantified the presence of nitrotyrosine in the lungs of patients with the adult respiratory distress syndrome (ARDS) and in the lungs of rats exposed to hyperoxia (100% O_2 for 60 h) using quantitative immunofluorescence. Fresh frozen or paraffin embedded lung sections were incubated with a polyclonal antibody to nitrotyrosine, followed by goat anti-rabbit IgG coupled to rhodamine. Sections from patients with ARDS (n=5), or from rats exposed to hyperoxia (n=3), exhibited a two fold increase of specific binding over controls. This binding was blocked by the addition of an excess amount of nitrotyrosine and was absent when the nitrotyrosine antibody was replaced with non-immune IgG. In additional experiments we demonstrated that significant nitrotyrosine formation could be demonstrated in lung tissues incubated with peroxynitrite, but not nitric oxide, or reactive oxygen species.

In the second series of experiments we investigated whether peroxynitrite injured surfactant protein A (SP-A), the most abundant surfactant apoprotein. The simultaneous generation of NO and superoxide by 3-morpholinosydnonimine (SIN-1; 0.1-2 mM) resulted in oxidation of dihydrorhodamine, a marker of peroxynitrite production, and a dose-dependent decrease in the ability of SP-A to enhance lipid aggregation. Western blot analysis of SIN-1 exposed SP-A samples, overlaid with a polyclonal antibody against nitrotyrosine, were consistent with nitration of SP-A tyrosine residues. Superoxide dismutase (100 U\ml), L-cysteine (5mM) xanthine oxidase (10 mU/ml) and xanthine (500 μM), or urate (100 μM), prevented the SIN-1 induced dihydrorhodamine oxidation, and injury to SP-A. NO alone, generated by S-nitroso-N-acetylpenicillamine plus 100 μM L-cysteine, or superoxide and hydrogen peroxide, generated by pterin and xanthine oxidase in the absence of iron, did not damage SP-A, or oxidize dihydrorhodamine. Taken together, this data suggests that toxic levels of peroxynitrite may be formed in the lungs of patients with acute lung injury, and that peroxynitrite may contribute to acute lung injury, by damaging key components of the pulmonary surfactant.

REFERENCES

Haddad, I.Y., Ischiropoulos, H., Holm, B.A., Beckman, J.S., Baker, J.R. and Matalon, S., 1993, Mechanisms of peroxynitrite-induced injury to pulmonary surfactants. *Am J Physiol 265 (Lung Cell Mol Physiol. 9)*: L555-L564.

Haddad, I.Y., Crow, J.P., Ye, Y., Beckman, J.S. and Matalon, S., 1994, Concurrent generation of nitric oxide and superoxide damages surfactant protein A (SP-A). *Am J Physiol (Lung Cell Mol Physiol)*, in press.

VITAMIN E ENHANCES ENDOTHELIAL FUNCTION

J. Matz, A. Stewart-Lee, T.L.G. Andersson, *T.O. Klemsdal, *K. Gjesdal, G.A.A. Ferns and E.E. Änggård

The William Harvey Research Institute
St Bartholomew's Hospital Medical College
Charterhouse Square
London EC1M 6BQ

The Department of Cardiology
Oslo University Hospital
Ullevål N-0407

Endothelial dysfunction is an early event in atherogenesis. Endothelial damage caused by oxidative stress during LDL-oxidation reduces endothelium-dependent (ED) vasodilation mediated by nitric oxide (NO) in larger arteries and resistance vessels. We have attempted to prevent and restore cholesterol (CH)-induced changes by using dietary supplementation of vitamin E (VIT E).

Method

Rabbits were fed a normal chow of 1% CH for 4 or 8 weeks. One group of rabbits received additional 0.2% VIT E. A second group were fed additional 0.2% VIT E only in the last half of the experiment. ED-vasodilation was studied *in vivo* by photoplethysmography (blinded study) and *in vitro* on isolated carotid rings and perfused hears and ears.

Results

Photoplethysmography: The relative height of the dicrotic notch (which predominantly indicates arterial tone in the larger vessels) was reduced by Ach in a dose-dependent manner, but in rabbits fed CH (n=10) this response was significantly decreased (P<0.01). Rabbits receiving concomitant VIT E (n=10) responded in a similar manner a controls. In a second set of experiments control rabbits, (n=5) were infused with the NO-synthase inhibitor L-nitro-arginine (L-NA). Infusion of L-NA caused an increase from baseline in the relative height of the dicrotic notch, and the responses to Ach were blunted. This indicates that the height of the dicrotic notch may in part be a NO-mediated phenomenon. *In vitro studies:* Ach-induced vasodilation was enhanced in preconstricted carotid arteries of rabbits fed CH + VIT E as compared to controls (P<0.01) and CH alone (P<0.001). Additional studies in which rabbits only received CH + VIT E in the second half of the experiment showed a significant restoration of ED-responses to Ach in isolated carotid rings, perfused rabbit hearts and perfused rabbit ears. Responses to endothelium-independent vasodilators were similar in all groups and preparations.

Conclusion

These data suggest that dietary VIT E can reduce the oxidative injury in early hypercholesterolemia that may cause impairment of NO-mediated responses. We also show that non-invasive inspection of peripheral pulse curves recorded by a photoplethysmographic method may provide independent prognostic information in respect to a vascular disorder.

HYDROGEN PEROXIDE TREATMENT UNMASKS A ROLE FOR SODIUM NITROPRUSSIDE IN MODULATION OF ENDOTHELIAL CELL BARRIER DYSFUNCTION

K.E. McQuaid, E.M. Smyth and A.K. Keenan

Department of Pharmacology
University College Dublin
Belfield, Dublin 4, IRELAND

Increased generation of reactive oxygen species (ROS) has been associated with many pathological situations such as inflammation, ischemia-reperfusion associated tissue injury *(Rubanyi, 1988)* and atherosclerosis *(Kozar, 1994)*. The first target of ROS is often the vascular endothelium, whose barrier function then becomes compromised. The role of NO in this oxidant-induced endothelial cell (EC) injury is unclear, as investigations have indicated both a cytoprotective *(Kim, 1992)* and a cytodestructive *(Palmer, 1992)* role.

In this study the role of NO in oxidant-induced EC barrier dysfunction was assessed using hydrogen peroxide (H_2O_2) as the oxidant, sodium nitroprusside (SNP) and glyceryl trinitrate (GTN) as NO donors, and an *in vitro* model of EC monolayer permeability.

Bovine pulmonary artery endothelial cells (BPAEC, passages 10-15) were seeded at a density of 2.3 x 10^5 cells/cm^2 onto fibronectin-treated (25 μg/ml) Costar Transwell polycarbonate membrane assemblies (6.5 mm diameter, 04 μm pore size). three days after seeding onto membranes, the monolayers were treated with test reagent(s) for 20 min. Permeability was then determined by incubating monolayers with 4% trypan blue labelled bovine serum albumin (TB-BSA) for 1 hr at membrane assembly, aliquots were sampled from the lower compartment after 1 hr and albumin transfer quantified by measuring absorbance at 590 nm. Results were calculated as the percentage clearance of TB-BSA (ratio of the absorbance in the lower chamber after 1 hr, to that added initially to the upper chamber) and are presented as fold increase in % clearance over basal values.

A significant increase in TB-BSA transfer was seen when BPAEC monolayers were incubated for 20 min with 0.2 mM H_2O_2. Assessment of permeability 48 hr after H_2O_2 exposure showed reestablishment of EC barrier integrity. The H_2O_2-induced alteration in BPAEC monolayer permeability was abolished by catalase (30 u/ml). A 20 min incubation with the NO donor SNP (100 μM) did not affect TB-BSA transfer, however when added in combination with H_2O_2 (0.2 mM), a significant increase in albumin transfer over that induced by H_2O_2 alone was observed ($p < 0.0001$). Furthermore, while the NO donor GTN (100 μM), did not itself alter permeability, in contrast to SNP it offered significant protection against the effect of H_2O_2 ($p < 0.001$).

These results suggest that the effect of NO on ROS-induced endothelial barrier dysfunction may be related to its concentration. SNP liberates NO spontaneously, while the enzymatic conversion of GTN to NO is tightly regulated and limited by the metabolic capacity of the EC[5]. Thus relatively large amounts of NO liberated from SNP may combine with H_2O_2 to form cytotoxic species, while smaller quantities of NO released from GTN are cytoprotective.

Supported by a grant from the Wellcome Trust.

REFERENCES

Kim, H. et al., 1992, *Biochem. Biophys. Res. Commun.* 189:670-676.
Kozar, R.A. et al., 1994, *J. Surg. Res.* 56:32-36.
Lopez-Belmonte, J. et al., 1993, *Br. J. Pharmacol.* 108, 73-78.
Palmer, R.M. et al., 1992, *Br. J. Pharmacol.* 105:11-12.
Rubanyi, G.N., 1988, *Free Radical Biol. Med.* 4:107-120.

EFFICIENT GENE TRANSFER INTO HUMAN UMBILICAL VEIN ENDOTHELIAL CELLS ALLOWS FUNCTIONAL ANALYSIS OF THE HUMAN TISSUE FACTOR GENE PROMOTER

Amit C. Nathwani, Karen M. Gale, Kieran D. Pemberton, David C. Crossman*, Edward G.D. Tuddenham and John H. McVey

Hameostasis Research Group
MRC Clinical Sciences Centre, RPMS
Du Cane Road, London W12 ONN, UNITED KINGDOM

*Department of Cardiology
Northern General Hospital
Herries Road
Sheffield S5 7AU, UNITED KINGDOM

Tissue factor (TF) is a cellular receptor and cofactor for factor VII/VIIa which inititates the blood coagulation cascade. We have investigated the role of 5'-flanking DNA sequences in regulating the expression of the human TF gene in human umbilical vein endothelial cells (HUVEC). Using a chloramphenicol acetylttransferase (CAT) reporter gene, we attempted to transfect primary cultured HUVEC (passage 3 to 4) with calcium phosphate coprecipitation, DEAE Dextran, lipopolyamine coated DNA or electroporation. Electroporation in HEPES buffered saline of 1×10^7 cells at 200 volts and 250 µF was found to be optimal. Using these conditions varying lengths of TF 5'-flanking sequences coupled to the CAT reporter gene were tested in transient expression studies. CAT expression corrected for variation in transfection efficiency and cell viability revealed that the sequences between -111 and +14 basic pairs are essential for minimal transcriptional activity. This region contains consensus sequences for a TATA box and three Sp1 binding sites. A domain from -382 to -111 bp, which containes two AP-1 consensus elements promoted high levels of gene expression. This transcriptional activity was repressed by 50% with constructs containing sequences between -550 and -382 bp. A further two fold drop in transcription activity was attributed to the region between -948 and -550 bp. These results suggest that the basal transcription of the human TF gene in HUVEC is mediated through at least two negative regulatory elements upstream of the proximal promoter domain. The proximal promoter region which contains two AP-1 sites is essential for efficient transcription.

FREE RADICAL MEDIATED ENDOTHELIAL INJURY, PROMOTE Z.549 CARCINOMA CELL INVASION

Eleftheria Missirlis[1] and William Orr[2]

[1]Department of Pharmacology
University of Patras Medical School
261 10 Rio Patras, Greece

[2]Department of Pathology
McMaster University
Hamilton, Ontario, Canada

Experiments *in vivo* have demonstrated that endothelial cell injury promotes the local arrest of circulating, intravascular cancer cells and the subsequent formation of metastatic tumors. Our hypothesis was that injury of the endothelium also causes damage to the adjacent vascular basement membrane, which in turn facilitates the passage of cancer cells across the vessel wall. To test our hypothesis we performed experiments on confluent monolayers of bovine pulmonary artery endothelial cells where they were incubated with ^{3}H-2-deoxyglucose or ^{3}H-proline to label the endothelial cells or the basement membrane respectively. After adding H_2O_2 or xanthine-xanthine oxidase to these cultures, damage of the endothelium and basement membrane was detected by release of the isotopes into the culture medium. The kinetics and magnitude of basement membrane degradation correlated with damage to the endothelial cells. Evidence for involvement of endothelial proteases in the basement membrane injury included identification of a 63Kd gelatinase in the culture medium, inhibition of injury by protease inhibitors, and the inability of H^2O_2 to cause^3 H-proline release when applied directly to basement membranes. Scanning electron microscopy demonstrated that greater numbers of A549 lung adenocarcinoma cells attached to the basement membrane and endothelium at points of endothelial retraction. However, this was not due to an increase in the adhesive properties of the basement membrane. The media from injured endothelial cultures stimulated the motility of A549 cells in a Boyden chamber assay. Furthermore, in a 24 hour invasion assay, greater numbers of A549 cells migrated through injured basement membranes than through control membranes. We conclude that endothelial cell injury can cause enzymatic damage to the underlying basement membrane and postulate that this can facilitate the transvascular passage of cancer cell *in vivo*.

ENDOGENOUS Ca^{2+} RELEASE INITIATES HYPERPERMEABILITY IN ENERGY-DEPLETED CORONARY ENDOTHELIAL MONOLAYERS

Thomas Noll, Andreas Muhs, Michael Besselmann and Hans Michael Piper

Institut fur Herz- und Kreislaufphysiologie
Heinrich-Heine-Universitat Düsseldorf
P.O.B. 10 10 07
D-40001 Düsseldorf, GERMANY

In the coronary microvascular bed, the barrier function of the endothelium against macromolecules is already compromised in the early phase of deep hypoxia or ischaemia. Failure of the endothelial barrier results in edema formation which can jeopardise the survival of the surrounding tissue. The underlying mechanism of the endothelial hyperpermeability during energy deprivation is only partly understood. Here, the causal relationship between the cytosolic calcium concentration (Ca_i; indicator Fura-2), and the macromolecule permeability for albumin (P) was studied in a cultured model of microvascular coronary endothelial monolayers from rat under energy depleting conditions (5 mM KCN plus 5 mM 2-deoxy-D-glucose (2DG)).

Upon addition of KCN/2-DG, increase of Caj and P were already initiated by a partial loss of the endothelial ATP-content (30% decrease of ATP reverses within the first minute). Within the first 5 min after onset of the metabolic blockade, Caj increased from 60 to 250 nM. This rise was transient, Caj decreased again within the following 30 min to a level of 180 nM which remained almost constant during further 30 min of observation. The initial increase of Caj was independent of extracellular calcium and could be abolished when intracellular calcium stores had been discharged by a 20-min pre-incubation in the presence of 300 nM thapsigargin (inhibitor of the microsomal calcium-ATPase). Within 1 min after addition of KCN/2-DG, P was increased 1.5-fold and approached a maximum (4-fold increase) after 20 min. This increase of P was independent of extracellular calcium and did not occur when endogenous calcium stores had been discharged before by 300 nM thapsigargin. The increase in P could be partially inhibited by 20 mM, 2,3-butadione monoxime (blocker of actin-myosin-interaction) suggesting an active mechanism of endothelial retraction.

Conclusions

Energy depletion of coronary endothelial cells instantaneously discharges calcium from intracellular stores and induced a steep increase of macromolecule permeability. The loss of barrier function, albeit triggered by a partial loss of energy, depends on the availability of residual energy reserves. The mechanism responsible for the endothelial barrier failure has the characteristics of a contractile process and may contribute to the early development of hyperpermeability of the coronary endothelium in the hypoxic or ischemic heart.

ROLE OF ENDOTHELIAL CELLS AS ANTIGEN PRESENTING CELLS IN TRANSPLANTATION AND INFLAMMATORY RESPONSES

Christopher Page, Magdi Yacoub and Marlene Rose

National Heart and Lung Institute
Harefield Hospital
Harefield, Middlesex
ENGLAND

Human capillary endothelial cells (EC), unlike rodent, constitutively express MHC class II antigens, raising the question of whether they can cause direct stimulation of resting T cells. Although previous reports have demonstrated an alloproliferative response between cultured human EC and resting T lymphocytes, they have not stringently proven the absence of contaminating leucocytes in both the stimulator and responder populations.

We quantitated the number of contaminating leucocytes in serially passaged human umbilical vein endothelial cells (HUVEC and examined the ability of HUVEC and an EC line (EAhy.926) to cause allostimulation of positively selected CD4+ and CD8+ T cells. Contaminating CD45+ leucocytes were found in primary cultures at a mean level of 0.43±0.49%; by passage 3 and 4 there were less than 1/10,000 cells. CE4+ and CD8+ T cells were purified by positive selection on antibody coated Dynabeads and residual DR+ cells were removed by complement mediated lysis. They were shown to be depleted of monocytes by their failure to proliferate in response to OKT3 and PHA. Nevertheless, the same cells gave an unequivocal response to cytokine treated, serially passaged HUVEC or EAhy.926 cells.

The response of CD4+, but not CD8+ T cells was dependent upon pretreatment of HUVEC or EAhy.926 with human interferon gamma. Neither CD4+ nor CD8+ T cells that recognized EC did not respond to spleen cells in the MLR. The results confirm that EC, unlike other non-bone marrow derived cells, have the ability to provide all the signals necessary for direct stimulation of resting allogeneic T cells.

INHIBITION OF NITRIC OXIDE SYNTHASE REVERSES THE ANTIANGIOGENIC AND CYTOTOXIC EFFECTS OF X-RAY IRRADIATION ON CHICK EMBRYO CHORIOALLANTOIC MEMBRANE

Stamatis Papaioannou*, Olga Hatjiconti***, Chrissa Katsarou***, Panagiota Ravazoula****, John Demopoulos*** and Michael Maragoudakis**

*Department of Molecular Pharmacology
School of Pharmacy
University of Patras
Patras, Greece

**Department of Pharmacology
School of Medicine
University of Patras
Patras, Greece

***Department of Radiology
School of Medicine
University of Patras
Patras, Greece

****Department of Pathology
School of Medicine
University of Patras
Patras, Greece

The detailed molecular mechanisms for the cytotoxic and antiangiogenic effects of ionizing radiation are poorly understood. Evidence indicates that eukaryotic cell damage is mediated by oxygen-derived free radicals. Earlier studies *(Hatjiconti et al., 1993;)* have shown that ionizing irradiation inhibits the active angiogenic process of the chick embryo chorioallantoic membrane (CAM), and enhances nitric oxide (NO) synthesis in mice *(Voevodskaya and Vamin, 1992)*. NO inhibits CAM angiogenesis *(Pipili-Synetos et al., 1993)* and is cytotoxic to tumor cells *(Hibbs et al., 1987)*. On the basis of the above data we have hypothesized that NO is involved in the antiangiogenic effect of ionizing irradiation.

The CAM model *(Folkman, 1985)* was used to study the effects and molecular mechanisms of X-ray irradiation on CAM angiogenesis. CAM angiogenesis was assessed morphologically by the Harris-Hooker method *(Harris-Hooker et al., 1983; Maragoudakis et al., 1988)* or a

computer assisted image analysis method allowing the determination of small (30-100 μm), medium (100-300 μm) and large (300-900 μm) diameter vessels. Collagenous protein biosynthesis was also determined as a biochemical index of angiogenesis *(Maragoudakis et al., 1988)*. Preliminary histological and immunohistochemical studies were also used to assess CAM tissue, endothelial cell and microvessel injury *Ravazoula et al., 1994)*. CAMs were irradiated with 10 Gy per cm^2 on the 9th or the 14th days of the chick embryo development showed that the small CAM vessels were more radiosensitive
than the medium and large ones. Maximum effects were observed 24 hours post-irradiation. The 14th day chick embryo data showed minimal or insignificant effects of the irradiation on CAM angiogenesis.

Harris-Hooker morphometric studies showed similar results. Ninth day chick embryo studies at 24 hours post-irradiation showed that N^G-nitro-L-arginine methyl ester (L-NAME 3.3 μg/CAM site) enhanced CAM vascular density by 28% compared to the non-irradiated control. When L-NAME was combinated with irradiation treatment it caused a complete reversal of the antiangiogenic effect of irradiation, resulting in not significant vascular density difference from that of the L-NAME treated CAM. D-NAME (3.3 μg/CAM site) showed no such a reversal effect.

Collagenous protein biosynthetic studies of the 9th day embryo CAM showed a 22% reduction of the ^{14}C-proline incorporation six hours post-irradiation. L-NAME (3.3 μg/CAM site) enhanced the incorporation by 25% compared to the non-irradiated control. When L-NAME was combined with the irradiation treatment, it caused a complete reversal of the incorporation, resulting in not significant incorporation difference from the L-NAME treated CAM. D-NAME (3.3 μg/CAMsite) was without effect on the ^{14}C-proline incorporation of the control or the irradiated CAM.

Preliminary studies indicated superoxide dismutase partially inhibited the irradiation-induced effect on the vascular density, and had no effect on the collagenous protein biosynthesis of the 9th day embryo CAM.

Histological studies *(Ravazoula et al., 1994)* showed that the irradiated 9th day embryo CAM preparations contained much lower stroma cell and microvessel densities that the non-irradiated CAM preparations. Immunohistochemical studies *(Ravazoula et al., 1994)* indicated that the irradiated CAM preparations did not stain with the anti-CD31 antibody, whereas the combined irradiation and L-NAME treatment resulted in positive stain of the CAM vascular endothelium.

We conclude that the irradiation-induced antiangiogenic and cytotoxic effects on the actively proliferating CAM are NO-dependent. Potential clinical implications of these findings were indicated.

REFERENCES

Folkman, J., 1985, Tumor angiogenesis. *Adv. Cancer Res.* 43:175-203.

Harris-Hooker, S.A., Gajdusek, C.M., Wight, T.N. and Schwartz, S.M., 1983, Neurovascular responses induced by cultured aortic endothelial cells. *J. Cell. Physiol.* 114:302-310.

Hatjiconti, O., Papaioannou, S., Haralabopoulos, G.C., Ravazoula, P., Demopoulos, J. and Maragoudakis, M.E., 1993 (in press), Effects of X-ray irradiation on angiogenesis and tube formation by endothelial cells in vitro. *In:* Proceedings of NATO Advanced Study Institute on Angiogenesis: Molecular Biology, Clinical Aspects. Ed. by Maragoudakis, M.E., Gullino, P., and Lelkes, P.I.

Hibbs, J.B., Jr., Vavrin, Z. and Taintor, R.R., 1987, L-Arginine is required for expression of the activated macrophage effector mechanism causing selective metabolic inhibition in target cells. *J. Immunol.* 138:550-565.

Maragoudakis, M.E., Sarmonica, M. and Panoutsacopoulou, M., 1988, Inhibition of basement membrane biosynthesis prevents angiogenesis. *J. Pharm. Exp. Ther.* 244:729-733.

Maragoudakis, M.E., Panoutsacopoulou, M. and Sarmonika, M., 1988, Rate of basement membrane biosynthesis as an index to angiogenesis. *Tissue Cell* 20:531-539.

Pipili-Synetos, E., Sakkoula, E. and Maragoudakis, M.E., 1993, Nitric oxide is involved in the regulation of angiogenesis. *Br. J. Pharmacol.* 108:855-857.

Ravazoula, P., Hatjiconti, O., Maragoudakis, M.E. and Papaioannou, S.E., 1994, Histological and immunohistochemical evidence that L-NAME protects CAM from X-ray irradiation-induced tissue injury. Concurrent presentation at the NATO ASI meeting on "Vascular Endothelium: Responses to Injury", Crete, Greece, June 22.

Voevodskaya, N.V. and Vamin, A.F., 1992, Gamma-irradiation potentiates L-arginine-dependent nitric oxide formation in mice. *Biochem. Biophys. Res. Commun.* 186:1423-1428.

ACTIVATION OF EXTRINSIC FIBRINOLYSIS DURING CARDIOTHORACIC OPERATIONS WITH AND WITHOUT CARDIOPULMONARY BYPASS (CPB)

Rachel Parratt, Beverly J. Hunt, Panny Kallis and Sajila Sheikh

Research Haemotology
Harefield Hospital
Middlesex UB9 6JH
UNITED KINGDOM

Cardiac surgery using CPB is known to induce fibrinolytic activation. However the extent to which CPB itself induces these changes is not clear. The aim of this study was to assess activation of extrinsic fibrinolysis in cardiothoracic operations with and without CPB. Blood samples were collected from 8 patients receiving coronary artery bypass grafts using CPB, and 7 patients receiving thoracic operations without CPB (4 thoracotamies and 3 two-stage oesophagectomies). Sequential blood samples were taken from the CPB group at time points 1. pre-operatively, 2. pre-CPB, 3. 5 minutes after the start of CPB, 4. end of CPB, 5. 20 minutes post-operative, 6. 24 hours post-operative, 7. 48 hours post-operative. Samples from the thoracic group were taken at equivalent time intervals.

Euglobulin clot lysis time (1) showed a significant shortening in both groups, with shortest times in the CPB group at the end of CPB. Plasma levels of t-Pa antigen (Biopool) increased significantly from preoperative levels 10.2 (2) ng/ml mean (standard error of mean) in the CPB group, and 6 (0.65) ng/ml in the thoracic group to peak at the end of the operation (time point 5) to 21.5 (4.4) ng/ml in the CPB group and 10.3 (2.1) ng/ml in the thoracic group. U-Pa antigen (Biopool) levels did not alter significantly from preoperative levels in either group. Plasma PAI-1 activity (Quadratech) in the CPB group decreased significantly from pre-op levels of 14.5 (3.1) Au/ml to 7.7 (1.9) Au/ml at time point 4. This was followed by a significant increase at 24 hours to 36.4 (2.7) Au/ml and remained elevated at 48 hours. D-dimer titres (Porton) increased significantly perioperatively reaching higher titre ($P<0.01$) in the thoracic group at time point 3, 6.8 mg/kl (1.6).

In conclusion activation of t-Pa occurred to a similar extent during thoracic operations whether CPB was used or not. This suggests that the surgical procedure itself, not CPB induced increased fibrinolytic activation (2).

REFERENCES

Mackie, I.J. and Machin, 1989, An account of laboratory techniques, *in:* Laboratory Hematology, Chanarin I. (ed) by Churchill Livingstone, London and New York.

Tanaka, K., Takao, M., Yade, I., Yuasa, H., Kusagawa, M. and Deguchi, 1989, Alterations in coagulation and fibrinolysis associated with CPB during open heart surgery, *J. Cardiothoracic Anesthesia 3:181-188.*

PERMEABILITY CHANGES IN THE HUMAN ENDOTHELIAL CELL LINE, SGHEC-7, CAUSED BY IL-1β, TNFα AND SUBSTANCE P

*+H.J. Parry, +G. St.J. Whitley and *P. Vallance

*Department of Pharmacology and Clinical Pharmacology
St. George's Hospital Medical School
Cranmer Terrace
Tooting, London SW17 ORE

+Department of Cellular and Molecular Sciences
St. George's Hospital Medical School
Cranmer Terrace
Tooting, London SW17 ORE

Vascular microvessels normally have low permeability to macromolecules. During an inflammatory reaction, however, this barrier function is lost and fluid leaks from the vessel into the tissues. Soluble mediators, such as cytokines an vasoactive hormones are thought to contribute to this process. Recent technical advances in endothelial cell (EC) culture and the production of porous membranes have allowed the study of EC permeability *in vitro (Albelda et al., 1988; Burke-Gaffney and Keenan, 1993)*. Permeability of EC monolayer grown on porous membranes has been shown to increase following stimulation with recombinant human interleukin-1α(rHuIL-1α) and -β and recombinant human tumor necrosis factor-α(rHuTNFα) *(Burke-Gaffney and Keenan, 1993)*. *In vivo* substance P dilates human arterial and venous beds in skeletal muscle *(McEwan et al., 1988)* and skin *(Foreman et al., 1983)*. Substance P has also been shown to dilate isolated human large vessels where the responses seem to be endothelium dependent in arteries *(Luscher and Vanhoutte, 1988)* but not in all veins *(Luu et al., 1992)* and may contribute to oedema formation by increasing vascular permeability.

We have used a human umbilical vein endothelial cell line, SGHEC-7, produced by electrotransfection in the presence of the plasmid pSV3neo *(Fickling et al., 1992)*, to study permeability changes in an endothelial cell monolayer following the addition of rHuIL-1β, rHuTNFα and substance P.

This assay involved the culture of SGHEC-7 cells on the porous membrane of Costar Transwell inserts. Gelatin coated Costar Transwell inserts (8μm pore size) were seeded with $2x10^5$ SGHEC-7 cells and cultured in medium 199:RPMI (Life Technologies Ltd.) supplemented with 10% (v/v) FCS (Advanced Protein Products), 2.5 μg/ml endothelial cell growth supplement (Sigma) and 0.09 mg/ml heparin (Sigma)(completed medium). These plates were then cultured for 48hr to allow growth into confluent monolayers.

Permeability was determined by incubating the monolayers with ^{125}[I]-labelled bovine serum albumin (ICN Flow, BSA) in complete medium. Cytokines (rHuIL-1β, Promega, Specific activity 2-5x10^7 Units/mg, rHuTNFα, R and D Systems, Specific activity 5 x 10^8 Reference Units/mg against NIBSC standard #88/552) or substance P (Sigma) were applied to the subluminal chambers at the specified concentrations and the luminal chambers at twice the concentrations immediately before the addition of the ^{125}[I]-labelled BSA. After varying incubation intervals the medium in the subluminal chamber was harvested and the radioactivity determined by γ-counting. The results are expressed as % diffusion across the membrane.

SGHEC-7 cells formed a confluent monolayer on Costar Transwell inserts forming a barrier against the diffusion of ^{125}[I]-labelled BSA, whereas inserts alone and inserts coated with gelatin did not present a barrier against diffusion. SGHEC-7 inhibited the diffusion by 40 and 45% over 30 mins, by 50 and 52% over 1 hr, by 56 and 60% over 2 hrs, by 69 and 66% over 4 hrs and by 79 and 78% over 24 hrs for Transwells alone and Transwells coated with gelatin respectively. The addition of rHuIL-1β and rHuTNFα at 100, 50, 10 and 1 ng/ml did not have any consistently significant effect on the permeability of SGHEC-7 monolayers compared with control monolayers. Similarly, substance P at 10^{-4}, 10^{5}, 10^{6} and 10^{-7} did not have any significant effect on the permeability of SGHEC-7 monolayers compared with control monolayers.

It has previously been reported that rHuIl-1β and rHuTNFα increases the permeability of human EC monolayers to ^{125}[I]-labelled BSA *(Burke-Gaffney and Keenan, 1993)*, however we found that these cytokines had no effect on SGHEC permeability. Receptors for IL-1β and TNFα are present on SGHEC-7 cells since these cytokines induce the expression of the adhesion molecules VCAM-1 and E-selectin and upregulate the expression of ICAM-1 *(Cartwright et al., 1994)*. We have, therefore, demonstrated differential effects in the response to cytokines between adhesion molecule expression and permeability changes. This may be peculiar to the SGHEC-7 cell line and we are now performing experiments with human umbilical vein EC to determine this. Substance P also had no effect on the permeability of SGHEC-7 monolayers. Studies are in hand to determine whether there are functional substance P receptors on the SGHEC-7 cell line.

REFERENCES

Albelda, S., Sampson, P.M., Haselton, F.R. et al., 1988, *J. Applied Physiol.* 64:308-322.
Burke-Gaffney, A. and Keenan, A.K., 1993, *Immunopharmacology* 25:1-9.
Cartwright, J., Whitley, G. and Johnstone, A., 1994 (submitted) *Immunology*.
Fickling, S.A., Tooze, J.A. and Whitley, G. StJ., 1992, *Exp. Cell Res.* 201-517-521.
Foreman, J.C., Jordan, C.C., Oeheme, P. and Renner, H., 1983, *J. Physiol.* 335:449-465.
Luscher, T.F. and Vanhoutte, P.M., 1988, *Trends Pharmacol. Sci.* 9:181-184.
Luu, I.H., Chester, A.H., O'Neil, G.S. et al., *1992, Br. Heart J.* 67:474-477.
McEwan, J.R., Benjamin, M., Larkin, S. et al., 1988, *Circulation* 77:1072-1080.

HUMAN BRAIN ENDOTHELIAL CELL-CONTRACTION INDUCED BY HEMOSTASIS AND FIBRINOLYSIS FACTORS, IN VITRO CELL CULTURE STUDY

Z. Nagy, K. Kolev*, M. Pék, É. Csorkia, M. Vastag, R. Machovich*

Stroke Centre
Semmeiweis Medical University
1083 Budapest, HUNGARY

*Department of Biochemistry
Semmeiweis Medical University
1083 Budapest, HUNGARY

Thrombo-emboli caused ischemic brain lesion could be prevented by fibrinolytic therapy in the acute phase. In the course of thrombus formation and during the lysis the endothelial cells are involved in different processes, but the exact mechanisms are not fully elucidated. We challenged endothelial cells cultured from microvessels of human brain with thrombin, plasmin, minimplasmin, PMN elastase, pancreatic elastase, urokinase streptokinase, fibrinogen, fibrin, heparin and protamine sulfate. The changes of cell morphology has been analyzed with computerized morphometry and cell perimeter area and the ratio were calculated. Fluorescein labelled thrombin, plasmin and miniplasmin binding to cell membrane has been visualized.

The cells cultured from the brain tissue were havily labelled with acetilated LDL characteristic for endothelial cells. Thrombin, plasmin, and miniplasmin binding were detected in both living and formalin fixed cells.

Rapid cell contraction could be observed following plasmin, miniplasmin treatment. Thrombin induced cell contraction after 60-90 mins of treatment. Elastase and protamin sulfate attracted cells contracted stepwise during the 120 mins observation period of time. Fibrin contraction was not significant. Urokinase and streptokinase do not affect the cell morphology observed with the phase contrast inverted microscope.

The hemostasis factors induced contraction resulted a spider like cell morphology characterized by reduced surface area and increases perimeter. The other form of contraction was tipical following protamin sulfate treatment, where cells has been rounded up, resulting reduced both perimeter and area. Cell contraction induced by thrombin plasmin, miniplasmin could be a factor in the junctional opening of endothelial cells in the genesis of vasogenic brain edema.

MODULATION OF ANGIOGENESIS IN THE CHICK CHORIOALLANTOIC MEMBRANE (CAM) BY STEROID AND NON-STEROID ANTIINFLAMMATORY DRUGS

Platon Peristeris, Paraskevi Andriopoulou, Eva Pipili-Synetos and Michael E. Maragoudakis

University of Patras Medical School
Department of Pharmacology
PATRAS-GREECE

Angiogenesis, a multistep process of great pathophysiological significance, is associated with many disease states such as cancer, psoriasis and arthritis *(Maragoudakis, 1993)*. In addition, it is one of the main features of the inflammation *(Folkman, 1985)*. Cells involved in the inflammatory process, such as macrophages, secrete angiogenic factors the production of which may be inhibited by antiinflammatory drugs *(Fantuzzi and Ghezzi, 1993; Hu et al., 1983)*.

In the present study, we investigated the modulation of angiogenesis by steroid (dexamethasone, hydrocortisone) and non-steroid (indomethacin) antiinflammatory drugs in the *in vivo* model of the chick chorioallantoic membrane (CAM) as described by Maragoudakis et al. *(Maragoudakis et al., 1988)*.

It is reported here, that the antiinflammatory drugs tested produce different responses due to modulation of distinct biological mechanisms, although the end result is inhibitory:

-Dexamethasone produced a biphasic effect on angiogenesis in the CAM. At low doses (0.1 - 20 μg/disc) it caused an increase in collagenous protein biosynthesis (19.6 ± 3.7% - 41.7 ± 6.4% of control). At 200 and 600 μg/disc dexamethasone caused a 24.8 ± 9.5% and 65.0 ± 4.9% decrease in collagenous protein biosynthesis. Hydrocortisone produced an inhibitory effect on angiogenesis ranging between -21.6 ± 5.8% of the control at 3 μg/disc and -35.6 ± 8.8% at 600 μg/disc. Increased doses of corticosteroids result in inhibition of angiogenesis, probably through the inhibition of th production of angiogenic factors by cells of the immune system.

-Indomethacin produced a potent inhibition of angiogenesis in the CAM, -43.1 ±6.4% at 1 μg/disc. In order to establish that the inhibitory effect of indomethacin is produced via 0.4 μg/disc), was added (0.4 μg/disc) in combination with 3 μg/disc of indomethacin. This partially reversed the antiangiogenic effect of indomethacin.

All the above steroid and non-steroid antiinflammatory drugs have been reported to inhibit tumor growth *(Sandstrom et al., 1990; Wright et al., 1969)* and this inhibition may, partially, be due to inhibition of angiogenesis.

Furthermore, these findings strengthen the concept that angiogenesis and inflammation are two procedures sharing common steps and that angiogenesis may comprise a "non-pathological" inflammatory status.

REFERENCES

Fantuzzi, G. and Ghezzi, P., 1993, Glucocorticoids as cytokine inhibitors: role in neuroendokrine control and therapy of inflammatory diseases. *Mediat. Inflamm.* 2:263-270.

Folkman, J., 1985, Towards an understanding angiogenesis: search and discovery. *Persp. in Biol. and Med.* 29:10-35.

Hu, D.E., Hor, Y. and Fan, T.P., 1983, Interleukin-8 stimulates angiogenesis in rats. *Inflammation* 17:135-143.

Maragoudakis, M.E., 1993, Angiogenesis, in: Annual of cardiac surgery, eds. M. Yacoub and J. Pepper (Current Science, London), p. 13-19.

Maragoudakis, M.E., Sarmonica, M. and M. Panoutsakopoulou, M., 1988, Rate of basement membrane biosynthesis as an index to angiogenesis. *Tissue Cell* 20:531-539.

Sandstrom, R., Gelin, J. and Lundholm, K., 1990, The effect of indomethacin on food and water intake, motor activity and survival in tumour bearing rats. *Eur. J. Cancer* 26:811-814.

Wright, R.L., Shaumba, B. and Keller, J., 1969, The effect of glucocorticoids on growth and metabolism of experimental glial tumours. *J. Neurosurg.* 30:140-145.

THROMBIN MODULATION OF PLASMINOGEN ACTIVATOR INHIBITOR-1 (PAI-1) IN CULTURED HUMAN VASCULAR SMOOTH MUSCLE CELLS

Prati L, Mussoni L, Sironi L, Colli S, Porta M and Tremoli E.

Institute of Pharmacological Sciences
University of Milan
Milan, ITALY

The effects of highly purified human thrombin (αThr) on fibrinolytic potential of cultured human vascular smooth muscle cells (HSMC) from saphenous, mammary and umbilical vessels were studied. Human mammary artery and saphenous vein were obtained from patients undergoing coronary bypass surgery. HSMCs from mammary artery, saphenous and umbilical veins, were cultured and used between second and sixth passages. For experimental purpose cells were made quiescent (48 hr in MEM containing 0.4% foetal calf serum) and then incubated in serum free medium with or without different concentrations of αThr. PAI-1 antigen levels were measured in conditioned medium (CM) and in extracellular matrix by ELISA. Experiments performed incubating HSMCs with thrombin for 16 hr showed that thrombin treatment increased PAI-1 antigen levels in CM of smooth muscle cells derived from saphenous veins and mammary arteries but not from umbilical veins. In particular, treatment of HSMCs with αThr (1.25 IU/ml) increased PAI-1 antigen levels in CM of cells derived from saphenous veins (+125 ± 18 SE % over control, n=4) and mammary arteries (+82 ± 20 SE% over control, n=3). PAI-1 antigen levels in the extracellular matrix showed a similar pattern, with an increase in PAI-1 associated with matrix from HSMCs obtained from saphenous and mammary vessels and no change in the umbilical vein after overnight incubation. The effect of αThr on these cells was dose-dependent in the range of 0.0125-1.25 IU/ml. Time course experiments showed that thrombin affected, with a different behavior, PAI-1 antigen release in HSMCs of all three districts considered. The effect of thrombin on HSMCs obtained from umbilical vein was evident only after 30-60 minutes incubation. On the other hand in HSMCs obtained from saphenous vein PAI-1 increased from 8 reaching a plateau after 16 hr incubation. Preincubation of αThr with hirudin or PPACK prevented the stimulatory effect on PAI-1 production, indicating that intact active site was required for this effect. The increase in PAI-1 release was accompanied by concomitant changes in PAI-1 mRNA levels in HSMCs from saphenous vein and mammary artery as determined by Northern blot analysis. In conclusion, at sites of vascular injury or denudation, smooth muscle cells interaction with thrombin may result in an increase of PAI-1 production, predisposing the vessels to thrombus formation.

REFERENCES

Berk, B.C. et al., *1991, Biochem. Journ.* 274:799-805.
Hiroko Noda-Heiny et al., 1993, *Circ. Res.* 72:36-43.
Lupu, L. et al., 1993, *Arter. and Thromb.* 13:1090-1100.
Wojta, J. Gallicchio et al., 1993, *Thromb. and Haemost.* 70(3):469-474.

IN VITRO MODELS FOR INVESTIGATION OF THE INTERACTION BETWEEN NEUTROPHILS AND THE VESSEL WALL

G.E. Rainger, S. Sheikh, G.B. Nash

Department of Haematology
The Medical School
University of Birmingham
Edgbaston, Birmingham, UNITED KINGDOM

We are interested in the role played by neutrophilic granulocytes in vascular disease, and particularly the mechanisms by which they become immobilized in the microcirculation. We have therefore developed *in vitro* models based on the flow of isolated neutrophils through coated microslides (glass tubes with rectangular cross-section and good optical qualities). Two types of model are under investigation, utilizing coatings of cultured human umbilical vein endothelial cells (HUVEC) or platelets (Cooke et al., Microvasc Res 45:33, 1993; Buttrum et al., Blood 82:1165, 1993). With HUVEC, increased interaction follows cytokine stimulation, and e.g., stationary adhesion and migration of neutrophils follows treatment with interleukin-1 for 4 hours. This process is dependent on E-selectin expressed on the HUVEC and an activating message passed to the neutrophils by the endothelial cells. For surfaces of immobilized platelets, continuous rolling adhesion of neutrophils is mediated by P-selectin. However, if an activating stimulus is supplied exogenously to the rolling cells, they rapidly become stationary but them gradually detach. These forms of adhesion are potential promoters of vascular injury and obstruction, and continuing collaborative studies seek to establish a link between the *in vitro* phenomena and pathological processes occurring, for example, following exercise-induced ischaemia in peripheral vascular disease.

HISTOLOGICAL AND IMMUNOHISTOCHEMICAL EVIDENCE THAT L-NAME PROTECTS CAM FROM X-RAY-IRRADIATION - ASSOCIATED TISSUE INJURY

P. Ravazoula*, O. Hatjiconti**, M.E. Maragoudakis** and S. Papaioannou***

*Department of Pathology
School of Medicine
University of Patras
Patras, Greece

**Department of Pharmacology
School of Medicine
University of Patras
Patras, Greece

***Department of Molecular Pharmacology
School of Pharmacy
University of Patras
Patras, Greece

We have followed up histologically and immunohistochemically recent evidence *(Papaioannou et al., 1994)* that the NO-synthase inhibitor N^G-nitro-N-arginine-methylester (L-NAME) *(Moncada et al., 1991)*, reverses the antiangiogenic effect of X-ray irradiation in chick embryo chorioallantoic membrane (CAM), in order to complement and verify these recent biochemical and pharmacological results *(Papaioannou et al., 1994)*.

Non-irradiated (control) CAM, at the ninth day of embryo development, were used in vivo as established in the literature *(Folkman, 1985; Maragoudakis et al., 1958)*. The membranes were irradiated with X-ray doses of 10 Gy.

Alternatively the CAM preparations were treated with L-NAME (3,3 μg/CAM site), or irradiated as above and treated immediately with L-NAME.

The control and the treated CAM preparations were incubated at 37°C(3,4) for 24 hours, fixed with 10% buffered formalin, and CAM tissues were embedded in parafin.

Thin horizontal sections of the parafin blocks were obtained and stained histochemically with hematoxylin-eosin *(Lee and Luna. 1960)* or immunohistochemically with factor VIII, or CD 31 (endothelial antigen) antibodies *(True, 1990; Miettinem et al., 1994)*.

The hematoxylin-eosin histological studies showed that irradiated CAM preparations contained much lower stromal cell and microvessel densities than the control CAM preparations. The L-NAME treated CAM showed no significant differences compared with the control CAM preparation. Combination treatment of CAM with X-rays and L-NAME resulted in stromal cell and microvessel densities similar to the control CAM preparations.

The immunohistochemical studies indicated that the irradiated CAM preparations did not stain, whereas, the irradiated and L-NAME treated CAM preparations stained positively at the endothelial cell layers.

The results indicate that, under the present experimental conditions, L-NAME protects CAM stomal cells, microvessels, and endothelial cells from the X-ray irradiation-induced tissue injury; implying that irradiation-induced NO plays an important role in such injury.

These data corroborate the recent biochemical pharmacological studies on the protective effect of L-NAME on the irradiated CAM vascularity *(Papaioannou et al., 1994)*.

REFERENCES

Folkman J., 1985, Tumor angiogenesis. *Adv. Cancer Res.* 43:175-203.

Lee G. and Luna, H.T. (ASCP), 1960, Armed Forces Institute of Pathology Manual Histological Staining Methods Third Edition pp. 38-39.

Maragoudakis, M.E., Panoutsacopoulou M. and Sarmonica M., 1958, Rate of basement membrane biosynthesis as an index to angiogenesis. *Tissue and Cell* 20:531-539.

Markku Miettinem, A. Earle Lindenmayer and Abbiject Chaubal, 1994, Endothelial Cell Markers CE31, CD34, and BHHg antibody to H-and Y-antigens. Evaluation of the specificity and sensitivity in the diagnosis of vascular tumors and comparison with von Willenbrand factor. *Modern Path.* 7:82-90.

Moncada, S. Palmer, R.M.J. and Higgs, E.A., 1991, Nitric oxide: Physiology, Pathophysiology, and Pharmacology. *Pharmacol. Rev.* 43:109-142.

Papaioannou, S., Hatjiconti, O., Demopoulos I., Katsarou, C., Ravazoula, P. and Maragoudakis, M., 1994, Inhibition of nitric oxide synthase reverses the antiangiogenic and cytotoxic effects of X-ray irradiation on chick embryo chorioallantoic membrane. Concurrent presentation at the NATO ASI meeting on "Vascular Endothelium: Responses to Injury, Crete, Greece, June 25.

True, L.W., 1990, Atlas of diagnostic Immunohistopathology. pp. 9.2-9.15.

VASCULAR ENDOTHELIAL CELLS EXPOSED TO SUSTAINED HYDROSTATIC PRESSURE: A THREE DIMENSIONAL EVALUATION OF CYTOSKELETAL REARRANGEMENT

S. Salwen[1], D.H. Szarowski[2], J.N. Turner[2], R. Bizios[1]

[1]Department of Biomedical Engineering
Rensselaer Polytechnic Institute
Troy, New York 12180, USA

[2]Wadsworth Center for Laboratories and Research
New York State Department of Health
Albany, New York 12201, USA

The response of vascular endothelium to the hemodynamic forces present under normal and pathological conditions is important in understanding how mechanical injury of the endothelium effects vascular function. During hypertension, increased blood pressure exposes the vascular cells to increased normal stresses. Endothelial cell cytoskeletal changes in response to normal stress (hydrostatic pressure) are explored in this study in three dimensions.

The morphological responses of bovine pulmonary arterial endothelial cells (ATCC cell line CCL-209; passage 20-25) to sustained hydrostatic pressure (1.5, 2.5, 5, 10, and 15 cm H_2O) for up to 7 days were investigated *in vitro*. The three dimensional aspects of cytoskeletal rearrangement and cell multilayering were evaluated using laser scanning confocal microscopy.

Cell bilayering was observed when the cells were exposed to pressure for 5 days. The endothelial cells showed evidence of rearrangement of the F-actin filaments of the cytoskeleton after 7 days of exposure to the hydrostatic pressures. The stress fibers of the cytoskeleton were observed primarily in upper level (primary monolayer) cells which had been exposed to 5, 10 and 15 cm H_2O for over 5 days. Microtubules showed reorganization in conjunction with the shape changes in elongated cells.

The present study provided visualization of the three dimension aspects of endothelial cell morphology and cytoskeletal rearrangement under sustained hydrostatic pressure.

THE L-ARGININE-NITRIC OXIDE PATHWAY IN THE VASCULAR SMOOTH MUSCLE: REGULATION AND PATHOPHYSIOLOGICAL SIGNIFICANCE

V.B. Schini

Zentrum der Physiologie
Klinikum der JWG-Universität
Theordor-Stern-Kai 7
D-60590 Frankfurt
GERMANY

Nitric oxide is a multifunctional effector in the vascular system. In normal healthy blood vessels, nitric oxide is produced predominantly by endothelial cells which have a constitutively expressed nitric oxide synthase. Endothelial-derived nitric oxide plays a key role in the regulation of blood flow and the perfusion of vital organs. A second source of nitric oxide can be elicited in the blood vessel wall by inflammatory mediators. The exposure of cultures of smooth muscle cells from rat aorta to cytokines such as interleukin-1 β and tumor necrosis factor alpha causes after a delay of three to six hour the expression of an inducible type of nitric oxide synthase as indicated by the appearance of inducible nitric oxide synthase mRNA and protein. It is also supported by the fact that the cytokines evoke a time-dependent release of nitrite (a stable oxidation product of nitric oxide) and the intracellular accumulation of cyclic GMP in the smooth muscle cells. Both responses are inhibited by nitro L-arginine, an inhibitor of nitric oxide synthase. Similarly to endothelium-derived nitric oxide, nitric oxide derived from smooth muscle cells inhibits vascular tone and also prevents the activation of platelets. In addition, the endogenous cytokine-evoked production of nitric oxide in vascular smooth muscle cells brakes their proliferative response to mitogens. The expression of inducible nitric oxide synthase in vascular smooth muscle cells is controlled in a coordinated manner by factors-derived from both blood and vascular cells. Platelet-derived growth factor, transforming growth factor β, insulin-like growth factor and thrombin inhibit while epidermal growth factor, basic fibroblast growth factor and plasmin potentiate the expression of nitric oxide synthase in cytokine-activated vascular smooth muscle cells. These factors may prevent the production of large amounts of nitric oxide at sites of injury which are cytotoxic for most types of mammalian cells. As the *in vivo* endothelial denudation of the rat carotid artery by the repeated passage of a balloon catheter is associated with a time-dependent production of nitric oxide by the injured blood vessel wall; it is likely that smooth muscle-derived nitric oxide is an important mediator of the vascular response to injury causing local dilatation, limiting the activation of platelets and thus the formation of the thrombus and also curtailing the mitogenic response of factors released by platelets.

STUDIES ON THE REVERSIBILITY OF PROTEIN S-THIOLATION IN HUMAN ENDOTHELIAL CELLS

Ina Schuppe-Koistinen, Peter Moldeus and Ian A. Cotgreave

Divison of Toxicology
Institute of Environmental Medicine
Karolinska Institute
Stockholm, SWEDEN

Exposure of human umbilical vein endothelial cells to oxidants, such as hydrogen peroxide and diamide, has been shown to induce protein-specific S-thiolation of cellular proteins. In this study we have now identified GSH as the major low molecular weight thiol that is bound to protein substrates in HUVE cells during oxidative stress and investigated the dose- and time-response relationship of diamide-and hydrogen peroxide-induced S-thiolation of HUVE cell protein. Intact HUVE cells are able to rapidly re-reduce S-thiolated proteins with almost quantitative re-appearance of reduced glutathione in the cells and protection from acute, lytic cytotoxicity. Additionally, studies were performed with detergent-solubilized cell extracts to determine the nature of the reductants operating in HUVE cells to maintain protein thiol homeostasis and the results clearly show the involvement of NADH-and NADPH-dependent systems. These data suggest that the reversible S-thiolation of proteins in these human endothelial cells may represent a significant cellular antioxidant and regulatory mechanism during oxidative stress.

VITAMINS C AND E ALTER THE RESPONSE TO CORONARY BALLOON INJURY IN THE PIG

Demetrios S. Sgoutas#, Gilberto L. Nunes*, Michael B. Gravanis#, Spencer B. King III*, Robert A. Redden*, Steven R. Sigman* and Bradford C. Berk*

*Departments of Medicine (Cardiology Division, Andreas Gruentzig Cardiovascular Center)

#Pathology
Emory University School of Medicine

Restenosis after percutaneous transluminal coronary angioplasty is the major limitation of the long-term success of this procedure *(Holmes et al., 1984)*. The process of restenosis involves repair of vascular injury and remodeling of vessel architecture *(Kuntz and Baim, 1993)*. Therapeutic interventions that improve vascular function may therefore be beneficial in the treatment of restenosis *(Kuntz et al.)* Antioxidants such as probucol and vitamins C and E have proved effective in improving endothelial function in hypercholesterolemia, inhibiting lipid accumulation in animal models of atherosclerosis and decreasing cardiovascular mortality in humans *(Halliwell, 1989)*.

In the present study we tested the hypothesis that the antioxidant vitamins C and E would favorably alter the vascular response to balloon injury in the swine coronary. Important parameters analyzed were: analysis of left anterior descendingt (LAD) and circumflex (CFX) artery separately; measurement of circulating lipid redox state and vascular production of oxygen radicals, and development of a method to anlayze morphologic changes in all segments of the injured artery.

Forty-two female domestic swine were divided into four study groups (control, n=12); vitamin C (500 mg/d, Group C, n=9); vitamin E (1000 U/d, Group E, n=10) and vitamin C and E (500 mg/d + 1000 U/d, Group C+E, n=11) before oversized balloon injruy of the LAD and CFX coronary arteries. Vitamins were administered 7 days before balloon injury and continued until the swine were killed 14 days after injury. Significant differences in morphometric parameters were present only in the C=E group with increases in vessel and lumen area in the segment with maximal injury. Although, there was no decrease in intima area nor maximal intima thickness, the ratio of intima area/vessel area and the lumen diameter were significantly educed, consistent with a positive effect in the C+E group. A graphic analysis of the relation between initial vessel injury (using internal elastic lamina fracture length/lumen parameter) and vessel response to injury (using intima area/vessel area) for all segments showed improved indices for C=E ($p < 0.005$). The beneficial effect of vitamins on the repair of the vessel correlated with changes in lipid and vessel redox state. Low density lipoprotein (LDL) thiobarbituric acid reactive substances showed approximately 70% decrease in all treatment groups and the lag phase for LDL conjugated diene a measure of the degree of oxidizability of LDL formation was significantly increased with C=E>E>C. Vessel superoxide production was increased 2-fold in injured vessels and was significantly decreased in the LAD of C=E animals.

The combination of vitamins C and E improved the vascular response to injury, due to a beneficial effect on vascular remodeling. The fact that C=E was better than E or C alone is consistent with the ability of C to improve the antioxidant effect of E, suggesting that the improved vessel response was due to a change in redox state. This study demonstrates the importance of oxygen radicals in the vascular response to injury and suggests that vascular remodeling as well as intimal proliferation is important to the restonotic process. Since the greatest morphologic benefit occurred in the LAD and this vessel exhibited the greatest reduction in superoxide production, a cause and effect relationship is suggested.

REFERENCES

Halliwell, B., 1989, Free radicals, reactive oxygen species and human disease: a critical evaluation with special reference to atherosclerosis. *Br. J. Exp. Pathol.* 70:737-757.

Holmes, D.R., Vlietstra, R.E., Smith, H.C., Vetrovec, G.W., Kent, K.M., Cowley, M.J., Faxon, D.P., Gruentzig, A.R., Kelsey, S.F., Detre, K.M., VanRaden, M.J. and Mock, M.B., 1984, Restenosis after percutaneous transluminal coronary angioplasty (PTCA): A report from the PTCA regisry of the National Heart, Lung and Blood Institute. *Am. J. Cardiol.* 53:77C-81C.

Kuntz, R.E. and Baim, D.S., 1993, Defining coronary restenosis. Newer clinical and angiographic paradigms. *Circulation* 88:1310-1323.

Kuntz, R.E., Safian, R.D., Carrozza, J.P., Fishman, R.F., Mansour, M. and Baim, D.S., The importance of acute luminal diameter in determining restenosis after coronary atheroectomy or stenting.

NITRIC OXIDE (NO) AND PEROXYNITRITE ($ONNO^-$) RELEASED BY ALVEOLAR MACROPHAGES PLAY A POTENTIAL ROLE IN LUNG INJURY OF CIGARETTE SMOKERS

J.C. Stavrides, V. Villiotou and G. Deliconstantinos

Department of Experimental Physiology
University of Athens Medical School
Athens, Greece 11527

Cigarette smoking causes serious health problems according to undeniable evidence. Still uncertain are the mechanisms involved in causing these health problems. We have recently demonstrated that the exhaled cigarette smoke contains large amounts of •NO and $ONNO^-$. Alveolar macrophages suffer oxidative stress when exposed to inhaled cigarette smoke. The causative stimuli for this macrophages stress are to a larger extent O and to a lesser extent CO. Oxidative stress of the macrophages activates the NO Synthase. This triggering mechanism results in NO release in a prolonged and amplified reaction. •NO inside the lungs is transformed into $ONOO^-$, reacting with superoxide ($O_2^{\cdot}$) released by alveolar macrophages. Oxidative stress also causes the production of hydrogen peroxide (H_2O_2). Both in vitro and in vivo measurements confirm all these findings. In vitro experiments were conducted with the aid of a special apparatus built in our lab, consisted of two chambers (A and B) separated by a tefflon membrane permeable to NO but not to ONOO and to NO_2^-. Mice lung macrophages placed in an HBSS solution were emptied into chamber A. An equal volume of Griess reagent was placed in the chamber B. The amount of NO released by the macrophages diffuses through the tefflon and was measured in chamber B. The amount of $ONOO^-$ and NO_2 contained inside the chamber A by the impervious tefflon membrane is also measured Oxidative stress of macrophages and human erythrocytes both exposed to cigarette smoke was estimated. All findings prove conclusively that, (a) macrophages exposed to cigarette smoke produce a dramatic increase of NO, NOx, $ONOO^-$ and H_2O_2 inside the lung. These substances stimulate the macrophages and perhaps even the endothelium of the alveolar vessels. They may cause indirect damage by triggering the production of more nitrogen compounds inside the lung in a vicious circle.

SINGLE VERSUS DOUBLE-LUNG INJECTION OF ^{3}H-BPAP IN HUMANS

[**]S.E. Orfanos, [*]D. Langleben, [*]J. Khoury, [*]R.D. Schlesinger, [*]L. Dragatakis, [**]Ch. Roussos, and [***]J.D. Catravas.

[*]Lady Davis Research Institute and Division of Cardiology Sir Mortimer B. Davis Jewish General Hospital, Mc Gill University, Montreal, Canada

[**]Critical Care Department, Evangelismos Hospital, University of Athens Medical School, Athens, Greece

[***]Vascular Biology Center, Medical College of Georgia, Augusta, GA, USA

ABSTRACT

The pulmonary vascular endothelium participates in various important physiologic and pharmacokinetic processes, such as synthesis and degradation of vasoactive peptides. Enzymes that catalyze many of these reactions are located on the luminal surface of the capillaries and their activities may be measured *in vivo* by means of *indicator-dilution* techniques. Among these enzymes, angiotensin-converting enzyme (ACE) has been extensively studied in animals: monitoring ACE activity under toxic conditions provides an early index of lung injury, whereas under normal conditions allows estimations of the dynamically perfused capillary surface area (PCSA), (*Orfanos et al., 1994*). The purpose of this study was: i) to validate the use of the aforementioned techniques as a tool of pulmonary endothelial function assessment in humans, and ii) to investigate if the assays of pulmonary endothelial ACE, under normal conditions provide estimations of PCSA in the human lung.

Six human volunteers (five males and one female), 34 to 71 years old, undergoing cardiac catheterization for coronary artery disease assessment, participated in our study. Participants did not have a medical history of pulmonary disease, they were not receiving ACE-inhibitor treatment, and exhibited normal pulmonary blood pressures. Single-pass transpulmonary hydrolysis of the specific ACE substrate ^{3}H-benzoyl-Phe-Ala-Pro (BPAP; 30μCi, 22.2 Ci/mmol) was measured by means of *indicator-dilution* techniques, and expressed as %metabolism (%M) and v= ln(1-M). BPAP was injected as a bolus i) into a main pulmonary artery, and ii) inside the right atrium, to assess ACE activity in one and both lungs respectively. We also calculated A_{max}/K_m, an index of PCSA. Pulmonary plasma flow (F_p) was determined immediately prior to each injection by thermodilution. F_p in one lung was estimated as 50% of total F_p (*Lipscomb and Pride, 1977; Wernly et al., 1980*). Similar values of %M (67±2.9 and 65±2.5), and v (1.13±0.09 and 1.07±0.07) were observed in both and one lung respectively, denoting homogeneous pulmonary capillary ACE concentrations and capillary transit times in both human lungs. A_{max}/K_m decreased from 3804±375 ml/min (both lungs) to 1899±189 (one lung), denoting a 50% reduction in PCSA, as expected. We conclude that i) pulmonary endothelial ectoenzyme activity and thus pulmonary endothelial function may be assessed in humans by means of *indicator-dilution* techniques and ii) this procedure can be used to quantify PCSA in the human lung. (Supported by Fonds de la Recherche en Sante du Quebec, and the National Health System of Greece).

REFERENCES

Lipscomb, D.J. and Pride, N.B., 1977, *Thorax* 32:720-725.
Orfanos, S.E. et al., 1994, *Toxicol. Appl. Pharmacol.* 124:99-111.
Wernly, J.A. et al., 1980, *J. Thorac. Cardiovasc. Surg.* 80:535-543.

ENDOGENOUS INHIBITORS OF THE INDUCTION OF THE INDUCIBLE ISOFORM OF NITRIC OXIDE SYNTHASE

Csaba Szabó, M.D., Ph.D.

The William Harvey Research Institute
St. Bartholomew's Hospital Medical College
Charterhouse Square, London EC1M 6BQ
UNITED KINGDOM

The induction of a distinct isoform of nitric oxide (NO) synthase (iNOS) in various organs and in the vascular smooth muscle occurs in a variety of diseases and is an important factor leading to therapy-resistant circulatory failure in shock *(Szabó and Thiemermann, 1993)*. The identification of endogenous factors that interfere with the induction of iNOS may have great potential. Here we overview the evidence supporting the view that glucocorticoids, polyamines, and IL-4 and IL-10 act as endogenous inhibitors of the induction of iNOS.

To study the role of endogenous glucocorticosteroids in the induction of iNOS by E. coli endotoxin (LPS), the responses of normal rats were compared with those of adrenalectomized (ADX) animals. In normal rats, LPS (10 mg/kg; i.v.) caused both an immediate and a delayed fall in blood pressure (MAP) and a reduction of the pressor responses to noradrenaline (NA) at 60-180 min. The fall in MAP and the reduction in pressor responses was inhibited by N^G-methyl-L-arginine or dexamethasone. Pronounced iNOS induction in the lung was observed at 180 after endotoxin application. This induction of iNOS was inhibited by dexamethasone pretreatment *(Szabó et al., 1993a)*. In normal Wistar rats, a low dose of LPS (0.1 mg/kg i.v.) did not alter MAP and pressor responsiveness over 120 min. In contrats, in ADX rats a profound hypotension (at 30-120 min) and hyporeactivity to NA (at 120 min) developed in response to the same dose of endotoxin. This was prevented by dexamethasone and significantly inhibited by NMA pretreatment. NOS induction in lung homogenates of ADX, but not of normal Wistar rats was found at 120 min after endotoxin *(Szabó et al., 1993b)*. Thus, endogenous glucocorticoids exert a tonic suppression against the induction of iNOS by LPS. The role of endogenous glucocorticoids is crucial in the development of endotoxin tolerance, since endotoxin tolerance is associated with increased plasma levels of glucocorticoids which accounts for the blunted cardiovascular response and reduced induction of iNOS in these animals *(Szabó et al., 1994a)*.

In the presence, but not in the absence, of fetal calf serum, spermine inhibits the production of nitric oxide (NO) in cultured J774.2 macrophages stimulated with LPS or with γ- interferon, showing that polyamines may act as suppressants of NO-mediated immune functions *(Szabó et al., 1994b)*. Other polyamines also inhibit the formation of NO in cultured J774.2 macrophages, with the order of potency being spermine > spermidine >> putrescine = cadaverine. Spermine is not a direct inhibitor of the activity of iNOS in induced J774.2 cells but it is an inhibitor of the induction of iNOS. The effect of spermine can be prevented by isoniacid or hydroxylamine, inhibitors of spermine oxidase as well as by phenylhydrazine, an aldehyde inhibitor. Spermine dialdehyde (SDA), but not malon dialdehyde inhibits nitrite formation in J774.2 cells in the absence of serum when given as a pretreatment but not when given 6h after stimulation. Thus, aldehyde metabolites of spermine, such as SDA, account for the inhibitor effect of polyamines on the induction of NOS *in vitro*. The inhibitory effect of polyamines on iNOS induction appears to be fairly specific to iNOS induction, for spermine does not inhibit LPS-induced production of prostaglandin $F_{2\alpha}$ or tumour necrosis factor *(Szabó et al., 1994c)*.

In peritoneal macrophages *ex vivo*, LPS (0.01-0.3 mg/kg i.p.) induces a dose-dependent increase in *ex vivo* nitrite production (a stable end-product of NO) at 24h. This is inhibited by IL-4 (0.1-1 µg i.v.). IL-4 also significantly inhibits the accumulation of nitrite when NO synthase was induced by endotoxin *in vitro* in peritoneal macrophages or in J774.2 cells *(Szabó et al., 1994d)*. IL-10 (0.01-1 µg i.v.) also inhibits NO production in endotoxin-stimulated peritoneal macrophages *ex vivo* but not *in vitro*.

Thus, endogenous glucocorticoids, polyamines and certain anti-inflammatory cytokines (e.g. IL-4 and IL-10) act as endogenous inhibitors of the induction of iNOS. Disturbances in the production of these factors may well contribute to the development of a variety of diseases associated with the induction of iNOS.

REFERENCES

Szabó, C. and Thiemermann, C., 1993, *Current Op. Develop. Drugs* 2:1-10.
Szabó, C., Mitchell J.A., Thiemermann, C. and Vane, J.R., 1993a, *Br. J. Pharmacol.* 108:786-792.
Szabó, C., Thiemermann, C. and Vane, J.R., 1993b, *Proc. Royal Soc. B.* 253:233-238.
Szabó, C., Thiemermann, C. Wu, C.C., Perretti, M. and Vane, J.R., 1994a, *PNAS* 91:271-275.
Szabó, C., Southan, G.J., Thiemermann, C. and Vane, J.R., 1994b, *Br. J. Pharmacol.* 112:355-356.
Szabó, C., Southan, G.J., Thiemermann, C. and Vane, J.R., 1994c, *Br. J. Pharmacol.* in press.
Szabó, C., Perretti, M., Thiemermann C., Flower, R.J. and Vane, J.R., 1994d, *Br. J. Pharmacol.* 112:519P.

THE PULMONARY VASODILATORY PROPERTIES OF PGE_1 ARE BLUNTED AFTER EXPERIMENTAL SINGLE LUNG TRANSPLANTATION

Sinikka Kukkonen, MD[a], Lasse Heikkilä, MD[b], Kalervo Verkkala, MD[b], Severi Mattila, MD[b], and Hannu Toivonen, MD, PhD[a]

Department of Anesthesiology[a]
Helsinki University Central Hospital
Helsinki, Finland

Department of Thoracic Surgery[b]
Helsinki University Central Hospital
Helsinki, Finland

Prostaglandin E^1 (PGE_1) is a short-lived powerful vasodilator, which modifies platelet and neutrophil function and exerts a cytoprotective effect. In search of improved pulmonary preservation techniques PGE_1 has been extensively studied as an adjunct to the crystalloid flush solutions or in donor pretreatment. Less attention has been paid to its possible usefulness when administered during the reperfusion period. Therefore, we evaluated the effects of PGE_1-infusion on pulmonary hemodynamics and gas exchange during the first four hours of reperfusion of a pig left lung allotransplantation.

The transplanted lung was preserved with modified Euro-Collins solution and stored immersed in a cold (+4 °C) saline. The range of the graft ischemia times was 130-226 minutes. To gain accurate control of right (nontransplanted) and left (transplanted) lung circulations we used extracorporeal circulation and total right heart bypass with individual cannulations of the right and left pulmonary arteries. To simulate normal clinical situation, systemic blood flow of 2 litres/min was divided between the lungs with two separate perfusion pumps to maintain equal perfusion pressures in each lung. Periodically, the blood flow was divided evenly to both lungs, 1 litre/min each, to measure pulmonary vascular resistance at constant flow conditions. Six animals received an increasing PGE_1 infusion (10, 25, 50, 100 ng/kg/min; 15 min each) and after two and a half hours of reperfusion (75 min without the drug) the infusion schedule with 25, 50, 100 ng/kg/min was repeated. The highest PGE_1-dose was selected on the basis of the pilot experiments to create a 30% decrease in the systemic mean arterial pressure (MAP). During the first infusion period PGE_1 decreased MAP by 25%, but an almost similar decreased occurred in the control animals. During the second infusion series a 28% decrease in MAP was observed only in the PGE_1 treated animals. None of the infusions was able to decrease pulmonary vascular resistance. Instead, PGE_1 diverted 2/3 of the pulmonary blood flow towards the native lung, and this manifested itself as a tendency to an earlier improvement of P_aO_2 in the drug treated animals. End-tidal $_2CO$ in the mixed expiratory gas was not significantly different from the values from each individual lung, but there was a distinct gradient of 14-20 mmHg (range) between the arterial and end-tidal CO_2 values. We conclude, that PGE_1 in doses tolerated by the systemic circulation, is ineffective in the treatment of the increased pulmonary vascular resistance after pulmonary transplantation. *This study was supported by the Finnish Academy.*

REFERENCES

Novick R., Reid K., Denning L., Duplan J., Menkis A. and McKenzie F., 1991, Prolonged preservation of canine lung allografts: the role of prostaglandins. *Ann. Thoracic Surg.* 51:853-9.

Prielipp R., McLean R., Rosenthal M. and Pearl, R., 1991, Hemodynamic profiles of prostaglandin E_1, isoproterenol, prostacyclin, and nifedipine in experimental porcine pulmonary hypertension. *Crit Care Med.* 19:60-7.

Ueno T., Yokomise H., Oka T. et al., 1991, The effect of PGE_1 and temperature on lung function following preservation. *Transplantation* 52:626-30.

MECHANISM OF BIOMINERALIZATION OF BOVINE PERICARDIUM TISSUE

Branko B. Tomazic

ADAHF Paffenbarger Research Center/NIST
Gaithersburg, Maryland 20899, USA

Pathologic biomineralization is the major cause of failure of tissue derived heart valve bioprosthesis *(Schoen, 1989)*. The formation of calcified deposits in the cardiovascular system is a very important process and if its mechanism is properly understood, it may greatly affect the technology of long-lasting bioprosthesis. The present study was to investigate the mechanism of *in vitro* mineralization of de-endotheliazed, glutaraldehyde-treated bovine pericardium (BP), the essential component of BP bioprostheses. The primary question is: does the mineralization *in vitro* proceed as surface (extrinsic) process *(Tomazic et al., 1991)* or as a subsurface (intrinsic) process recorded *in vivo (Schoen et al., 1985)*? The BP mineralization *in vitro* was investigated by using a diffusion cell with two compartments that contained calcium phosphate solutions of different concentrations, separated by de-endotheliazed BP membrane. Diffusion of Ca and P through the membrane was monitored over extended periods under specified thermodynamic conditions. This permitted predicting which calcium phosphate compound could form at the membrane/solution interface (extrinsic mineralization), but more complexity occurs within the membrane corresponding to intrinsic mineralization. The analysis of transport of Ca and P through the BP membrane indicated the retention of Ca and P at/within membrane. Histology, scanning electron microscopy (SEM) and energy dispersive x-ray microanalysis (EDX) provided good evidence of subsurface mineralization, in addition to surface mineralization. The latter was not documented by histology, only with EDX. The EDX data compiled by line scan analyses indicated the Ca/P heterogeneity of the early precursor formed in the membrane environment that was indicated by Ca and P x-ray mapping of subsurface calcified regions. The majority of Ca/P ratios appeared higher than those of any stiochiometric phase. This result indicates that the "three phasic mechanism" *(Urist, 1973)* may be prevailing in the formation of early precursor of biological apatites, in addition to a transient precursor, octacalcium phosphate *(Brown et al., 1987)*. One may expect that the presence of endothelium that acts as a protective barrier would greatly decrease the extent of intrinsic mineralization; therefore, successful re-endothelization may be a key factor in the longevity of tissue-derived bioprostheses. *(Supported in part by NHLBI grant HL30035)*.

REFERENCES

Brown, W.E., Eidelman, N. and Tomazic, B., 1987, Octacalcium phosphate as a precursor in biomineral formation. *Adv. Dent. Res.* 1:306-313.

Schoen, F.J., 1989, Interventional and surgical cardiovascular pathology. *Clinical Correlations and Basic Principles*. W.B. Saunders Company, Philadelphia.

Schoen, F.J., Levy, R.J., Nelson, A.C., Bernhard, W.F., Nashef, A. and Hawley, M., 1985, Onset and progression of experimental bioprosthetic heart valve calcification. *Lab Invest.* 52:523-532.

Tomazic, B.B., Siew, C. and Brown, W.E., 1991, A comparative study of bovine pericardium mineralization. *Cells Mater* 1:231-241.

Urist, M.R., 1973, Biologic initiators in calcification, *In:* Biological Mineralization, Zipkin, I. (ed), John Wiley & Sons, New York.

IMAGING OF ARTERIOSCLEROSIS AND THROMBOSIS

Irene Virgolini, M. Banyai, F. Rauscha, A. Kurtaran, P. Angelberger, et al.

Department of Nuclear Medicine
University of Vienna
Vienna, AUSTRIA

Over the past decade several groups have used radiolabelled low density lipoprotein (LDL) in order the investigate liver LDL receptor function on the basis of LDL kinetic studies, and also to demonstrate LDL uptake in human atherosclerotic lesions. The influx of radiolabelled LDL into the vessel wall has allowed visualization of some active forms of atherosclerotic plaques. We could visualize atherosclerotic plaques over 48 hours after injection of autologous ^{111}In-labelled LDL. The highest uptake of ^{111}In-LDL was found at 2 hours after injection for both the carotid and femoral artery plaques. The kinetic pattern of LDL uptake by the vessel wall could be an indicator for the integrity of endothelial lining: deendothelialized arterial segments seem to retain the maximum of radioactivity. We also found that plaques enriched with foam cells take up ^{123}I-LDL/^{111}In-LDL. In a further approach we have investigated ^{111}In-LDL liver uptake in normolipemic subjects and in patients with heterozygous familial hypercholesterolemia (FH). In these studies, FH patients were shown to possess a decreased accumulation of LDL in the liver due to LDL receptor deficiency.

Many objective diagnostic procedures fail to satisfy the criteria of simplicity and noninvasiveness in the clinical diagnosis of thrombi. Radiolabelled platelets, fibrinogen, plasmin or monoclonal antibodies directed against platelets and fibrin have been used as clot imaging agents. We have used ^{123}I-labelled rh-tPA as clot imaging agent in patients with deep venous leg thrombosis, Cimino shunts thrombosis or vena cava superior thrombosis. The *in vivo* biodistribution and fibrin localizing capacity of ^{123}I-rh-tPA were investigated. After i.v.-injection to patients (200 MBq/1 mg) organ distribution studies demonstrated that the early rapid uptake of ^{123}I-rh-tPA occurred exclusively in the liver. In all patients with deep venous leg thrombosis or Cimino shunt thrombosi ^{123}I-rh-tPA scanning provided excellent imaging of venous thrombi, whereas thrombi of the vena cava superior could not be detected. ^{123}I-rh-tPA rapidly accumulated in clots within 5 minutes after injection. The scintigraphic findings correlated well with ultrasound and/or phlebography. None of other isotope techniques (platelet imaging, ^{123}I-plasmin or fibrinogen-scanning) was superior to ^{123} I-rh-tPA scanning. We conclude that ^{123}I-rh-tPA scanning could be a safe and rapid method for the determination of venous thrombi.

EVIDENCE FOR INCREASED VASCULAR MONOCYTE UPTAKE IN PATIENTS WITH HETEROZYGOUS FAMILIAL HYPERCHOLESTEROLEMIA *IN VIVO*

Irene Virgolini, Shu-Ren Li, Walter Pirker, Peter Fitscha, Graziana Lupatteli, Ross G. Gerrity, and Helmut Sinzinger

Department of Nuclear Medicine
and Wilhelm Auerswald Atherosclerosis Research Group
Vienna, AUSTRIA

Policlinic Monteluce
University of Peruggia
ITALY

Department of Anatomic Pathology
Medical College of Georgia
Augusta, Georgia, 30912, USA

Considerable attention has been focused on the potential pathogenetic role of monocytes in the atherogenic process widely regarded as being the result of tissue reaction after injury to the vascular endothelium. There is strong evidence that one of the earliest key events during atherogenesis is the adherence of circulating monocytes to the endothelial lining of the artery, followed by migration across the endothelial barrier into the subendothelial space where they may appear as lipid-loaden foam cells. Various experimental studies have shown that hypercholesterolemia induces an increased adhesion of monocytes *in vitro*, and reinjection studies using labelled autologous monocytes in hyperlipemic animals have demonstrated their localization in atherosclerotic lesions and lesion-susceptible areas in aorta. Whereas there are abundant data on *in vitro* monocyte adhesion to the arterial wall, and evidence for increased adhesion in experimentally induced hypercholesterolemia, nearly nothing is known for *in vivo* involvement and the predisposing factors in human. We approached this problem by developing the "monocyte scintigraphy" which uses 10^8 autologous monocytes radiolabelled with 0.42 MGq^{111} In-oxine (viability > 95%). Following transient imaging of the lungs during the first 15 minutes after reinjection, the predominant target organ of labelled monocytes was the spleen. Small "hot spots" over the thigh or lower leg arteries became visible within 2 to 8 hours after reinjection. In patients with heterozygous familial hypercholesterolemia (FH) monocyte uptake was found in 68% of lesions that were positive upon angiography (defined as a 30-60% luminal stenosis of the artery). In contrast, in normolipidemic patients monocyte accumulation was noted only in 42% of lesions that were positive upon angiography. The monocyte uptake ratio (MUR; i.e. the cpm measured in a region of interest (ROI) inserted into a lesion showing monocyte uptake in relation to the cpm measured in a ROI inserted into an area without visible (hot spot) monocyte uptake of the contralateral side) was significantly (p,0.001) higher in FH patients as compared with controls. Successful lipid-lowering intervention (medication and LDL-apheresis) in 10 FH patients resulted in a decreasing MUR (1.35 ± 0.07 vs 1.20 ± 0.09; $p<0.001$ at 48 hours p.i.) after a 2-month intervention period. Taken together, these imaging results in human *in vivo* provide evidence for an enhanced vascular monocyte adhesion in patients with FH, and open a new research area into human atherosclerosis.

CROSS-COMPETITION BETWEEN VASOACTIVE INTESTINAL PEPTIDE (VIP) AND SOMATOSTATIN (SST) FOR BINDING TO SPECIFIC RECEPTORS EXPRESSED ON ENDOTHELIAL CELLS, BASOPHILS, MAST CELLS AND MONOCYTES: PATHOPHYSIOLOGY AND CLINICAL IMPLICATIONS

Irene Virgolini, Q. Yang, S.R. Li, H. Agis, A. Kurtaran, S. Kapiotis, W. Sperr, P. Angelberger and Peter Valent

Departments of Nuclear Medicine and Hematology
University of Vienna
A-1090 Vienna, AUSTRIA

Increasing evidence exists that growth and proliferation as well as the functional properties of various cells are regulated by a complex network of cytokines and by peptide interactions. VIP is a 28-amino acid neuroendocrine mediator with a broad range of biologic activities in diverse cells and tissues. VIP is a vasodilatory substance responsible for the watery diarrhea syndrome in patients with VIP-secreting tumors. More recent studies suggest that in addition, VIP promotes cell growth and proliferation. The effects of VIP are mediated via cell surface membrane receptors which are widely distributed throughout the gastrointestinal tract, but are also found on various immune cells. SST is a 14-amino acid peptide acting as a neurotransmitter or as a hormone, depending on the site of action and target cell type. In contrast to VIP, SST inhibits cell growth and proliferation. We have analyzed the in vitro (and in vivo) biodistribution and receptor binding of ^{123}I-labelled VIP and of ^{123}I-labelled octreotide (long-acting SST analog, OCT). Furthermore, possible interactions between VIP and SST (and its analogs) were analyzed. Significant cross-competition between VIP and SST/OCT for binding onto endothelial cells (HUVEC), basophils (KU812 cells), mast cells (HMC1 cells), peripheral blood monocytes and a variety of primary tumor cells and tumor cell lines (PANC1, A431, HT29, COLO320) was observed. The rank-order of potency for displacement of ^{123}I-VIP binding to intact endothelial cells was VIP (3.8 nM) > OCT (IC_{50}=12.9 ± 3.2 nM) = SST > substance P = secretin (IC_{50} = 1 µM). Binding of ^{123}I-Tyr-3-OCT to endothelial cells, in turn, was inhibited by OCT = Tyr-3-OCT (IC_{50}=3.1 ± 1.9 nM) = SST > VIP (IC_{50}=8.2 ± 3.0 nM). This rank-order of potency was also obtained for all other normal and malignant cells investigated. Furthermore, SST and OCT inhibited VIP-induced 3H-thymidine incorporation, cAMP-formation and tyrosine kinase activity with IC_{50} values <10 nM. In vivo biodistribution studies indicated that the lung is the primary organ of ^{123}I-VIP (200 MBq, 300 pmol) uptake. At 5 minutes after i.v. -injection >90% of the radioactivity was cleared from the blood, excretion occurred via the urinary tract. In patients with ^{123}I-VIP positive tumors (adenocarcinomas of the intestine, lymphomas, endocrine tumors) the tumor masses and/or metastases were visualized after i.v.-injection. In contrast to ^{123}I-VIP, ^{123}I-Tyr-3-OCT (200 MBq, 10µg) was found in the liver and spleen, and excreted into the bile. VIP-receptor positive endocrine tumors were also visualized with ^{123}I-OCT. Together, our data provide evidence for functional interactions between SST and VIP on various circulating and tissue-fixed cells. These interactions may involve peptide cross-competition at cellular binding sites and may have implications for the biology and pathophysiology of respective cells and disease states.

ANGIOGENESIS IN DELAYED REVASCULARIZATION MODEL IS ACCELERATED BY LOW MOLECULAR WEIGHT HYALURONAN

Lees VC[1] and West, DC[2]

[1]Department of Pharmacology
University of Cambridge, UNITED KINGDOM

[2]Department of Immunology
University of Liverpool, UNITED KINGDOM

A delayed revascularization model *(Lees and Fan, 1992)*, has been used to assess the angiogenic activity of hyaluronan fragments. Revascularization of single full thickness skin autografts in rats was delayed by subjecting isolated grafts to a sublethal cryoinjury prior to implantation. Hyaluronan fragments were delivered to the grafts using slow release pellets. Rates of release were measured *in vitro* by ELISA. The angiogenic response to the application of 100µg of low (1-4kDa) molecular weight by hyaluronan and 100µg of medium (33kDa) molecular weight hyaluronan was tested in separate groups of 15 rats, and compared to unstimulated cryoinjured controls (n+30). Return of graft blood flow was measured on anaesthetized animals over a 10 day period using laser Doppler flowmetry and ^{133}Xe clearance. Quantitative histological assessment of graft vessel profile numbers was performed on days 3,7 and 10 after grafting.

Low molecular weight by hyaluronan fragments increased blood flow ($p<0.001$) as measured by both flow measuring techniques, and increased graft vessel profile numbers assessed histologically at each of the three time points samples ($p<0.05$). Medium molecular weight fragments had no such effect and the measured responses were not significantly different from the control.

Further investigation of the effect of application of the angiogenic low molecular weight hyaluronic acid fragments on endogenous graft hyaluronic acid content was undertaken in a separate histological study, using a biotinylated probe to stain for hyaluronic acid.

The hypothesis that there may be physiological regulation of angiogenesis by hyaluronic acid and its metabolites is supported by the results of these studies. The data provides further evidence of the utility of the cryoinjured graft model for the study of *in vivo* angiogenesis.

REFERENCES

Lees, V.C. and Fan, T-P.D., 1992, Development of a new wound healing model for the quantitative study of angiogenesis in rats. *Int. J. Microcirc.* 11(1): S183.

ANTI-ENDOTHELIAL CELL ANTIBODIES IN THROMBOTIC THROMBOCYTOPENIC PURPURA AND HAEMOLYTIC URAEMIC SYNDROME

Wickham, N.W.R.^*, Crisp, S.J., Rose, M.L., Dunn, M.J., Rowley, M.* and Bevan, D.H.*

Department of Transplant Immunology
Heart Science Centre
Harefield Hospital
Middlesex

Department of Haematology
St George's Hospital*
London

^Present address:
Department of Clinical Oncology
Prince of Wales Hospital
Chinese University of Hong Kong
Shatin, Hong Kong

INTRODUCTION

Thrombotic thrombocytopenic purpura (TTP) and haemolytic-uraemic syndrome (HUS), are rare and poorly understood disorders involving multiorgan microvascular platelet thrombus formation and microangiopathic haemolytic anaemic *(Kaplan et al., 1992; Kaplan 1992; White et al., 1988)*. Only in HUS has an initiating pathologic stimulus been identified in the form of toxins associated with specific enteric infections (i.e., E Coli, Shigella)*(Karmali et al., 1983; Karmali, 1989; Kavi and Wise, 1989)*. Because of the effectiveness of plasma exchange therapy *(Moake, 1991; Rock et al., 1992; Bell et al., 1991)* workers have looked for circulating factors that might explain the therapeutic response. Several have been detected, most notably, abnormally large high molecular weight von Willebrand multimers *(Rose et al., 1984)* a calcium-dependent cysteine protease *(Murphy et al., 1987)* and a platelet aggregating protein (p37) *(Siddiqui et al., 1985)*. All these factors reflect responses to a damaging stimulus and are implicated in platelet activation, aggregation and deposition *(Moake and McPherson, 1989; Moore et al., 1990; Lian et al., 1991)*.

RATIONALE

Anti-endothelial cell antibodies (AECA) have been implicated in a number of other disorders involving immune/inflammatory damage to the vessel wall, but to what extent they play a causal role is not known (Editorial 1991; *Pearson, 1991*). Lytic anti-endothelial cell antibodies have been detected in some patients with TTP or HUS *(Leung et al., 1988; Burns and Zucker-Franklin, 1982; Foster and Anderson, 1979)* and these may form part of an initial damaging stimulus in a manner analogous to that of Vero cytotoxin *(Kavi et al., 1987)*. In a previous study that demonstrated lytic/cytotoxic AECA in cases of TTP and HUS, it was not possible to demonstrate binding to endothelial cell proteins *(Leung et al., 1988)*. However, this phenomenon has been demonstrated by Dunn and colleagues in sera from patients with AECA associated with coronary artery disease post cardiac-transplantation, with a sensitive technique using one-dimensional sodium dodecyl sulphate polyacrylamide gel electrophoresis and western blotting *(Dunn et al., 1992)*. Therefore, we decided to examine stored serum from TTP\HUS patients using this technique.

STUDY METHODS

Nine adult patients with TPP/HUS, admitted to St George's Hospital between 1988 and 1992, had had serum/plasma samples stored (at -70°C) on the day of admission and, in some cases, during subsequent treatment and follow-up. Samples were analyzed for the presence of endothelial cell antibodies using the established human endothelial cell line EA.hy 926 *(Edgell et al., 1983)* as a source of endothelial cell antigens *(Crisp et al., 1994)*. Tests were carried out both with and without prior activation of the endothelial cells using gamma-interferon. Samples were also tested against an epithelial cell line (A549), monocytes and platelets to assess the specificity of any antibodies detected. Cytotoxicity of the samples against endothelial cells was tested for by a modification of a complement-mediated cell lysis assay as described by *Martin et al, 1991*.

RESULTS

AECA were detected in all patients, and distinct patterns could be discerned in the western blots that correlated with the patients' clinical presentation. All sera contained IgG against a 96kD protein; in one patient with chronic relapsing TTP, no other bands could be detected; in one patient with acute relapsing TTP, an additional band at 120 kD was seen; samples from two patients with viral prodromes, adult HUS and acute renal failure also contained IgG against a 35 kD protein, whilst three patients with post-viral TTP contained IgM against a 35 kD protein; two patients with atypical TTP/HUS associated with another underlying autoimmune disorder (MCTD; seronegative lupus) also contained strong bands (IgM and IgG respectively) against a 69 kD protein. In those cases who had follow-up samples, there was no change in the pattern of antibodies that could be related with response. Indeed, there was either no change in the antibodies detected or further antibodies were found. In 4 out of 7 patients re-tested, a 69 kD antibody appeared (in 3 cases IgG, 1 case IgM). Reactivity was not affected by prior gamma-interferon endothelial cell activation and no lytic activity could be demonstrated. Band specificities and reactivity with different cell types showed that the 96 kD antibody had strong reactivity (+++) with endothelial cells but weak reactivity (±) against monocytes and epithelial cells; the 69 kD band showed moderate activity against endothelial cells, monocytes and platelets, but none against epithelial cells, whilst the 35 kD band showed moderate activity (+) against endothelium and platelets only.

CONCLUSIONS

AECA were detected in all cases of HUS/TTP. They were relatively specific, non-lytic and did not disappear with successful therapy. The apparent association between different patterns of antibodies is an intriguing finding that is in keeping with the heterogeneity of these disorders, and suggests that whilst the presence of AECA is an epiphenomenon rather than a cause, these patterns may be clue to the initial pathologic stimulus. Elucidation of the nature of the target antigens will be an important step towards determining the significance of these findings.

REFERENCES

Burns, E.R. and Zucker-Franklin, D., 1982, Pathologic effects of plasma from patients with thrombotic thrombocytopenic purpura on platelets and cultured vascular endothelial cells. *Blood* 60:1030.

Bell, W.R., Braine, H.G., Ness, P.M. and Kickler, T.S., 1991, Improved survival in thrombotic thrombocytopenic purpura-hemolytic uremic syndrome. *N. Engl. J. Med.* 325:398.

Crisp, S.J., Dunn, M.J., Rose, M.L., Barbir, M., and Yacoub, M.H., 1994, Antiendothelial antibodies after heart transplantation: the accelerating factor in transplant-associated coronary artery disease? *J. Heart Lung Transplant* 13:81.

Dunn, M.J., Crisp, S.J., Rose, M.L., Taylor, P.M. and Yacoub M.H., 1992, Anti-endothelial antibodies and coronary artery disease after cardiac transplantation. *Lancet* 339(8809):1566.

Edgell, C.J.S, McDonald, C.C. and Graham, J.B., 1983, Permanent cell line expressing human factor VIII-related antigen established by hybridization. *Proc. Nat. Acad. Sci. USA* 80:3734.

Editorial., 1991, Antibodies to endothelial cells. *Lancet* 337 (March 16):649.

Foster, P.A. and Anderson, J.C., 1979, Effects of plasma from patients with thrombotic thrombocytopenic purpura on cultured endothelial cells. *Blood* 54:240.

Kaplan, B.S., Trompeter, R.S. and Moake, J.L., 1992, Haemolytic Uremic Syndrome and Thrombotic Thrombocytopenic Purpura, 558, New York: Marcel Dekker.

Kaplan, B.S., 1992, Commentary on the relationship between hemolytic uremic syndrome and thrombotic thrombocytopenic purpura, in Hemolytic Uremic Syndrome and Thrombotic Thrombocytopenic Purpura, Dekker. Kaplan, Trompeter, Moake (eds). 29-39, New York.

Karmali, M.A., Steele, B.T., Petric, M. and Lim, C.,1983, Sporadic cases of haemolytic-uraemic syndrome associated with faecal cytotoxic and cytotoxin-producing Eschericia Coli in stools: *Lancet* (i): 619.

Karmali, M.A., 1989, Infection by Verocytotoxin-producing Eschericia coli. *Clin Microbiol Revs* 2:15.

Kavi, J. and Wise, R., 1989, Causes of the haemolytic uraemic syndrome. *Br Med J.* 298:65.

Kavi, J., Chant, I., Maris, M. and Rose, P., 1987, Cytopathic effect of verotoxin on endothelial cells: *Lancet* (ii):1035.

Leung, D.Y.M., Moake, J.L., Havens, P.L., Kim, M. and Pober, J.S., 1988, Lytic anti-endothelial cell antibodies in haemolytic-uraemic syndrome: *Lancet* (July 23):183.

Lian, E.C., Siddiqui, F.A., Jamieson, G.A. and Tandon, N.N., 1991, Platelet agglutinating protein p37 causes platelet agglutination through its binding to membrane glycoprotein IV. *Thromb Haemost* 65(1):102.

Martin, S., Benchley, P.E., Postlethwaite, R.J., Johnson, R.W.G. and Dyer, P.A., 1991, Detection of anti-epithelial antibodies in association with pediatric renal transplant failure using a novel microcytotoxicity assay. *Tissue Antigens* 37:152.

Moake, J.L., 1991, TTO - desperation, empiricism, progress. *N. Engl. J. Med.* 325(6):426.

Moake, J.L. and McPherson, P.D., 1989, Abnormalities of von Willebrand factor multimers in thrombotic thrombocytopenic purpura and the hemolytic-uremic syndrome. *Am. J. Med.* 87:9.

Moore, J.C., Murphy, W.G. and Kelton, J.G., 1990, Calpain proteolysis of von Willebrand factor enhances its binding to platelet membrane glycoprotein IIb/IIIa: an explanation for platelet aggregation in thrombotic thrombocytopenic purpura. *Br. J. Haematol.* 74(4):457.

Murphy, W.G., Moore, J.C. and Kelton, J.G., 1987, Calcium-dependent cysteine protease activity in the sera of patients with thrombotic thrombocytopenic purpura. *Blood* 70(5):1683.

Pearson, J.D., 1991, Autoantibodies to endothelial cells, in Vascular Endothelium. Catravas, Callow, Gillis, Ryan (eds), 1991, 117, Plenum Press, New York.

Rock, G., Shumak, K., Kelton, J., Blanchette, V.S., Buskard, N., Nair, R. and Spasoff, R., 1992, Thrombotic thrombocytopenic purpura: outcome in 24 patients with renal impairment treated with plasma exchange. Canadian Apheresis Study Group: *Transfusion* 32(8):710.

Rose, P.E., Enayat, S.M., Sunderland, R., Short, P.E., Williams, C.E. and Hill, F.G.H. , 1984, Abnormalities of factor VIII related protein multimers in the haemolytic uraemic syndrome. *Arch Dis Child* 59:1135.

Siddiqui, F.A., and Lian, E.C., 1985, Novel platelet agglutinating protein from a thrombotic thrombocytopenic purpura plasma. J Clin Invest 76:1330.

White, D.J., Yong, F. and McKendrick, M.W., 1988, Haemolytic syndrome in adults. Br Med J 296:899.

ENDOTHELIAL FUNCTION AND MALIGNANT COAGULOPATHY

*Wickham, N.W.R., **Bradshaw, A., Evans, J., Bevan, *D.H. and Mansi, J.

Clinical Oncology
St George's Hospital
London

*Department of Haematology
St George's Hospital
London

**Department of Haematology
Hammersmith Hospital
London

`Present address:
Department of Clinical Oncology
Prince of Wales Hospital
Chinese University of Hong Kong
Shatin, Hong Kong

INTRODUCTION

Patients with cancer suffer an increased incidence of thrombo-embolic events. This is particularly true of breast, lung, gastro-intestinal, gynecological and prostatic tumors *(Jacobs and Yeager, 1986)*. This risk is further enhanced by metastatic disease and chemotherapy *(Goodnough et al., 1984)*. Many coagulation abnormalities have been detected in these patients, but none have been shown to be predictive of subsequent thrombo-embolism *(Nand et al., 1987; Bick, 1978; Nand and Messmore, 1990; Miller and Heilmann, 1988; Zurbon et al., 1990)*. The fact that thrombo-embolic events often occur or recur despite apparently adequate anticoagulation with warfarin is further evidence of the unsatisfactory nature of conventional assays in indicating thrombotic status in these patients.

RATIONALE

One of the more consistently abnormal findings in malignancy is deranged fibrinolysis, particularly with respect to an increase in the endothelial cell product, plasminogen activator inhibitor-1 *(Zurbon et al., 1990; Mannucci et al., 1990)*. Furthermore, both interleukin-1-beta (IL-1b) and tissue necrosis factor-alpha (TNFa) activate the endothelial cell to become procoagulant *(Kobayashi et al., 1990; Bevilacqua et al, 1984)*, with an increase in PAI-1 as well as an increase in tissue factor expression *Bevilaxqua et al., 1986; Nachman et al., 1986; Dosne et al., 1988)*. The production of endothelin-1 (ET-1), a vasoactive peptide with growth factor activity *Ohnishi et al., 1989)*, also may be increased by the action of IL-1 and other cytokines on the endothelial cell *(Casey et al., 1991)*. It is also a marker of possible endothelial cell damage in a wide variety of disorders including DIC *(Ishibashi 1991)*, arterial vascular disease *(Lerman et al., 1991)*, cardiogenic shock *(Cernacek and Stewart, 1989)* and trauma *(Koller et al., 1991)*.

PILOT STUDY DESIGN

To test the hypothesis that endothelial activation, or factors known to cause endothelial activation, may be increased in cancer patients to a degree that would correlate with subsequent thrombosis, 18 patients with cancer were recruited into a pilot study between October 1992 and April 1993, either at diagnosis or during therapy/follow-up in the cancer day-unit at St George's Hospital. Fully-informed consent was obtained from all patients who agreed to provide an extra 30 mls of blood every 2-4 weeks at the time of their routine testing to avoid unnecessary venepuncture. The sample consisted of 10 mls in trisodium citrate, 10 mls in EDTA, and 10 mls in lithium heparin + 200 µl Trasylol. All samples were centrifuged and the plasma separated within 20 minutes of collection and stored frozen at -70°C. Tests to be performed included PT, APTT, fibrinogen, PAI-1, IL-1b, TNFa and ET-1 (ET-1 assay performed by Dr M Ghatei, Hammersmith Hospital). Twenty four duplicate samples were analyzed from 10 patients - 5 who were thought to have had new thrombotic events after enrollment in the study and 5 with no signs or symptoms suggesting thrombosis.

RESULTS

The following tests were within the normal range in all samples - PT, APTT, fibrinogen, IL-1b and ET-1. TNFa was raised only in samples from two out of the three patients known to have widespread metastases and was associated with abnormal elevation of PAI-1. PAI-1 was raised in 7 of the 10 patients. In 5 of these patients, there was evidence of clinically/radiologically significant thrombosis. In addition, if a graph was plotted of the PAI-1 levels against date of assay (plotted as days prior to detection of thrombosis) then there was an exponential rise (correlation coefficient r=0.84) from day -70, which exceeded the upper limit of normal about 20 days before clinical presentation.

CONCLUSIONS

The procoagulant and pro-inflammatory cytokine IL-1b was not increased in any of the cancer patients tested. Likewise, ET-1 remained within its normal range in all samples. TNFa, however, was most likely to be raised in patients with widespread metastases and was all ways associated with a raised PAI-1. The added procoagulant effect of TNFa in addition to the raised PAI-1 may explain why patients with metastatic cancer are even more prone to thrombosis. Thus TNFa may be a marker for a particularly high risk group. A raised PAI-1 level was detected in 7 out of 10 patients, and was associated with clinically significant thrombosis in 5 out of the 7. There was evidence that a rising PAI-1 may possibly be predictive of a clinical thrombosis, suggesting that routine serial PAI-1 estimations may serve as a method of detecting those who may benefit from preemptive full anticoagulation. Thus, this study confirms the findings of others, that PAI-1 is commonly abnormally elevated in patients with cancer. It also suggests that an increasing PAI-1 level may have a predictive value for subsequent thrombosis, especially in association with a raised TNFa. However, a larger prospective study will be necessary to determine the sensitivity and specificity of serial PAI-1 estimations in predicting subsequent thrombosis in a routine clinical setting, and to confirm the postulated additional consequences of an elevated plasma TNFa.

REFERENCES

Bevilacqua, M.P., Pober, J.S., Majeau, G.R., Fiers, W., Cotran, R.S. and Gimbrone, M.A.J., 1986, Recombinant tumor necrosis factor induces procoagulant activity in cultured human vascular endothelium. *Proc. Natl. Acad. Sci. USA* 83(12):4533.

Bevilacqua, M.P., Pober, J.S., Majeau, G.R., Cotran, R.S., and Gimbrone, M.A.J., 1984, Interleukin 1 (IL-1) induces biosynthesis and cell surface expression of procoagulant activity in human vascular endothelial cells. *J. Exp. Med.* 160(2):618.

Bick, R.L., 1978, Alterations of hemostasis associated with malignancy: etiology, pathophysiology, diagnosis and management. *Semin. Thromb. Hemost.* 5(1):1.

Casey, M.L., Word, R.A. and MacDonald, P.C., 1991, Cytokine stimulated endothelin release from endothelial cells. *Life Sci.* 48(14):1379.

Cernacek, P. and Stewart, D.J., 1989, Immunoreactive endothelin in human plasma: marked elevations in patients in cardiogenic shock. *Biochem. Biophysic. Res. Commun.* 161:562.

Dosne, A.M., Dubor, F., Lutcher, F., Parant, M. and Chedid, L., 1988, Tumor necrosis factor (TNF) stimulates plasminogen activator inhibitor (PAI) production by endothelial cells and decreases blood fibrinolytic activity in the rat. *Thromb. Res. Suppl.* 8:115.

Endo, S., Inooka, H., Ishibashi, Y., Kitada, C., Mizuta, E. and Fujino, M., 1990, Interleukin 1 increases the production of endothelin-1 by cultured endothelial cells. *Biochem. Biophys. Res. Commun.* 166(1):324.

Goodnough L.T., Saito, H., Manni, A., Jones, P.K. and Pearson, O.H., 1984, Increased incidence of thromboembolism in stage IV breat cancer patients treated with a five-drug chemotheraphy regimen. A study of 159 patients. *Cancer* 54(7): 1264.

Ishibashi, M., Saito, K., Watanabe, K., Uesugi, S., Furue, H. and Yamaji, T., 1991, Plasma Endothelin-1 levels in patients with disseminated intravascular coagulation. *N. Engl. J. Med.* 324(21):1513.

Jacobs M.B. and Yeager M., 1986, Idiopathic deep vein thrombosis in an apparently healthy patient as a premonitory sign of occult cancer. *Cancer* 57(9):1846.

Kobayashi, T., Shimada, K., and Ozawa, T. , 1990, Human recombinant interleukin-1 beta- and tumor necrosis factor alpha- mediated suppression of heparin-like compounds on cultured porcine aortic endothelial cells. *J. Cell Physiol.* 144(3):383.

Koller, J., Mair, P., Wieser, C., Pomnaroli, A., Puschendorf, B. and Herold, M., 1991, Endothelin and big endothelin concentrations in injured patients. *N. Engl. J. Med.* 325(21):1518.

Lerman, A., Edwards, B.S., Hallett, J.W., Heublein, D.M., Sandberg, S.M. and Burnett Jr, J.C., 1991, Circulating and tissue endothelin immunoreactivity in advanced atherosclerosis. *N. Engl. J. Med.* 325(14):997.

Mannucci, P.M., Cugno, M., Bottasso, B., Marongiu, F., Maniezzo, M., Vaglini, M., and Cascinelli, N., 1990, Changes in fibrinolysis in patients with localized tumors. *Eur. J. Cancer* 26(2):83.

Miller, B. and Heilmann, L., 1988, Hemorheologic variables in breast cancer patients at the time of diagnosis and during treatment. *Cancer* 62(2):350.

Nachman, R.L., Hajjar, K.A., Silverstein, R.L. and Dinarello, C.A., 1986, Interleukin induces endothelial cell synthesis of plasminogen activator inhibitor. *J. Exp. Med.* 163(6):1595.

Nand, S., Fisher, S.G., Salgia, R. and Fisher R.I., 1987, Hemostatic abnormalities in untreated cancer: incidence and correlation with thrombotic and hemorrhagic complications. *J. Clin. Oncol.* 5(12):1998.

Nand, S. and Messmore, H., 1990, Hemostasis in malignancy. *Am. J. Hematol.* 35(1):45.

Ohnishi, A., Yamaguchi, K., Kusuhara, M., Abe, K. and Kimura, S., 1989, Endothelin potentiates growth factor-stimulated DNA synthesis in Swiss 3T3 cells. *Jpn. J. Cancer Res.* 80(4):302.

Zurbon, K.H., Duscha, H., Gram, J., and Bruhn, H.D., 1990, Investigations of coagulation system and fibrinolysis in patients with disseminated adenocarcinomas and non-Hodgkin's lymphomas. *Oncology* 47(5):376.

PROTECTION FROM APOPTOSIS OF ENDOTHELIAL CELLS BY ALBUMIN

H. Zoellner, M. Hofler, R. Beckmann, P. Hufnagl and B.R. Binder

Clinical Experimental Physiology
Department of Medical Physiology
Schwarzspanierstrasse 17, A-1090
Vienna, AUSTRIA

Apoptosis is programmed cell death and is thought to be responsible for the removal of excess cells in tissues *(Gerschenson and Rotello, 1992)*. Endothelial cells are known to undergo apoptosis in-vivo *(Walker et al., 1989; Walker and Gobe, 1987)*. Similarly, apoptosis has been observed in cultured HUVEC deprived of serum and or attachment *(Araki et al., 1990; Araki et al., 1990; Meredith et al., 1993)*. We have found that bovine (BSA) and human (HSA) serum albumin both block the loss of attachment and DNA fragmentation typical of HUVEC apoptosis. There is no difference in the dose response of the two albumins in their ability to prevent loss of attachment, and the response is maximal at concentrations approaching those in blood. However, albumin is not able to rescue HUVEC after 8 hr of serum starvation, while before this time it is only partially effective in rescuing HUVEC. Serum on the other hand is able to rescue serum starved cells at all time points tested (up to 18 hr). This suggests that serum contains other factors blocking apoptosis such as fibroblast growth factor *(Araki et al., 1990)*. BSA is partially inactivated by CNBr, while HSA is unaffected. Reduction of both albumins results in inactivation. We suggest that the effect of albumin in blocking apoptosis may provide a mechanism for the removal of nonfunctional vessels *in-vivo*.

REFERENCES

Araki, S., Simada, Y., Kaji, K., and Hayashi, H., 1990, *Biochem. Biophys. Res. Commun.* 168:1194.
Araki, S., Simada, Y., Kaji, K. and Hayashi, H., 1990, *Biochem. Biophys. Res. Commun.* 172:1081.
Gerschenson, L.E. and Rotello, R.J., 1992, *FASEB J.* 6:2450.
Meredith, J.E. Jr, Fazeli, B. and Scwartz, M.A., 1993, *Mol. Biol. Cell.* 4:953.
Walker, N.I., Bennett, R.E. and Kerr, J.F., 1989, *Am J. Anat.* 185:19.
Walker, N.I. and Gobe, G.C., 1987, *J. Pathol.* 153:333.

VII. PARTICIPANTS

PARTICIPANTS

Steffan G. Anderson
Department of Physiological Sciences
College of Veterinary Medicine
Oklahoma State University
Stillwater, OK 74078-0350
USA

Frederick Andreasen, M.D.
University of Aarhus
The Bartholin Building
DK-8000 Arhus C
DENMARK

Paraskevi Andriopoulou
University of Patras
Department of Pharmacology
26 110 Rio
Patras
GREECE

Maria Bastaki
Department of Pharmacology
University of Patras
26 110 Rio
Patras
GREECE

Stephan Baydanoff, M.D.
Department of Biology and Immunology
University School of Medicine
5800 Pleven
BULGARIA

Panayotis Behrakis, M.D.
Department of Experimental Physiology
University of Athens School of Medicine
11527 Youdi
Athens
GREECE

B. R. Binder, M.D.
Clinical Experimental Physiology
Department of Medical Physiology
Schwarzspanierstraβe 17
A-1090 Vienna
AUSTRIA

Rena Bizios, Ph.D.
Biomedical Engineering Department
Rensselaer Polytechnic Institute
Troy, NY 12180-3590
USA

Julie Borland
Heart Science Centre
Harefield Hospital
Harefield, Middlesex UB9 6JH
UNITED KINGDOM

Robert W. Caldwell, Ph.D.
Department of Pharmacology & Toxicology
Medical College of Georgia
Augusta, GA 30912-2300
USA

Ruth B. Caldwell, Ph.D.
Department of Cellular Biology and Anatomy
Medical College of Georgia
Augusta, GA 30912-2300
USA

Allan D. Callow, M.D., Ph.D.
Boston University School of Medicine
80 E Concord Street
NEB 606
Boston, MA 63110
USA

Sean Carroll
Biomedical Engineering Department
Rensselaer Polytechnic Institute
Troy, NY 12180-3590
USA

Judith E. Cartwright
Division of Immunology
Department of Cellular and Molecular Sciences
St. George's Hospital Medical School
Cranmer Terrace
London SW17 ORE
UNITED KINGDOM

John D. Catravas, Ph.D.
Vascular Biology Center
Medical College of Georgia
Augusta, GA 30912-2500
USA

Romeo Cecchelli, Ph.D.
Institute Pasteur De Lille
SERLIA et U. 325 INSERM
1 rue du Prof. Calmette
59019 Lille
FRANCE

Adrian Chester, Ph.D.
Heart Science Centre
Harefield Hospital
Harefield, Middlesex UB9 6JH
UNITED KINGDOM

Allessandro Chinellato
Department of Pharmacology
University of Padova
Largo E. Meneghetti, 2
1-35131 Padova
ITALY

Eric Choi, M.D.
Department of Surgery
Washington University Medical Center
4906 Audobon Avenue
St. Louis, MO 63110
USA

Alan D. Collins
Department of Cardiothoracic Surgery
Harefield Hospital
Harefield, Middlesex UB9 6JH
UNITED KINGDOM

James D. Crapo, M.D.
Department of Pulmonary Medicine
Duke University Medical Center
Box 3177
349 Bell Building
Durham, NC 27707-3162
USA

Luis M. Cunha-Ribeiro, M.D.
Department of Physiology
Porto Medical School
Hosp. S. Joao
4200 Porto
PORTUGAL

Attila Cziraki, M.D.
Vascular Biology Center
Medical College of Georgia
Augusta, GA 30912-2500
USA

Mariella De Biasi
Department of Pharmacology
University of Padova
Largo E. Meneghetti, 2
1-3531 Padova
ITALY

M.P. Dehouck, Ph.D.
Institute Pasteur De Lille
SERLIA et U. 325 INSERM
1 rue du Prof. Calmette
59019 Lille
FRANCE

Elisabetta Dejana, Ph.D.
Department of Biologie-INSERM U 217
17 rue des Martyrs
38054 GRENOBLE Cedex 9
FRANCE

George Deliconstantinos, Ph.D.
Department of Experimental Physiology
University of Athens
Micras Asias 75
115 27 Goudi Athens
GREECE

Ramazan Demir, M.D.
Department of Histology and Embryology
Akdeniz University
07070 Kampus
Antalya
TURKEY

Christiana Dimitropoulou
Department of Pharmacology
University of Patras Medical School
26 110 Rio Patras
GREECE

K. Dirkranian, M.D.
Christiana Dimitropoulou
University of Patras Medical School
Department of Pharmacology
26 110 Rio Patras
GREECE

Dr. Thomas Douraki
Cardiac Surgery
Iatrikon Kentron
Maroussi
GREECE

Salwa A. Elgebaly, Ph.D.
Department of Surgery
Hartford Hospital
80 Seymour Street
Hartford, CT 06102-5037
USA

Suzanne G. Eskin, Ph.D.
Texas Biotechnology Corporation
7000 Fannin, Suite 1920
Houston, TX 77030
USA

John W. Fenton II, Ph.D.
State of New York Department of Health
P.O. Box 509
Albany, NY 12201-0509
USA

John A. Frangos, Ph.D.
The Pennsylvania State University
150 Penske Laboratory
University Park, PA 16802-4400
USA

Peter I. Ghenev, M.D.
Department of Pathology
Medical University Varna
Marin Drinov STR 55
9002 Varna
BULGARIA

Adel Giaid, M.D.
Montreal General Hospital
1650 Cedar Avenue
Montreal H3G1A4
CANADA

Carolyn Go
Vascular Biology Center
Medical College of Georgia
Augusta, GA, 30912-2500
USA

Keith J. Gooch
Pennsylvania State University
158 Fenske Laboratory
University Park, PA 16802-4400
USA

Howard P. Greisler, M.D.
Section of Peripheral Vascular Surgery
Loyola University Chicago
2160 South First Avenue
Maywood, IL 60153
USA

Gail H. Gurtner, M.D.
New York Medical College
Division of Pulmonary and Critical
Care Medicine
Valhalla, NY 10595
USA

Alison W. Halliday, M.D.
St. Thomas' Hospital
Division of Surgery
Lambeth Palace Road
London SE1 7EH
UNITED KINGDOM

George Haralabopoulos
University of Patras
Department of Pharmacology
26 100 Rio Patras
GREECE

Medora M. Hardy, Ph.D.
G.D. Searle, Inc.
800 North Lindbergh Blvd.
St. Louis, MO 63167
USA

Vanessa J. Harrison
The William Harvey Research Institute
St. Bartholomew's Medical College
Charterhouse Square
London EC1M 6BQ
UNITED KINGDOM

Dorian O. Haskard, M.D.
Hammersmith Hospital Du Cane Road
London W12 ONN
UNITED KINGDOM

Olga Hatzikonti
University of Patras
Department of Pharmacology
26 110 Rio Patras
GREECE

Maria A. Hildago
Surgical Research Department
Clinical Research Centre
Northwick Park Hospital
Harrow, Middlesex HA1 3UJ
UNITED KINGDOM

Kun Huang, Ph.D.
Department of Immunology
XOMA Corporation
2910 7th Street
Berkeley, CA 94710
USA

Martino Introna, Ph.D.
Instituto Mario Negri
Via Eritrea 62
20157 Milano
ITALY

Alan K. Keenan, Ph.D.
Department of Pharmacology
University College of Dublin
Foster Avenue
Blackrock Co, Dublin
IRELAND

Nicholas A. Kefalides, M.D., Ph.D.
Connective Tissue Research Section
University City Science Center
3624 Market Street
Philadelphia, PA 19104
USA

Matthew Konneh
William Harvey Research Institute
Charterhouse Square
London EC1M, 6BQ
UNITED KINGDOM

David Langleben, M.D.
Jewish General Hospital
Division of Cardiology
3755 Chemin De La Cote Ste-Catherine
Montreal, Quebec H3T E2
CANADA

Paul S. Lidbury, Ph.D.
St. Bartholomew's Hospital
Medical College
Charterhouse Square
London EC1M, 6BQ
UNITED KINGDOM

Peter Lockey, Ph.D.
Rhone Poulenc Rorer
Rainham Road South
Dagenham, Essex RM10 7XS
UNITED KINGDOM

Enrico Lupia
Intituto di Nefro-Urologia
Ospedale Maggiore S. Giovanni
Battista-Molinette
Corso Dogliotti, 14
10126 Torion
ITALY

Thin Luu
Heart Science Centre
Harefield Hospital
Harefield, Middlesex UB9 6JH
UNITED KINGDOM

Savvas C. Makrides, Ph.D.
Advanced Biotechnology Building
Room 509
Boston University Medical Center
700 Albany Street
Boston, MA 02118
USA

Michael Maragoudakis, Ph.D.
University of Patras
Department of Pharmacology
26 110 Rio Patras
GREECE

Nandor Marczin, M.D.
Department of Anesthesia and
Intensive Therapy
Medical University of Pecs
Ifjusag Str 13, POB 99
HUNGARY

Sadis Matalon, Ph.D.
University of Alabama at Birmingham
619 South 19th Street
Birmingham, AL 35233-1924
USA

Jorgen Matz
The William Harvey Research Institute
St. Bartholomew's Hospital Medical
College
Charterhouse Square
London EC1M 6BQ
UNITED KINGDOM

Rhoda M. McDouall
Heart Science Centre
Harefield Hospital
Harefield, Middlesex UB9 6JH
UNITED KINGDOM

Karen McQuaid
Department of Pharmacology
University College of Dublin
Foster Avenue, Blackrock Co/Dublin
IRELAND

John H. McVey
Hemostasis Research Group
Clinical Research Centre
Watford Road
Harrow, Middlesex HA1 3UJ
UNITED KINGDOM

Eleftheria Missirlis, Ph.D.
University of Patras
Department of Pharmacology
26 110 Rio Patras
GREECE

Karoly Nagy, M.D.
Hungarian Heart Center
Department of Internal Medicine
8231 Balantonfured, Gyogy t.2
HUNGARY

Zolton Nagy, M.D.
Stroke Unit of Semmeweis Medical
University
Balassa u.6.
1083 Budapest
HUNGARY

Thomas Noll, M.D.
Institute fur Herz-und Kreislaufphysiologie
Heinrich-Heine-Universitat
POB 10 10 07
D-4001 Dusseldorf
GERMANY

Suzanne Oparil, M.D.
Department of Medicine
University of Alabama at Birmingham
1034 Zeigler Building, 703 19th Street
Birmingham, AL 35294-0007
USA

Stylianos Orfanos, M.D., Ph.D.
University of Athens Medical School
Evangelismos General Hospital
45-47, Ipsilandou St.
106 76 Athens
GREECE

Christopher Page, Ph.D.
Heart Science Centre
Harefield Hospital
Harefield, Middlesex UB9 6JH
UNITED KINGDOM

Luisa Pandolfo, M.D.
Department of Pharmacology
University of Padova
Largo E. Meneghetti, 2
1-35131 Padova
ITALY

Staniatis Papaioannou
University of Patras School of
Pharmacy
Department of Molecular Pharmacology
Patras
GREECE

Andreas Papapetropoulos
Vascular Biology Center
Medical College of Georgia
Augusta, GA 30912-2500
USA

Rachel Parratt
Heart Science Centre
Harefield Hospital
Harefield, Middlesex UB9 6JH
UNITED KINGDOM

Heather J. Parry, Ph.D.
Department of Pharmacology and
Clinical Pharmacology
St. George's Hospital Medical School
Cranmer Terrace
London SW17 PRE
UNITED KINGDOM

Marta Pek, M.D.
Stroke Unit of Semmeweis
Medical University
Balassa u.6.
1083 Budapest
HUNGARY

Markku Pentikainen
Wilhuri Research Institute
Kalliolinnantie 4
00140 Helsinki
FINLAND

Platon Peristeris, Ph.D.
University of Patras
Department of Pharmacology
26 110 Rio Patras
GREECE

Livia Prati, M.D.
Institute of Pharmacological Sciences
University of Milan
Via Balzaretti 9
20133 Lilan
ITALY

Scott K. Pruitt, Ph.D.
Duke University Medical Center
Department of Surgery
Box 31147
Durham, NC 27710
USA

G.E. Rainger, M.D.
University of Birmingham Medical
School
Department of Hematology
Edgbaston, Birmingham B15 2TT
UNITED KINGDOM

Stuart Rich, M.D.
Department of Cardiology
University of Illinois Hospital
840 S Wood
Chicago, IL 60612
USA

Werner Risau, M.D.
Max-Planck-Institut
W.G. Kerckhoff-Institute
Parkstraße 1
D-61231 Bad Nauheim
GERMANY

Marlene L. Rose, Ph.D.
Reader in Transplant Immunology
Heart Science Centre
Harefield Hospital
Harefield, Middlesex UB99 6JH
UNITED KINGDOM

Jean-Jacques Rouby, M.D.
Surgical Critical Care Unit
Department of Anesthesiology
Hopital de la Pitie
75013 Paris
FRANCE

Una S. Ryan, Ph.D.
T-CellS, Inc.
115 4th Avenue
Needham, MA 02194
USA

Ravazoula Giota
Department of Pathology
Patras Medical School
Rio Patras
GREECE

Sarah A. Salwen
Biomedical Engineering Department
Rensselaer Polytechnic Institute
Troy, NY 12180-3590
USA

Valerie B. Schini, Ph.D.
Zentrum der Physiologie
Klinikum der JWG-Universitat
Theodor-Stern-Kai 7
D-60590 Frankfurt/Main
GERMANY

Ina Schuppe-Koistinen
Institute of Environmental Medicine
Division of Toxicology
Box 210
S-171 77 Stockholm
SWEDEN

Cathy Schyns
Heart Science Centre
Harefiel Hospital
Harefield, Middlesex UB9 6JH
UNITED KINGDOM

Demetrios S. Sgoutas, Ph.D.
Department of Pathology and
Laboratory Medicine
Emory University School of Medicine
Atlanta, GA 30322
USA

Inan Soydan, M.D.
Department of Cardiology
Edg University Medical Faculty
Bornova/Izmir
TURKEY

Saliha Soydan, M.D.
Ege University Medical Faculty
Bornova/Izmir
TURKEY

John Stavridis, M.D.
Department of Experimental Physiology
University of Athens
Micras Asias 75
115 27 Goudi Athens
GREECE

D.J.A. Sutherland, M.D.
Medical Oncology
Toronto-Bayview Regional
Cancer Centre
Ontario Cancer Treatment and
Research Foundation
Toronto M4N 3M5
CANADA

Csaba Szabo, M.D.
St. Batholomew's Hospital Medical
College
Charterhouse Square
London EC1M 6BQ
UNITED KINGDOM

Patricia M. Taylor, Ph.D.
Harefield Hospital
Harefield, Middlesex UB9 6JH
UNITED KINGDOM

Michael Theodorakis, M.D.
Department of Medicine
University of Manchester
Manchester Royal Infirmary
Oxford Road, Manchester M13 9WL
UNITED KINGDOM

Stergios Theodoropoulos, M.D.
Cardiac Surgery
Iatrikon Kentron
Maroussi
GREECE

Hannu Toivonen, M.D.
Department of Anesthesiology
Helsinki University Hospital
Haartmaninkatu 4
SF-00290 Helsinki
FINLAND

Branko B. Tomazic, Ph.D.
American Dental Association Health
Foundation
Paffenbarger Research Center
National Institute of Standards and
Technology
Gaithersburg, MD 20899
USA

Anthony T. Tu, Ph.D.
Department of Biochemistry and
Molecular Biology
Colorado State University
Fort Collins, CO 80523
USA

Meral Tuncer, M.D.
Hacettepe University
Faculty of Medicine
Department of Pharmacology
06100 Ankara
TURKEY

Vassiliki Vallidtou
Department of Experimental Physiology
University of Athens
Micras Asias 75
115 27 Goudi Athens
GREECE

Rocco H. van den Berg
Department of Nephrology
University Hospital Leiden
Rijnsburgerweg 10, Leiden
THE NETHERLANDS

Victor W.M. van Hinsbergh, M.D.
Department of Endothelial
Pathophysiology
Gaubius Lab IVVO-TNO
PO Box 430
Leiden
THE NETHERLANDS

Monika Vastaq, M.D.
Stroke Unit of Semmeweis
Medical University
Balassa u.6.
1083 Budapest
HUNGARY

Irene Virgolini, M.D.
University Clinic for Nuclear Medicine
Allgemeines Krankenhaus Der Stadt
Wein
Wahringer Gurtel 18-20
Vienna A-1090
AUSTRIA

Peter A. Ward, M.D.
MS240 MSI, Box 0602
University of Michigan Medical School
1301 Catherine
Ann Arbor, MI 48109-0602
USA

David C. West, Ph.D.
Department of Immunology
Royal Liverpool University Hospital
P.O. Box 147
Liverpool, L69 3BX
UNITED KINGDOM

Nicholas W.R. Wickham, Ph.D.
Department of Clinical Oncology
The Chinese University of Hong Kong
Prince of Wales Hospital
Shatin, N.T.
HONG KONG

Johann Wojta, M.D., Ph.D.
Clinical Experimental Physiology
Department of Medical Physiology
Schwarzpanierstraße 17
A-1090 Vienna
AUSTRIA

Liz Wood
The William Harvey Research Institute
St. Bartholomew's Hospital Medical
College
Charterhouse Square
London EC1M 6BQ
UNITED KINGDOM

Sir Magdi Yacoub, FRCS
Department of Cardiothoracic Surgery
Harefield Hospital
Harefield, Middlesex UB9 6JH
UNITED KINGDOM

M. Yanagisawa, M.D.
University of Texas
Southwestern Medical Center at Dallas
5323 Harry Hines Boulevard
Room Y5.210
Dallas, TX 52351
USA

Vassilis I. Zannis, Ph.D.
Department of Medicine
Boston University
80 East Concord Street
Boston, MA 02118-2394
USA

Hans Zoellner, M.D., Ph.D.
Clinical Experimental Physiology
Department of Medical Physiology
Schwarzspanierstraße 17
A-1090 Vienna
AUSTRIA

VIII. INDEX

INDEX

acute phase proteins 142
acute respiratory failure 181-186, 188-190, 197, 200, 202, 203
acute pulmonary hypertension 197
adheren junctions 167, 171
adhesion 28, 30, 47, 62, 145-154
adhesion molecules 131, 134, 136
allotransplantation 3, 9
Alzheimer's disease 73, 75, 89, 90, 99, 124
angiogenesis 157-159, 161-165
angioplasty 46, 56, 61, 66-68
angiotensin converting enzyme 11, 14, 20, 21, 205, 208, 211, 230-233, 235-238
apoE deficiency 75, 94-98, 111
apolipoprotein E 75, 90, 115, 119-123, 125-128
ARDS 19, 131, 187-191, 193-199, 201-203
arthritis 157, 162
atherogenic diets 89, 110, 113
atherosclerosis 35-37, 39, 43, 49, 64, 66, 67, 73, 75, 85-87, 89, 90, 92, 96, 98, 103, 110-112, 114, 118, 119, 121, 123, 126, 127
basement membrane 158, 159, 164
cadherins 168, 169, 171
cancer 161, 162, 164
capillaries 14, 15, 18, 65, 132, 134, 157, 158, 161, 163, 171
cholesterol 34, 37, 41, 63, 64, 69-72, 74, 76, 78-80, 82-84, 86, 88, 92, 93, 95, 96, 98, 103, 110-114, 116-119, 121, 124, 125, 127, 128
coagulation 45-47, 49, 56, 59, 61, 62, 64, 65, 160, 161, 181
communicating junctions 168
complement 3, 10, 60, 61, 64-68, 131-134, 136, 137
coronary 5, 13-15, 17-20, 33, 34, 36, 40, 43, 61-63, 65-68, 71, 93, 98, 114, 115, 124, 128, 163, 180, 207, 217, 234
critically ill patients 191, 195, 197
cytokines 28, 53, 62, 65, 131, 134-136, 141-144, 149, 151, 153, 154
endothelial cells 8, 10-12, 24, 27, 30, 38, 40, 47-60, 62, 63, 65, 66, 139, 141-145, 147, 149, 151-154
endothelium 167-170
endotoxin 53, 59, 67, 110, 127, 201, 203
differential screening 139
fibrin 39, 40, 45-47, 49-51, 54, 55, 60, 62, 161, 163, 165
fibrinolysis 45, 47, 49, 50, 53-55, 58, 60
growth factors 158, 159, 161, 164
HUVEC 149, 153, 161
hyperacute rejection 67
hyperoxia 11, 23, 30, 194
hypoxia 20, 167, 203, 206-212, 230, 233-235
imaging 150, 152
immune complex 131, 133-135, 137
immunoglobulins 169
inflammation 10, 23, 28, 29
inflammation 55, 64, 68, 148, 150, 151, 153, 154
inhaled nitric oxide 181, 200-203
integrins 147, 148, 152, 153, 168, 170, 171
interleukin-1 148
intimal hyperplasia 61, 64, 66, 67
ischemia-reperfusion 133

leukocytes 25, 29, 62, 65, 68, 145-148, 150, 151, 153, 154
lipoprotein remnants 69-74, 76, 78-80, 82-84, 86, 89, 91-93, 95-98, 101, 103, 105, 110-118, 120, 121, 123-128
lipoproteins 53, 69-71, 73-77, 83, 87, 89-93, 101, 103, 111, 112, 114-121, 125-127
lung injury 131-137
macrophages 27, 28, 30, 51, 52, 60, 62, 72, 74, 94, 96, 114, 122, 123, 131, 135, 157, 162, 163, 195, 206, 218, 222, 232, 238
monocytes 141-144
mRNA 53, 55, 57, 58, 73, 90, 93-97, 103, 115, 120, 123, 128, 206, 212-221, 227, 233, 235, 237
nitric oxide 159, 162, 164
occluding junctions 168
PECAM 168-171
pentraxins 140, 141
permeability 4, 9, 23, 30, 43, 58, 167, 168, 170, 171, 186, 194
plasminogen activator 45, 47-50, 54-59, 139, 143, 158
plexogenic arteriopathy 177
protein kinase 161, 164, 165
pulmonary circulation 19, 20, 188-190, 193, 194, 197, 200, 201, 209
pulmonary hypertension 175-180
rejection 3-10, 17, 64, 65, 67, 157, 163
renin-angiotensin system 62, 205-207, 215, 217-220, 222-225, 227-232, 234-236
restenosis 33-37, 39-43, 46, 49, 56, 61-68
retinopathy 157, 163
selectins 131, 132, 134-137, 145, 146, 154
selective pulmonary vasodilation 181, 202
serum amyloid P component 141, 143, 144
smooth muscle 25, 26, 27, 29, 37-40, 42, 43, 45, 47, 49, 51, 52, 54, 56, 58, 61-65, 67, 68, 83, 118, 157, 158, 160, 161, 169, 176, 178, 180, 186, 188, 193-195, 200, 205, 218, 219, 221, 222, 224, 231, 232, 235, 237
soluble complement receptor type 1 3, 10, 67
therapy 3, 5, 8, 30, 42, 63, 64, 66, 67
thrombin 158, 161-163, 165
thrombomodulin 45-48, 57
tumor necrosis factor 53, 58, 60, 148, 152-154
type III hyperlipoproteinemia 73, 75, 89, 92, 93, 95, 96, 98, 101, 121, 122
vascular permeability 30, 170
vascular permeability factor 38, 42, 43
vasculogenesis 157, 160
vasodilators 177, 178
VE-cadherin 168
wound healing 157, 161, 163
xenotransplantation 3, 7, 9

MIX
Papier aus verantwortungsvollen Quellen
Paper from responsible sources
FSC® C105338

If you have any concerns about our products,
you can contact us on
ProductSafety@springernature.com

In case Publisher is established outside the EU,
the EU authorized representative is:
Springer Nature Customer Service Center GmbH
Europaplatz 3, 69115 Heidelberg, Germany

Printed by Libri Plureos GmbH
in Hamburg, Germany